An Introduction to Thermodynamics

Revised Edition

Y. V. C. RAO

Universities Press

Universities Press (India) Private Limited

Registered Office
3-6-747/1/A & 3-6-754/1, Himayatnagar, Hyderabad 500 029 (A.P), India
e-mail: *www.universitiespress.com*

Distributed by
Orient Blackswan Private Limited

Registered Office
3-6-752, Himayatnagar, Hyderabad 500 029 (A.P), India

Other Offices
Bangalore / Bhopal / Bhubaneshwar / Chennai / Ernakulam / Guwahati
Hyderabad / Jaipur / Kolkata / Lucknow / Mumbai / New Delhi / Patna

First published in 1993 by Wiley-Eastern Limited.
Reprinted 1997, 1998, 2001

This revised edition published by Universities Press (India) Private Limited 2004
Reprinted 2005, 2009, 2010, 2012

ISBN: 978 81 7371 461 0

Typeset by
AICRA, Hyderabad 500 020

Printed at
Charita Impressions
Hyderabad 500 020

Published by
Universities Press (India) Private Limited
3-6-747/1/A & 3-6-754/1, Himayatnagar,
Hyderabad 500 029 (A.P), India

Dedicated to
my wife Ramana and children Srinivas and Sawjanya
for standing by me through some very trying times

Contents

Preface

Thermodynamics is a basic science that deals with equilibrium, energy and its transformation, and the laws governing such transformation. These laws are of wide applicability and are used in several branches of engineering and science. It is essential that every engineer should have a thorough knowledge of thermodynamics and hence thermodynamics has been an essential part of Engineering curricula all over the world. The merit of a book lies in its well balanced coverage of physical concepts and the demonstration of the underlying principles and concepts through practical situations. This introductory text is intended for use in undergraduate engineering courses. The traditional classical or macroscopic approach is used in this text. This approach is easier for students for learning the subject matter. A majority of students often consider thermodynamics a difficult subject. On the contrary, it is a simple subject and students should have no difficulty in understanding it. Future engineers should have a clear understanding of the basic principles in order to apply the laws of thermodynamics for analyzing real life problems.

Particular emphasis is placed on precise and logical presentation of the concepts and principles which are essential for understanding thermodynamics. The first two chapters of the book are devoted to the introduction of the basic definitions and concepts. The properties of fluids are discussed in the third chapter. Then, the laws of thermodynamics are discussed in Chapters 4 to 7. The application of the first law of thermodynamics to chemically reacting systems is included in Chapter 6. An important task in thermodynamics is to develop relationships for the estimation of non-measurable properties in terms of measurable properties. Therefore, Chapter 8 is devoted to this purpose. The Jacobian method of deriving the required thermodynamic relations is emphasized in Chapter 8. Power cycles and refrigeration cycles are presented in Chapter 9. Lastly, gas–vapor mixture and psychrometry are discussed in Chapter 10.

As is well known, different universities and engineering colleges follow slightly different syllabi for thermodynamics. Even within the same syllabus, emphasis on certain topics may change from university to university. Moreover, a student may like to study thermodynamics more rigorously or less rigorously depending on his/her career goal. Therefore, to cater to the requirements of all universities or engineering colleges as well as the individual aptitudes or aspirations of the students, this book covers all the important topics. The material presented in this book can be easily covered in one semester (of 14 working weeks at the rate of 3 lectures per week).

Every chapter begins with a set of learning objectives outlining the chapter's content. At the end of each chapter, a short summary of that chapter is included for a quick overview of the basic concepts. Thermodynamics is a quantitative subject and can be best imparted by demonstrating the applications of the principles to a few selected problems. Moreover, the readers' confidence and ability to apply the principles of thermodynamics to solve real life problems will be greatly improved if they work out a considerable number of problems. Keeping this objective in mind, numerous solved examples are included in this book. The text contains more than 550 unsolved problems and the instructor can assign them, without repetition, as tutorial problems for four years. It is impossible for any textbook to provide enough data through tables or charts to satisfy every reader. As this book is intended for undergraduate students, sufficient data is included in the appendices. SI system of units only has been adopted in this book. I hope that teachers and students will find the material presented in this text both useful and stimulating. It is also hoped that this text would contribute to effective teaching of thermodynamics to undergraduate students, and that students would derive maximum benefit from it. Some errors might have inadvertently crept into the book, and I would greatly appreciate if readers would bring them to my attention.

The financial support provided by the Curriculum Development Cell, Quality Improvement Programme (IIT Kanpur) of the Ministry of Human Resource Development for the preparation of the manuscript is gratefully acknowledged, Finally, I wish to sincerely thank the Publishers, Universities Press (India) Pvt. Ltd., for bringing out this book in its present form.

Kanpur
June 2003

Y.V.C. Rao

Nomenclature

A	Cross sectional area
A	Helmholtz free energy or potential
a	Acceleration
a	Molar (or specific) Helmholtz free energy
a	van der Waals constant
B	Second virial coefficient
B'	Second virial coefficient
b	Depth
b	van der Walls constant
C	Number of components
C	Third virial coefficient
C'	Third virial coefficient
COP_R	The coefficient of performance of a refrigerator
COP_{HP}	The coefficient of performance of a heat pump
C_P	Molar (or specific) heat capacity at constant pressure
C_{Pf}	Specific heat at constant pressure for the saturated liquid
C_{Pg}	Specific heat at constant pressure for saturated vapor
C_v	Specific heat at constant volume
C_{Ps}	Specific heat at constant pressure for solid
d	Time unit day
e	Specific energy
E	Energy of the system
E	Emf of a reversible cell
e_e	Specific energy of material at the exit port of control volume
e_i	Specific energy of material at the inlet port of control volume
$e(t)$	Specific energy of material inside control volume at time t
F	Degrees of freedom of a thermodynamic system
F	Force
F_K	Generalized force
G	Gibbs free energy
g	Acceleration due to gravity
g	Molar (or specific) Gibbs free energy
H	Enthalpy
H	Magnetic field
h	Molar (or specific) enthalpy
h	Time unit hour

h^*	Enthalpy of air–water vapor mixture expressed on the basis of dry air
h_e	Specific enthalpy of material at the exit port of a control volume
h_f	Molar (or specific) enthalpy of saturated liquid
h_{fg}	Latent heat of vaporization ($= h_g - h_f$)
h_g	Molar (or specific) enthalpy of saturated vapor
h_i	Molar (or specific) enthalpy of component *i*
h_i	Specific enthalpy of material at the inlet port of a control volume
h_{sf}	Latent heat of fusion
I	Irreversibility
I	Moment of inertia about the centre of mass of a molecule
i	Current
J	Joule, unit of energy
K	Constant
K	Degree kelvin
kg	Kilogram
L	Length
m	Mass of a particle
min	Minute, time unit
M	Magnetization
M	Molar mass
M_i	Molar mass of the constituent *i*
m	Metre
$\dot{m}$	Mass flow rate through a control volume
m_a	Mass of dry air in the air–water vapor mixture
$\dot{m}_a$	Mass flow rate of dry air
$\dot{m}_e$	Rate of mass leaving the control volume
m_f	Mass inside control volume at the end of the transient flow process
m_f	Mass of saturated liquid
m_g	Mass of saturated vapor
m_i	Mass of component *i*
$\dot{m}_i$	Rate of mass entering control volume
m_0	Mass inside control volume at the start of the transient flow process
m_w	Mass of water vapor in the air–water vapor mixture
$\dot{m}_w$	Mass flow rate of water
$m(t)$	Mass inside control volume at time *t*
N	Newton
N	Number of moles of a substance
N_i	Number of moles of component *i*
P	Number of phases present in a system
P	Pressure
Pa	Pascal
P_c	Critical pressure
P_e	Pressure at the exit port of the control volume
P_i	Pressure at the inlet port of the control volume
P_m	Mean effective pressure

P_a	Ambient pressure
P_r	Reduced pressure
P_s	Saturation pressure
p_a	Partial pressure of air
p_i	Partial pressure of component *i*
p_w	Partial pressure of water vapor
Q	Energy transferred as heat
$\dot{Q}$	Rate of energy flow as heat into the control volume
Q_e	Electric charge
Q_H	Energy rejected as heat to the high temperature thermal reservoir
Q_L	Energy absorbed as heat from the low temperature thermal reservoir
q	Energy transferred as heat per unit mass
R	Universal gas constant
r_c	Cut off ratio
r_0	Compression ratio in an Otto or Diesel cycle
r_P	Pressure ratio (Brayton cycle)
r_{vp}	Constant volume pressure ratio (Dual cycle)
S	Entropy
S_G	Entropy generation
$\dot{S}_G$	Rate of entropy generation
$S(t)$	Entropy of matter in the control volume at time *t*
s	second
s	Molar (or specific) entropy
s_e	Specific entropy of matter at the exit port of control volume
s_f	Molar (or specific) entropy of saturated liquid
s_{fg}	Change in the specific entropy due to vaporization
s_g	Molar (or specific) entropy of saturated vapor
s_i	Molar (or specific) entropy of component *i*
s_i	Specific entropy of matter at the inlet port of control volume
$s(t)$	Specific entropy of matter inside the control volume at time *t*
T	Temperature
T_c	Critical temperature
T_{DB}	Dry-bulb temperature
T_H	Temperature of the high temperature thermal reservoir
T_L	Temperature of the low temperature thermal reservoir
T_0	Ambient temperature
T_r	Reduced temperature
T_{WB}	Wet-bulb temperature
t	Time
U	Internal energy
u	Molar (or specific) internal energy
u_f	Molar (or specific) internal energy of material inside control volume at the end of transient flow process
u_f	Molar (or specific) internal energy of saturated liquid
u_{fg}	Change in the specific internal energy due to vaporization ($= u_g - u_f$)

u_g	Molar (or specific) internal energy of saturated vapor
u_i	Molar (or specific) internal energy of component *i*
$\mathcal{V}$	Velocity
V	Volume
$\mathcal{V}_e$	Flow velocity at the exit port of control volume
$\mathcal{V}_i$	Flow velocity at the inlet port of control volume
v	Molar (or specific) volume
v_c	Critical volume
v_e	Specific volume of material at the exit port of control volume
v_f	Molar (or specific) volume of saturated liquid
v_{fg}	Change in molar volume due to vaporization ($= v_g - v_f$)
v_g	Molar (or specific) volume of saturated vapor
v_{fg}	Change in specific volume due to vaporization
v_g	Specific volume of saturated vapor
v_i	Molar (or specific) volume of component *i*
v_i	Specific volume of material at the inlet port of control volume
v_r	Reduced volume
v_r'	Pseudo-reduced volume
W	Watt
W	Work done
$\dot{W}$	Rate of work done
W_s	Shaft work
$\dot{W}_s$	Rate of shaft work done
W_{net}	Net work done by a heat engine
X	Quality of a wet mixture
X_i	Mole fraction of component *i* in a liquid phase
X_k	Generalized displacement
Y'	Fraction of steam extracted from the turbine at an intermediate state in a regenerative cycle
Z	Compressibility factor
Z	Elevation
Z_c	Critical compressibility factor
Z_e	Elevation of the exit port of a control volume above a datum plane
Z_i	Elevation of the inlet port of a control volume above a datum plane

Greek Symbols

β	Coefficient of volume expansion
γ	Ratio of specific heats
ε	Strain
η	Thermal efficiency of a heat engine
η_P	Isentropic efficiency of a pump
η_T	Isentropic efficiency of a turbine
θ	Thermodynamic temperature
κ	Isothermal compressibility
μ	Chemical potential

μ_{JT}	Joule–Thompson coefficient
μ_i	Chemical potential of component i
ρ	Density
σ	Stress
τ	Tension
ϕ	Relative humidity
ϕ	Availability function for a non-flow process
ϕ_i	Mass fraction of component i
ω	Angular velocity
ω	Specific humidity or humidity ratio

1 Introduction

Learning Objectives

This chapter will enable readers to:

- Know the basic questions an engineer faces in his/her professional career.
- State the essence of the four laws of thermodynamics.
- Know the macroscopic and microscopic approaches adopted in the study of thermodynamics.
- Gain familiarity with the use of SI units.

1.1 Introduction

Thermodynamics, which deals with energy and its transformation, is one of the most fascinating branches of science. Historically, thermodynamics originated as a result of man's endeavour to convert heat into work. In the eighteenth century it was primarily associated with the development of heat engines. Later, Gibbs (1839–1903) developed thermodynamics into a science of such broad scope and generality that presently there is no field in physical, chemical and biological sciences which is not influenced by thermodynamics.

The application of the principles of thermodynamics in practical design tasks—may be that of a simple pressure cooker or of a complex chemical plant—involves examination of the resources—space, time, energy and matter—and ensuring their optimum use. For example, take the design of a chemical reactor. The endeavour of the engineer will be to design it in such a way that it occupies least space, consumes least energy and converts major fraction of the raw materials into the desired products in least time. Such a reactor is obviously optimally designed. An engineer's knowledge of the behavior of different materials under varying conditions, gained from the study of physical sciences, enables him to identify the constraints coming in the way of ensuring optimum use of the resources and compensate for the constraints suitably.

At this point let the author be permitted to digress to touch upon an aspect, namely, the shaping of the social consciousness of the engineer, which has assumed much importance of late. While the principal task of an engineer is ensuring optimum use of the resources, he/she should be equally concerned about the ultimate benefits which the society would derive from the systems/processes/products he/she designs and implements. It is possible to produce highly dangerous compounds (like killer gas) or highly desirable compounds (like life-saving drugs). And every engineer, sooner or later, will be called upon to make his choice: Whether to design systems that would produce killer gas or to design systems that would produce life-saving drugs. Bright youngsters, who have now embarked on an ambitious career- building task, should occasionally start reflecting on such and related questions. It is to help engineers to develop a reflective frame of mind, thereby to keep themselves sensitive to the needs of the society and relate themselves better to the society, that a preliminary course in social sciences has been introduced in the engineering curriculum. The feel for the society and its individual members which budding engineers imbibe through the study of social sciences would not only help them to discharge their social responsibilities better but also enable them to effectively deal with people in the man-management positions they will occupy in future.

Reverting to the topic, the main activities of an engineer still remain as: (a) design and development of new systems and processes, and (b) improvement of the existing systems and processes. The rest of this book concentrates on these main activities of an engineer to help him ensure optimum use of the resources.

Let us consider a specific example—production of ammonia which is a basic ingredient for a variety of fertilizers. From a knowledge of basic chemistry, we know that ammonia can be synthesized from nitrogen and hydrogen at high pressure and temperature, according to the following reaction:

$$N_2 + 3H_2 \longrightarrow 2NH_3$$

The important steps involved in the production of ammonia are presented in Fig. 1.1. Before an engineer takes up the task of producing ammonia, he would like to know answers to the following questions in detail:

1. Is the proposed reaction feasible?
2. Does the reaction go to completion, or does it proceed only to a certain extent?
3. What factors (operating variables like pressure, temperature, ratio of reactants, etc.) govern the extent of the reaction?
4. How much energy is required to compress and raise the temperature of the reactants to the desired level?
   (a) How much energy is required to carry out the synthesis reaction?
   (b) How much cooling (rate of refrigeration) is required to condense the ammonia and to separate it from the unreacted material?
5. What is the source of the required raw materials? If there are several sources, which one is to be preferred?

Let us consider the availability of raw materials first. Nitrogen is available in abundance in the atmosphere. Air can be liquefied and distilled into nitrogen and oxygen. To liquefy air, it has to be compressed to a sufficiently high pressure and then cooled. The question that needs to be answered is: What are the energy requirements for the compression and liquefaction?

Next, let us look into the availability of hydrogen. There are several ways for producing industrial hydrogen. Some of the important processes are:

1. Electrolysis of water.

2. Steam-carbon reaction:

$$C + H_2O \longrightarrow CO + H_2$$
$$CO + H_2O \longrightarrow CO_2 + H_2$$

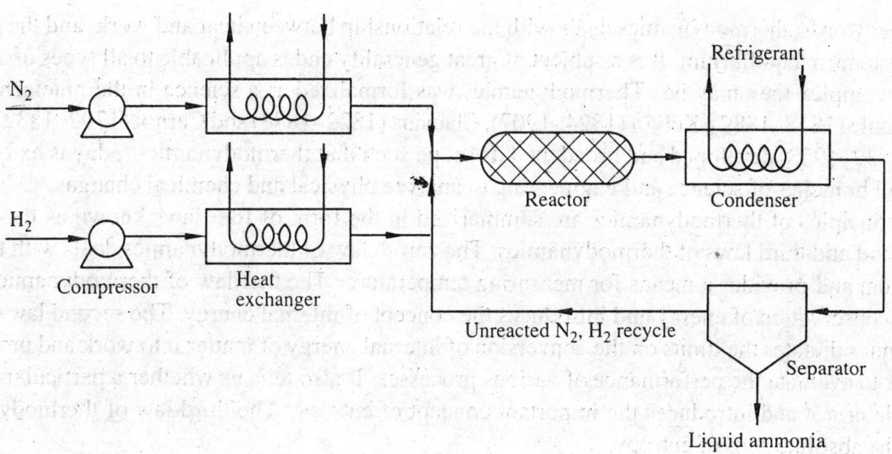

Fig. 1.1 *Schematic diagram showing the important steps involved in the production of ammonia*

3. Methane-steam reaction:

$$CH_4 + H_2O \longrightarrow CO + 3H_2$$
$$CO + H_2O \longrightarrow CO_2 + H_2$$

From the reaction products CO_2 can be easily separated by absorbing it in monoethanolamine solution, and later recovered by stripping. The CO_2 can also be removed by washing the product gas mixture with a weak alkaline solution, producing a carbonate liquor. Depending upon the end use of CO_2, any of the above processes can be adopted. Of the three processes indicated for the production of hydrogen, which process should be preferred? Obviously , the process which requires spending the least amount of energy should be preferred.

In the above example, and in general, the engineer quite often comes across the following questions:

1. Is a certain change, physical or chemical, feasible or not?
2. How much energy is required for the desired change to take place?
3. To what extent does the change take place? Is there any limit to the extent of change under the specified conditions?

If the engineer is in a position to find answers to the above questions, he will be able to design and develop new processes and/or improve upon the existing ones. Thermodynamics—the study of energy and its transformation—enables engineers to answer the above questions and hence its importance.

1.2 What is Thermodynamics?

Thermodynamics is the science dealing with:

- Energy and its transformation,
- Feasibility of a process, and
- Equilibrium processes.

In other words, thermodynamics deals with the relationship between heat and work, and the properties of systems in equilibrium. It is a subject of great generality and is applicable to all types of systems however complex they may be. Thermodynamics was formalized as a science in the nineteenth century by Joule (1818–1889), Kelvin (1824–1907), Clausius (1822–1888) and Carnot (1792–1832). Later Gibbs (1839–1903) developed and broadened its scope such that thermodynamics today is extensively used in all branches of science and engineering to analyze physical and chemical changes.

The principles of thermodynamics are summarized in the form of four laws known as the zeroth, first, second and third laws of thermodynamics. The zeroth law of thermodynamics deals with thermal equilibrium and provides a means for measuring temperature. The first law of thermodynamics deals with the conservation of energy and introduces the concept of internal energy. The second law of thermodynamics dictates the limits on the conversion of internal energy of matter into work and provides a yardstick to evaluate the performance of various processes. It also tells us whether a particular change is feasible or not and introduces the important concept of entropy. The third law of thermodynamics defines the absolute zero of entropy.

The four laws of thermodynamics are based on logical reasoning and were deduced from experimental results gathered over centuries. And so far no experimental evidence is available which casts any doubt on the correctness of these laws and hence they are accepted universally. It is necessary to emphasize here that these laws cannot be deduced or proved directly as we prove theorems in mathematics. Their validity, sturdiness and beauty lie in that they enable scientists and engineers to explain most of the phenomena they come across; solve practical problems; predict the behavior of materials and systems under varying conditions; and design beautiful systems that work wonders—automobiles, refrigerators, air-conditioners, to name a few common ones—which add to the comforts of human beings.

1.3 Two Different Approaches

In the study of thermodynamics one can adopt two different approaches, namely, macroscopic and microscopic. Suppose a certain amount of gas is trapped in a container, say, in a circular cylinder. One can measure the volume occupied by the gas by measuring the diameter and height of the cylinder. The pressure exerted by the gas can be measured with the help of a pressure gauge; and its temperature can be measured with the help of a thermometer or a thermocouple. The composition of the gas can also be measured. Then the state of the gas can be described by specifying the pressure (P), volume (V) and the temperature (T). The values of these variables (P, V, T) which can be measured, are, in essence, space-averaged characteristics or properties of the gas under consideration at a particular instant. Thus, in macroscopic approach, the state of the matter under consideration is specified by only a few of its measurable parameters. In classical thermodynamics, we adopt the macroscopic approach and the relations we obtain by this approach are independent of the nature of the matter under consideration. Summarizing, in the macroscopic approach:

- The structure of matter is not considered,
- Only a few variables are used to describe the state of the matter under consideration, and
- The values of the variables used to describe the state of the matter are easily measurable.

On the other hand, the same gas can be considered as consisting of a large number of small particles each of which moves at random with independent velocity. The state of each particle can be specified in terms of the position coordinates (x_i, y_i, z_i) and the momentum components (p_{xi}, p_{yi}, p_{zi}), or alternatively, by specifying their complete wave function $\psi(x, y, z, t)$. If we consider a gas occupying a volume of one cm^3 at ambient temperature and pressure, the number of particles present in it are of the order of 10^{20} and the same number of position coordinates and momentum components are required to specify the state of the gas. Such a description, or specification, constitutes the essence of the microscopic approach. Since the state of individual particles constituting the gas (system) varies continuously with time because of the large number of collisions among them, the properties of the system normally measured are its time-averaged properties. To deduce time-averaged properties, statistical mechanics approach is normally employed. In statistical thermodynamics, the microscopic approach is adopted and the type of statistics employed depends upon the nature of the particles dealt with. Summarizing, in the microscopic approach:

- A large number of variables are required for a complete specification of the state of matter under consideration,
- The variables used to describe the state of matter cannot be measured easily and precisely, and
- A knowledge of the structure of matter under consideration is essential.

Since this book is intended to serve as an introduction to the study of thermodynamics the macroscopic approach is adopted in presenting the principles.

1.4 Units of Measurement

A thorough knowledge of the units of measurement of the physical quantities commonly encountered is essential for the understanding of thermodynamics. Although the system of units employed in thermodynamics is identical with that employed in mechanics, a brief review is presented below.

1.4.1 Base units

All physical quantities are divided into two groups—primary quantities and secondary quantities. Primary quantities are measured in terms of the base units. The secondary quantities (all physical quantities other than the primary quantities) are measured in terms of the derived units. The international system of units (abbreviated as SI units from *Systeme International d' Unites*) have been used in this book. The SI system was adopted at the Eleventh *Conference Generale des Poids et Mesures* (abbreviated as CGPM) in 1960. The basic units of mass, length, time and temperature in the SI units are described below:

(i) Mass

The basic SI unit for mass is kilogram (abbreviated as kg). A standard alloy block of platinum and iridium maintained at the International Bureau of Weights and Measures at Sevres, Paris, is taken as the base unit of mass. Since all materials are known to consist of atoms, molecules, etc., the quantity of a substance can also be conveniently expressed in terms of the number of such particles present in the matter. In thermodynamics, quite often, the quantity of matter is expressed in terms of the unit mole (abbreviated as mol). The *mole is the amount of substance of a system which contains as many elementary entities as there are atoms in 0.012 kg of carbon-12*. When the mole is used the elementary entities must be specified and may be atoms, molecules, ions, electrons, other particles or specified groups of such particles. This unit mole was adopted by the Fourteenth *Conference Generale des Poids et Measures* (CGPM) in 1971.

(ii) Length

The basic SI unit for length is the metre (abbreviated as m). For many years the accepted basic unit, metre, was the distance between two marks on a platinum-iridium bar, kept at the International Bureau of Weights and Measures, Sevres, France, when measured at 0°C. In 1960, the Eleventh *Conference Generale des Poids et Measures* (CGPM 1960) held in Paris redefined the metre in terms of a particular wavelength of light. The *metre is defined as the length equal to 1,650,763.73 wavelengths in vacuum of the radiation corresponding to the transition between the levels* $2p_{10}$ *and* $5d_5$ *of the krypton-86 atom.*

(iii) Time

The basic SI unit for time is the second (abbreviated as s). For many years, the accepted basic unit, second, was defined as 1/86,400 of the mean solar day. Later, it was recognized that the duration of the mean solar day can vary and the measurement of time with respect to the rotation of the earth, with reference to a fixed star, is tedious and requires elaborate measurements. The development of the atomic clock made precise measurement of frequency possible. Nowadays, it is possible to measure frequencies with an accuracy of 2 parts in 10^{11} and hence, time can be measured with high accuracy. Consequently, in 1967, the Thirteenth *Conference Generale des Poids et Measures* defined the second as the *duration of 9,192,631,770 periods of the radiation corresponding to the transition between the two hyperfine levels of the ground state of the cesium-133 atom.* Quite often, other units of time like minute (min), hour (h) and day (d) are used though they are not the basic units of time.

(iv) Temperature

The basic SI unit for temperature is kelvin (abbreviated as K). In 1967, the Thirteenth *Conference Generale des Poids et Measures* defined kelvin as the fraction 1/273.16 of the thermodynamic temperature of the triple point of water. (The thermodynamic temperature is defined in Chapter 7 based on the laws of thermodynamics).

Quite often, the Celsius temperature (symbol °C) is used to express the temperature of a substance. The Celsius temperature is defined as:

$$°\text{C} = \text{K} - 273.15$$

1.4.2 Derived units

The secondary quantities are expressed in terms of the derived units, which in turn are formed from the base units. The relation between the derived units and the base units depends on a law or a definition. For example, velocity ($\mathcal{V}$) is defined as $\mathcal{V} = \frac{dL}{dt}$; where L is length and t is time, and acceleration a is given by $a = \frac{d\mathcal{V}}{dt}$. Therefore, the units of velocity and acceleration are m/s and m/s^2, respectively. The units of force can be derived with the aid of Newton's second law ($F = ma$, where F is force) as kg m/s^2. In honour of Newton, the unit of force has been named newton and is abbreviated as (N). Newton is the force required to produce an acceleration of one m/s^2 in a body of mass one kg. The units of pressure, energy and power in the SI units are pascal (Pa), joule (J) and watt (W). It was internationally agreed to abbreviate the units named after the scientists by capital letters, and all other quantities in lower case letters. These units should not be represented in plural form and no period (full stop) should be placed at the end of the abbreviated form. The dimensions and units of some of the physical quantities commonly used in thermodynamics are presented in Table 1.1.

Table 1.1 *Dimensions and units of a few selected physical quantities*

Quantity	Dimensions	Units	Name	Abbreviation
Length	L	metre	–	m
Time	T	second	–	s
Mass	M	kilogram	–	kg
Velocity	LT^{-1}	m/s	–	–
Acceleration	LT^{-2}	m/s^2	–	–
Force	MLT^{-2}	kg m/s^2	newton	N
Pressure	$MT^{-2}L^{-1}$	kg/m s^2 (N m^{-2})	pascal	Pa
Energy	ML^2T^{-2}	kg m^2 s^{-2} (N m)	joule	J
Power	ML^2T^{-3}	kg m^{2}s^{-3} (J s^{-1})	watt	W

The SI unit of pressure (pascal) Pa is very small in magnitude. For convenience, pressure is expressed commonly in kilopascal and megapascal. Similarly, other units can be expressed in terms of appropriate multiples of SI units. The choice of the decimal multiple or submultiple of an SI unit is governed by convenience. It is internationally accepted to use the prefixes given in Table 1.2 to form names and symbols of multiples of the SI units.

Table 1.2 *SI unit prefixes*

Factor	Prefix	Symbol	Factor	Prefix	Symbol
10^{18}	exa	E	10^{-18}	atto	a
10^{15}	peta	P	10^{-15}	femto	f
10^{12}	tera	T	10^{-12}	pico	p
10^{9}	giga	G	10^{-9}	nano	n
10^{6}	mega	M	10^{-6}	micro	µ
10^{3}	kilo	k	10^{-3}	milli	m
10^{2}	hecto	h	10^{-2}	centi	c
10	deca	da	10^{-1}	deci	d

The following conventions are recommended by CGPM to express the SI units and quantities:

- The symbol of a prefix is to be combined with the singular form of the unit symbol. For example, 500 000 Pa can be written as either 500 kPa or 0.5 MPa.

- Compound prefixes should not be used to express a quantity. For example; 10^{-9} m should not be expressed as 1 mμ m (millimicro metre) but should be expressed as 1 nm (nanometre).
- The unit symbol should not be used in plural form. For example, 250 joules should be expressed as 250 J and not as 250 Js.
- Lower case letters should be used to write the SI units in full. For example; newton, pascal, joule, etc. On the other hand, Newton, Pascal and Joule refer to the scientists.
- No period (.) should be used after a symbol of SI unit, unless the sentence ends with the unit.
- When a unit is obtained by multiplication or division of two or more units, they can be indicated as N m for newton × metre; m s for metre × second; m/s or m s^{-1} for metre ÷ second, etc.
- Care should be taken to avoid confusion between the SI unit prefix milli (m) and the base unit metre (m) while expressing a quantity. For example, mN stands for millinewton whereas, m N stands for metre × newton. Similarly, ms is used for millisecond and m s is used for metre × second.
- Avoid using ×, ⋆ or . in between two SI units to indicate their product, but leave a gap between the two SI unit symbols. For example indicate the product of metre and second as m s and not as m × s; m ⋆ s or ms.
- The base unit for mass, kilogram, contains the prefix name *kilo*. Therefore, the names of the multiples of the unit mass are formed by adding the prefix to the word *gram* like milligram, mg or microgram, µg.
- The SI prefixes should be selected such that the numerical value lies between 0.1 and 1000. For example, 101 325 Pa can be written as 101.325 kPa or 0.101 325 MPa. However, when the values are presented in a tabular form, for the same quantity it is advisable to use the same SI prefix throughout even if some of the numerical values lie outside the range of 0.1 to 1000.
- The unit symbol should be placed after the numerical value, leaving a space in between the numerical value and the unit symbol.
- While writing a quantity with several digits, to facilitate reading, they should be separated into groups of three, counting from the decimal point towards the right and left. These groups should be separated by a small gap and should not include a comma.

Sometimes, pressure is expressed in terms of an independent unit, the standard atmosphere (1 atm). One standard atmosphere is defined as equal to the pressure exerted by a column of 760 mm of mercury at 0°C when density of mercury is 0.013 595 1 kg/cm^3, and acceleration due to gravity is equal to 980.665 cm/s^2. The pressure exerted by one millimetre column of mercury under these conditions is named as Torr, in honour of the Italian physicist Torricelli (1608–1647), who first constructed the barometer to measure pressure.

$$1 \text{ atm} = 760 \text{ Torr} = 101\ 325 \text{ Pa}$$

$$1 \text{ Torr} = 0.1333 \text{ kPa}$$

In engineering thermodynamics, quite often the pressure is expressed in terms of the unit bar (1 bar = 100 kPa). A simple and widely used instrument to measure the pressure difference is a U tube manometer. Because of the use of manometers, pressure is quite often expressed in terms of millimeters of mercury column.

$$1 \text{ mm Hg} = 0.1333 \text{ kPa}$$

A pressure which is less than atmospheric pressure is often referred to as vacuum. That is, a pressure which is less than atmospheric pressure by 600 mm Hg is called a vacuum of 600 mm Hg and it is expressed as 600 mm Hg vac. Most of the pressure measuring devices like Bourdon pressure gage, measure the difference in pressure between the fluid and the atmosphere. This pressure difference (or the pressure indicated by the gage) is called *gage pressure*. The absolute pressure, P, of the fluid is then given by

$$P = P_{\text{atm}} + P_{\text{gage}}$$

$$P = P_{\text{atm}} - P_{\text{vac}}$$

The relationships between the absolute pressure, gage pressure, vacuum and atmospheric pressure is shown in Fig. 1.2.

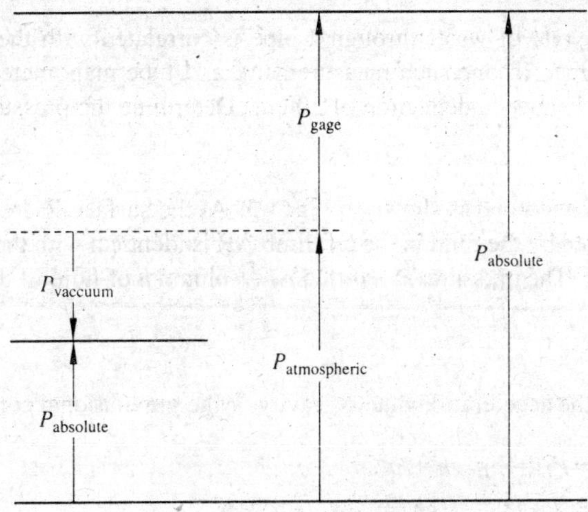

Fig. 1.2 *Relation between absolute pressure, gage pressure, vacuum and atmospheric pressure*

There are certain units which are outside the SI and are recognized by the *Comit' International des Poids et Measures* (CIPM). In view of their practical importance, these units are also used alongwith the SI units in engineering thermodynamics. The non-SI units recognized by CIPM are presented in Table 1.3.

Table 1.3 *Non-SI units recognized by CIPM*

Quantity	Unit	Symbol	Relation with SI unit
Time	Minute	min	1 min = 60 s
	Hour	h	1 h = 3600 s
	Day	d	1 d = 8.64×10^4 s
Volume	Litre	l	1 l = 1×10^{-3} m^3
Mass	Tonne	t	1 t = 1000 kg
Energy	Electronvolt	ev	1 ev = $1.602\ 19 \times 10^{-19}$ J
Pressure	Bar	bar	1 bar = 100 kPa

***Example* 1.1** A bell jar 50 cm in diameter is made to rest on a flat plate and is evacuated with the help of a vacuum pump till the pressure inside the jar reduces to 25 Pa. If the atmospheric pressure is 101.325 kPa, determine the force required to lift the bell jar off the plate.

Solution Net pressure acting on the bell jar $P = 101\,325 - 25 = 101\,300$ Pa

Force acting on the plate is given by

$$F = PA = P\frac{\pi D^2}{4} = 101\,300 \times \frac{\pi(0.5)^2}{4} = 19.89 \text{ kN}$$

Hence the minimum force required to lift the bell jar off the plate = 19.89 kN.

***Example* 1.2** The flow rate of water through a pipe is correlated with the pressure drop across a specified length of the pipe. In one such measurement, a U-tube manometer filled with mercury of density 13.6×10^3 kg/m^3 shows a deflection of 20 cm. Determine the pressure drop if the density of water is 1000 kg/m^3.

Solution The U-tube manometer is shown in Fig. 1.3. At the surface R–R, the pressure is uniform. Hence, the pressure exerted by the fluid in the left limb AR is identical with the pressure exerted by the fluid in the right limb BR. The pressure, P, exerted by a column h of fluid of density ρ is given by

$$P = \rho h g$$

where, $g = 9.81$ m/s^2 is the acceleration due to gravity or the gravitational constant. Then

$$P_A + \rho_w g(h + l) = P_B + \rho_{Hg} g h + \rho_w g l$$

$$\text{or} \quad P_A - P_B = \rho_{Hg} g h - \rho_w g h = (\rho_{Hg} - \rho_w) g h$$

$$= (13.6 - 1) \times 10^3 \times 9.81 \times 0.2 = 24.721 \text{ kPa}$$

where, P_A, P_B = pressure at locations A and B, respectively, ρ_w = density of water, and, ρ_{Hg} = density of mercury.

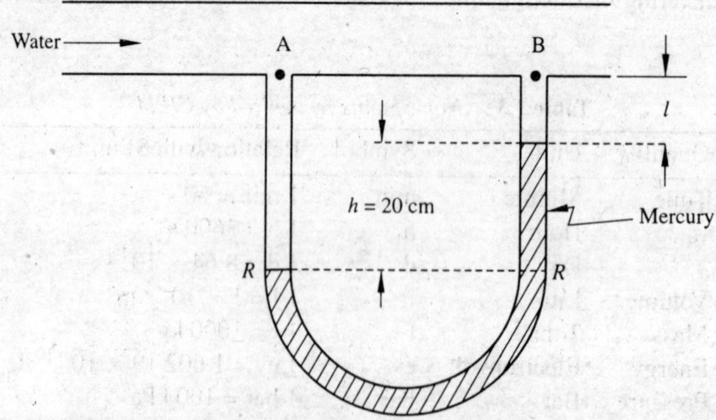

Fig. 1.3 *Sketch for Example 1.2*

***Example* 1.3** The cylinder–piston assembly shown in Fig. 1.4 contains 0.1 m^3 of a gas at a pressure of 101.325 kPa. At this stage the spring is touching the piston but exerts no force on it. The gas is heated till the volume is doubled. During this process the force exerted by the spring is proportional to the displacement of the piston. If the spring constant is 50 kN/m and the cross-sectional area of the piston is 0.05 m^2, determine the final pressure of the gas in the cylinder.

Solution A force balance on the piston gives

$$PA = P_a A + kx = P_a A + \frac{k(V - V_0)}{A}$$

$$\text{or, } P = P_a + \frac{k(V - V_0)}{A^2}$$

where, P_a = atmospheric pressure
A = cross-sectional area of piston
k = spring constant
V = volume of the gas at any instant during expansion
V_0 = initial volume of the gas.

Substituting the given values in the above equation, we get

$$P = 101.325 + \frac{50(0.2 - 0.1)}{(0.05)^2} = 2101.325 \text{ kPa} = 2.1013 \text{ MPa}$$

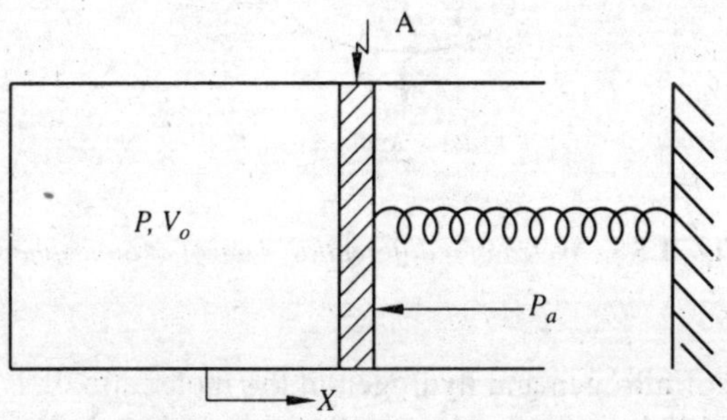

Fig. 1.4 *Sketch for Example 1.3*

***Example* 1.4** Assume that the atmosphere is locally isothermal, that is, the variation of the pressure with the specific volume of the atmospheric air follows the relation $Pv = P_0 v_0$ where the subscript zero denotes the conditions at the surface of the earth. Show that the pressure variation with the height in such an atmosphere is given by:

$$P = P_0 \exp\left(\frac{-gh}{P_0 v_0}\right)$$

where h is the height above the earth's surface and g the acceleration due to gravity.

Solution A force balance on an elemental portion of the atmosphere shown in Fig. 1.5 gives:

$$(P + dP)A + \frac{Agdh}{v} = PA$$

or,

$$dP = -\frac{gdh}{v} \quad \text{or} \quad \frac{dP}{P} = -\frac{gdh}{P_0 v_0}$$

Integrate the above relation from $h = 0(P = P_0)$ to $h = h(P = P)$ to obtain:

$$\ln \frac{P}{P_0} = -\frac{gh}{P_0 v_0} \quad \text{or} \quad P = P_0 \exp\left(\frac{-gh}{P_0 v_0}\right)$$

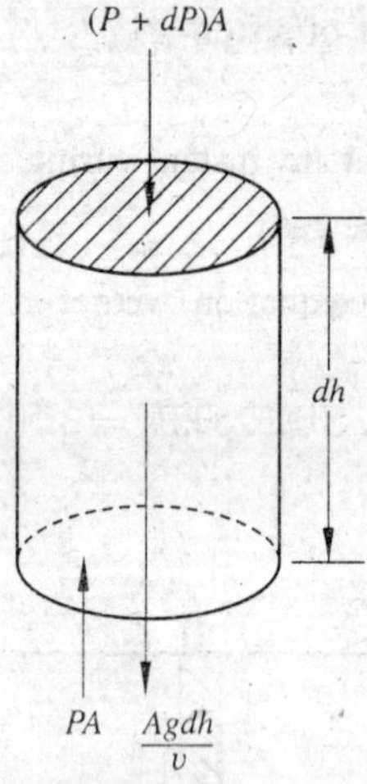

Fig. 1.5 *Sketch of a differential element of atmosphere*

***Example* 1.5** A mixture of nitrogen and hydrogen in the mole ratio of 1:3 enters an ammonia synthesis reactor at the rate of 100 kg/min. Express the flow rate in terms of kmol/min.

Solution

1 mol gas mixture contains 0.25 mol N_2 and 0.75 mol H_2

0.25 mol N_2 = 0.25 (molar mass of N_2) = $0.25(28 \times 10^{-3}) = 7 \times 10^{-3}$ kg

0.75 mol H_2 = $0.75\ (2 \times 10^{-3}) = 1.5 \times 10^{-3}$ kg

Therefore, one mol mixture = $(7 + 1.5)10^{-3} = 8.5 \times 10^{-3}$ kg

Hence, 100 kg mixture/min = $\dfrac{100}{8.5 \times 10^{-3}} = 11.765$ kmol/min

1.5 Summary

Thermodynamics is a science of broad scope and great generality and is applicable to all types of systems however complex they may be. Thermodynamics deals with the study of energy and its transformation and enables engineers to determine the feasibility of a given process, the energy required to carry out the process and the limit or the extent to which the process/change can take place.

In the study of thermodynamics one can adopt either a macroscopic approach or a microscopic approach. In macroscopic approach the state of the matter under consideration is specified by only a few of its measurable parameters like P, v and T. This approach is adopted in classical thermodynamics. On the other hand, in microscopic approach a large number of variables which cannot be measured directly, are used to specify the state of matter under consideration. This approach is used in statistical thermodynamics.

A knowledge of the units of measurement of the physical quantities commonly encountered is essential for the understanding of thermodynamics. SI units are briefly reviewed. The units of force, pressure, energy and power in the SI units are newton (N), pascal (Pa), joule (J) and watt (W). These units should not be represented in plural form and no period (full stop) should be placed at the end of the abbreviated form. The units can be expressed in terms of appropriate multiples of SI units.

1.6 Review Questions

1. Industrial grade methanol can be produced from CO and H_2 according to the reaction $CO + 2H_2 \longrightarrow CH_3OH$. If an engineer has to design a plant for the production of methanol by making use of the above reaction, list the various questions for which he/she needs complete answers.
2. State the essence of the four laws of thermodynamics.
3. What are the different approaches adopted in the study of thermodynamics and which approach is adopted in the study of classical thermodynamics?
4. What are the base units in SI units?
5. Define the term mole.
6. Distinguish between Joule and joule.
7. Name the SI units which are named after famous scientists.
8. What are the important conventions adopted in the use of SI units?
9. What is a bar and what is a Torr?
10. What is the value of one standard atmosphere pressure in SI units?
11. What are the SI unit symbols for millisecond and metre $\times$ second?
12. Distinguish between absolute pressure and gage pressure.

1.7 Problems

1.1 Atmospheric pressure is usually measured with the help of a barometer shown in Fig. 1.6. On a particular day at a particular location where $g = 9.7$ m/s^2, if a barometer reads 735 mm Hg, determine the atmospheric pressure in kPa and in bars.

1.2 The pressure gages, in common use, are usually calibrated in terms of kg/cm^2 (the pressure exerted by 1 kg mass on an area of 1 cm^2). If a pressure gage connected to a gas chamber

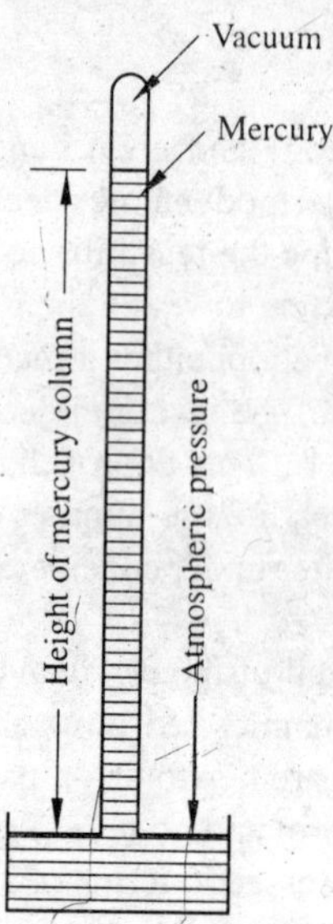

Fig. 1.6 *Schematic diagram of a barometer*

reads 5 kg/cm^2, what is the absolute pressure (in bars) of the gas in the chamber? Assume that $g = 9.78$ m/s^2 and the atmospheric pressure is 98.6 kPa.

1.3 A vacuum gage connected to a partially evacuated chamber reads 690 mm Hg vac. Calculate the absolute pressure of the gas in the chamber when a barometer reads 740 mm Hg at the same location where $g = 9.75$ m/s^2.

1.4 A silicone oil manometer is used to measure the pressure of a gas in a test cell. One limb of the manometer is evacuated and sealed and the other limb is connected to the test cell. The density of the silicone oil is 800 kg/m^3. If the difference in the height of the silicone oil in the two limbs is 30 cm, determine the pressure of the gas in the test cell.

1.5 A mercury manometer is used to measure the pressure drop of water flowing in a pipeline. If the difference of the mercury levels in the two limbs reads 35 cm, determine the pressure drop. The density of mercury is 13.6×10^3 kg/m^3 and the acceleration due to gravity $g = 9.81$ m/s^2

1.6 A hemispherical bulb of 10 cm diameter is held against the ceiling of a room, and the space between the ceiling and the bulb is evacuated to a pressure of 0.325 kPa. Safety considerations require that not more than 60% of the mass which separates the bulb from the ceiling can be suspended to a hook attached to the bulb. Determine the mass that can be safely held by the bulb. The atmospheric pressure is 101.325 kPa.

1.7 A cylinder-piston assembly shown in the Fig. 1.4 contains 0.1 m^3 of gas at a pressure of 5 bar. The spring exerts a force which is proportional to the displacement of the piston. The atmospheric pressure is 1 bar. The gas is heated until the volume is doubled at which point the pressure of the gas is 20 bar. Determine the spring force constant. The cross-sectional area of the piston is 20 cm^2.

1.8 Assume that the atmosphere is adiabatic, that is, the pressure variation with the specific volume of the air follows the relation $Pv^\gamma = P_0 v_0^\gamma$ where the subscript zero denotes the conditions at the surface of the earth. Derive an expression to estimate the pressure as a function of height above the surface of the earth.

1.9 Assuming that the atmosphere is locally isothermal and at 27°C, estimate the pressure at a hill station which is 3000 m above the sea level. It is known that the pressure at the sea level is 1 bar and assume that g is constant and is equal to 9.8 m/s^2.

1.10 Rework Problem 1.9 assuming that the atmosphere is adiabatic.

1.11 Liquefied Petroleum Gas (LPG) is sold in small cylinders for use as cooking gas. If one LPG cylinder contains 15 kg gas, express the quantity of gas contained in the cylinder in moles. Treat LPG as butane (C_4H_{10}) only.

1.12 Determine the number of moles in one kg of each of the following: H_2; O_2; CH_4 and C_4H_{10}.

1.13 A gas mixture has the following composition:

Component	CO_2	He	O_2	H_2
Mass %	20	40	10	30

Determine the number of moles of gas in 100 kg of the mixture.

1.14 A particular spring stretches 3 mm for every newton of applied force. When an object of unknown mass is suspended from the spring a deflection of 3 cm is observed. Assuming that the acceleration due to gravity at that location is 9.81 m/s^2, determine the mass of the object.

1.15 A home cooler has a fan of 170 W rating and a water recirculating pump of 30 W rating. If the cooler operates for a period of 8 hours, determine the energy consumed by the cooler.

1.16 A domestic refrigerator has a compressor of 200 W rating. The compressor runs for 40% of the time only to maintain the inside of the refrigerator at the required temperature. Estimate the energy required to operate the refrigerator for one month. If electrical energy costs Rs 3.75 per kWh, estimate the monthly expenditure for operating the refrigerator.

1.17 A submarine is located at a depth of 300 m in sea. If the average density of sea water is 1040 kg/m^3 and the local acceleration due to gravity is 9.7 m/s^2, estimate the pressure on the outer surface of the submarine.

2 Concepts and Definitions

Learning Objectives

This chapter will enable readers to:

- Define the terms system and surroundings.
- Understand thermodynamic property.
- Distinguish between extensive and intensive properties.
- Know kinetic energy, potential energy and internal energy.
- Understand the criteria of thermal equilibrium, mechanical equilibrium and chemical equilibrium.
- State the thermodynamic definition of work.
- Distinguish between heat and work.
- Calculate the work done by a system.

2.1 System

The study of any branch of science or engineering requires a thorough understanding of the words which are used in explaining its principles. A sound knowledge of the definitions of the terms at the outset will eliminate any possible misunderstanding. Therefore, it is essential to build the vocabulary or terminology needed in the study of thermodynamics. In this chapter we shall discuss the basic concepts and definitions of some terms which are widely used in thermodynamics.

In any scientific or engineering analysis it is essential to clearly identify the material on which we focus our attention. In mechanics, when we want to study the behavior of a body with the help of Newton's laws of motion, the body is isolated from its surroundings. This is done by imagining that the body or a part of the body is separated from the surroundings by an assumed boundary. Such an assumed boundary sometimes passes through the body; sometimes it coincides with the geometric surface of the body. Similarly, in thermodynamic analysis, we clearly identify the subject of discussion or analysis by enclosing it within a boundary which separates the subject of analysis from its surroundings. For this purpose, we use the term thermodynamic system or just system, for simplicity. A thermodynamic system is the material under consideration and it is defined by the analyst depending upon the particular problem at hand.

In other words, a thermodynamic system refers to a definite quantity of matter bounded by some closed surface. The definition of a system and its boundary is completely arbitrary. Depending upon the problem, the choice of the system may differ from person to person performing the analysis. In other words, for the same problem there can be several alternative systems. The choice of a system and its boundary is largely dictated by convenience and not by any principle. The boundary surface may change in shape as well as in size when the system undergoes a change. For example, if we wish to analyse the state of a gas enclosed in an elastic balloon, the physical boundary of the balloon can be treated as the system boundary. The balloon, which is initially spherical in shape, may be deformed into a cylindrical shape or some other geometrical shape, or it may get squeezed to a reduced volume. Therefore, the boundary need not retain the same shape or size during a given change. The boundary need not necessarily be real either. It can be imaginary. A system can be very simple, such as a given amount of matter, or it can be a very complex one such as a complete thermal power plant. However, the quantity of the material constituting the system must remain constant. To emphasise the fact that the system contains a definite quantity of matter, sometimes we use the term *control mass* for a system. When the changes in the properties of matter are to be estimated, we usually use control-mass analysis or system analysis.

2.2 Surroundings

The combination of matter and space outside the system, that is, the material which is not included in the system boundary, constitutes the surroundings. Whenever a system undergoes a change, its effect will be seen in the immediate surroundings. However, at a distance, far removed from the system, the effect will not be detectable and hence, for all practical purposes that part of the surroundings where the effects are not detectable need not be considered.

A system and its surroundings are separated from each other by some boundary. The boundary or the wall, employed to separate a system from its surroundings can be classified as:

(a) Movable or rigid depending upon whether it allows a change in volume of the system. A rigid wall is said to be restrictive with respect to volume whereas a movable wall is said to be nonrestrictive with respect to volume.
(b) Diathermal or adiabatic depending upon whether it allows the exchange of energy in the form of heat. An adiabatic wall is said to be restrictive with respect to energy flow in the form of heat whereas a diathermal wall is said to be nonrestrictive with respect to energy flow in the form of heat.

A system can be made to undergo a change by manipulating the nature of the wall enclosing the system. A system which is enclosed by an adiabatic wall can exchange energy in the form of work with the surroundings. A system which is enclosed by a rigid and adiabatic wall cannot exchange energy with its surroundings either in the form of heat or work. Such a system is usually referred to as an *isolated system*. In classical thermodynamics, the changes in the properties of the system can be estimated without paying any attention to what happens to the individual particles (molecules or atoms) that constitute the system.

2.3 Property

We have learnt that a thermodynamic system refers to a definite quantity of matter. A thermodynamic system may consist of various species. Suppose a mixture of nitrogen, hydrogen and ammonia is contained in a reactor at a given temperature and pressure. If we are interested in predicting the change of state of the gas mixture when the reaction conditions are altered, we would like to choose the gas mixture as our system. This system consists of the species nitrogen, hydrogen and ammonia. To describe such a system and to predict the changes that occur in the system because of its interaction with the surroundings, a knowledge of the properties of the materials constituting the system is essential. What is a property? Sometimes we specify the properties of water as colorless, odorless and tasteless. When our interest lies in recommending water for potable purposes, certainly these properties of water are relevant. However, since thermodynamics deals with energy and its transformation, any property that is not associated with energy transfer is irrelevant for the purpose of thermodynamic analysis. Therefore, only those properties which are associated with energy and its transformation are relevant for thermodynamic analysis. That is, a property in relation to a thermodynamic system is a characteristic of the system, which can be quantitatively evaluated and is associated with energy and its transformation. We describe a system by specifying its properties. The essential features of a property are:

(a) A property should have a definite value when the system is in a particular state, and
(b) The value of the property should be determinable irrespective of how the system is brought to that particular state.

Since the value of a property of a system is independent of the process or path followed by the system in reaching a particular state, the change in the value of the property depends only on the initial and the final states. Mathematically, if Z is a property of the system, then the change in the property in going from the initial state i to the final state f is given by:

$$\int_i^f dZ = Z_f - Z_i \tag{2.1}$$

If $Z = Z(x, y)$, then

$$dZ = \left(\frac{\partial Z}{\partial x}\right)_y dx + \left(\frac{\partial Z}{\partial y}\right)_x dy = Mdx + Ndy \tag{2.2}$$

where

$$M = \left(\frac{\partial Z}{\partial x}\right)_y \quad \text{and} \quad N = \left(\frac{\partial Z}{\partial y}\right)_x$$

If $\left(\frac{\partial M}{\partial y}\right)_x = \left(\frac{\partial N}{\partial x}\right)_y$, then dZ is said to be an exact differential, and Z is a point function. The change in the value of Z in reaching the state f, starting from the state i can be written as:

$$\Delta Z = \int_i^f dZ = Z_f - Z_i$$

Therefore, *a thermodynamic property is a point function and not a path function.* Pressure, temperature, volume/mole are some of the quantities which satisfy these requirements and, therefore, they are used to describe the state of a thermodynamic system.

***Example* 2.1** The differential for the molar volume, (v) of a gas is given by one of the following expressions:

(a) $dv = \frac{R}{P}dT - \frac{RT}{P^2}dP$
(b) $dv = \frac{2R}{P}dT - \frac{RT}{2P^2}dP$

where P is pressure, T the temperature and R, a constant. Identify the correct equation and find the relation between P, v and T for the gas.

Solution

(a) Given $v = v(P, T)$, Then

$$dv = \left(\frac{\partial v}{\partial T}\right)_P dT + \left(\frac{\partial v}{\partial P}\right)_T dP = MdT + NdP$$

This is of the form $dZ = Mdx + Ndy$. Since volume is a property, its differential must be exact. To find the correct expression, let us check whether the given differentials are exact. The test for exactness is:

$$\left(\frac{\partial M}{\partial P}\right)_T = \left(\frac{\partial N}{\partial T}\right)_P$$

$$\left(\frac{\partial M}{\partial P}\right)_T = \frac{-R}{P^2} = \left(\frac{\partial N}{\partial T}\right)_P$$

Hence (a) is an exact differential.

(b) $$\left(\frac{\partial M}{\partial P}\right)_T = \frac{-2R}{P^2} \quad \text{and} \quad \left(\frac{\partial N}{\partial T}\right)_P = \frac{-R}{2P^2}$$

Hence (b) is not an exact differential. Therefore (a) is the correct differential equation for volume.

Now, to obtain the explicit relation $v = v(T, P)$, integrate the first term of (a) partially with respect to T and obtain

$$v = \frac{RT}{P} + f(P)$$

where $f(P)$ is a function of P only. Then

$$\left(\frac{\partial v}{\partial P}\right)_T = \frac{-RT}{P^2} + \frac{\partial f(P)}{\partial P} = N$$

Therefore, $\frac{\partial f(P)}{\partial P} = 0$. Hence $f(P)$ = Constant. Then $v = \frac{RT}{P}$ + const.

2.3.1 Intensive and extensive properties

Suppose a homogeneous system, say air at a specified temperature and pressure, shown in Fig. 2.1, is divided into several parts by inserting a few partitions. Then, the mass of the whole system is equal to the sum of the masses of the various parts. The volume of the whole system is also given by the sum of the volumes of the constituent parts. On the other hand, the temperature of the whole system is not equal to the sum of the temperatures of the constituent parts. The temperature of the whole system is equal to the temperature of any part. Similarly, the pressure of the whole system is not equal to the sum of the pressures of the constituent parts, but it is equal to the pressure of any part.

$$m = m_A + m_B + m_C + m_D$$

$$V = V_A + V_B + V_C + V_D$$

where, V = volume of the whole system,

m = mass of the whole system,

V_A, V_B, V_C, V_D = volume of parts A, B, C and D, respectively,

m_A, m_B, m_C, m_D = mass of parts A, B, C, and D, respectively.

A	B	C	D
m_A	m_B	m_C	m_D
V_A	V_B	V_C	V_D

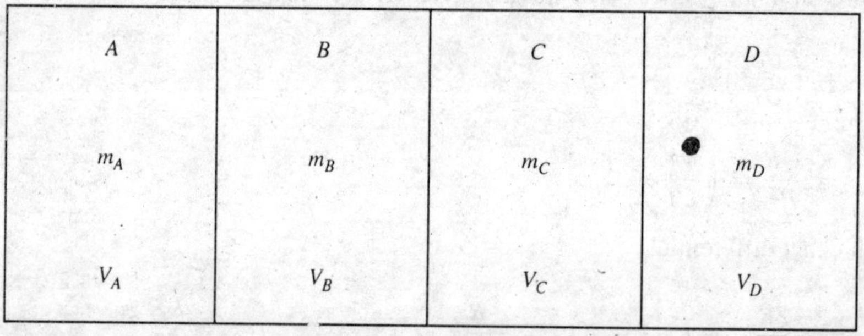

Fig. 2.1 *A homogeneous system divided into several parts A, B, C, D*

Thus we notice that there are certain properties which depend upon the size or extent of the system, and there are certain properties which are independent of the size or extent of the system. The properties like volume which are additive, that is, which depend on the size of the system are called *extensive properties*. The properties like temperature and pressure which are independent of the mass of the system are called *intensive properties*. The ratio of the extensive property to the mass (or property per unit mass) is called the specific value of that property.

Specific volume $= v = V/m$

Specific internal energy $= u = U/m$

where, U and m are internal energy and mass of the system respectively. Similarly, the molar properties are defined as the ratios of the properties to the mole number (N) of the substance.

Molar volume $= v = V/N$

Molar internal energy $= u = U/N$

We shall use the capital letters V, U, etc. to denote the extensive property and the lower case letters v, u etc. to denote the specific property or molar property. It should be clearly noted that the same lower case letters are used to represent the specific property as well as molar property, and the units will clearly indicate whether the property under consideration refers to the specific property or molar property.

Units of specific volume, $v = \text{m}^3/\text{kg}$

Units of molar volume, $v = \text{m}^3/\text{mol}$

2.4 Energy

2.4.1 Macroscopic modes of energy

Kinetic energy

The concept of energy was developed very slowly over a period of several hundred years, and has finally led to the establishment of the general principle of conservation of energy. From Newton's second law of motion, we know that when a body of mass m is acted upon by a force F, its acceleration a is related to the force by the relation $F = ma$. With the help of definition of work in mechanics as the product of force and displacement in the direction of force, we can evaluate the kinetic energy and the potential energy of a body. If during a small time interval dt, the displacement of the body is dL, then the work done by the force is given by:

$$dW = FdL = madL = m\left(\frac{d\mathcal{V}}{dt}\right)dL = m\left(\frac{dL}{dt}\right)d\mathcal{V} = m\mathcal{V}d\mathcal{V} \qquad (2.3)$$

If the body is accelerated from an initial velocity $\mathcal{V}_1$ to a final velocity $\mathcal{V}_2$ the total work done on the body is

$$W = \int_1^2 m\mathcal{V}d\mathcal{V} = \frac{m(\mathcal{V}_2^2 - \mathcal{V}_1^2)}{2} = \frac{\Delta\mathcal{V}^2}{2} \qquad (2.4)$$

The quantity $(m\mathcal{V}^2/2)$ has been given the name kinetic energy (KE) by Lord Kelvin in 1856.

The work done on a body in accelerating it from its initial velocity $\mathcal{V}_1$ to a final velocity $\mathcal{V}_2$, is equal to the change in the kinetic energy of the body. If the body is decelerated from a velocity $\mathcal{V}_2$ to a velocity $\mathcal{V}_1'$ by the application of a resisting force, the work done by the body is equal to the decrease in its kinetic energy. The work done on the body in the first instance can be recovered by reversing the direction of the process in bringing it back to its original condition. Therefore, the body, by virtue of its velocity, has the ability to do this work. The work done on the body in accelerating it has been stored in the body by virtue of its having a higher velocity and can be recovered at a later stage. The energy possessed by a body by virtue of its motion is called kinetic energy.

Example 2.2 A glider of mass 150 kg, which is initially at rest, is accelerated to a speed of 60 km/h by a vehicle. Calculate the work done on the glider by the vehicle. In course of time, if the glider's speed reduces to 10 km/h, calculate the change in the kinetic energy of the glider.

Solution Speed of glider $\mathcal{V}_2 = 60\ \text{km/h} = \dfrac{60 \times 10^3}{3600} = 16.67\ \text{m/s}$

Work done on the glider, W is given by

$$W = \frac{m(\mathcal{V}_2^2 - \mathcal{V}_1^2)}{2} = \frac{150\{(16.67)^2 - 0\}}{2} = 20.842 \text{ kJ}$$

Final velocity, $\mathcal{V}_f = \dfrac{10 \times 10^3}{3600} = 2.78$ m/s

Change in the kinetic energy, $\Delta(KE)$ is given by

$$\Delta(KE) = \frac{m(\mathcal{V}_f^2 - \mathcal{V}_i^2)}{2} = \frac{150\{(2.78)^2 - (16.67)^2\}}{2} = -20.263 \text{ kJ}$$

(ii) Potential energy

If a body of mass m is raised from an initial elevation Z_1 to a final elevation Z_2, in a gravitational field, the force exerted on the body is given by the relation $F = mg$, where g is the acceleration due to gravity. This force has moved a distance $(Z_2 - Z_1)$. Therefore, the work done on the body in moving it from Z_1 to Z_2 is:

$$W = F(Z_2 - Z_1) = mg(Z_2 - Z_1) = \Delta(mgZ) \tag{2.5}$$

The quantity (mgZ) has been given the name potential energy (PE) by Rankine. The work done on the body in raising it from one elevation to another is equal to the change in the potential energy of the body. The work done on the body can be recovered by lowering the body from elevation Z_2 to elevation Z_1. Therefore, the body has the ability to do work, and this ability to do work is called energy. The work done on the body in elevating it from Z_1 to Z_2 has been stored in the body by virtue of its elevation. Similarly, one can stretch a rubberband by doing work on it and the work done on the body is stored as energy in the body. This energy can be recovered at a later stage by allowing the rubber band to come to its original state. The energy possessed by a body by virtue of its elevation or configuration is called potential energy. These examples demonstrate that the work done on the body in accelerating it or in raising it, can be recovered later by reversing the direction of the process. These forms of energy, namely Kinetic Energy and Potential Energy, can be easily estimated in terms of macroscopically observable quantities like mass, velocity and elevation, and hence, they are known as macroscopic modes of energy..

2.4.2 Microscopic modes of energy

From a microscopic point of view, we can consider all materials as consisting of a large number of particles (molecules, atoms, electrons, etc.). We know that these particles are in continuous motion, and therefore possess energy in several forms. These forms of energy cannot be calculated in terms of macroscopically observable quantities and hence are called microscopic modes of energy. The molecule, as a whole, moves through space and possesses kinetic energy. The kinetic energy associated with the translational motion of the molecule is given by $m\mathcal{V}^2/2$, where m is the mass of the molecule and $\mathcal{V}$ is its velocity. This energy is termed as *translational energy* of molecule. The diatomic and polyatomic molecules rotate about the center of mass and possess energy due to rotational motion. This energy, called *rotational energy*, is given by $I\omega^2/2$, where I is the moment of inertia about the center of mass, and ω is the angular velocity. In addition to this, the atoms of the molecule vibrate back and forth about their common centre of mass. Because of this vibrational motion, the molecule has potential and kinetic energies also. This mode of energy is called the *vibrational energy* of the molecule. At low temperature the energy of a gas molecule is largely due to the translational and rotational modes of motion only. But, at high temperatures the vibrational modes contribute towards the energy of the molecule. It is not possible to evaluate the energy associated with the translational, rotational and vibrational modes of motion in terms of macroscopically measurable quantities. The energy possessed by all these microscopic modes is clubbed together and is called *internal energy U*.

The electrons continuously move around the nucleus and they have a certain amount of kinetic energy associated with this motion. The amount of this energy depends upon the orbit. Electrons in the lower orbits have smaller energies and the electrons in the outer orbits have higher energies. Usually, the electrons tend to be in the lower energy orbits. However, when the molecules are excited the electrons may move to high energy orbits. Then, the electrons also spin, and hence possess some spin energy. Some of these microscopic modes of energy are shown in Fig. 2.2. The electrons and nucleus are held together through coulombic interactions which are responsible for keeping the electrons in the orbits, and some potential energy is associated with these interactions. In addition, molecules are held together by molecular binding forces and there is a certain amount of potential energy associated with these intermolecular forces.

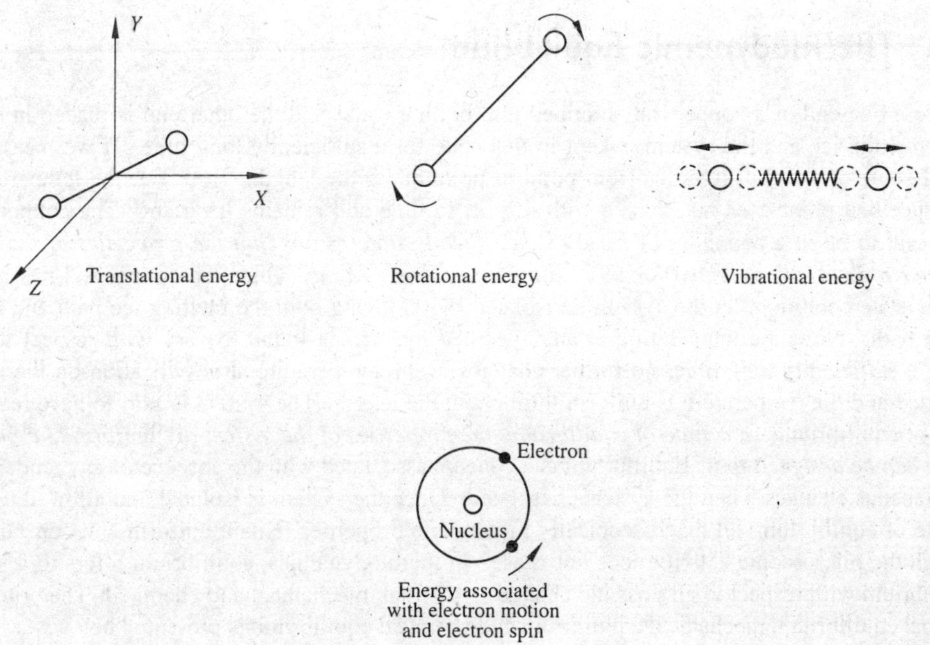

Fig. 2.2 *A few microscopic modes of energy*

In a molecule, the atoms are held together by chemical bonds. The chemical reactions are associated with the breaking of existing bonds and the formation of new bonds. The chemical reactions are accomplished by evolution or absorption of energy because the energy associated with the product molecules is different from the energy associated with the reactant molecules. The difference in energy of the molecules arises due to the difference in the structure of the molecules. The energy associated with the chemical bonds is called the *chemical energy, E_{ch}*. In an atom, the nucleus consists of neutrons and protons and these particles are often referred to as nucleons. These nucleons are bonded together in a nucleus. The energy associated with the nuclear bonds is called the *nuclear energy, E_{nu}*. Thus, we find that a system possesses energy in several forms and the energy of a system is given by

$$E = KE + PE + U + E_{ch} + E_{nu}$$

In absence of nuclear and chemical reactions, the above relation reduces to

$$E = KE + PE + U \tag{2.6}$$

We know that kinetic energy of a system is by virtue of its velocity, and that potential energy is by virtue of its configuration. These forms of energy can be readily converted into work and, hence, are available in the most useful or 'organized' form. On the other hand, microscopic modes of energy, which are due to random motions, or disorganized motions of molecules, cannot be readily converted into work. When dealing with energy and its transformation in thermodynamics, the major task involved is devising means for converting the disorganized internal energy into useful or organized work.

2.5 Thermodynamic Equilibrium

Suppose one end of a copper rod is dipped into boiling water and the other end is placed in contact with melting ice and the system is kept in that state for a sufficiently long time. If we measure the temperature of the rod, it varies from point to point along the length. However, the temperature at any specified point does not change with respect to time and remains invariant. The copper rod is then said to be in a condition of steady state. *Steady state means that the properties of the system (copper rod) at any specified location, are independent of time.* Once the copper rod has reached steady-state condition, let the system be isolated by removing both the melting ice bath and boiling water bath. Now, the temperature at any specified location is found to vary with respect to time. After a sufficiently long time, no further changes in the temperature at any location on the rod are detected and the temperature is uniform throughout the length. The system is said to have reached a state of equilibrium. *In a state of equilibrium, the properties of the system are uniform and only one value can be assigned to it.* Equilibrium is a concept associated with the absence of any tendency for spontaneous changes when the system is isolated. Once the system is isolated and allowed to reach a state of equilibrium, all macroscopically measurable properties remain uniform and constant even though the microscopic activity does not cease. In thermodynamics, equilibrium refers to a state of equilibrium with respect to all possible changes—thermal, mechanical and chemical. The criteria for thermal equilibrium, mechanical equilibrium and chemical equilibrium is presented below.

2.5.1 Thermal equilibrium

Consider two blocks of copper, one hot and the other cold. When these two blocks are brought into contact with each other, observe their temperatures continuously. It can be observed that the temperature of both the blocks keep changing, but after some time there are no further changes. Both the blocks ultimately attain the same temperature. At this stage, these blocks are said to be in a state of thermal equilibrium. In general then, if any two given systems are in a state of thermal equilibrium, they have the same temperature. Conversely, if the temperatures of two bodies are equal, they will be in thermal equilibrium when placed in diathermal contact. A state of thermal equilibrium can be described as one in which the temperature of the system is uniform.

2.5.2 Mechanical equilibrium

If a force acts on a system, stresses develop within the system to balance the applied force. If the system is isolated, the stresses continue to develop till certain level, which then remain constant with respect to time. This state in which the applied force and the developed stresses are balanced, is called the state of mechanical equilibrium. In that state the system would not have any tendency to change its mechanical state. For thermodynamic purposes, a system in a state of mechanical equilibrium will have uniform pressure as long as the effect of gravity is not significant, otherwise a pressure variation can exist as in the case of a liquid in a long vertical tube.

2.5.3 Chemical equilibrium

Suppose a small crystal of potassium permanganate ($KMnO_4$) is dropped in a glass beaker containing water. Immediately after dropping the crystal, the color of water in the immediate vicinity of the crystal will be dark purple and in other places the water will be colorless, which indicates that the concentration of $KMnO_4$ in water is very large in the immediate vicinity of the crystal, and at other places the concentration of $KMnO_4$ is zero. As time progresses, we find a gradient in the intensity of the color of water from dark purple in the vicinity of crystal to light purple in the top layer. If we wait for a sufficient period of time, we shall find that all the water in the beaker has the same uniform color. The potassium permanganate diffuses from the region of high concentration to the region of low concentration. When the concentration is uniform throughout, diffusion or mass transfer ceases and the system attains a state of chemical equilibrium. In a homogeneous or single phase system, if the concentration is uniform and there is no tendency for mass transfer or diffusion, the system is said to be in a state of chemical equilibrium.

Consider a tank containing a mixture of nitrogen and hydrogen at a sufficiently high temperature and pressure, where the chemical reaction between H_2 and N_2 is feasible leading to the formation of NH_3. If we start with a mixture of pure N_2 and H_2 (in the presence of a catalyst), as time progresses we notice the formation of NH_3. The composition of the mixture undergoes a change continuously with respect to time. If the system is isolated and left to itself, after a sufficiently long time, the composition of the mixture reaches a steady state value and there is no further change in the composition of the mixture. The system is now said to be in a state of chemical equilibrium. However, at microscopic level, the chemical reaction does not cease, but proceeds in the forward as well as in the backward directions, such that at macroscopic level there is no detectable change in the composition. In other words, if a system is in a state of chemical equilibrium, there is no detectable change in the chemical composition of the system.

If the composition of a system does not undergo any change because of chemical reactions and diffusion, and the system has no tendency for any chemical change when it is isolated, then the system is said to be in a state of chemical equilibrium. If two systems, each consisting of several components, are in a state of chemical equilibrium, then the chemical potential μ_i (which is defined later) of each component must be the same in both the systems. Therefore, the criterion for chemical equilibrium is the equality of chemical potential.

$$\mu_i^A = \mu_i^B$$

where μ_i is the chemical potential of component i, and the superscripts A and B refer to the systems A and B.

We have noticed that the criteria for equilibrium are:

- Equality of pressure—Mechanical equilibrium
- Equality of temperature–Thermal equilibrium
- Equality of chemical potential—Chemical equilibrium

We have learned that temperature and pressure are intensive properties, and the criterion for thermal equilibrium and mechanical equilibrium is expressed in terms of these intensive properties. The chemical potential is also an intensive parameter. Therefore, we observe that the criteria of equilibrium is expressed in terms of intensive parameters only.

Thus a system which is simultaneously in a state of mechanical equilibrium, thermal equilibrium and chemical equilibrium is said to be in a state of thermodynamic equilibrium.

2.6 Stability of Equilibrium States

The equilibrium states of a system can be classified as:

- Stable equilibrium
- Neutral equilibrium
- Unstable equilibrium
- Metastable equilibrium.

Considering a mechanical analogy as an example, these equilibrium states are illustrated in Fig. 2.3.

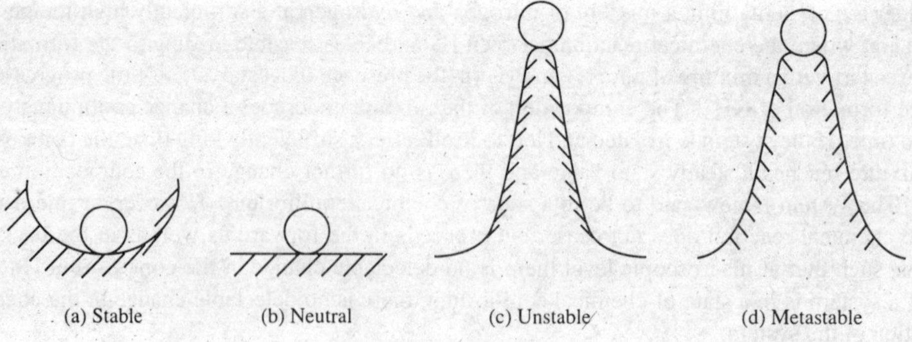

Fig. 2.3 *States of equilibrium*

The equilibrium condition shown in Fig. 2.3(a) is called stable equilibrium because the sphere always returns to its original position after it has been disturbed from its equilibrium position irrespective of how large the disturbance may be. The equilibrium condition shown in Fig. 2.3(b) is called neutral equilibrium, because the sphere assumes a new position where it remains indefinitely after it is disturbed. Fig. 2.3(c) shows an unstable equilibrium condition, where the sphere leaves the position and does not return to its original position after it is disturbed, even though the magnitude of the disturbance may be very small. Metastable equilibrium position is shown in Fig. 2.3(d) in which the sphere returns to its original position if subjected to small disturbances, but when the disturbance exceeds a certain magnitude, the sphere moves to a new state of equilibrium.

2.6.1 Examples of metastable states

(a) Metastable states in phase transition

Usually, metastable states exist along the phase boundaries - vapor–liquid, liquid–solid and solid–vapor transition lines. With careful cooling it is possible to produce a supercooled liquid, that is, a liquid at a temperature below which it normally solidifies. With care, water can be cooled to a temperature well below 0°C without solidification. Any disturbance, or addition of nucleation sites, initiates the process of solidification and the supercooled water immediately freezes. It is possible to have an undercooled vapor, that is, vapor at a temperature at which it normally condenses. Similarly, it is possible to produce a supersaturated liquid, that is, a liquid above its boiling point. Water can be heated to reach a temperature above 100°C without boiling. The introduction of a rough surface which provides nucleation sites, leads to immediate vapor formation. The metastable states in solids can last for long periods of time and can also be insensitive to large disturbances. For example, objects made of iron in Egypt about 6000 years ago contain cementite—a metastable phase of iron and carbon—and do not exhibit graphite which would appear in them in stable equilibrium. The metastable equilibrium states are also of some importance in metallurgy.

Figure 2.4(a) shows on a temperature–volume diagram the process paths for attaining metastable states by heating or cooling at constant pressure. If the liquid in state a is heated at constant pressure, it reaches a saturated state b and vaporization occurs. The path followed in such a case is $ab2$. However, if it is heated with care in a smooth glass container which does not provide any nucleation sites, the path followed will be $ab1$. The state 1 is a supersaturated liquid. Any disturbance, or insertion of a rough surface which provides nucleation sites, returns the system to state 2 which is a state of thermodynamic equilibrium. Similarly, constant-pressure cooling with care can result in the production of an undercooled vapor, and the path followed by the system is $cd3$. The state 3 represents undercooled vapor and d represents saturated vapor where condensation occurs. Any disturbance shifts the system to the thermodynamic equilibrium state 4.

(b) Metastable states in reactive systems

A mixture of hydrogen and oxygen at ambient conditions of temperature and pressure is an example of metastable equilibrium state. The mixture can remain in that state for a sufficient time. However, if the mixture is disturbed by introducing an electric spark, chemical reaction occurs spontaneously, leading to the formation of water vapor.

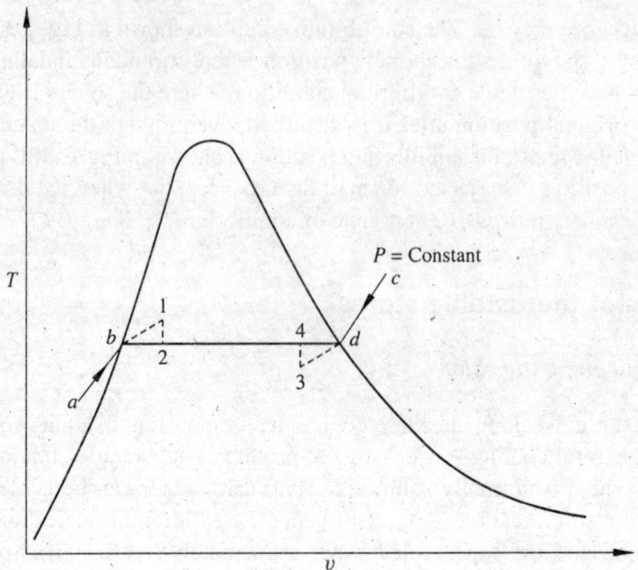

Fig. 2.4 *T–v diagram showing constant pressure heating/cooling to attain metastable states during phase transition*

2.7 Process

In thermodynamics we are mainly concerned with systems which are in a state of equilibrium. Whenever a system undergoes a change in its condition from one equilibrium state to another equilibrium state, the system is said to undergo a process. Consider a certain amount of gas enclosed in a piston–cylinder assembly as our system. Suppose the piston moves under such conditions that the opposing force is always infinitesimally smaller than the force exerted by the gas. Then the process followed by the system is said to be *quasi-static*. On the other hand, if the piston moves under such conditions that the forces are exactly balanced, the system is said to undergo a *reversible process*. Then the direction of the process can be reversed by an infinitesimal change in the pressure across the boundary. For such a process the states of the system can be located on a *P–V* diagram. A reversible process is always a quasi-static process, but the converse is not necessarily true (reversible and irreversible processes are discussed in detail in Chapter 7). It can be stated here without much explanation that *a process is said to be reversible if the system and its surroundings are restored to their respective initial states by reversing the direction of the process.* The process is irreversible if it does not fulfill this criterion.

All actual processes are non-quasistatic because they take place with finite temperature difference or pressure difference between a system and its surroundings. For the purpose of thermodynamic analysis, we idealize several processes as reversible. Many processes are characterized by the fact that some property of the system remains constant during the process. A process in which the volume of a system remains constant is called a *constant-volume or isochoric or isometric* process. If the pressure of a system remains constant during a process, it is called a *constant-pressure or isobaric* process. A process in which the temperature of the system remains constant is called *isothermal.* A process carried out by a system while it is enclosed by an adiabatic wall is called an *adiabatic process.*

2.8 Work

It has already been stated that thermodynamics deals with interactions between a system and its surroundings, or between two systems. These interactions are of two kinds: (a) Work interaction, and (b) Heat interaction.

A precise understanding of the meaning of these terms, namely heat and work, is essential in the thermodynamic analysis of a process because the major task of an engineer is to devise means to convert the highly disorganized internal energy of matter, into work. Historically, thermodynamics originated with man's ceaseless efforts aimed at converting internal energy of matter into work.

While discussing the kinetic energy and potential energy, we evaluated the work done on a body with the aid of the definition of work, as applied in mechanics.

In mechanics, work is defined as the product of the force (F) and the displacement (dL) in the direction of the force.

$$dW = FdL \tag{2.7}$$

To elaborate, consider a certain amount of gas enclosed in a piston–cylinder assembly shown in Fig. 2.5 as our system. Let A represent the cross-sectional area of a piston and dL its displacement during a small time interval dt. The infinitesimal work done by the system is

$$dW = \text{force} \times \text{distance} = PA(dL) = PdV \tag{2.8}$$

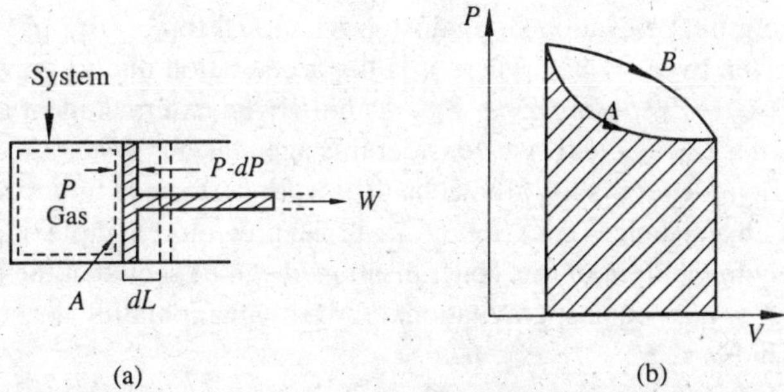

Fig. 2.5 *(a) Example of work done during a reversible expansion process (b) Illustration showing work done is a path function*

If the gas is allowed to expand (reversibly) from the initial pressure P_1 to the final pressure P_2, the total work done by the gas during the expansion process is

$$W = \int_1^2 PdV \tag{2.9}$$

During the expansion process the external pressure is always infinitesimally smaller than the gas pressure. The expansion process can be reversed at any time by increasing the external pressure infinitesimally more than the gas pressure. In other words, during the expansion process the forces on the moving boundary are balanced and the system undergoes a reversible process. This process can be represented on a P–V diagram. The area under the P–V curve represents the work done during the process. To evaluate the work done we should integrate PdV and, hence, we should know the relation between P and V, or the path of the process. It is clear from Fig. 2.5 that for the given initial and final states, the work done by the system in following path B is different from the work done by the system if it follows path A. Therefore, we observe that the work done by the system depends on the path followed by the system in reaching the given final state from the specified initial state. That is, *the work done by a system is a path function, and not a point function.* Therefore, work is not a property of the system, and it cannot be said that work is possessed by the system. It is an interaction across the boundary. What is stored in the system is energy, but not work. A decrease in the energy of the system appears as work done. Therefore, *work is energy in transit and can be identified only when the system undergoes a process.*

2.8.1 Thermodynamic definition of work

Let us consider a storage battery as our system. The terminals of the battery are connected to a resistance through a switch. When the switch is closed, current flows through the resistor. The circuit external to the system constitutes our surroundings. We find the resistor (surroundings) becoming warmer and the charge of the storage battery (system) decreasing. Obviously, the system has interacted with the surroundings. Can this interaction be called work? If we restrict ourselves to the definition of work in the sense it is applied in mechanics, the answer is no because no force has moved through any distance.

Before attempting a thermodynamic definition of work, let us consider the raising of a mass m from an initial elevation Z_1 to a final elevation Z_2 against gravitational force. To raise this mass, the force acting on the mass is given by $F = mg$, where g is the acceleration due to gravity. The work done on this body is $W = F(Z_2 - Z_1) = mg(Z_2 - Z_1)$. Definitely an external agent acted on the mass to raise it through the distance $(Z_2 - Z_1)$. If we consider this agent as our system, then the work done by the system is $mg(Z_2 - Z_1)$. Generalising, all mechanical work performed by a system is equivalent to raising a mass (m) through a distance (ΔZ). Reverting to the previous example of a gas expanding in a piston–cylinder assembly in the light of this generalisation, it can be seen that the expansion of the gas gets reduced to raising of a mass against gravitational force by manipulating the surroundings external to the system as shown in Fig. 2.6.

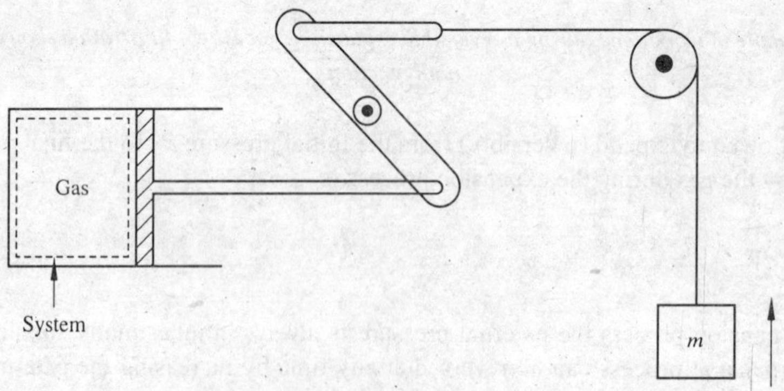

Fig. 2.6 *Visualization of work done by a system as equivalent to raising of a mass*

In the light of the above observations, we can now proceed to state a thermodynamic definition of work which has a wider scope than that used in mechanics. In thermodynamics, *work done by a system on the surroundings during a process is defined as that interaction whose sole effect, external to the system, could be viewed as the raising of a mass through a distance against gravitational force.* This definition of work is based on the definition first suggested by M. Planck and later developed by J.H. Keenan. This definition does not state that the mass is actually raised by the application of the force: what it states is that the sole effect, external to the system, could be the raising of a mass through a distance against gravitational force.

To show that the thermodynamic definition of work is more general than that used in mechanics, let us consider the interaction of a storage battery (system) with a resistor. According to the definition of work in mechanics, this interaction cannot be called work. However, according to the thermodynamic definition of work, this interaction can be termed as work. Let us modify the external circuit by replacing the resistor by a motor which draws the same amount of current as the resistor. The motor can wind a string and thereby raise the mass which is suspended, as shown in Fig. 2.7. When the switch is closed, the motor runs and the mass is raised. In this case, external to the system (battery), the only effect is the raising of a mass and hence the interaction is called work. As far as the battery (system) is concerned, the situation is identical to the previous case of discharging through a resistor. Therefore, the interaction of a battery with a resistor is called work, though no mass has been raised through any distance. By manipulating the environment that is external to the system, the effect can be reduced to raising of a mass against gravitational force, and that is the only effect in the surroundings.

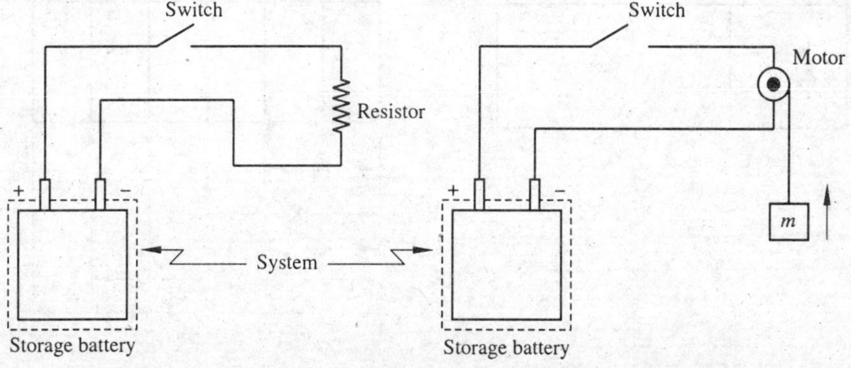

Fig. 2.7 *Visualization of battery interaction as work done*

Situation in which $W \neq \int Pdv$

Suppose a vessel is divided into two compartments by a rigid partition. One compartment of volume V_1 contains a gas at a given initial pressure (P_1) and the second compartment is evacuated. If the partition is removed, the gas expands and occupies both the compartments. The gas is said to have undergone a free expansion, that is, an expansion which is not restrained by an opposing force. Let the pressure be P_2 and volume V_2 in the final state. One can locate (P_1, V_1) and (P_2, V_2) on a P–V diagram. What is the work done by the gas during the expansion process? Is it equal to $\int PdV$? In other words, will it be equal to the area under the curve on a P–V diagram? To determine the area under the curve, we should know the path followed by the system or the trace of the process on the P–V diagram. The process can be carried out in steps as shown in Fig. 2.8. Suppose we place a few partitions a, b, c, d, e in the second compartment, and remove the partitions one at a time. Once the first partition a is removed, the gas pressure and volume can be measured (say P_b, V_b). Similarly, the gas pressure and volume can be measured after removing each partition, which provide us with the intermediate values of P and V. These values can then be plotted on the P–V diagram. One can draw a smooth dotted curve passing through the points $1, b, c, d, e, 2$, as shown in Fig. 2.9, and estimate the area under the curve. Does the area under the curve represent the work done by the system? The answer is no, because, the process is not reversible. The expansion of the gas is not restrained by an equal and opposing force at the moving boundary. Therefore $W \neq \int PdV$ in this case. Since no external force has moved through any distance in this case, the work done is equal to zero. An expansion which is not restrained is called a free expansion and the work done by a gas during free expansion is zero. We observe that, for a reversible process

$$W = \int PdV \tag{2.10}$$

and for all irreversible processes

$$W \neq \int PdV \tag{2.11}$$

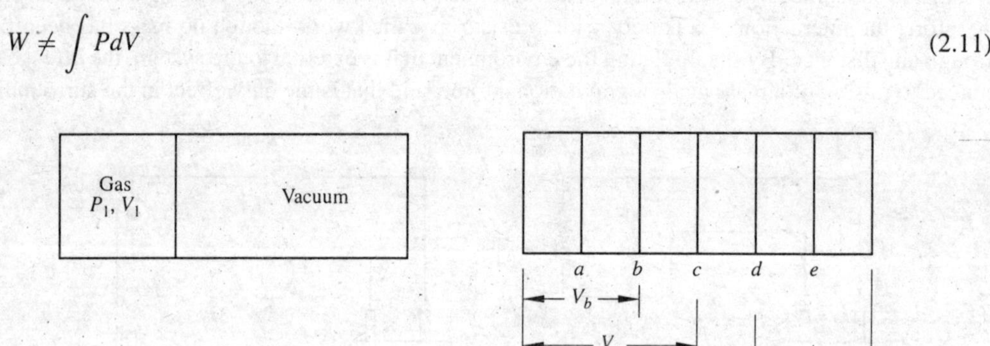

Fig. 2.8 *Schematic representation to show that work done during free expansion is equal to zero*

Let us consider the gas enclosed in a piston–cylinder assembly as shown in Fig. 2.10. The piston is held rigid with the help of latches. A paddle wheel is inserted into the gas space and made to run for a while. Since the latches hold the piston in position, there is no change in the volume of the gas ($dV = 0$). Is there any work done? The answer is yes, because the work is done on the system, and the amount of work done on the gas is equal to the decrease in the potential energy of the mass m. This is an example of a situation where $dV = 0$ and yet dW is not equal to zero. Stirrer or paddle wheel, work can be performed on a system, without causing any change in its volume. Note that this form of work can be done in one direction only, that is, work is done on the system by the surroundings.

Example 2.3 Consider a gas enclosed in a piston–cylinder assembly as the system. The gas is initially at a pressure of 500 kPa and occupies a volume of 0.2 m^3. The gas is taken to the final state

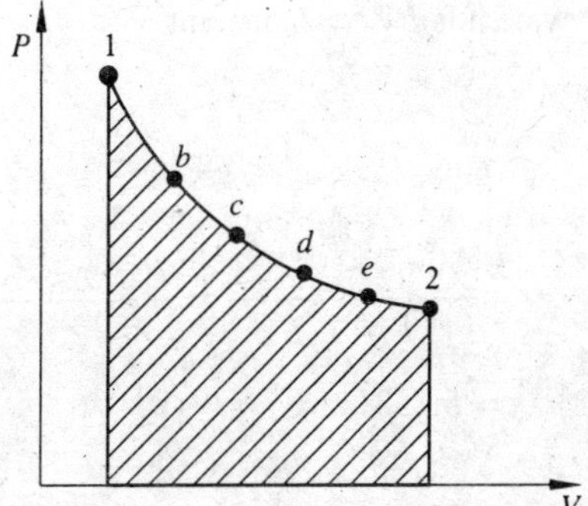

Fig. 2.9 *P–V diagram for free expansion when carried out in successive steps*

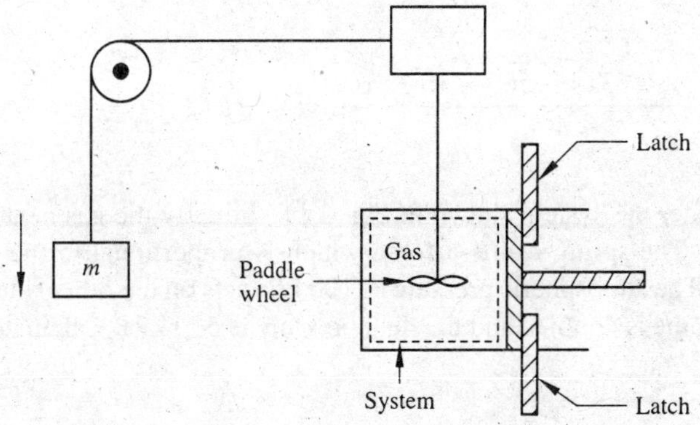

Fig. 2.10 *A process in which $dV = 0$ and yet $dW \neq 0$*

where $P_2 = 100$ kPa by the following two different processes. Calculate the work done by the gas in each case.

(a) The volume of the gas is inversely proportional to the pressure.
(b) The process follows the path PV^{γ} = Constant, where $\gamma = 1.4$.

Solution (a) The path followed by the system is $V \propto \frac{1}{P}$ or PV = Constant.

$$P_1V_1 = P_2V_2 = PV$$

$$P = \frac{P_1V_1}{V}$$

Work done, W is given by

$$W = \int_1^2 PdV = P_1V_1 \int_1^2 \frac{dV}{V} = P_1V_1 \ln \frac{V_2}{V_1}$$

$$= P_1V_1 \ln \frac{P_1}{P_2} = 500 \times 10^3 \times 0.2 \ln \frac{500}{100} = 160.94 \text{ kJ}$$

(b) The path followed by the system is PV^γ = Constant

$$P_1V_1^\gamma = P_2V_2^\gamma = K = PV^\gamma$$

or $P = \frac{K}{V^\gamma}$

$$W = \int_1^2 PdV = \int_1^2 \frac{KdV}{V^\gamma} = \frac{K(V_2^{1-\gamma} - V_1^{1-\gamma})}{1-\gamma} = \frac{P_2V_2 - P_1V_1}{1-\gamma}$$

$$P_1V_1^\gamma = 500 \times 10^3 \times (0.2)^{1.4} = 5.253 \times 10^4 = P_2V_2^{1.4}$$

Therefore,

$$V_2 = \left(\frac{5.253 \times 10^4}{100 \times 10^3}\right)^{\frac{1}{1.4}} = 0.6314 \text{ m}^3$$

Hence,

$$W = \frac{100 \times 10^3 \times 0.6314 - 500 \times 10^3 \times 0.2}{1 - 1.4} = 92.15 \text{ kJ}$$

***Example* 2.4** Consider the system shown in Fig. 2.11. Initially the gas is at 200 kPa and occupies a volume of 0.1 m^3. The spring exerts a force which is proportional to the displacement from its equilibrium position. The atmospheric pressure of 100 kPa acts on the other side of the piston. The gas is heated until the volume is doubled and the final pressure is 500 kPa. Calculate the work done by the gas.

Solution

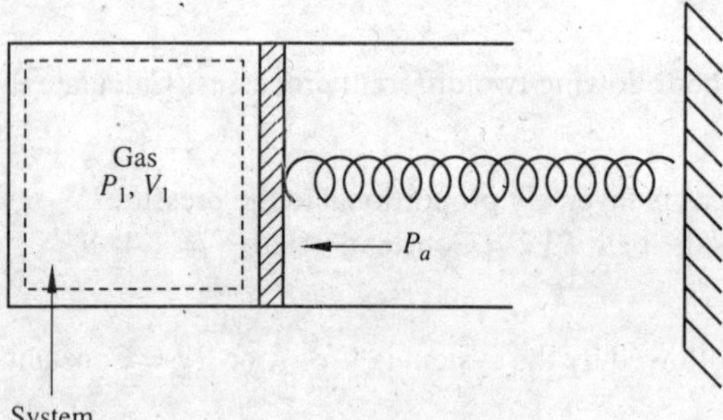

Fig. 2.11 *Sketch for Example 2.4*

Given that $P_1 = 200$ kPa, $V_1 = 0.1$ m^3; $P_2 = 500$ kPa, $V_2 = 0.2$ m^3

The force exerted by the spring, $F_s \propto x$ or $F_s = Kx$ where x is the displacement of the spring from its equilibrium position. Let A be the cross-sectional area of the piston. A force balance on the piston gives:

$$PA = P_aA + F_s = P_aA + Kx = P_aA + \frac{K(V - V_0)}{A}$$

or

$$P = P_a + \frac{K(V - V_0)}{A^2}$$

where P_a is the atmospheric pressure and V_0 is the volume of the gas when the spring is in its natural length.

Work done W is given by:

$$W = \int_1^2 PdV = \int_1^2 \left(P_a + \frac{K(V - V_0)}{A^2}\right) dV$$

$$W = P_a(V_2 - V_1) + \frac{K(V_2 - V_0)^2 - (V_1 - V_0)^2}{2A^2}$$

or,

$$W = P_a(V_2 - V_1) + \frac{K}{2A^2}(V_2 - V_0 + V_1 - V_0)(V_2 - V_0 - V_1 + V_0)$$

$$= P_a(V_2 - V_1) + \frac{K}{2A^2}(V_2 + V_1 - 2V_0)(V_2 - V_1)$$

$$W = \left(P_a + \frac{K}{2A^2}(V_2 + V_1 - 2V_0)\right)(V_2 - V_1) \tag{A}$$

Force balance across the piston, in the initial state gives:

$$P_1A = P_aA + \frac{K}{A}(V_1 - V_0)$$

or

$$P_1 = P_a + \frac{K}{A^2}(V_1 - V_0) \tag{B}$$

Similarly

$$P_2 = P_a + \frac{K}{A^2}(V_2 - V_0) \tag{C}$$

Adding Eqs. (B) and (C) and dividing by 2, we get:

$$\frac{P_1 + P_2}{2} = P_a + \frac{K}{2A^2}(V_2 + V_1 - 2V_0) \tag{D}$$

Substituting Eq. (D) in Eq. (A), we get:

$$W = \frac{P_1 + P_2}{2}(V_2 - V_1) = \frac{(200 + 500) \times 10^3}{2}(0.2 - 0.1) = 35 \text{ kJ}$$

The pressure versus volume diagram for the gas is shown in Fig. 2.12. It can be observed from Fig. 2.12 that the work done is given by the area of trapezium A12B.

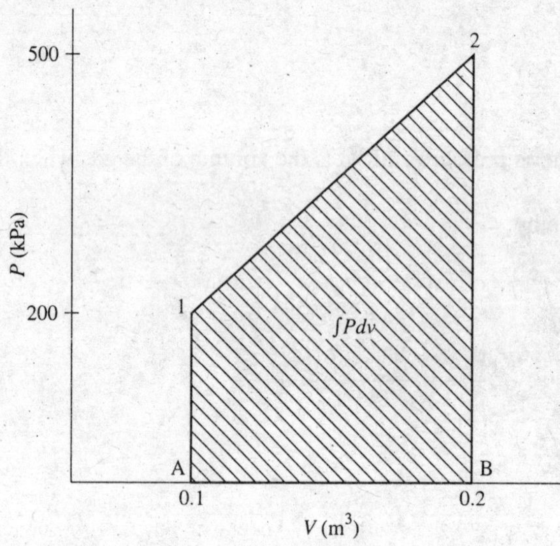

Fig. 2.12 *P–V diagram for Example 2.4*

Example 2.5 A spherical balloon of 1 m diameter contains a gas at 150 kPa. The gas inside the balloon is heated until the pressure reaches 450 kPa. During the process of heating the pressure of the gas inside the balloon is proportional to the cube of the diameter of the balloon. Determine the work done by the gas inside the balloon.

Solution Initial conditions: $P_1 = 150$ kPa and $D_1 = 1$ m

$P \propto D^3$ or $P = KD^3$

Therefore, $P_1 = 150 \times 10^3 = K(1)^3$ or $K = 150$ kPa/m^3

$P_2 = KD_2^3$ or $450 \times 10^3 = 150 \times 10^3 D_2^3$ or $D_2 = 1.4417$ m

$$W = \int_1^2 PdV = \frac{KD^3 \pi D^2 dD}{2} = \frac{\pi K(D_2^6 - D_1^6)}{12}$$

$$= \frac{\pi(150 \times 10^3)\left((1.4417)^6 - 1^6\right)}{12} = 313.352 \text{ kJ}$$

2.9 Other Modes of Work

So far we have discussed the compression and expansion (PdV) mode of work only. The work done is given by $dW = PdV$ in such systems. For most engineering purposes, attention is confined to systems which are described in terms of the coordinates P and V, in which the pressure (P) is the intensive coordinate and volume (V) is the extensive coordinate. There are a number of other modes of work frequently encountered in thermodynamic analysis of systems. A brief discussion of some of the common modes of work is presented below.

2.9.1 Extension of an elastic bar

Consider a solid rod of cross-section A as shown in Fig. 2.13. Let the normal stress applied on the rod in the x direction be σ. Then the force exerted on the solid is σA. Let dL represent the deformation in the solid rod. Then the work done on the rod is $dW = -\sigma A dL$. Negative sign is introduced to take into account the fact that when dL is positive (extension of the bar) work must be done on the system. The sign convention for heat and work is introduced in detail in Chapter 5. For the present, let it suffice to say that the work done on the system is considered as negative. The strain (ε) is defined as the deformation per unit length.

$$d\varepsilon = \frac{dL}{L} = \frac{AdL}{V} \tag{2.12}$$

Therefore,

$$dW = -V\sigma d\varepsilon \tag{2.13}$$

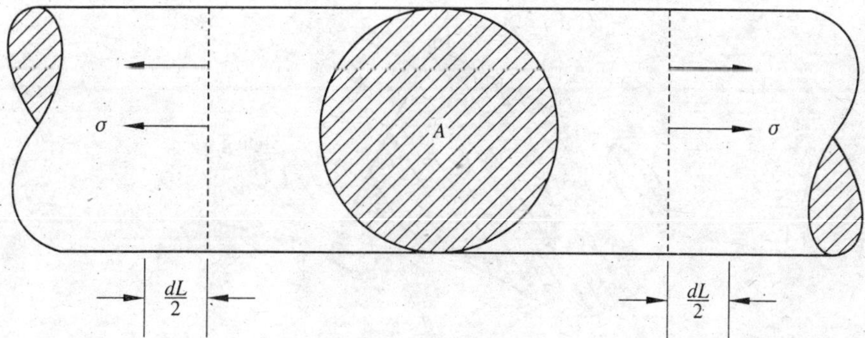

Fig. 2.13 *Extension of an elastic rod*

For an elastic bar, the intensive coordinate is stress (σ) and the extensive coordinate is strain (ε), and the work done per unit volume of the rod is:

$$\frac{dW}{V} \doteq -\sigma d\varepsilon \tag{2.14}$$

2.9.2 Stretching of a wire

Let the tension applied on a wire be τ, which results in a change of length dL of the wire. Then the work done on the wire is given by:

$$dW = -\tau dL \tag{2.15}$$

where, τ is the intensive coordinate and L is the extensive coordinate. Negative sign is introduced to take into account the fact that for a positive dL (extension of the wire) work has to be done on the system. The expression can be integrated once the relation between the tension (τ) and L is known.

2.9.3 Stretching of a liquid film

Consider a liquid film held between the wire frame, one side of which can be moved with the aid of the sliding wire as shown in Fig. 2.14. The interfacial tension or surface tension is the force per unit length normal to a line on the surface. The total surface tension force exerted on the wire is equal to $2\tau b$, where b is the width of the liquid film. The factor 2 appears because of the two surfaces of the film. If the sliding wire is moved through a distance dL, then the work done is given by

$$dW = -2\tau b dL \tag{2.16}$$

The total change in the surface area of the film = $dA = 2bdL$
Therefore,

$$dW = -\tau dA \tag{2.17}$$

Negative sign has been introduced for the reasons explained earlier.

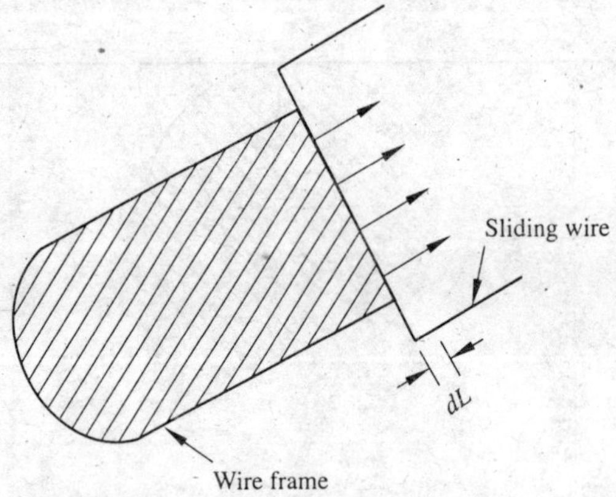

Fig. 2.14 *Stretching of a liquid film*

2.9.4 Electrical work

Consider an electrolytic cell of emf E, shown in Fig. 2.15. The cell is connected to an external circuit through which an electric current i is flowing.

Let dQ_e represent the change in the charge of the cell because of the flow of electric current. Then, the work done by the cell is given by:

$$dW = -E dQ_e \tag{2.18}$$

Negative sign is introduced to take into account the fact that when the cell discharges, dQ_e is negative and work is done by the cell. The expression for the work done by the cell can be put in a convenient form if we recall that the current (i) is given by:

$$i = \frac{dQ_e}{dt} \tag{2.19}$$

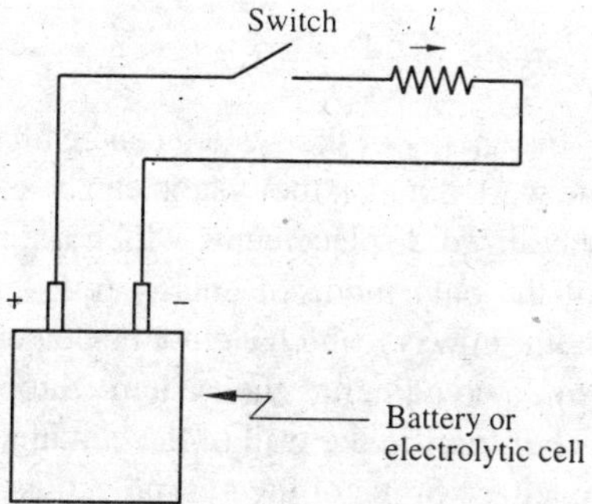

Fig. 2.15 *Electrolytic cell*

where t is time. Therefore,

$$dW = -Eidt \tag{2.20}$$

It can be noticed that in all the above cases the work done by a system is equal to the product of an intensive coordinate and the differential of an extensive coordinate. Now we can generalize and express the work done as:

$$dW = F_k dX_k \tag{2.21}$$

where F_k is the generalized force and dX_k is the generalized displacement. The generalized forces and the generalized displacements for various work modes are summarized in Table 2.1.

Table 2.1 *Generalized forces and displacements for various modes of work*

System	Generalized force	Generalized displacement	Work done
Compression/ Expansion	Pressure (P)	Volume (dV)	PdV
Solid rod (elastic)	Stress (σ)	Strain ($d\varepsilon$)	$-\sigma d\varepsilon$
Stretching of wire	Tension (τ)	Extension(dL)	$-\tau dL$
Liquid film	Surface tension (τ)	Surface area (dA)	$-\tau dA$
Reversible cell	emf (E)	Charge (dQ_e)	$-EdQ_e$
Polarization	Voltage (V)	Charge (dQ_e)	$-VdQ_e$
Magnetization	Magnetic field (H)	Magnetization (dM)	$-HdM$

2.10 Heat

Depending on the type of system, the energy of the system can be changed by performing work on the system by one or more modes. The work done on the system can be easily estimated from a knowledge of the generalized forces and generalized displacements which are macroscopically observable. We shall see shortly that work is not the only mode of energy transfer. It is also possible to transfer energy to a system (or from a system) in ways which cannot be detected as a macroscopic work mode. Consider a certain amount of gas in a container as the system. Suppose the container is brought into thermal contact with a hot body. The atoms in the wall of the container interact with the gas molecules which move with independent velocities. Some of the gas molecules receive energy from the atoms of the wall and move faster. The interaction between the atoms of the wall of the container and the gas molecules results in an increase in the energy of the system. At a microscopic level, work is done by the atoms of the wall on the gas molecules in increasing their molecular activity (the translational energy, rotational energy, vibrational energy, etc. of the gas molecules increases). However, we fail to observe any work interaction at macroscopic level and, therefore, the macroscopically observable work is zero. The energy of the system changes without any macroscopically observable energy transfer in the form of work. Such a mode of energy transfer, which cannot be accounted as work from a macroscopic point of view is called heat interaction. Such energy transfer occurs because of the temperature difference across the boundary. Therefore, *energy transfer between a system and its surroundings can be called as heat if it occurs by virtue of the temperature difference across the boundary*. Energy is transferred across the boundary from a system at a higher temperature to its surroundings which are at a lower temperature.

The two modes of energy transfer, namely work and heat, depend on the choice of the system. Consider a vessel containing some water, which is heated by an immersion heater connected to a storage battery. If we choose the water as our system and the rest of the universe as its surroundings, the interaction between the system and the surroundings is called heat because the energy transfer takes place by virtue of the temperature difference between the heating coil and the water. On the other hand, if we consider the water and the immersion heater as our system, the interaction is called work because the energy transfer is due to the interaction of the battery. Hence, whenever the interactions are named as work or heat, they are with reference to choice of the particular system.

Suppose we consider the water contained in a vessel as our system. The energy of the system can be increased through work interaction by rotating a paddle wheel in it as shown in Fig. 2.16(a). The energy of the system can be increased also through heat interaction by bringing it in contact with a hot body as shown in Fig. 2.16(b). From a knowledge of the initial energy and the final energy of the system (which can be estimated by measuring the temperature and the quantity of water), it is not possible to specify the mode of energy transfer through which the energy of the system has changed. It is so because the same change can be effected either by work interaction or by heat interaction. Therefore, once the process is completed, it is not possible to find out the mode through which the energy transfer has taken place. It can be identified only while the system is interacting with its surroundings.

Heat, like work, is energy in transit. It can be identified only at the boundary of the system. Heat is not stored in the body, but energy is stored in the body. Heat, like work, is not a property of the system and hence its differential is not exact. Heat and work are two different ways of transferring energy across the boundary of a system.

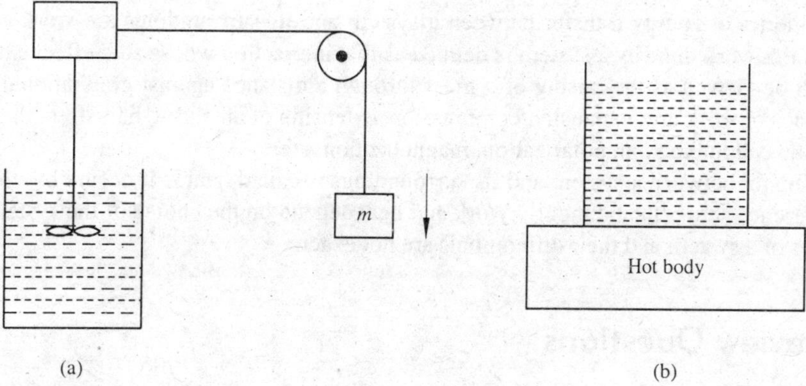

Fig. 2.16 *Schematic arrangement to show energy transfer through (a) work interaction (b) heat interaction*

2.11 Summary

In thermodynamics a system is defined as a definite quantity at matter bounded by some closed surface–real or imaginary. The boundary surface may change in shape as well as in size as the system undergoes a change. The combination of matter and space external to the system is called its surroundings. The boundary employed to separate a system from its surroundings can be classified as: movable or rigid, diaterminal or adiabatic. A system which is enclosed by a rigid and adiabatic wall is called an isolated system.

The state of a thermodynamic system is described by a set of parameters called properties. A thermodynamic property is a characteristic of the system and it should have a definite value when the system is in a particular state. It is a point function and its differential is exact. The extensive properties depend on the mass of the system whereas the intensive properties are independent of the mass of the system.

A thermodynamic system possesses energy in several forms. The kinetic energy and potential energy of a system can be evaluated in terms of macroscopically observable quantities and hence are called macroscopic modes of energy. The system possesses other forms of energy associated with translational, rotational and vibrational motion of its constituent molecules. These energy terms cannot be evaluated in terms of macroscopically observable quantities and hence are called microscopic modes of energy.

In thermodynamics we deal with systems in equilibrium. The criteria for equilibrium are:

- Equality of temperature - Thermal equilibrium
- Equality of pressure - Mechanical equilibrium
- Equality of chemical potential - Chemical equilibrium

A system which satisfies simultaneously the criteria for thermal, mechanical and chemical equilibria is in a state of thermodynamic equilibrium.

If a thermodynamic system undergoes a change in its state it does so by means of a process. In a reversible process the system is always in a state of equilibrium. If a process is carried out in such a way that at every instant the system departs infinitesimally from an equilibrium state, the process is called quasistatic. A reversible process is always a quasistatic process, but the converse is not necessarily true.

The two modes of energy transfer between a system and its surroundings are work and heat. In thermodynamics work done by a system is defined as that interaction whose sole effect external to the system could be viewed as the raising of a mass through a distance against gravitational force. The different modes of work are: expansion/compression, extension of an elastic bar, stretching of a liquid film, reversible cell interaction, polarization, magnetization, etc.

Energy transfer between a system and its surroundings is called heat if it occurs by virtue of temperature difference across the boundary. Work and heat depend on the choice of the system. They are not properties of a system and their differentials are not exact.

2.12 Review Questions

1. Define a thermodynamic system.
2. Distinguish between an adiabatic system and isolated system.
3. Is it possible to change the energy of an isolated system? Explain.
4. What is meant by surroundings and universe?
5. Define thermodynamic property.
6. What are the essential features of a thermodynamic property?
7. What is a point function? Give some examples which are not point functions.
8. Distinguish between extensive property and intensive property.
9. Name a few macroscopic modes of energy and a few microscopic modes of energy.
10. What is internal energy?
11. Which forms of energy are included in internal energy and which forms are not included in it?
12. Distinguish between steady state and equilibrium.
13. State the criteria which a system should satisfy for it to be in a state of thermodynamic equilibrium.
14. What is a quasi-static process and how does it differ from a reversible process?
15. What is the criterion to judge whether a process is reversible or not?
16. State the thermodynamic definition of work done by a system.
17. Is it possible to define the work done on a system? If not, explain why.
18. Define the term heat.
19. Which characteristics are common to both work and heat?
20. Is it possible to name the interaction, as work or heat, by knowing the initial and final energies of a system? Give an example.

2.13 Problems

2.1 Three thermodynamic quantities A, B, and C are defined as $dA = Pdv$, $dB = vdP$ and $dC = Pdv - vdP$, where $Pv = RT$. Which of the quantities can be used as properties?

2.2 Given $Pv = RT$, where P is pressure, v is molar volume, T is temperature and R is a constant, determine whether the following quantities can be used as properties:

(a) $(Pdv + vdP)$

(b) $\left(\frac{dT}{T} + \frac{Pdv}{v}\right)$

2.3 The differential for the volume V of a particular body is expressed as $dV = 2\pi RhdR + \pi R^2 dh$. Test whether the differential is exact. If it is exact, find the relation between V, R and h.

2.4 An equation of state gives one of the following expressions for the differential of pressure. Identify the correct expression and determine the equation of state.

$$dP = \frac{2(v-b)}{RT}dv + \frac{(v-b)^2}{RT^2}dT$$

$$dP = -\frac{RT}{(v-b)^2}dv + \frac{R}{v-b}dT$$

2.5 Classify the following properties of a system: volume, specific volume, pressure, temperature, density, concentration expressed in terms of moles of the constituent species and internal energy.

2.6 A scooter of mass 90 kg is moving at a speed of 60 km/h. Estimate the kinetic energy of the scooter.

2.7 An aircraft with a mass of 25 000 kg flies at a speed of 1000 km/h at an altitude of 10 km. Calculate the kinetic energy and potential energy of the aircraft.

2.8 A vehicle of mass 90 kg moves at a speed of 50 km/h. Determine the work to be done by an opposing force on the vehicle to reduce its speed to 20 km/h.

2.9 An object of mass 10 kg is dropped from the roof of a building which is 100 m above the earth's surface. Calculate the velocity with which the object hits the earth's surface if $g = 9.81$ m/s^2.

2.10 In a particular fountain, water is ejected vertically up through a nozzle. If the velocity of the water at the tip of the nozzle is 30 m/s, determine the height to which the water can reach in the atmosphere.

2.11 A hollow sphere of mass m and volume V is immersed in a liquid of density ρ. If the sphere is raised through a distance of h in the liquid by an external agent, determine the work done by the external agent. Is there any energy transfer as work between the sphere and the liquid? What happens to the energy of the fluid?

2.12 A gas is compressed reversibly from the initial state (P_1, V_1) to the final state (P_2, V_2). During the compression process the pressure and volume are related by the equation PV^γ = constant. Determine the work done by the gas.

2.13 A balloon which is initially collapsed and flat, is slowly filled with helium from a cylinder, forming the balloon into a sphere of 5 m in diameter. The ambient pressure is 100 kPa. During the filling process the temperature of helium inside the cylinder remains constant at 300 K. Determine the work done by the cylinder–balloon system.

2.14 Initially nitrogen is contained in a cylinder of volume 0.01 m^3 at a pressure of 15 MPa and 300 K. The nitrogen slowly leaks into the atmosphere until the gas pressure in the cylinder is reduced to 10 MPa. The temperature of the cylinder and the gas remains at 300 K all the time. If the atmospheric pressure and temperature are 100 kPa and 300 K respectively, is there any energy transfer as work? If so, determine the work done by the nitrogen. Nitrogen obeys the relation $PV = NRT$ where $R = 8.314$ kJ/kmol K and N is the number of moles.

2.15 A particular gas obeys the van der Waals equation of state. Calculate the work done per kmol of the gas if the gas is compressed reversibly at constant temperature from the initial volume v_1 to the final volume v_2.

2.16 Consider the system shown in Fig. 2.11. Initially the gas is at 500 kPa and occupies a volume of 0.2 m^3. The spring exerts a force which is proportional to the displacement from its equilibrium position. The ambient pressure is 100 kPa. The gas is heated until the volume is doubled at which point the pressure of the gas is 1 MPa. Determine the work done by the gas.

2.17 A spherical balloon of 1 m diameter contains a gas at 200 kPa and 300 K. The gas inside the balloon is heated until the pressure reaches 500 kPa. During the process of heating the pressure is proportional to the diameter of the balloon. Determine the work done by the gas inside the balloon.

2.18 A gas is compressed reversibly in a piston–cylinder assembly, from the initial state of 2 bar and 0.2 m^3 to a final state of 10 bar and 0.04 m^3. The pressure–volume relationship during the compression process is $P = a + bV$. Show the path followed by the gas on a pressure versus volume diagram and calculate the work done on the gas.

2.19 A gas is contained in a piston–cylinder assembly. Initially, the gas is at 10 bar and 0.05 m^3. The gas is allowed to expand reversibly to a final pressure of 1 bar. Calculate the work done by the gas for the following reversible processes: (a) The product PV is constant. (b) The product PV^2 is constant.

2.20 A linear spring, with a spring constant $K = 10$ kN/m has a natural length of 20 cm. Determine the work required to compress the spring (a) from its natural length to 17 cm, (b) from 18 cm to 16 cm, and (c) from 16 cm to 15 cm.

2.21 A linear spring with a spring constant $K = 100$ kN/m is compressed from its natural length to a final length of 10 cm. If the work required for compressing the spring is 500 J, determine the natural length of the spring.

2.22 It is required to stretch a steel bar of cross-section 2 cm × 2 cm from its original length of 1 m to 1.01 m. Determine the work required if the Young's modulus of steel is 2×10^{11} N/m^2.

2.23 A parallel plate condenser in a D.C circuit is charged by increasing the voltage across the condenser from 0 to 220 V. Under this condition, the relation between the electric charge (Q_e) and the voltage (V) is given by $Q_e = CV$, where C is the capacitance of the condenser. Calculate the work done on the condenser if $C = 50 \times 10^{-6}$ coulombs/volt.

2.24 An electric potential of 220 V is applied across an electrical resistance of 50 Ω for a period of 10 minutes. Determine the work done.

2.25 A particular gas is made to undergo a process in a cylinder–piston assembly. The initial conditions of the gas are: $P_1 = 1$ MPa and $V_1 = 1$ m^3. The final volume is 2 m^3. Calculate the work done by the gas for the following processes: (a) pressure remains constant, (b) the product of pressure and volume remains constant, and (c) the pressure is proportional to the square of the volume.

2.26 In a bomb calorimeter the sealed bomb is loaded with 10 g fuel oil and then pressurized with oxygen to a pressure of 1 MPa. The calorimeter is immersed in a water bath. The oil is ignited by an electric spark. Immediately after the ignition, there is no detectable change in the temperature of the water bath. After a period of 2 minutes, the temperature of the water bath rises by 5°C. The mass of the water bath is 10 kg. State whether or not work was done, energy transferred as heat, internal energy has changed, energy has changed at the following two instants: (a) immediately after the ignition process, and (b) after the water bath registered a temperature rise.

2.27 In each of the following cases state whether work is done, energy transferred as heat, E and U changed or not: (a) A viscous liquid is stirred by a paddle wheel in an insulated container thereby raising the temperature of the liquid; (b) An insulated vessel is divided into two compartments by a partition. One compartment contains strong sulfuric acid at a given temperature and the second compartment contains water at the same temperature. The partition is ruptured and the temperature of the mixture rises; (c) An electrical resistor is placed in an insulated container having 10 kg water. A steady electric current flows through the resistor and the temperature of the water rises. (i) Consider the water as the system, and (ii) consider the resistor and water as the system.

3 Thermodynamic Properties of Fluids

Learning Objectives

This chapter enables the readers to:

- Define a pure substance and a phase.
- Define a simple compressible substance.
- Understand the widely used methods of presenting the thermodynamic properties of substances.
- Define an ideal gas.
- Understand the van der Waals equation of state.
- Determine the parameters of an equation of state in terms of critical properties.
- Define compressibility factor.
- Use the generalized compressibility chart for estimating the properties of real gases.
- Understand the various thermodynamic diagrams like T–v; P–v; P–T, etc.
- Use the Mollier diagram in estimating the properties of steam.
- Gain familiarity with the use of steam tables.

The heat and work interactions across the boundaries of the system are path functions and depend on the process followed by the system. To evaluate these interactions a knowledge of the properties of the system is essential. That is, in order to apply the principles of thermodynamics to the problems of interest, it is essential to have a knowledge of the properties of the materials that constitute the system, as well as the relationships between these properties.

Most useful properties of matter in the study of thermodynamics are :

1. Properties like pressure, volume and temperature which can be measured directly.
2. Properties like internal energy, enthalpy, entropy, etc. which cannot be measured directly.

The properties are point functions as discussed in Chapter 2, and do not depend on the path followed by the system. Moreover, properties have unique values when the system is in a particular state. Hence the properties which cannot be measured directly can be evaluated as a function of the measurable properties P, v and T. For example, enthalpy of a gas can be tabulated as a function of temperature and pressure. Therefore, it is necessary to have relationships between these two groups of properties. Before we look into these relationships, let us consider the relationships among the measurable properties P, v and T.

3.1 Pure Substance

We shall concentrate mainly on the properties of substances which are uniform in composition. A pure substance is one, which consists of a single chemical species. Mostly we deal with systems that contain a specified mass of a substance of fixed chemical composition (like air which is a mixture of nitrogen and oxygen of fixed chemical composition) which undergoes a given process. Suppose air is cooled to sufficiently low temperature where it can be present in both the liquid and vapor states. Then the air cannot be treated as a pure substance, because the chemical composition of the liquid is different from that of the vapor.

3.2 Concept of Phase

Substances can be found in different states of physical aggregation. Ice, water and steam are the three different physical states of the same species H_2O. Based on the physical states of aggregation the substances are generally classified into three states: (a) solid (b) liquid and (c) gas.

Solids can exist in different allotropic forms or states. At atmospheric pressure sulfur can be found in rhombic form (lemon yellow in color) as well as in monoclinic form (orange in color).The properties of the allotropic states differ from each other. A microscopic examination of these allotropic forms reveals that their crystalline structure is different from each other. Details of these crystalline structure can be studied through X-ray diffraction methods. Similarly, several solids exist in different allotropic forms, and they are of importance. The different states in which the pure substances can exist are called phases. If a system consists of a mixture of several phases, the phases are separated from each other by a phase boundary. The thermodynamic properties of the substance change abruptly at the phase boundary, even though the intensive properties like pressure and temperature are identical. For example, when ice and water coexist in equilibrium at 0°C and 101.325 kPa, the solid phase (ice) is clearly separated from the liquid phase (water). The properties of ice are much different from that of water. It is also known that liquid helium can exist in two different liquid phases called liquid helium I and liquid helium II.

Mixtures also exist in different phases. For example, a mixture of alcohol and water can exist in both liquid and vapor phases. The chemical composition of the vapor phase is generally different from that of the liquid phase.

Generalizing the above observation, it can be said that *a system which is uniform throughout both in chemical composition as well as in physical state is called a homogeneous substance or a phase.*

All substances—solids, liquids and gases—change their specific volume when they are subjected to a change in pressure and temperature. Depending on the nature of the substance, and on the range of the applied pressure and temperature, the change in the specific volume may be large, or negligible. When the change in the specific volume is negligible as in the case of liquids and solids, the substance may be treated as *incompressible.* The isothermal compressibility κ of a substance is defined as

$$\kappa = -\frac{1}{V}\left(\frac{\partial V}{\partial P}\right)_T \tag{3.1}$$

and the coefficient of volume expansion β of a substance is defined as

$$\beta = \frac{1}{V}\left(\frac{\partial V}{\partial T}\right)_P \tag{3.2}$$

Hence, if κ is negligible, then the fluid is said to be incompressible. The specific volume of a gas varies greatly with pressure and temperature. Hence, gases are considered as compressible fluids. Since the properties of compressible fluids vary to a large extent, we shall concentrate mostly on the study of their pressure–volume–temperature relationships.

3.3 Simple Compressible Substance

Instead of attempting to study the thermodynamic properties of a 'general substance' which is capable of exchanging energy through several work modes, let us consider a simple substance—a substance which can interact through only one relevant reversible work mode. If the relevant reversible work mode is the compression/expansion (Pdv) work mode, then the substance is called a simple compressible substance.

Thermodynamic properties of a substance may be presented in the form of mathematical relations, charts or diagrams and tables. These properties are usually experimentally determined at selected temperatures and pressures, and are interpolated using exact thermodynamic relations to tabulate the data in fine intervals. At high temperatures and in some special cases they are estimated by the methods of statistical thermodynamics.

3.4 Equations of State

The most convenient method of representing the properties or the behavior of a simple compressible substance is by a mathematical expression that is, an equation which represents the P–v–T behavior of a fluid. A general form of such an expression is

$$f(P, v, T) = 0 \tag{3.3}$$

which is known as an equation of state. This is the most convenient way of condensing a large amount of data in a form which is highly useful.

Equation (3.3) can be rewritten as

$$v = v(T, P)$$

or

$$dv = \left(\frac{\partial v}{\partial T}\right)_P dT + \left(\frac{\partial v}{\partial P}\right)_T dP$$

or

$$\frac{dv}{v} = \frac{1}{v}\left(\frac{\partial v}{\partial T}\right)_P dT + \frac{1}{v}\left(\frac{\partial v}{\partial P}\right)_T dP \tag{3.4}$$

Substituting the definitions of κ and β from Eqs. (3.1) and (3.2) in Eq.(3.4), we get:

$$\frac{dv}{v} = \beta dT - \kappa dP \tag{3.5}$$

One can experimentally measure the values of κ and β, and use Eq. (3.5) to determine the equation of state.

3.4.1 Ideal gas

Experimental methods were first used in developing the equation of state for gases. In 1662 Boyle observed that if the temperature of a gas is held constant, its molar volume is inversely proportional to the pressure. This relation is known as Boyle's law. That is

$$v \propto \frac{1}{P} \quad \text{or} \quad Pv = \text{constant at constant temperature} \tag{3.6}$$

In 1787 Charles observed that (a) if the pressure is held constant, the molar volume of a gas is directly proportional to the temperature of the gas, i.e.,

$$v \propto T \text{ at constant pressure} \tag{3.7}$$

and (b) if the molar volume is held constant, the pressure exerted by a gas is directly proportional to the temperature, i.e.,

$$P \propto T \text{ at constant volume} \tag{3.8}$$

Boyle's law and Charle's law can be combined to yield the following equation of state.

$$Pv = RT \tag{3.9}$$

where v is the molar volume of the gas and R is the universal gas constant, which is equal to 8.314 J/mol K or 8.314 kJ/kmol K. This 'law' is only an approximation to the actual behavior of the gases and Eq. (3.9) is known as the *ideal gas law.* At high densities, the deviations from this law are large. In general, as the density of the gas is decreased, that is at sufficiently low pressure or at high temperature, all gases tend to behave like ideal gas.

Based on kinetic theory arguments, the ideal gas law can be derived by assuming that: (1) the molecules do not occupy any volume, and (2) there are no attractive forces between the molecules. It can be shown that for an ideal gas, the internal energy is a function of temperature only. That is, for an ideal gas

$$u = u(T) \tag{3.10}$$

In 1843, Joule demonstrated this fact through the following experiment. Two containers were connected through a valve and immersed in a water bath as shown in Fig. 3.1. The water bath was kept in an insulated container. One of the vessels was filled with air at 22 atm. pressure (1 atm = 101.325 kPa) and the other vessel was evacuated. The initial temperature of the water bath was measured.

The valve connecting the two containers was opened allowing the air to occupy both the containers. The temperature of the water bath was measured again and it was found that the temperature had not changed. That is, there was no change in the internal energy of the gas. The pressure and volume of the gas changed and yet the internal energy did not change. Hence, he concluded that the internal energy is independent of pressure and volume. That is

$$\left(\frac{\partial u}{\partial P}\right)_T = 0 \quad \text{and} \quad \left(\frac{\partial u}{\partial v}\right)_T = 0 \tag{3.11}$$

Since the internal energy of the gas did not change when the temperature was held constant, he concluded that the internal energy is a function of temperature only.

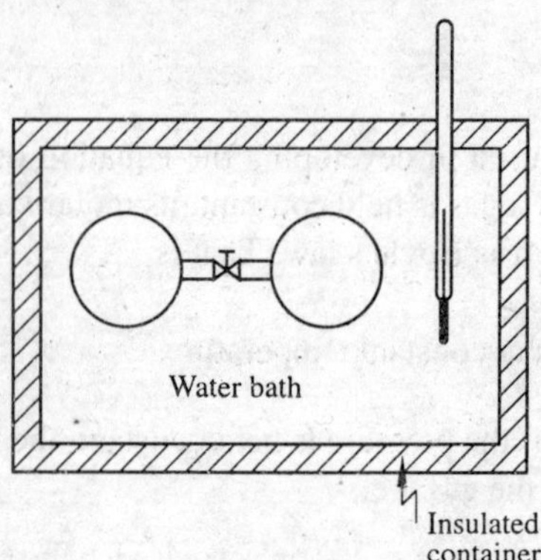

Fig. 3.1 *Schematic arrangement of Joule's experiment to show that $u = u(T)$ only for an ideal gas*

In engineering thermodynamics, one deals with fluids over a wide range of temperatures and pressures. The behavior of many fluids in the region of high temperature and low pressure can be approximated by the relationship $Pv = RT$. Hence, a knowledge of the ideal gas behavior is very useful in the study of the P–v–T behavior of fluids. The behavior of all gases at sufficiently low densities can be approximated as that of an ideal gas.

3.4.2 Van der Waals equation of state

The assumptions made in deriving the ideal gas law cannot be justified for a real gas (non-ideal gas). Many attempts have been made to represent the P–v–T data accurately. Clausius pointed out that the free volume available for molecular motion is $(v - b)$, where b is a measure of the volume occupied by the gas molecules themselves. Then making use of the correction in the ideal gas law, he obtained the following equation of state

$$P(v - b) = RT \tag{3.12}$$

which is known as the Clausius equation of state.

In 1873, van der Waals (Johannes D. van der Waals, 1837–1923) proposed an equation of state which can be regarded as a modification of the ideal gas law. It represents one of the earliest attempts to develop an equation of state which is more accurate than the ideal gas law. It takes into account the effect of the volume occupied by the gas molecules as well as the effect of attractive forces between the molecules. As the density of the gas increases the number of molecules per unit volume increases. The free volume available for the motion of the molecules is reduced by the volume occupied by the molecules. Therefore, v in the ideal gas equation is modified as $(v - b)$ where b is the volume occupied by the gas molecules.

Suppose a gas molecule is considered as a hard sphere of radius r, the centre-to-centre distance between two molecules at the instant of collision is $2r$. Therefore, a molecule cannot reach another molecule such that its centre lies within a distance of $2r$ from the centre of the other molecule. That is, the centre of a molecule is excluded from a spherical region called the sphere of exclusion of radius $2r$ from other molecules as shown in Fig. 3.2. Then, the total volume, which is not available for molecular motion of N_0 molecules (Avagadro's number) is called excluded volume b and is given by

$$b = \frac{N_0}{2} \times \frac{4}{3}\pi(2r)^3 = \frac{16}{3}N_0\pi r^3 \tag{3.13}$$

The factor 2 is included in Eq. (3.13) to avoid counting each pair twice.

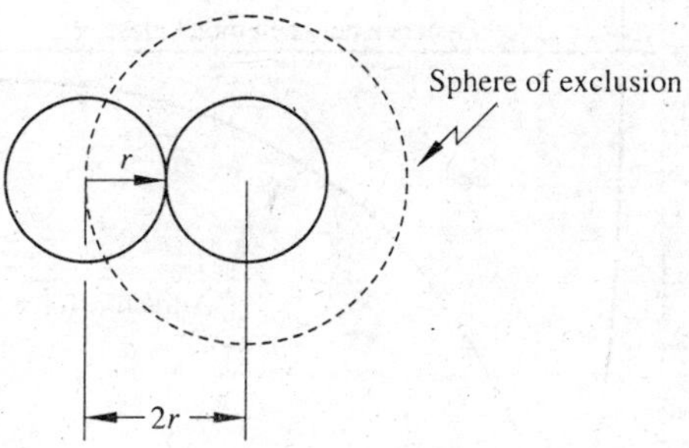

Fig. 3.2 *Sphere of exclusion*

In real gases, there are forces of attraction and repulsion between a pair of molecules. The intermolecular forces of attraction and repulsion between a pair of molecules are of the form shown in Fig. 3.3. If the molecules attract one another, a molecule in the centre of the gas is attracted equally on all sides and its motion is not affected. For a molecule which is very close to the wall of the vessel, all the gas molecules are on one side of it only. Because of the intermolecular attractive forces, the molecule near the wall of the container is attracted by the other gas molecules and hence exerts less force on the wall. Therefore, the observed pressure is less than the pressure in absence of the attractive forces. The reduction in pressure is proportional to the number of molecules per unit volume ($\frac{N_0}{v}$) in the layer adjacent to the wall, and to the number of molecules per unit volume in the next layer which are attracting. Therefore, the reduction in pressure ($P' - P$) is given by

$$P' - P = a'\left(\frac{N_0}{v}\right)^2$$

or

$$P' = P + a'\left(\frac{N_0}{v}\right)^2 = P + \frac{a}{v^2} \tag{3.14}$$

where, P' = pressure exerted by the gas in absence of attraction forces
P = observed pressure
a = constant.

If the corrections for the excluded volume and attraction forces are incorporated in the ideal gas equation of state, one obtains:

$$\left(P + \frac{a}{v^2}\right)(v - b) = RT \tag{3.15}$$

which is known as the van der Waals' equation of state.

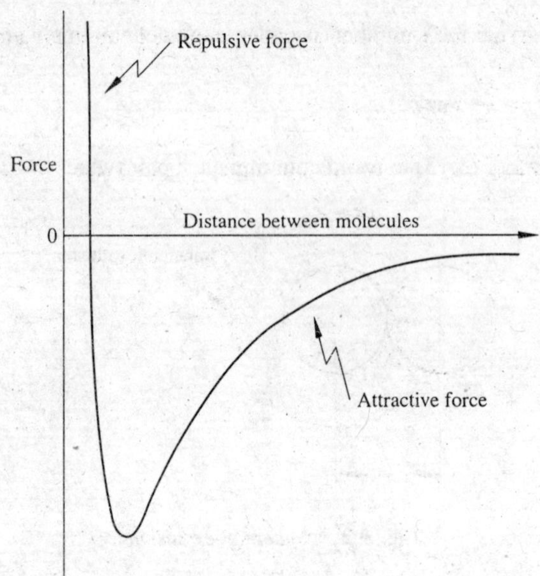

Fig. 3.3 *Intermolecular forces of attraction and repulsion between a pair of molecules*

Since, van der Waals equation takes into account the intermolecular forces, it is expected to hold good for the liquid phase also. Nowadays, the van der Waals equation is used to predict the phase equilibrium data. It is likely to be used extensively in future to predict the phase equilibrium data for pure substances as well as mixtures. The van der Waals constants a and b can be determined from experimental P–v–T data as described below.

Rearranging the Eq. (3.15) we get

$$P = \frac{RT}{v-b} - \frac{a}{v^2} \tag{3.16}$$

Therefore, a plot of P versus T at constant volume gives a straight line with a slope $R/(v-b)$ and an intercept $-a/v^2$. The van der Waals constants for some gases are given in Table 3.1. An interesting feature of the van der Waals equation is that it is cubic in volume, and in general, there are three values of v for a given set of P and T. However, in some range of values of P and T, there are three real values of v and in some range of values of T there is only one real value and two imaginary values of v. For $T > T_c$ only one real value of v exists and for $T < T_c$ three real values of v exist, where T_c is the critical temperature of the gas.

Figure 3.4 shows the isotherms of a van der Waals gas. It can be seen from Fig. 3.4 that at $T > T_c$, for a given value of P there is only one value of v. When $T < T_c$, there are three values of v for a given value of P. Suppose an experiment is performed with a particular gas in which the pressure and volume are measured at a specified temperature. These experimental points appear as shown by the points in Fig. 3.4. At $T > T_c$ the agreement between the experimental and theoretical values is good. However, at temperature less than the critical temperature the theoretical values depart from the experimental values. In particular, the curve ABDEF as predicted by the van der Waals equation of state is not observed. Suppose the substance is in the state indicated by point B. If the volume of the system is increased, the system moves along the curve BDE which is associated with an increase in pressure also. This is not physically feasible and the system is unstable. The curve ABDEF is practically not

Table 3.1 *van der Waals constants for a few selected gases*

Gas	Constant $a \times 10^3$ Pa $(m^3/mol)^2$	Constant $b \times 10^3$ (m^3/mol)
Air	135.653	0.037
Argon	136.282	0.032
Ammonia (NH_3)	422.546	0.037
Carbon dioxide (CO_2)	363.077	0.043
Carbon Monoxide(CO)	150.475	0.040
Carbon tetrachloride	2066.017	0.138
Chlorine	657.936	0.056
Chloroform	1537.100	0.102
Ethane	556.200	0.064
Ethylene	453.046	0.057
Ethyl ether	1761.029	0.134
Heptane (n)	3192.751	0.265
Hydrogen (H_2)	24.764	0.266
Hydrogen chloride	371.559	0.041
Mercury	820.023	0.017
Methane (CH_4)	228.296	0.043
Neon	21.349	0.017
Nitrogen (N_2)	140.849	0.039
Octane (n)	3789.000	0.237
Oxygen (O_2)	137.802	0.032
Propane	877.923	0.084
Sulphur dioxide	680.330	0.056
Sulphur trioxide	831.286	0.060
Water	553.667	0.300

realizable. Based on the stability criteria of thermodynamic systems, the physically feasible curve is the straight line path from A to F. It can be shown that the straight line AF is such that the area ABD is exactly equal to the area DEF for a stable system.

3.4.3 Critical constants

Every equation of state must satisfy the following conditions.

1. It must reduce to the ideal gas law as the pressure tends to zero.
2. The critical isotherm must show a point of inflection at the critical point.

Stated mathematically,

$$\left(\frac{\partial P}{\partial v}\right)_{T_c} = 0 \quad \text{and} \quad \left(\frac{\partial^2 P}{\partial v^2}\right)_{T_c} = 0 \tag{3.17}$$

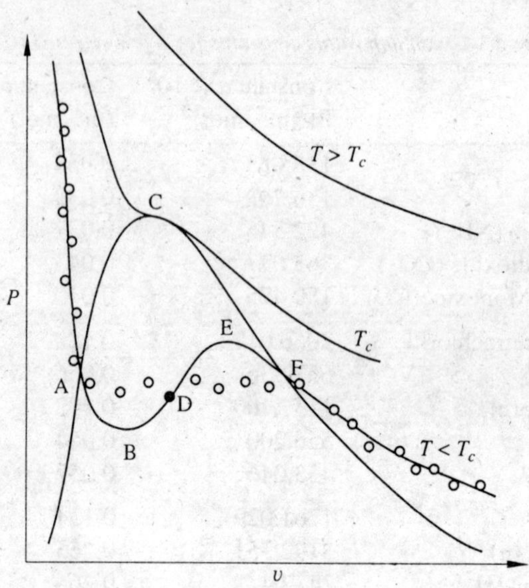

Fig. 3.4 *$P - v$ diagram for a pure substance showing isotherms in the region of critical point. Solid lines represent the values predicted by the van der Waals equation of state. Points represent the experimental values.*

One can use this criterion to obtain the critical values, or to determine the constants a and b of van der Waals equation. For a van der Waals gas

$$\left(\frac{\partial P}{\partial v}\right)_{T_c} = -\frac{RT_c}{(v_c - b)^2} + \frac{2a}{v_c^3} = 0 \tag{3.18}$$

$$\left(\frac{\partial^2 P}{\partial v^2}\right)_{T_c} = \frac{2RT_c}{(v_c - b)^3} - \frac{6a}{v_c^4} = 0 \tag{3.19}$$

At the critical point, van der Waals equation is

$$\left(P_c + \frac{a}{v_c^2}\right)(v_c - b) = RT_c \tag{3.20}$$

where P_c and v_c are the critical pressure and critical volume respectively. One can solve the three Eqs. (3.18–3.20) to obtain the critical constants and they are given by

$$v_c = 3b; \quad T_c = \frac{8a}{27Rb} \quad \text{and} \quad P_c = \frac{a}{27b^2} \tag{3.21}$$

Alternatively, one can calculate the van der Waals constants a and b in terms of the critical constants. Since it is possible to experimentally measure the critical temperature and critical pressure, a and b can be evaluated from a knowledge of P_c and T_c through the following relations.

$$a = \frac{27R^2T_c^2}{64P_c} \quad \text{and} \quad b = \frac{RT_c}{8P_c} \tag{3.22}$$

At the critical point, one can see that

$$\frac{P_c v_c}{RT_c} = \frac{3}{8} = 0.375 \tag{3.23}$$

3.5 Other Equations of State

Several attempts have been made to formulate the equations of state to represent the P–v–T data more accurately. Many equations of state have been proposed so far and many more are likely to be introduced in the future. No attempt is made, in this book, to discuss the various equations of state available in the literature. Most of the equations of state are empirical and only a few of them are in wide use. Some of the equations of state, commonly used in engineering analysis, are presented below.

3.5.1 Beattie–Bridgeman equation of state

Beattie–Bridgeman equation of state is given by

$$Pv^2 = RT\left(1 - \frac{c}{vT^3}\right)\left(v + B_0 - \frac{bB_0}{v}\right) - A_0\left(1 - \frac{a}{v}\right) \tag{3.24}$$

The constants A_0, B_0, a, b, and c are characteristic of a gas. These constants for some substances are given in Table 3.2.

Table 3.2 *Constants of the Beattie-Bridgeman equation of state*

Gas	A_0 $\left(\frac{\text{Pa m}^3}{\text{mol}^2}\right)$	$B_0 \times 10^6$ (m^3/mol)	$a \times 10^6$ (m^3/mol)	$b \times 10^6$ (m^3/mol)	c $\left(\frac{\text{m}^3\text{K}^3}{\text{mol}}\right)$
Air	0.1318	46.11	19.31	−11.010	43.40
Ammonia	0.2425	34.15	170.31	19.112	4768.70
n - Butane	1.8030	246.20	121.61	94.230	3500.00
Carbon dioxide	0.5073	104.76	71.32	72.350	660.00
Ethane	0.5958	94.00	58.61	19.150	900.00
Ethylene	0.6234	121.56	49.64	35.970	226.80
Helium	0.00219	14.00	59.84	0.000	0.04
Hydrogen	0.0200	20.96	−5.06	−43.590	5.04
Methane	0.2307	55.87	18.55	−158.700	128.30
Neon	0.0215	20.60	21.96	0.000	1.01
Nitrogen	0.1362	50.46	26.17	−6.910	42.00
Oxygen	0.1511	46.24	25.62	4.208	48.00
n - Pentane	2.8634	394.00	150.99	139.600	4000.00
Propane	1.2078	181.00	73.21	42.930	1200.00

3.5.2 Redlich–Kwong equation of state

Redlich–Kwong (RK) equation of state given below, is widely used for engineering calculations.

$$P = \frac{RT}{v-b} - \frac{a}{v(v+b)} \tag{3.25}$$

The constants a and b of the Redlich–Kwong equation of state can be estimated from the critical constants by the following relations.

$$a = \frac{0.42748 R^2 T_c^{2.5}}{P_c T^{1/2}} \tag{3.26}$$

$$b = \frac{0.0867 R T_c}{P_c} \tag{3.27}$$

3.5.3 Benedict–Webb–Rubin equation of state

The Benedict–Webb–Rubin (BWR) equation of state is given by

$$P = \frac{RT}{v} + \frac{1}{v^2}\left(RT(B_0 + \frac{b}{v}) - (A_0 + \frac{a}{v} - \frac{a\alpha}{v^4}) - \frac{1}{T^2}\left(C_0 - \frac{c}{v}(1 + \frac{\gamma}{v^2})\exp(\frac{-\gamma}{v^2})\right)\right) \tag{3.28}$$

where A_0, B_0, C_0, a, b, c, α and γ are constants for a given fluid. The BWR constants for a few selected gases can be found in Perry's Chemical Engineer's Handbook. This equation of state is quite complex and contains eight constants, and is able to predict the P–v–T data with higher accuracy compared to many other equations of state.

3.5.4 Virial equation of state

The word virial stems from the Latin word *Vis* (plural *vires*) meaning force, thus it refers to the interaction forces between molecules. In 1901 Kamerlingh Onnes suggested the virial equation of state which is expressed as a power series in reciprocal volumes; it is given by

$$\frac{Pv}{RT} = 1 + \frac{B}{v} + \frac{C}{v^2} + \frac{D}{v^3} \tag{3.29}$$

where B, C, D, etc. are known as second virial coefficient, third virial coefficient etc. Virial coefficients express the deviations from ideality in terms of intermolecular forces. These virial coefficients are functions of temperature only. The advantage of virial equation of state is that it may be made to represent the experimental P–v–T data as accurately as required by increasing the number of constants. The values of the second virial coefficients have been determined experimentally for a number of gases. Whenever the data is not available it is possible to estimate the second virial coefficients from the intermolecular potential functions. The third virial coefficients are not known for many substances and much less information is available beyond the third virial coefficient. Moreover, the virial equation of state with more than three terms is difficult to handle. It may be mentioned that the virial equation of state and the ideal gas law have a strong theoretical base, and they can be derived through statistical mechanical methods. All other equations of state are empirical or semi-empirical.

The virial equation of state is sometimes written as a power series in the pressure as

$$\frac{Pv}{RT} = 1 + B'P + C'P^2 + D'P^3 + \cdots \tag{3.30}$$

where the coefficients B', C', D', etc. are functions of temperature only. The coefficients B', C', D', etc. are related to the virial coefficients B, C, D, etc. by the following relations:

$$B' = \frac{B}{RT} \tag{3.31}$$

$$C' = \frac{C - B^2}{(RT)^2} \tag{3.32}$$

$$D' = \frac{D - 3BC + 2B^3}{(RT)^3} \tag{3.33}$$

It has been found that the virial equation (3.29) adequately represents the experimental data over a wide range of pressures, compared to the virial equation (3.30) when both these equations are truncated after the third term.

***Example* 3.1** It is necessary to store one kmol of methane at 300 K and 60 MPa. Determine the volume of the cylinder that is required for storage by each of the following: (a) Ideal gas law (b) van der Waals equation and (c) Redlich–Kwong equation.

Solution (a) Ideal gas law

$$v = \frac{RT}{P} = \frac{8.314 \times 10^3 \times 300}{60 \times 10^6} = 0.0416 \text{ m}^3$$

(b) The van der Waals constants for methane are: $a = 228.296 \times 10^{-3}$ Pa(m^3/mol)^2 and $b = 0.043 \times 10^{-3}$ m^3/mol

$$\left(P + \frac{a}{v^2}\right)(v - b) = RT$$

$$\left(60 \times 10^6 + \frac{228.296 \times 10^{-3}}{v^2}\right)(v - 0.043 \times 10^{-3}) = 8.314(300)$$

Assume $v = 0.065 \times 10^{-3}$ m^3/mol. Then

$$\left(60 \times 10^6 + \frac{228.296 \times 10^{-3}}{(0.065 \times 10^{-3})^2}\right)(0.065 \times 10^{-3} - 0.043 \times 10^{-3}) = 2508.8$$

This value is very close to the value 8.314(300) = 2494.2. Since the assumed value is very close, $v = 0.065 \times 10^{-3}$ m^3/mol.

(c) Redlich–Kwong equation:

$$a = \frac{0.42748R^2T_c^{2.5}}{P_cT^{1/2}} \quad \text{and} \quad b = \frac{0.0867RT_c}{P_c}$$

For methane $T_c = 190.70$ K and $P_c = 46.41$ bar

$$a = \frac{0.42748(8.314)^2(190.70)^{2.5}}{46.41 \times 10^5(300)^{1/2}} = 0.1846 \text{ m}^6 \text{ Pa/mol}^2$$

$$b = \frac{0.0867(8.314)190.70}{(46.41 \times 10^5)} = 2.9619 \times 10^{-5} \text{ m}^3/\text{mol}$$

Assume $v = 0.054 \times 10^{-3}$ m^3/mol. Then

$$P = \frac{RT}{v - b} - \frac{a}{v(v + b)}$$

$$P = \frac{8.314(300)}{0.054 \times 10^{-3} - 2.9619 \times 10^{-5}} - \frac{0.1846}{0.054 \times 10^{-3}(0.054 \times 10^{-3} + 2.9619 \times 10^{-5})}$$
$$= 61.42 \text{ MPa}$$

This value is very close to the given value 60 MPa. Therefore, $v = 0.054$ m^3/kmol.

***Example* 3.2** One kmol of ethylene is contained in a 0.6 m^3 steel vessel immersed in a constant temperature bath at 200°C. Calculate the pressure developed by the gas by each of the following: (a) Ideal gas law (b) van der Waals equation and (c) Redlich–Kwong equation.

Solution (a) Ideal gas law:

$$P = \frac{RT}{v} = \frac{8.314 \times 10^3 \times 473}{0.6} = 6.554 \text{ MPa}$$

(b) van der Waals constants for ethylene are $a = 453.046 \times 10^{-3}$ Pa(m^3/mol)^2 and $b = 0.057 \times 10^{-3}$ m^3/mol

$$P = \frac{RT}{v - b} - \frac{a}{v^2} = \frac{8.314 \times 473}{(0.6 - 0.057) \times 10^{-3}} - \frac{453.046 \times 10^{-3}}{(0.6 \times 10^{-3})^2} = 5.984 \text{ MPa}$$

(c) The constants appearing in the Redlich–Kwong equation of state are given by Eqs. (3.26) and (3.27). For ethylene $P_c = 51.17$ bar and $T_c = 283.1$ K. Therefore,

$$a = \frac{0.42748R^2T_c^{2.5}}{P_cT^{1/2}} = \frac{0.42748(8.314)^2(283.1)^{2.5}}{(51.17 \times 10^5)(473)^{1/2}} = 0.358 \text{ Pa m}^6/\text{mol}^2$$

$$b = \frac{0.0867RT_c}{P_c} = \frac{0.0867(8.314)283.1}{51.17 \times 10^5} = 3.988 \times 10^{-5} \text{ m}^3/\text{mol}$$

Therefore,

$$P = \frac{RT}{v - b} - \frac{a}{v(v + b)}$$

$$P = \frac{8.314(473)}{0.6 \times 10^{-3} - 3.988 \times 10^{-5}} - \frac{0.358}{0.6 \times 10^{-3}(0.6 \times 10^{-3} + 3.988 \times 10^{-5})} = 6.088 \text{ MPa}$$

3.6 Compressibility Chart

Equations of state to represent the P–v–T behavior of fluids have been discussed in Section 3.4. We have noticed that the ideal gas law holds good at sufficiently low densities, but as the density increases the behavior of the real gas deviates from the ideal gas law. To represent the P–v–T data more accurately, the ideal gas equation of state has been modified and several equations of state like Redlich–Kwong, Benedict–Webb-Rubin, and Beattie–Bridgemann equations of state have been proposed. To obtain accurate values it is always advisable to use an equation of state to predict the thermodynamic properties, though it involves calculations. However, with the availability of high speed digital computers the computational difficulty is easily overcome. For approximate values of the properties which are needed for quick estimate, one would like to use an equation of state which is simple and yet which predicts the properties with reasonable accuracy. In view of its simplicity the ideal gas law, if modified, is quite appealing for the above purpose. Hence it is necessary to know to what extent the behavior of the real gases deviates from that of an ideal gas.

To quantify the deviation of real gas behavior from the ideal gas behavior, we introduce the new term called the compressibility factor. The compressibility factor Z, is defined as the ratio of the actual volume to the volume predicted by the ideal gas law at a given temperature and pressure.

Z = Actual volume/Volume predicted by the ideal gas law

$$Z = \frac{v}{(RT/P)} = \frac{Pv}{RT} \tag{3.34}$$

If the gas behaves like an ideal gas, $Z = 1$ at all temperatures and pressures. Therefore, a plot of Z as a function of temperature and pressure (compressibility chart), should reveal the extent of deviation from the ideal gas law. Fig. 3.5 shows a plot of Z as a function of temperature and pressure for nitrogen.

3.6.1 Reduced equation of state

One can observe from Fig. 3.5 that $Z \rightarrow 1$ as $P \rightarrow 0$. That is, at low pressures, the gas behaves like an ideal gas, but at low temperatures and high pressures the behavior of the gas deviates from the ideal gas behavior. Similar charts can be constructed for various substances. Several attempts have been made to obtain a generalized compressibility chart to represent the behavior of all gases on a single chart. The constants a and b of the van der Waals equation of state have been determined in terms of the critical constants. It was observed that $P_c v_c / RT_c = 0.375$. That is, at the critical point, the compressibility factor (Z_c) for all substances obeying the van der Waals gas is 0.375. The value of Z_c for several fluids deviates from 0.375.

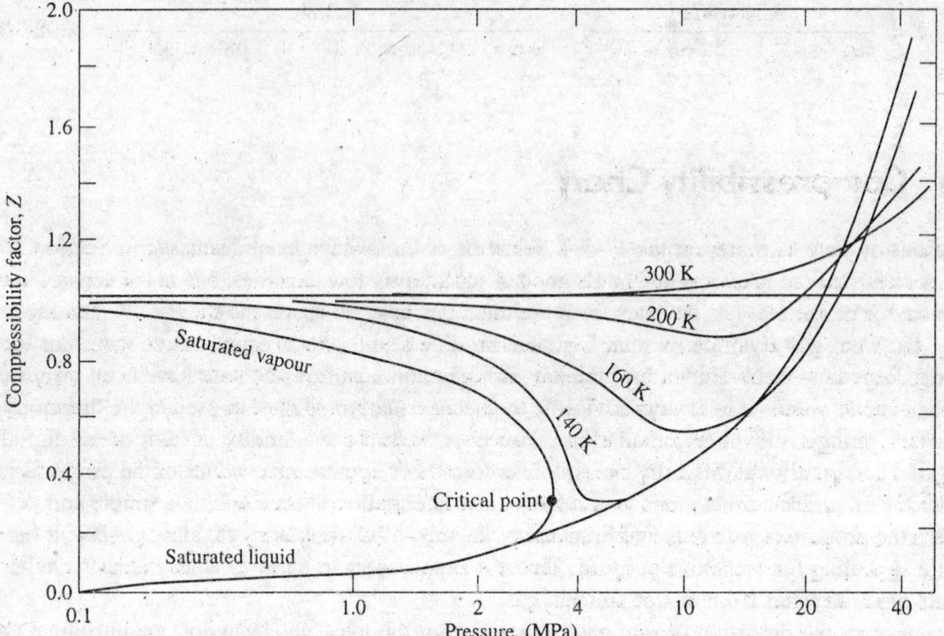

Fig. 3.5 *Compressibility chart for nitrogen*

The reduced parameters are defined as

$$\text{Reduced pressure,}\quad P_r = \frac{P}{P_c} \tag{3.35}$$

$$\text{Reduced temperature,}\quad T_r = \frac{T}{T_c} \tag{3.36}$$

$$\text{Reduced volume,}\quad v_r = \frac{v}{v_c} \tag{3.37}$$

The van der Waals equation of state can also be expressed in terms of reduced parameters as

$$\left(P_r + \frac{3}{v_r^2}\right)(3v_r - 1) = 8T_r \tag{3.38}$$

or

$$\left(Z + \frac{27P_r}{64ZT_r^2}\right)\left(1 - \frac{P_r}{8ZT_r}\right) = 1 \tag{3.39}$$

where a and b have been substituted in terms of Eq. (3.22) and P, v, T are expressed in terms of reduced parameters. This equation does not contain any constants characteristic of individual gases. It is a universal equation of state which is applicable to all van der Waals gases. An interesting feature of this equation of state is that it leads to the law of corresponding states which reads: *All gases when considered at the same value of reduced pressure and reduced temperature will have the same value of the compressibility factor.*

3.6.2 Generalized compressibility chart

Compressibility charts (which show the compressibility factor Z as a function of pressure P and temperature T) of several gases show a lot of similarities, but they differ one from the other. In an attempt to find the generalized compressibility chart which adequately represents the P–v–T behavior of several substances, one makes use of the law of corresponding states. Though the reduced equation of state does not quantitatively represent the P–v–T behavior of the gases, it can be expected that the compressibility factor can be expressed as a function of the reduced parameters as $Z = Z(P_r, T_r)$ for all gases. While preparing the compressibility chart, it was found to be convenient to use pseudo-reduced volume v_r' instead of reduced volume. The pseudo-reduced volume is defined as:

$$v_r' = \frac{v}{RT_c/P_c} = \frac{P_c v}{RT_c} \tag{3.40}$$

The compressibility chart is presented in Appendix Figs. A1–A2. It is observed from the generalized compressibility chart that there are two regions:

(a) Region in which $Z < 1$. Since $P = ZRT/v$, as $Z < 1$, the actual gas pressure is less than the pressure exerted by an ideal gas. That is, the forces of attraction between the gas particles play a dominant role in this region (at low temperature).
(b) Region in which $Z > 1$. The volume occupied by the gas is $v = ZRT/P$. When $Z > 1$, the actual volume is greater than the ideal gas volume. Hence in the region where $Z > 1$ (at all pressures above $P_r = 10$), the actual volume of the gas particles play a dominant role.

Though convenient to use, the generalized compressibility chart is not very accurate for predicting the thermodynamic properties of real gases. It is useful in predicting the properties of substances for which a more precise equation of state is not available.

The value of Z at $P_r = 1$ and $T_r = 1$ as read from the compressibility chart given in Appendix Fig. A1 is 0.27. That is, the critical compressibility factor $Z_c = 0.27$. This value is different from 0.375 as predicted by the van der Waals equation. A closer look at the critical constants of several substances indicates that Z_c varies from gas to gas, whereas the compressibility chart gives $Z_c = 0.27$ for all substances. Therefore, the generalized compressibility chart is necessarily not very accurate in representing the P–v–T behavior of gases. Several attempts have been made to improve the compressibility chart by including a third parameter in addition to P_r and T_r. Pitzer found that the compressibility factor for the gases argon, krypton, xenon, etc., the molecules of which are spherical are equal when all these gases are at the same reduced temperature and reduced pressure. That is, these gases are in corresponding states. Pitzer introduced the acentric factor as a measure of the deviation of the molecule from sphericity. The *acentric factor* is defined such that its value is zero for the gases argon, krypton and xenon. He found that all gases with the same acentric factor when considered at the same reduced temperature and reduced pressure deviate to the same extent from the ideal gas law. This observation led him to propose the three parameter corresponding states principle, and hence the three parameter generalized compressibility chart. The predicted values even with the improved compressibility charts differ by three percent for non polar gases, and by as much as ten percent or more for polar gases. The errors are still large for gases which exhibit association effects. The generalized compressibility chart does not provide reasonably good results for gases like hydrogen and helium. Therefore, the generalized compressibility chart should be used with caution and only when reliable experimental P–v–T data or an equation of state for the substance under consideration is not available. Moreover, it can be seen that the saturated liquid line is not shown in the compressibility chart. Hence the chart cannot be used to predict the properties of the liquid phase. The use of the compressibility chart is illustrated in the following examples.

***Example* 3.3** It is desired to store one kmol of methane at a temperature of 300 K and at a pressure of 60 MPa. Using the generalized compressibility chart determine the volume of the cylinder that is necessary for storage.

Solution For methane $P_c = 46.41$ bar and $T_c = 190.7$ K.

$$P_r = \frac{P}{P_c} = \frac{60 \times 10^6}{46.41 \times 10^5} = 12.93$$

$$T_r = \frac{T}{T_c} = \frac{300}{190.7} = 1.57$$

Read $Z = 1.34$ at $T_r = 1.57$ and $P_r = 12.93$ from the generalized compressibility chart as shown in Fig. 3.6. Therefore,

$$v = \frac{ZRT}{P} = \frac{1.34(8.314 \times 10^3)300}{60 \times 10^6} = 0.0557 \text{ m}^3/\text{kmol}$$

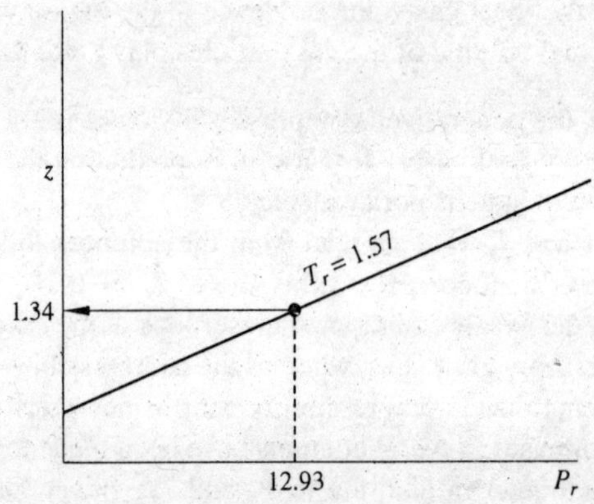

Fig. 3.6 *Sketch for Example 3.3*

***Example* 3.4** One kmol of ammonia is filled in a 0.1 m^3 vessel at a temperature of 200°C. Using the generalized compressibility chart determine the pressure which the ammonia exerts.

Solution For ammonia $T_c = 405.5$ K and $P_c = 112.77$ bar

$$T_r = \frac{T}{T_c} = \frac{473}{405.5} = 1.17$$

Since the pressure of the gas is not known, it is not possible to evaluate the reduced parameter P_r to read the value of Z. Therefore, an indirect procedure is to be applied to determine Z. The following method is simpler to evaluate Z and hence the pressure.

$$P_r = \frac{P}{P_c} \quad \text{or} \quad P = P_r P_c = 112.77 \times 10^5 P_r$$

$$Z = \frac{Pv}{RT} = \frac{112.77 \times 10^5 P_r (0.1)}{8.314 \times 10^3 \times 473} = 0.287 P_r$$

Draw a line with a slope of 0.287 on the compressibility chart as shown in Fig. 3.7 and locate the point of intersection of this line with the curve for $T_r = 1.17$ and obtain $Z = 0.55$ and $P_r = 1.9$. Therefore

$$P = 112.77 \times 10^5 P_r = 112.77 \times 10^5 \times 1.9 = 21.43 \text{ MPa}$$

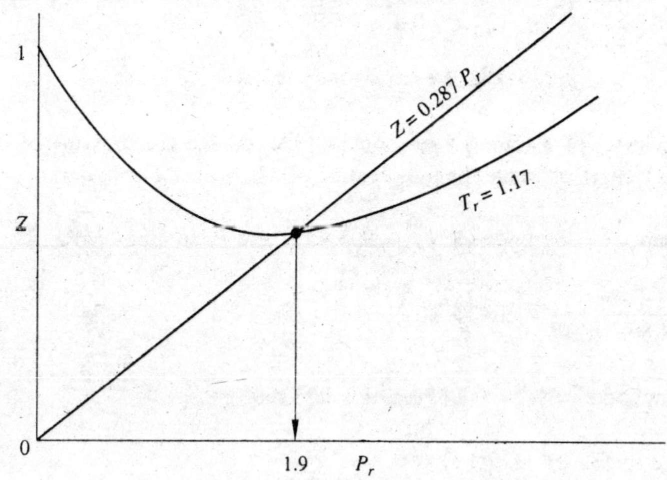

Fig. 3.7 *Sketch for Example 3.4*

***Example* 3.5** Rework Example 3.4 using the pseudo-reduced volume.

Solution For ammonia, read the values of T_c and P_c from Appendix A as:

$$T_c = 405.5\ K \quad \text{and} \quad P_c = 112.77 \text{ bar}$$

$$T_r = \frac{T}{T_c} = \frac{473}{405.5} = 1.17$$

$$v_r' = \frac{P_c(V/n)}{RT_c} = \frac{112.77 \times 10^5 \times (0.1/1 \times 10^3)}{8.314 \times 405.5} = 0.334$$

Then read $P_r = 1.9$ at the intersection of $T_r = 1.17$ and $V_r' = 0.334$ from Appendix Fig. A1 as shown in Fig. 3.8. Therefore,

$$P = P_r P_c = 1.9 \times 11.277 \times 10^6 = 21.43 \text{ MPa}$$

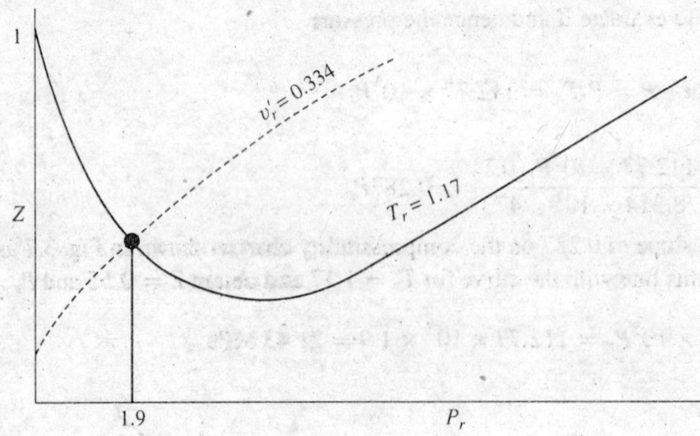

Fig. 3.8 *Sketch for Example 3.5*

Example **3.6** A cylinder of volume 0.1 m^3 contains 7 kg ethane at a pressure of 70 bar. Making use of the compressibility chart, estimate the temperature of ethane in the cylinder.

Solution For ethane, $M = 30$ kg/kmol, $T_c = 305.43$ K and $P_c = 4.884$ MPa.

$$P_r = \frac{P}{P_c} = \frac{70 \times 10^5}{4.884 \times 10^6} = 1.433$$

$$\text{Moles of ethane in cylinder} = \frac{7}{30} = 0.233 \text{ kmol} = 233 \text{ mol}$$

$$v_r' = \frac{P_c v}{RT_c} = \frac{4.884 \times 10^6 \times (0.1/233)}{8.314 \times 305.43} = 0.826$$

Draw a vertical line at $P_r = 1.433$ on the compressibility chart as shown in Fig. 3.9 and locate the point of intersection of this line with $v_r' = 0.826$ curve, and find that $T_r = 1.4$ curve passes through this point of intersection. Therefore,

$$T = T_r T_c = 1.4 \times 305.43 = 427.6 \text{ K}$$

3.7 Graphical Representation of Data

A convenient method of representing the behavior of a pure substance is by means of a graph or thermodynamic diagram. The commonly used thermodynamic diagrams are:

1. Temperature versus volume ($T - v$)
2. Pressure versus volume ($P - v$)
3. Pressure versus temperature ($P - T$)
4. Temperature versus entropy ($T - s$)
5. Enthalpy versus entropy ($h - s$, known as *Mollier diagram*)
6. Pressure versus enthalpy ($P - h$)

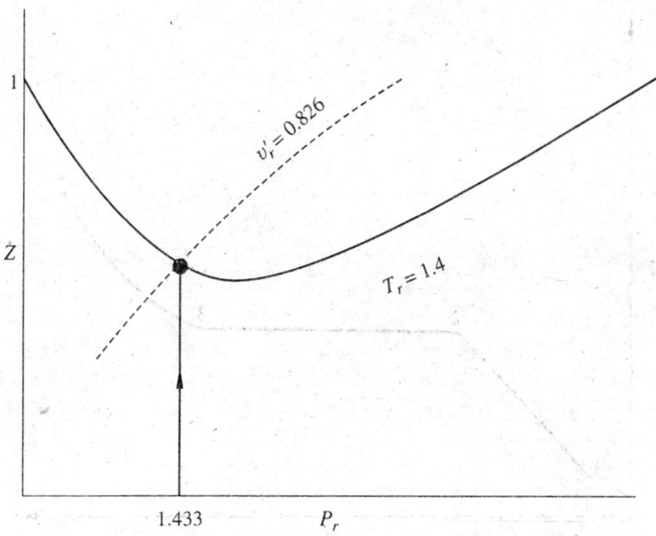

Fig. 3.9 *Sketch for Example 3.6*

A brief description of some of the important diagrams is given below.

3.7.1 Temperature–volume diagram

Let us consider constant pressure heating of liquid water in a cylinder–piston assembly. If the water is initially at state 1 say at 25°C and 101.325 kPa, on heating at constant pressure the temperature and the specific volume of the water increases and follows path 1–2 as shown in Fig. 3.10. At point 2, the water reaches a temperature of 100°C and starts vaporizing. During the vaporization process, the temperature and pressure remain constant till all the liquid is vaporized. This vaporization process is represented by path 2–3. The phase change from liquid to vapor starts at point 2 and ends at point 3. A *state at which a phase change begins or ends is called a saturated state.* The point 2 represents saturated liquid and the point 3 represents saturated vapor. Further heating produces superheated vapor, that is, vapor above its saturation temperature at the specified pressure. The final state (4) of the system is shown in Fig. 3.10. If the constant pressure heating is done at several different pressures the resulting temperature–volume diagram is as shown in Fig. 3.11.

The curve AC represents the saturated liquid line and the curve BC represents the saturated vapor line. Along the curves AC and BC both the liquid and vapor phases coexist in a state of equilibrium. To the left of the curve AC, the fluid exists in the liquid phase only and to the right of the curve BC, the fluid exists in the vapor phase only. In the region covered by the dome ACB, both liquid and vapor phases coexist in a state of equilibrium. The temperature at which phase transition from liquid to vapor begins or ends, at a specified pressure is called *saturation temperature.* For water, the saturation temperature at 101.325 kPa is 100°C. At a specified temperature, the pressure at which phase transition begins or ends is called *saturation pressure or vapor pressure.* The saturation pressure for water at 100°C is 101.325 kPa. A point in the two-phase region, shown as M in Fig. 3.11, represents a mixture of liquid and vapor. We know that volume is an additive property. Therefore, the volume of a mixture of liquid

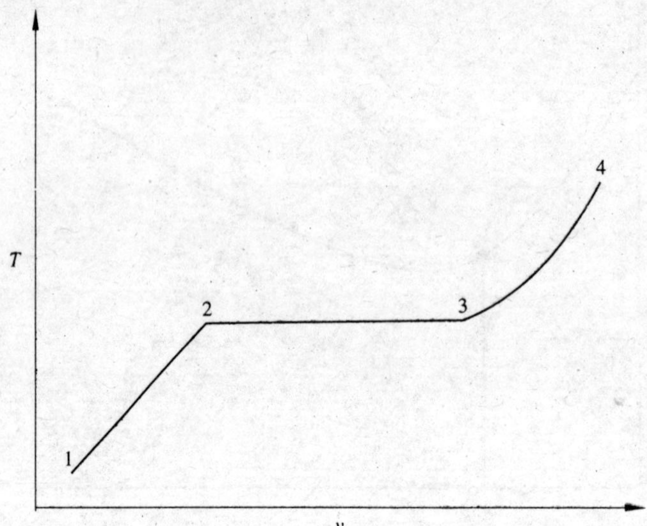

Fig. 3.10 *$T-v$ diagram showing constant pressure heating of a pure substance*

and vapor is equal to the sum of the volumes occupied by these phases. Hence,

$$(m_g + m_f)v = mv = m_g v_g + m_f v_f \tag{3.41}$$

or

$$v = \frac{m_g}{m} v_g + \frac{m - m_g}{m} v_f \tag{3.42}$$

where m = mass of mixture.
m_g = mass of saturated vapor.
m_f = mass of saturated liquid.
v = specific volume of mixture.
v_g = specific volume of saturated vapor.
v_f = specific volume of saturated liquid.

Usually, a mixture of saturated liquid and saturated vapor is described by specifying its quality X, which is defined as the ratio of the mass of saturated vapor to the total mass of mixture. That is,

$$X = \frac{\text{Mass of saturated vapor}}{\text{Total mass of mixture}} = \frac{m_g}{m} \tag{3.43}$$

Then, Eq. (3.42) can be rewritten as:

$$v = Xv_g + (1 - X)v_f = v_f + Xv_{fg} \tag{3.44}$$

where $v_{fg} = v_g - v_f$ = change in specific volume due to vaporization. Equation (3.41) can be rearranged as:

$$\frac{m_f}{m_g} = \frac{v_g - v}{v - v_f} \tag{3.45}$$

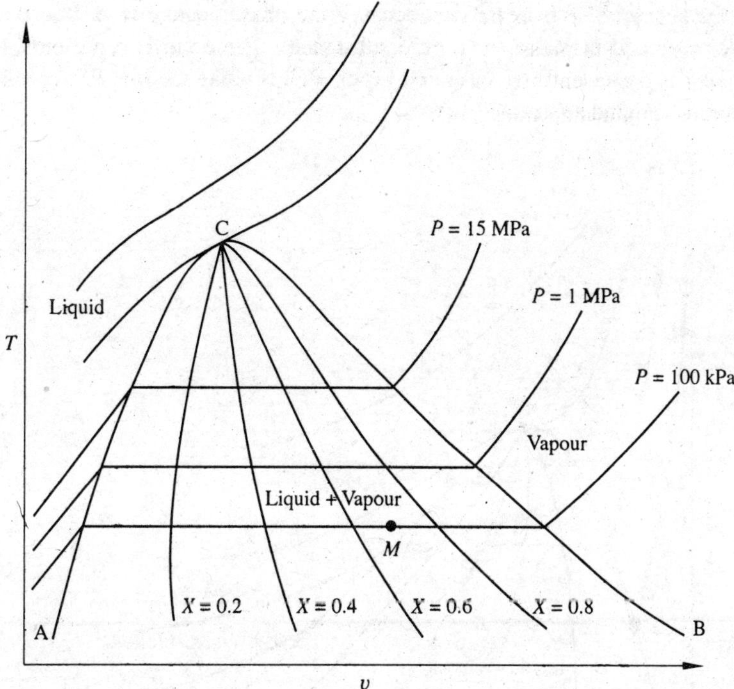

Fig. 3.11 *T − v diagram for a pure substance*

which is commonly known as the *lever rule*. The constant quality lines are shown in the two-phase region in Fig. 3.11.

3.7.2 Pressure–volume diagram

Suppose we plot P versus v for a large number of isotherms, we obtain a pressure versus volume diagram as shown in Fig. 3.12. The saturated liquid curve and the saturated vapor curve merge at point C, which is called the critical point. The critical point represents the highest temperature and pressure at which both the liquid phase and the vapor phase can coexist in equilibrium. At the critical point, the specific volumes and all other thermodynamic properties of the liquid phase and the vapor phase are identical. The corresponding temperature T_c, and the pressure P_c are called the critical temperature and critical pressure, respectively.

The isotherm T_1 is at a temperature greater than the critical temperature T_c. The isotherms T_2 and T_3 are at temperatures less than the critical temperature and they cross the phase boundary. The point C represents the critical point. The curve AC is called the saturated liquid line, and the curve BC is called the saturated vapor line. Area under the curve ACB is the two phase region where both the liquid and vapor are present. The region to the left of the curve AC is the liquid region and that to the right of the curve BC is the vapor region. The isotherms T_2 and T_3 appear in three segments (DE, EF, FG). The segment DE is almost vertical because the change in the volume of liquid is very small for large change in pressure. The segment FG is less steep because the vapor phase is highly compressible. In between these two limits the segment EF is horizontal because the phase change from liquid to vapor occurs at constant temperature and pressure for a pure substance. The point E represents 100% saturated liquid and the point F represents 100% saturated vapor. Points along the line EF represent all possible mixtures of saturated liquid and saturated vapor.

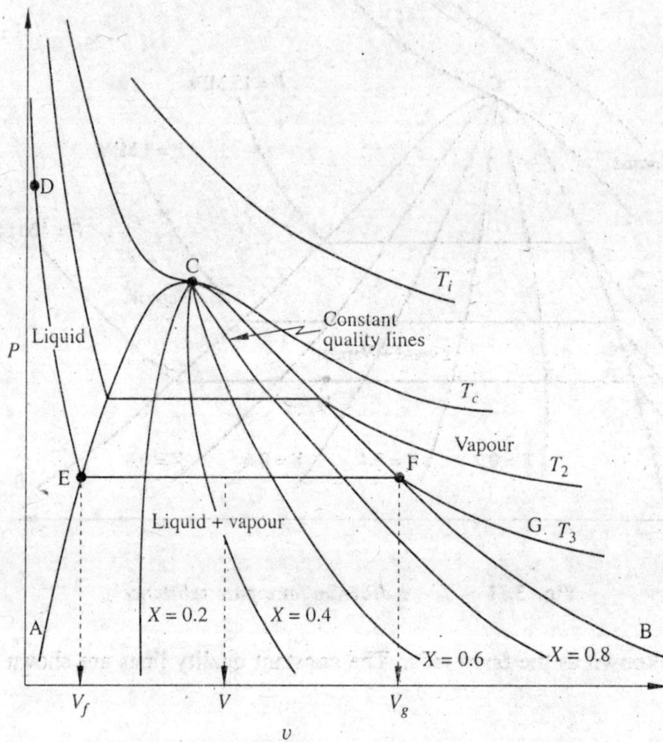

Fig. 3.12 *P–v diagram for a pure substance*

It can be seen from Fig. 3.12 that the horizontal segments become short as the temperature increases. Ultimately, at critical temperature the horizontal segment reduces to a point. At the critical point the specific volumes and all other thermodynamic properties of the liquid phase and the vapor phase are identical.

To understand more about the critical point, let us consider a constant volume process involving a phase change. Suppose a constant volume bulb is initially filled with a large quantity of saturated liquid and a small quantity of saturated vapor whose state is shown by point 1 on the *P–v* diagram in Fig. 3.13. If it is heated at constant volume, it follows the path 1–2–3 on the *P–v* diagram (see Fig. 3.13). With heating, the liquid expands and occupies the whole container and reaches state 2 where it is present as a saturated liquid. That is, the interface separating the liquid and vapor can be observed to move up. On further heating the liquid reaches state 3. On the other hand, suppose the constant volume bulb is initially filled with a small quantity of saturated liquid and a large quantity of saturated vapor. The state is represented by point 4 on the *P–v* diagram. If the mixture is subjected to a constant volume heating process, the path followed by the system is 4–5–6. That is, on heating the liquid vaporizes and the interface separating the two phases moves down. Once the liquid is completely vaporized, the state is represented by point 5, at which all substance is in saturated vapor form. Further heating takes the system to state 6 which is in the vapor region.

Between the above two extreme cases, it is possible to fill the vessel with a mixture of saturated liquid and saturated vapor such that the initial state is represented by point 7 on the *P–v* diagram. On heating the mixture at constant volume, the path followed by the system is 78C9. That is, the mixture passes through the critical point. In this case large changes in the position of the interface cannot be observed till the system reaches the critical point. At the critical point the interface suddenly disappears. As the system approaches the critical point the difference between the properties of saturated liquid and saturated vapor becomes less. Therefore, the interface separating the liquid and vapor phases becomes less distinct. At the critical point, the properties of the saturated liquid and saturated vapor are identical and the interface disappears. On further heating the system attains state 9.

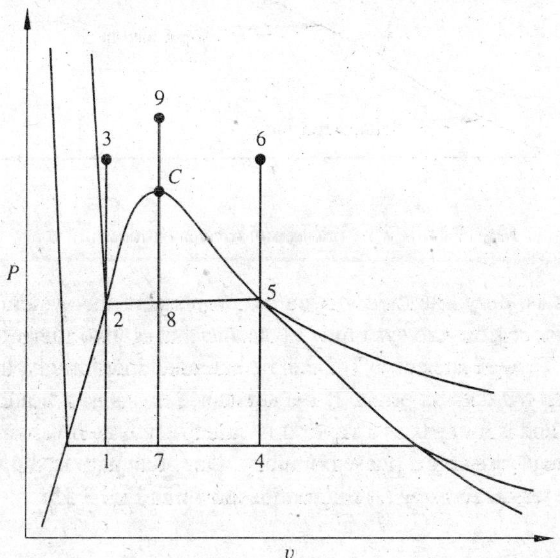

Fig. 3.13 *P–v diagram showing constant volume heating process for a pure substance*

3.7.3 Pressure–temperature diagram

Suppose one measures the vapor pressure of a substance for both the solid and liquid phases as a function of temperature. These measurements provide vapor pressure curves of the substance. A typical pressure–temperature diagram is shown in Fig. 3.14.

The segment 1–T represents the sublimation curve on which the solid and vapor phases coexist in equilibrium. The segment 2–T represents the fusion curve on which the solid and liquid phases coexist in equilibrium. The vaporization curve on which the liquid and vapor phases coexist in equilibrium is represented by T–C. The curve 1–T separates the solid and vapor regions. The curve 2–T separates the solid and liquid regions and the curve T–C separates the liquid and vapor regions. These three curves meet at point T, which is called the triple point, where all the three phases—solid, liquid and vapor coexist in equilibrium. At the triple point no thermodynamic property of the system can be varied independently. The system is said to be invariant. Along any one of the curves 1–T, 2–T or T–C the

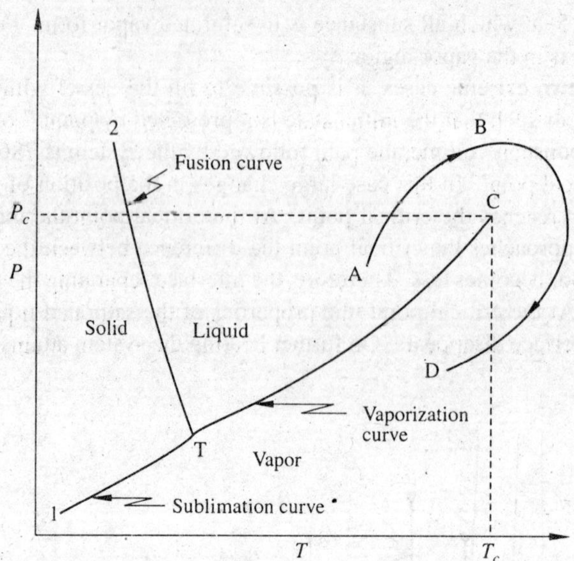

Fig. 3.14 *P – T diagram for a pure substance*

system is univariant, that is, only one thermodynamic property of the system can be varied independently. In the single-phase region, the system is bivariant, that is, two thermodynamic properties can be varied independently. Though the curve T–2 can be extended indefinitely, the curve T–C terminates at point C, which is called the critical point. If the substance exists as a liquid on the curve T–C, it is called a saturated liquid, and if it exists as a vapor it is called a saturated vapor. It is possible to convert the liquid into vapor without crossing a phase boundary. One such path is shown in Fig. 3.14 by ABD.

A typical temperature versus entropy (s) diagram is shown in Fig. 3.15.

3.7.4 Mollier diagram

The thermodynamic properties of a substance are quite often shown on an enthalpy–entropy diagram. This diagram was introduced into engineering thermodynamics by Richard Mollier (1863–1935) and was named after him as Mollier diagram. Enthalpy h is defined as:

$$h = u + Pv \tag{3.46}$$

While discussing the first law of thermodynamics in a later chapter it will be shown that the heat interaction for an isobaric process is equal to the change in the enthalpy of the system and in a flow process the work done by an adiabatic device is equal to the decrease in the enthalpy of the fluid flowing through the device. It will also be shown that a throttling process is an isenthalpic (constant enthalpy) process. Similarly, it will be shown that a reversible adiabatic process is an isentropic (constant entropy) process. Isobaric, isenthalpic and isentropic processes can be conveniently traced on an enthalpy- entropy diagram. Therefore, Mollier diagram can be conveniently used in the thermodynamic analysis of processes.

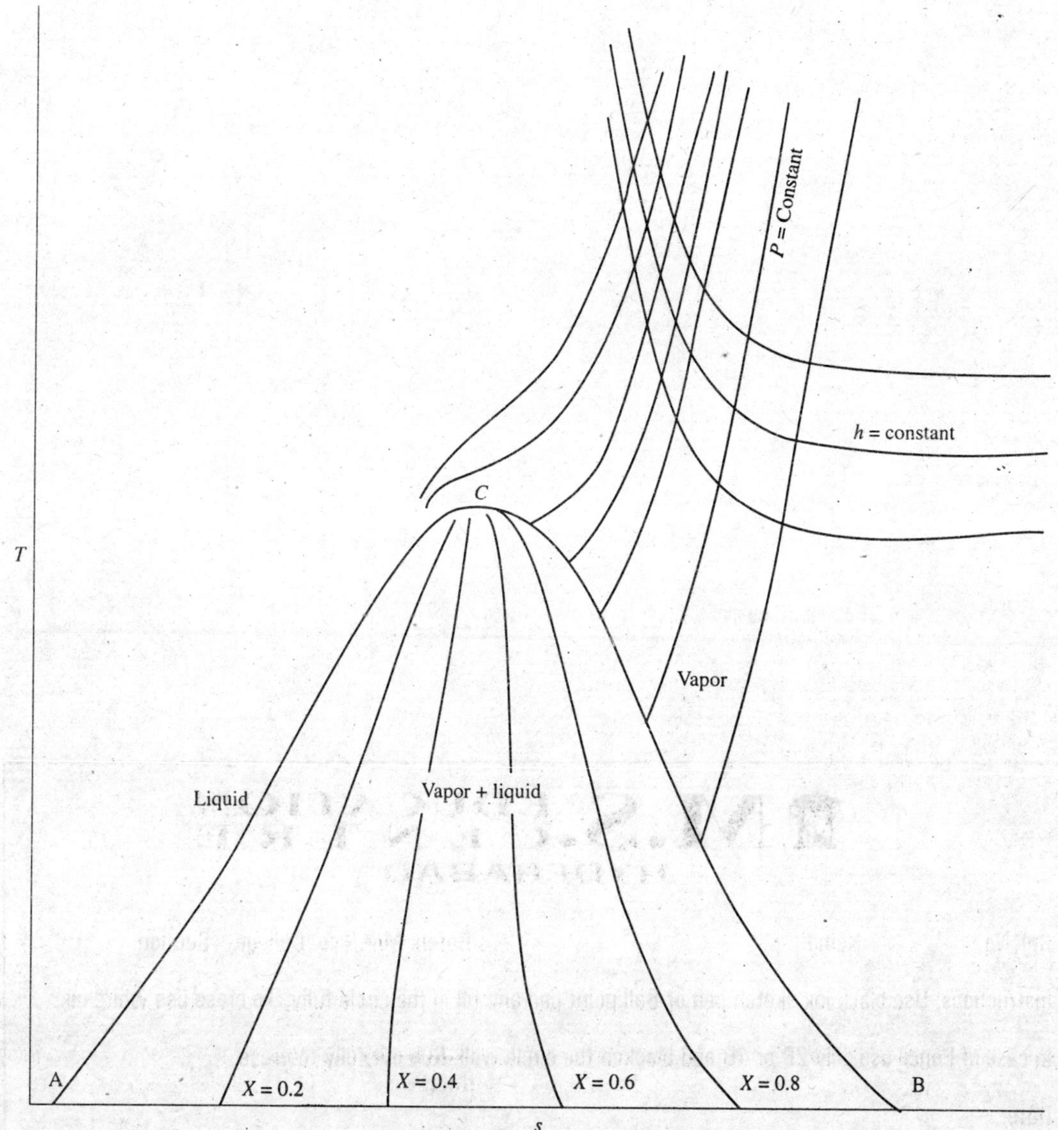

Fig. 3.15 *Temperature–entropy diagram for a pure substance*

A Mollier diagram for water is presented in Fig. 3.16. It consists of a family of constant pressure lines, constant temperature lines, constant quality lines and constant volume lines plotted on enthalpy–entropy coordinates. The saturation curve is shown as a dome which separates the two-phase region from the vapor phase region. It may be noted that the critical point does not exist at the top of the dome in an $h - s$ diagram. The lines of constant temperature and constant pressure coincide in the two-phase region.

M.S. EDUCATION CENTRE
HYDERABAD

Roll No.: Name: Batch: Mor./Eve./College Section:

Instructions: Use black ink sketch pen or Ball point pen and fill in the circle fully. To erase use white ink.

In case of Pencil use only 2B or 4B and blacken the circle well. Rub out fully to erase.

Date:

Marks obtained: Botany: Zoology: Physics: Chemistry: Total:

Rank:

Fig. 3.16 *Mollier diagram*

Since enthalpy (h) and entropy (s) are also extensive properties, equations similar to Eq. (3.44) can be written to estimate the mixture properties in terms of quality X.

$$h = Xh_g + (1 - X)h_f = h_f + Xh_{fg} \tag{3.47}$$

$$s = Xs_g + (1 - X)s_f = s_f + Xs_{fg} \tag{3.48}$$

where, h_g = specific enthalpy of saturated vapor,
h_f = specific enthalpy of saturated liquid,
$h_{fg} = h_g - h_f$ = latent heat of vaporization,
s_g = specific entropy of saturated vapor,
s_f = specific entropy of saturated liquid,
$s_{fg} = s_g - s_f$ = entropy change due to vaporization.

***Example* 3.7** Making use of the Mollier diagram, determine the saturation pressure of water at 150°C and the saturation temperature of water at 10 bar.

Solution On the Mollier diagram locate T = 150°C curve and follow it till it touches the saturation curve as shown in Fig. 3.17. The point of intersection of T = 150°C curve and the saturation curve is shown as point A in the Fig. 3.17. Now, notice that P = 5 bar curve passes through the point A. Hence, the saturation pressure is 5 bar.

Locate P = 10 bar curve on the Mollier diagram and follow it till it intersects (shown as point B in Fig. 3.17) the saturation curve. Then, notice that the T = 180°C curve passes through the point B. Hence, the saturation temperature is 180°C.

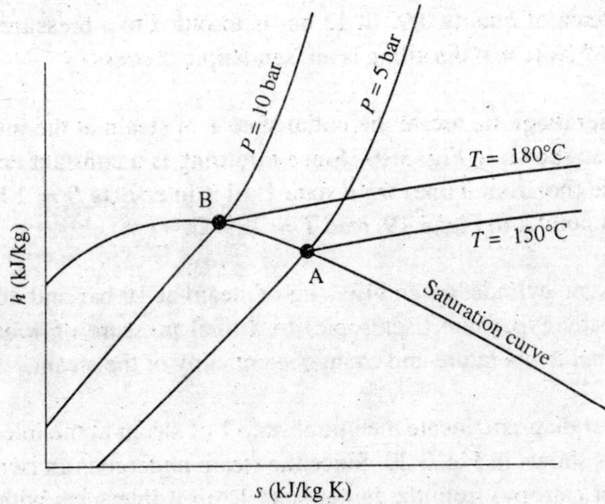

Fig. 3.17 *Sketch for Example 3.7*

***Example* 3.8** Using the Mollier diagram, determine the specific enthalpy and specific entropy of wet steam of quality 0.85 at 2 bar pressure.

Solution On the Mollier diagram, locate the constant quality line $X = 0.85$ and follow it till it intersects $P = 2$ bar curve at point A, as shown in Fig. 3.18. Then read the following values corresponding to point A.

$$h = 2380 \text{ kJ/kg} \quad s = 6.29 \text{ kJ/kg K}$$

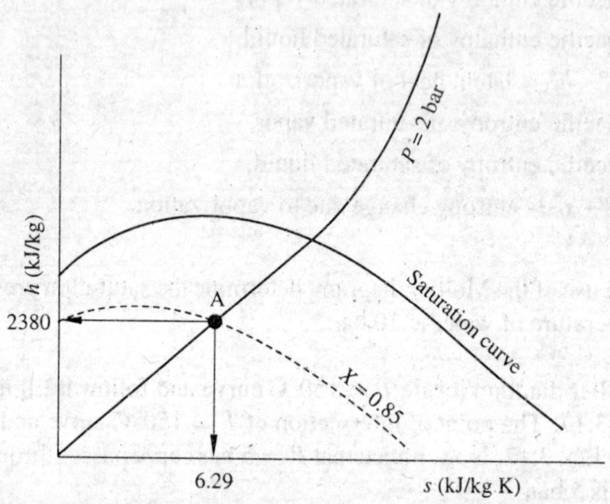

Fig. 3.18 *Sketch for Example 3.8*

***Example* 3.9** If wet steam of quality 0.97 at 15 bar is throttled to a pressure of 1 bar, what will be the resulting temperature? Note that throttling is an isenthalpic process.

Solution On the Mollier diagram, locate the initial state 1 of steam at the intersection of $X = 0.97$ and $P = 15$ bar curves, as shown in Fig. 3.19. Since throttling is a constant enthalpy process, follow the constant enthalpy line (horizontal line) from state 1 till it intersects $P = 1$ bar curve. At the point of intersection, shown as point 2 in Fig. 3.19, read $T = 129°\text{C}$.

***Example* 3.10** In a piston–cylinder assembly, 1 kg of steam at 10 bar and 400°C is allowed to undergo a reversible-adiabatic expansion (isentropic) to a final pressure of 1 bar. Using the Mollier diagram determine the final temperature and change in enthalpy of the steam.

Solution On the Mollier diagram, locate the initial state 1 of steam at the intersection of $P = 10$ bar and $T = 400°\text{C}$ curves as shown in Fig. 3.20. Since the steam undergoes an isentropic process, follow the vertical line (constant entropy) from the initial state 1, till it intersects with $P = 1$ bar curve and locate the final state 2 at the point of intersection. Then, find the constant temperature curve $T = 120°\text{C}$ that passes through point 2. Hence, the final temperature of steam is 120°C. Then read:

$$h_1 = 3275 \text{ kJ/kg} \quad h_2 = 2725 \text{ kJ/kg}$$

$$\Delta h = h_2 - h_1 = 2725 - 3275 = -550 \text{ kJ/kg}$$

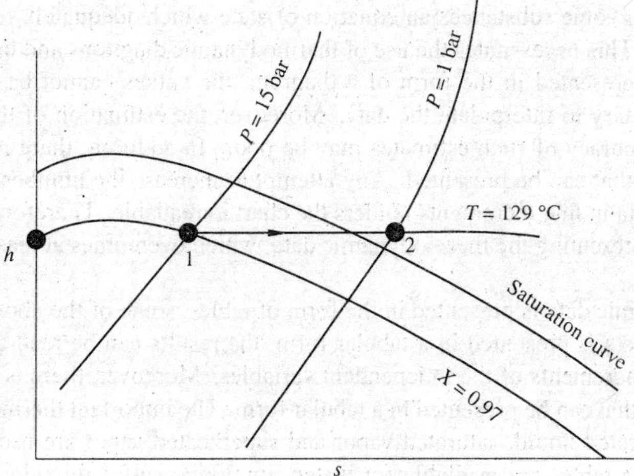

Fig. 3.19 *Sketch for Example 3.9*

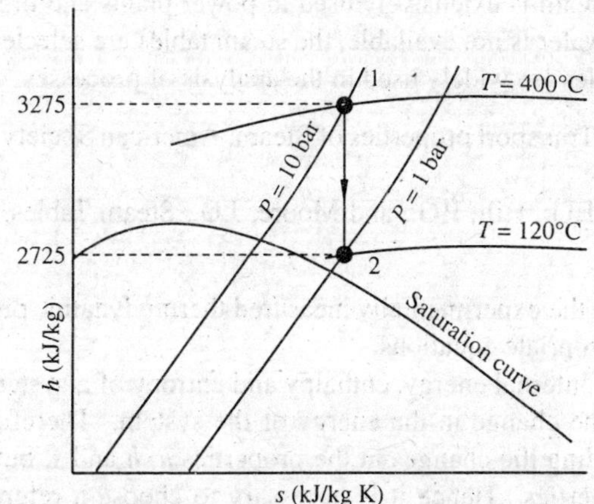

Fig. 3.20 *Sketch for Example 3.10*

3.8 Tabular Representation of Data

The mathematical and graphical representation of *P–v–T* behavior of fluids has been discussed in Sections 3.4 to 3.7. It was pointed out that the mathematical representation of data is very convenient to use, and accurate values can be predicted from the appropriate equation of state of the substance under consideration. Quite often, the other thermodynamic properties like internal energy, enthalpy, entropy, etc. of the substance, in addition to the *P–v–T* data, are extensively used in the analysis of processes. These properties can be estimated with a knowledge of the *P–v–T* data. The methods of estimation of these properties are discussed in Chapter 8. To estimate the thermodynamic properties, a knowledge of the partial derivatives of *P*, *v* and *T* is essential, which can be obtained from an equation of state. However, for some substances an equation of state which adequately represents the P–v–T data, is not available. This necessitates the use of thermodynamic diagrams and tabulated data.

When the data is presented in the form of a diagram, the values cannot be read accurately and sometimes it is necessary to interpolate the data. Moreover, the estimation of the partial derivatives is difficult and the accuracy of such estimates may be poor. In addition, there is a limitation on the number of properties that can be presented. Any attempt to increase the number of properties and/or the presentation of data in fine increments renders the chart unreadable. Therefore, there should be an alternative way of representing the thermodynamic data, which overcomes at least a few of the above drawbacks.

When thermodynamic data is presented in the form of tables, some of the above drawbacks can be avoided. When the data is presented in a tabular form, the results can be read accurately. The data can be listed in fine increments of the independent variables. Moreover, there is no limitation on the number of properties that can be presented in a tabular form. The important thermodynamic properties $v, u, h,$ and s for saturated liquid, saturated vapor and superheated vapor are usually provided in the form of a table. Such tables are available for water, ammonia, sulfur dioxide, carbon dioxide and Freons, etc. which are commonly used as working fluids in heat engines and refrigerators. In general, all tables of thermodynamic properties are similar. Therefore, it is necessary to discuss a representative table for one fluid. Since steam is extensively used in power plants and in several industries, and the exact equation of state for water is not available, the steam tables are selected for discussion.

The following steam tables are widely used in the analysis of processes:

1. Thermodynamic and Transport properties of Steam, American Society of Mechanical Engineers, 1967.
2. Keenan, J.H.,Keyes, F.G., Hill, P.G., and Moore, J.G., Steam Tables, John Wiley & sons, Inc., New York, 1969.

The steam tables present the experimentally measured thermodynamic properties of steam after the data is smoothed using appropriate equations.

Since the properties like internal energy, enthalpy and entropy of a system cannot be directly measured, they are related to the change in the energy of the system. Therefore, the experimental data provides information regarding the changes in the properties u, h and s, but does not provide the absolute values of these properties. Hence it is necessary to choose a reference state to which these properties can be assigned some numerical values. Then the properties of the system can be evaluated with reference to this reference state. For convenience, an arbitrary reference state is chosen in evaluating the thermodynamic properties. This reference state is the triple point of water ($T = 0.01°C$, $P = 0.6113$ kPa) where the internal energy and entropy of saturated liquid water are assigned a value of zero. This choice of the reference state is immaterial because we are concerned with the changes in the thermodynamic properties of the substance while it undergoes a process. For example, the heat and work interactions across the system boundaries are related to the change in the energy of the system.

3.8.1 Steam tables

The properties of steam are given in three different tables: Saturated steam - temperature table, saturated steam - pressure table and superheated steam table. In the saturated steam tables the properties of the saturated liquid, in equilibrium with the saturated vapor are presented. It can be observed either from the P–v diagram or the T–v diagram that during the phase transition, the temperature and pressure remain constant and are termed as saturation temperature and saturation pressure respectively. It has already been shown that along the vaporization curve on a *P–T* diagram the system is univariant, that is, only one thermodynamic property (either *P* or *T*) can be varied independently. Hence, it is possible to choose either the saturation temperature or the saturation pressure as an independent variable to specify the state of the system. When temperature is chosen as the independent variable and thermodynamic properties are tabulated, that table is referred to as the saturated steam-temperature table. For the saturated steam-pressure table, the independent variable is pressure. The information contained in these two tables is exactly the same. The choice of the table to be used is merely a matter of convenience. When the substance is in a single phase region (vapor region—for example superheated steam) it is necessary to specify two independent variables like temperature and pressure to describe its state. These tables are self explanatory, and do not require any discussion.

In the saturated steam tables, the values of specific volume, specific enthalpy and specific entropy of saturated liquid and saturated vapor are presented. In some steam tables additional data like internal energy, latent heat of vaporization and changes in specific volume and specific entropy are also tabulated. However, this additional data is redundant, since they can be calculated from the relations:

$$u = h - Pv$$

$$h_{fg} = hg - h_f$$

$$v_{fg} = v_g - v_f$$

$$s_{fg} = s_g - s_f$$

An extract of the saturated steam-temperature table is shown in Table 3.3. The second column in the Table gives saturation pressure.

In the steam tables, the thermodynamic properties of saturated liquid and saturated vapor are presented from the triple point (0.01°C and 0.6113 kPa) to the critical point (374.15°C and 221.2 bar). It can be seen from the table that the saturation pressure of water at 100°C is 1.0133 bar and at 200°C the saturation pressure is 15.549 bar. It can be observed from Table 3.3 that $v_f = v_g$; $h_f = h_g$ and $s_f = s_g$ at the critical point. An extract of the saturated steam- pressure table is shown in Table 3.4. The second column in this Table gives saturation temperature.

Table 3.3 *Extract from saturated steam-temperature table*

Temp.	Pressure	*Specific volume*		*Specific enthalpy*		*Specific entropy*	
T	P	v_f	v_g	h_f	h_g	s_f	s_g
°C	bar	m^3/kg	m^3/kg	kJ/kg	kJ/kg	kJ/kg K	kJ/kg K
0.01	0.006 113	0.001 000	206.15	0.000 71	2501.7	0.0000	9.1578
–	–	–	–	–	–	–	–
–	–	–	–	–	–	–	–
100	1.0133	0.001 043 7	1.6730	419.06	2676.0	1.3069	7.3554
–	–	–	–	–	–	–	–
200	15.549	0.001 156 5	0.1272	852.37	2790.9	2.3307	6.4278
–	–	–	–	–	–	–	–
374.15	221.200	0.003 17	0.003 17	2107.40	2107.40	4.4429	4.4429

Table 3.4 *Extract from saturated steam-pressure table*

Pressure	Temp.	Specific volume		Specific enthalpy		Specific entropy	
		v_f	v_g	h_f	h_g	s_f	s_g
bar	°C	m^3/kg	m^3/kg	kJ/kg	kJ/kg	kJ/kg K	kJ/kg K
0.006 113	0.01	0.001 000	206.15	0.000 71	2501.7	0.0000	9.1578
0.01	6.98	0.001 000 1	129.20	29.34	2514.4	0.1060	8.9767
–	–	–	–	–	–	–	–
1.0	99.63	0.001 043 4	1.694	417.54	2675.4	1.3027	7.3598
–	–	–	–	–	–	–	–
221.20	374.15	0.003 17	0.003 17	2107.4	2107.4	4.4429	4.4429

An extract from the superheated steam table is shown in Table 3.5. A superheated vapor is a vapor at a temperature which is greater than the saturation temperature, at a specified pressure. For example, at a pressure of 1 bar, the saturation temperature for water is 99.63°C (see Table 3.4) and hence steam at 1 bar and 200°C is called superheated steam. Some times, the difference between the steam temperature and the saturation temperature at the given pressure is called the *degree of superheat*.

Table 3.5 *Extract from superheated steam tables*

Pressure (bar)		Temperature (°C)					
(Sat. temp. °C)		100	200	300	600	700	800
0.2	v	8.585	10.507	13.219	20.146	22.455	24.762
(60.09)	h	2686.3	2879.2	3076.4	3705.4	3928.7	4158.7
	s	8.1261	8.5839	8.9618	9.8416	10.0836	10.3085
–	–	–	–	–	–	–	–
20	v	–	–	0.1255	0.1995	0.2232	0.2467
(212.37)	h	–	–	3025.0	3689.2	3916.5	4149.4
	s	–	–	6.7696	7.7022	7.9485	8.1763
–	–	–	–	–	–	–	–
1000	v	–	–	–	0.002 668	0.003 536	0.004 341
	h	–	–	–	2857.5	3324.4	3714.3
	s	–	–	–	5.1505	5.6579	6.0397

In the superheated steam tables, the saturation temperature is shown in parenthesis under pressure. It can be observed from the Table 3.5 that the saturation temperature for water at 20 bar is 212.37°C and hence superheated steam can exist at a temperature greater than 212.37°C. Therefore, the table indicates blanks at all temperatures less than 212.37°C for $P = 20$ bar. The critical pressure for water is 221.20 bar and water cannot exist in both liquid and vapor phases above the critical pressure. Therefore, saturation temperature is not indicated, in Table 3.5, at pressures greater than 221.20 bar. The thermodynamic properties of dichloro-difluoro methane (Freon-12) are presented in Appendix Table A5. The use of the thermodynamic property tables is illustrated in the following examples.

***Example* 3.11** Making use of the steam tables, determine (a) the saturation pressure and the latent heat of vaporization of water at 200°C and (b) the specific internal energy of saturated vapor at 200°C.

Solution (a) The independent variable used for the specification of the state of water, in this example, is temperature. Hence, it is convenient to use the saturated steam-temperature table. From Appendix Table A3, we read the following values at $T = 200°\text{C}$:

$$P = 15.549 \text{ bar} \quad h_f = 852.37 \text{ kJ/kg} \quad h_g = 2790.9 \text{ kJ/kg}$$

Therefore, saturation pressure = 15.549 bar. Latent heat of vaporization, $h_{fg} = h_g - h_f = 2790.9 - 852.37 = 1938.53$ kJ/kg.

(b) Read the following data at 200°C, from the Appendix Table A3.

$$P = 15.549 \text{ bar} \quad v_g = 0.1272 \text{ m}^3/\text{kg} \quad h_g = 2790.9 \text{ kJ/kg}$$

We know that $h = u - Pv$. Therefore, the specific internal energy of saturated vapor is given by:

$$u_g = h_g - Pv_g = 2790.9 \times 10^3 - 15.549 \times 10^5 \times 0.1272 = 2593.12 \text{ kJ/kg}$$

***Example* 3.12** Calculate the specific volume, specific enthalpy and specific internal energy of wet steam of quality 0.85 at 2 bar pressure.

Solution The independent variable used for the specification of the state of steam is pressure. Hence, it is convenient to use the saturated steam-pressure table given in Appendix Table A4. Read the following data at $P = 2$ bar from the Appendix Table A4.
$P = 2$ bar; $v_f = 0.001\,060\,8 \text{m}^3/\text{kg}$; $v_g = 0.8854 \text{ m}^3/\text{kg}$
$h_f = 504.70$ kJ/kg and $h_g = 2706.3$ kJ/kg

We know that the specific volume of a mixture of saturated liquid and saturated vapor is given by:

$$v = Xv_g + (1 - X)v_f = 0.85 \times 0.8854 + (1 - 0.85) \times 0.001\,060\,8 = 0.7527 \text{ m}^3/\text{kg}$$

Similarly, we can calculate the specific enthalpy of the steam.

$$h = Xh_g + (1 - X)h_f = 0.85 \times 2706.3 + (1 - 0.85) \times 504.70 = 2376.06 \text{ kJ/kg}$$

The specific internal energy is given by:

$$u = h - Pv = 2376.06 \times 10^3 - 2 \times 10^5 \times 0.7527 = 2225.52 \text{ kJ/kg}$$

***Example* 3.13** Using the steam tables, estimate the specific volume and specific enthalpy of saturated vapor at 32 bar.

Solution In the saturated steam-pressure table (see Appendix C), we find that the data is provided at $P = 30$ bar and $P = 35$ bar and no data is listed at $P = 32$ bar. Therefore, it is necessary to interpolate the data. One can use linear interpolation to estimate the data at 32 bar as shown below:

P (bar)	v_g (m^3/kg)	h_g (kJ/kg)
30	0.066 63	2802.3
35	0.057 03	2802.0

At 32 bar:

$$v_g = 0.066\,63 + \frac{(0.057\,03 - 0.066\,63)}{35 - 30} \times (32 - 30) = 0.062\,79 \text{ m}^3/\text{kg}$$

$$h_g = 2802.3 + \frac{(2802.0 - 2802.3)}{35 - 30} \times (32 - 30) = 2802.18 \text{ kJ/kg}$$

***Example* 3.14** If the specific enthalpy of steam at 2 bar is 2600 kJ/kg, determine its state.

Solution The state of steam is specified by describing the pressure and specific enthalpy. In such instances, it is desirable to first consult the saturated steam tables. From Appendix Table A4, we find at $P = 2$ bar:

$$h_f = 504.70 \text{ kJ/kg} \quad \text{and} \quad h_g = 2706.3 \text{ kJ/kg}$$

The given value of specific enthalpy 2600 kJ/kg lies in between the values of h_f and h_g and hence, it represents wet steam, that is, a mixture of saturated vapor and saturated liquid at 2 bar. We know that the specific enthalpy of wet steam of quality X is given by:

$$h = Xh_g + (1 - X)h_f$$

Therefore,

$$2600 = 2706.3X + (1 - X)504.70 \quad \text{or} \quad X = 0.952$$

Hence, the state of steam is a wet steam of quality 0.952 at 2 bar pressure.

The Mollier diagram can also be conveniently used to determine the state of the steam. On the Mollier diagram, locate the point of intersection of $h = 2600$ kJ/kg line and $P = 2$ bar curve as shown in Fig. 3.21. The point of intersection (shown as A) passes through $X = 0.95$ curve. Hence, the quality of steam is 0.95.

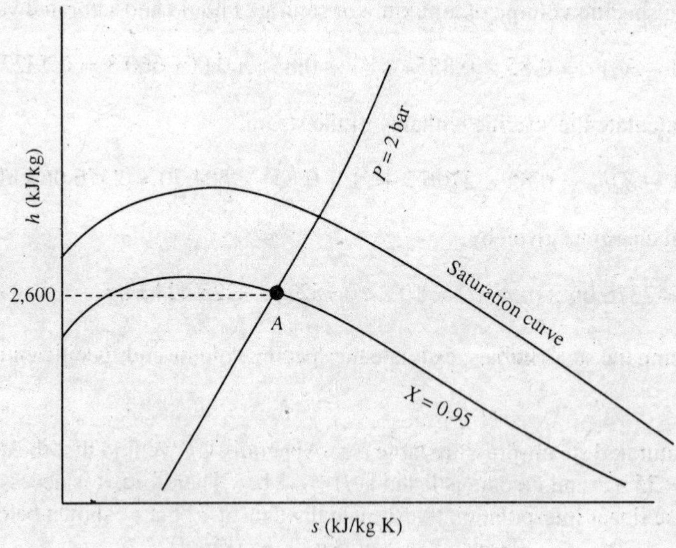

Fig. 3.21 *Sketch for Example 3.14*

***Example* 3.15** Estimate the specific volume and specific enthalpy of superheated steam at 25 bar and 350°C.

Solution The superheated steam tables in Appendix Table A5 does not list the properties at the required pressure and temperature. The data is listed at 24 bar and 26 bar and at 300°C and 400°C. Read the following data from the superheated steam tables:

P (bar)	T°C	v(m^3/kg)	h(kJ/kg)
24	300	0.103 36	3013.4
26	300	0.094 83	3007.4
24	400	0.125 22	3242.3
26	400	0.115 26	3239.0

First let us find the values of v and h at P = 25 bar at 300°C and 400°C by linear interpolation as:
At P = 25 bar:

	300°C	400°C
h(m^3/kg)	0.099 095	0.120 240
h(kJ/kg)	3010.4	3240.65

Then, by linear interpolation between 300°C and 400°C, find the values of v and h at 350°C as: v = 0.109 668 m^3/kg and h = 3125.53 kJ/kg.

***Example* 3.16** The domestic pressure cooker is a common kitchen appliance. Particularly they are useful in hill stations where the atmospheric pressure is less than 100 kPa. The weight placed on the steam vent, under normal conditions, ensures that the pressure inside the cooker never exceeds 200 kPa. If water is boiled in the pressure cooker, determine its boiling point.

Solution When the water boils, the pressure maintained in the cooker is 200 kPa. Therefore, the saturation temperature at a pressure of 200 kPa is the boiling point of water. The saturation temperature at 200 kPa as read from the saturated steam-pressure table is 120.23°C. Therefore, the boiling point of water at 200 kPa is 120.23°C. This information can also be obtained from the Mollier diagram as shown in Fig. 3.22. The intersection of the P = 2 bar line with the saturation curve has been located (point A) which corresponds to the saturation temperature of 120°C (interpolation between two constant temperature lines is sometimes required).

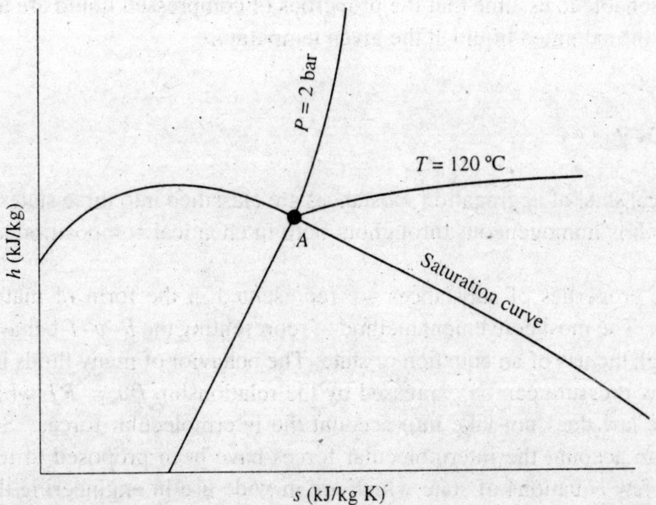

Fig. 3.22 *Sketch for Example 3.16*

***Example* 3.17** A rigid vessel is filled with a mixture of saturated liquid and saturated vapor at a pressure of 100 kPa. Determine the proportions by volume of liquid and vapor to be filled so that on heating the vessel, the state of the water passes through the critical state.

Solution Let V be the volume of the vessel and m be the total mass of water initially present in the vessel. Then the initial specific volume of the mixture $v_1 = V/m$. Since the process is constant volume heating, and the total mass is also remaining constant, the final specific volume is equal to the initial specific volume. $v_2 = v_1$. Since on heating the system should pass through the critical state, $v_1 = v_2 = v_c$.

From steam tables $v_c = 0.003\ 17$ m^3/kg.

At 100 kPa, $v_f = 0.001\ 043\ 4$ m^3/kg and $v_g = 1.694$ m^3/kg

$v_c = v_1 = Xv_g + (1 - X)v_f$, where X is the initial quality of the mixture.

$$0.003\ 17 = X(1.694) + (1 - X)(0.001\ 043\ 4) \quad \text{or} \quad X = 1.256 \times 10^{-3}$$

Volume of liquid $= (1 - X)v_f = 1.042 \times 10^{-3}$ m^3

Volume of vapor $= Xv_g = 2.128 \times 10^{-3}$ m^3

Ratio of liquid to vapor by volume $= \dfrac{(1 - X)v_f}{Xv_g} = 0.4897$

At a given pressure if a liquid is available at a temperature, which is less than its saturation temperature at that pressure, then the liquid is called a *subcooled liquid.* At a given temperature, if a liquid is available at a pressure, which is greater than the saturation pressure at that temperature, then the liquid is called a *compressed liquid.* The state of a subcooled liquid or compressed liquid lies in the region marked *Liquid* to the left of the saturated liquid line on P–v; T–v; and T–s diagrams, and in the region enclosed by the fusion curve and vaporization curve on the P–T diagram. The properties of the subcooled liquid or compressed liquid can in principle be obtained from steam tables. However, quite a few available steam tables do not provide the properties of subcooled liquid or compressed liquid. The variation of properties of compressed liquid with pressure is usually less than one percent. Therefore, it is reasonable to assume that the properties of compressed liquid are approximately equal to the properties of the saturated liquid at the given temperature.

3.9 Summary

Based on the physical state of aggregation substances are classified into three states as solid, liquid and gas. A system which is homogeneous throughout both in chemical composition and physical state is called a phase.

Thermodynamic properties of substances are represented in the form of mathematical relations, diagrams and tables. The most convenient method of representing the P–v–T behavior of compressible substances is through the use of an equation of state. The behavior of many fluids in the region of high temperature and low pressure can be expressed by the relationship $Pv = RT$ which is known as the ideal gas law. This law does not take into account the intermolecular forces. Several equations of state which take into account the intermolecular forces have been proposed to represent the P–v–T data accurately. A few equations of state which are in wide use in engineering thermodynamics are reviewed.

The compressibility factor, which quantifies the deviation of the real gas behavior from the ideal gas behavior, is defined as $Z = \frac{Pv}{RT}$. The reduced temperature and reduced pressure are defined as $T_r = \frac{T}{T_c}$ and $P_r = \frac{P}{P_c}$. When the van der Waals equation of state is expressed in terms of reduced parameters we obtain the reduced equation of state. This reduced equation of state suggests the law of corresponding states which in turn suggests that $Z = Z(P_r, T_r)$. The generalized compressibility chart is discussed and its use is illustrated through several examples.

The properties of simple substances can be conveniently represented on a thermodynamic diagram. The commonly used thermodynamic diagrams are: T versus v; P versus v; P versus T and h versus s (Mollier diagram). These thermodynamic diagrams are discussed and their use is illustrated through several examples.

At the triple point all the three phases—solid, liquid and vapor—coexist in equilibrium. No property of the system can be varied at the triple point. The critical point represents the highest temperature and pressure at which both the liquid and vapor phases coexist in equilibrium.

The quality (X) of a mixture is defined as the ratio of the mass of saturated vapor to the total mass of th mixture. That is, $X = \frac{m_g}{m}$. The specific volume of a mixture of quality X can be estimated by the relation

$$v = Xv_g + (1 - X)v_f = v_f + Xv_{fg}$$

Similar relations can be used for the estimation of any extensive property of the mixture.

The drawbacks associated with the graphical representation of thermodynamic data are discussed. Tables can be conveniently used for presenting thermodynamic data. The most frequently used properties $P, v.T, u, h$ and s of substances are presented in thermodynamic tables. Since steam is extensively used in several industries, steam tables are discussed. The reference state for steam tables is the triple point of water where $u_f = 0$ and $s_f = 0$. The use of steam tables is illustrated through several examples.

3.10 Review Questions

1. What are the thermodynamic properties which can be directly measured?
2. Name a few thermodynamic properties of a substance, which cannot be measured directly.
3. What is a pure substance?
4. What is a phase?
5. What is a compressible substance and what is a simple compressible substance?
6. What are the usual methods of presenting the thermodynamic data of substances?
7. Define the coefficient of volume expansion and the isothermal compressibility.
8. Define an ideal gas.
9. What is an equation of state?
10. What are the limitations of the ideal gas equation of state?
11. What is Boyle's law?
12. State the Charles law.
13. What assumptions are made in deriving the ideal gas law from kinetic theory of gases approach?
14. What is Clausius equation of state and how does it differ from ideal gas law?
15. Which factors are taken into account by van der Waals in developing the equation of state?
16. Explain how the attraction forces are related to $\frac{a}{v^2}$ in the van der Waal's equation of state.

17. What is meant by excluded volume?
18. Draw an isotherm at temperature, which is less than the critical temperature, for a van der Waals gas and indicate the zone of the isotherm which is not physically realizable.
19. What is a point of inflection?
20. How are the van der Waals constants related to critical temperature and critical pressure?
21. Name a few equations of state, which are in wide use in engineering thermodynamics.
22. Is it possible to estimate the parameters of a two parameter equation of state in terms of critical constants? If so explain the method.
23. Define the compressibility factor.
24. Define the reduced parameters and pseudo-reduced volume.
25. Sketch the generalized compressibility chart, which is applicable at low pressures and show the lines of constant T_r and constant v_r'.
26. What is the reference state adopted in tabulating the properties of steam in the steam tables and which properties are arbitrarily assigned a zero value?
27. What is the reference state adopted in tabulating the properties of Freon-12 and which properties are arbitrarily assigned a numerical value of zero?
28. What is a saturated state?
29. Define the quality of a mixture.
30. What is saturation pressure and saturation temperature?
31. What is a subcooled liquid?
32. What is a compressed liquid?
33. What is a critical point?
34. Sketch the phase (P–T) diagram for a substance, which contracts on freezing.
35. Sketch the phase diagram (P–T) for a substance which expands on freezing.
36. Is it possible to convert a liquid into vapor phase without ever observing the phase transition? If so sketch the process on P–T diagram.
37. Sketch the P–V diagram for a pure substance and show the isotherms and constant quality lines on it.
38. How does one estimate the specific volume and specific enthalpy of a mixture of saturated liquid and saturated vapor?
39. What is lever rule?
40. Sketch the Mollier diagram and show the isotherms, isobars and constant quality lines on it.

3.11 Problems

3.1 It is desired to design a tank to store 10 kmol of carbon dioxide at 500 kPa and 500 K. Determine the size of the tank using the following methods. (a) Ideal gas equation. (b) van der Waals equation. (c) Redlich–Kwong equation.

3.2 Determine the molar volume of n-Butane at 500 K and 8 MPa by each of the following methods. (a) Ideal gas law. (b) The virial equation of state

$$Z = 1 + \frac{B}{v} + \frac{C}{v^2}$$

with the experimental values of $B = -0.265 \times 10^{-3}$ m^3/mol and $C = 0.3025 \times 10^{-7}$ m^6/mol^2.

3.3 A kmol of methane is stored in a steel vessel of volume 0.6 m^3 at 200°C. Determine the pressure developed by the gas by each of the following methods. (a) Ideal gas law and (b) The Redlich–Kwong equation.

3.4 One mol of an ideal gas ($\gamma = 1.4$) undergoes the following reversible processes. The gas is compressed adiabatically from the initial state 100 kPa and 300 K to 1 MPa pressure. It is then cooled at constant pressure to 300 K. Finally the gas is allowed to expand isothermally to the final pressure of 100 kPa. Determine the work done for each of the processes.

3.5 One hundred kg of CO_2 is to be stored in a 0.5 m^3 container at 100°C. Compare the required pressure using the ideal gas law and the van der Walls equation of state.

3.6 The Berthelot equation of state is given by

$$\left(P + \frac{a}{Tv^2}\right)(v - b) = RT$$

where a and b are constants. Making use of the fact that the critical isotherm shows a point of inflexion show that $a = 9Rv_cT_c^2/8$ and $b = v_c/3$

3.7 The Dieterici equation of state is given by

$$P(v - b)\exp(a/RTv) = RT$$

where a and b are constants. Determine the values of a and b in terms of P_c and T_c.

3.8 Derive relations for the coefficient of thermal expansion β, and the coefficient of isothermal compressibility κ for (a) and ideal gas and (b) a Van der Waals gas.

3.9 It is required to store 90 kg ethane (C_2H_6) at 124°C and 73 bar. Using the generalized compressibility chart, estimate the volume of the tank.

3.10 Determine the compressibility factor for steam at 100 bar and 400°C. Also evaluate the compressibility factor from the data given in steam tables.

3.11 A gas cylinder contains 10 kg ethane when it is filled to a pressure of 57 bar at 335 K. Determine the pressure to which the cylinder is to be filled with ethylene at the same temperature, if the cylinder has to hold 10 kg ethylene. Use the compressibility chart to solve the problem.

3.12 It is required to store 100 kg ethylene at 65°C and 10 MPa. Making use of the compressibility chart, estimate the size of the tank.

3.13 If 100 kg methane is held in a tank of volume 1 m^3 at 110°C, estimate the pressure developed using (a) ideal gas law (b) van der Waals equation of state and (c) compressibility chart.

3.14 Making use of the compressibility chart, estimate the pressure to which a tank of volume 0.2 m^3 is to be charged with 50 kg ethylene at 100°C.

3.15 A cylinder of volume 0.05 m^3 contains 30 kg nitric oxide at 0°C. Make use of the compressibility chart to estimate the pressure exerted by the gas. Also determine the pressure, which is likely to be exerted if the temperature of the gas raises to 40°C.

3.16 A cylinder of volume 0.5 m^3 contains 60 kg ethane at 74 bar. Make use of the compressibility chart and estimate the temperature of the gas.

3.17 Calculate the specific volume of propane at 150°C and 70 bar using the (a) ideal gas law and (b) the compressibility chart.

3.18 It is required to store 100 kg propane at 150°C in a tank of volume 0.62 m^3. Making use of the compressibility chart, estimate the pressure to which the tank is to be charged.

3.19 If a tank of volume 0.62 m^3 is filled with 100 kg propane at 70 bar, what will be the temperature of the gas in the tank. Use the compressibility chart to solve the problem.

3.20 Sketch a temperature versus volume diagram for water and indicate the various zones on it. Show constant pressure lines for pressures below critical pressure and above critical pressure.

3.21 Wet steam of quality 0.85 at 5 bar is heated at constant pressure in a piston-cylinder assembly till it is converted into superheated steam at 300°C. Sketch the process on a P–v diagram.

3.22 Making use of the Mollier diagram, determine the specific volume, specific enthalpy and specific entropy of wet steam of quality 0.9 at 5 bar.

3.23 Determine the quality of a liquid-vapor mixture of water at

(a) 200°C if the specific volume is 0.1 m^3/kg.
(b) 2 bar if the specific volume is 0.75 m^3/kg.
(c) 150°C if the specific enthalpy is 2650 kJ/kg.
(d) 1.5 bar if the specific entropy is 6.5 kJ/kg K.

3.24 Making use of the steam tables, determine the volume occupied by 5 kg of steam at 20 bar and 350°C.

3.25 Calculate the volume occupied by 5 kg of Freon-12 at −20°C with a quality of 0.80.

3.26 A tank of volume 1 m^3 contains wet steam of quality 0.75 at 300°C. Determine the mass of saturated liquid and the mass of saturated vapor present in the tank.

3.27 A small cylinder of 3 litres volume contains Freon-12 of quality 0.8 at 2 bar. Determine the mass of Freon-12 in the cylinder and the fraction of total volume occupied by vapor phase.

3.28 Wet steam of quality 0.9 at 5 bar is heated in a piston-cylinder assembly, at constant pressure, till it reaches a state of superheated steam at 300°C. Sketch the process on a Mollier diagram and determine the changes in specific volume and specific enthalpy.

3.29 Wet steam with a dryness fraction of 0.98 at 2 bar is throttled to a pressure of 0.5 bar. Making use of the Mollier diagram, estimate the final temperature of steam. Note that throttling is an isenthalpic process.

3.30 When wet steam at 5 bar is throttled to a pressure of 0.2 bar, the resulting steam temperature is 65°C. Make use of the Mollier diagram to estimate the quality of wet steam.

3.31 Determine the minimum quality of wet steam at 20 bar that can be measured by throttling it to a pressure of 1 bar, if the resulting steam is to have a temperature of 10°C above its saturation temperature. Make use of the Mollier diagram, to determine the value.

3.32 A cylinder-piston assembly contains 1 kg wet steam of quality 0.91 at 1 bar pressure. The piston is held rigid and the steam is heated till the pressure rises to 2 bar. Sketch the process on a Mollier diagram and determine the change in specific enthalpy of steam and the final temperature of steam.

3.33 A steam radiator at a particular instant contains superheated steam at 1 bar and 110°C. The inlet and outlet valves of the radiator are closed. Due to energy losses as heat to the surroundings, the pressure of steam in the radiator reduces to 70 kPa. Making use of the Mollier diagram, determine the final state of steam in the radiator and the change in specific enthalpy of steam.

3.34 A rigid and closed cylinder is initially filled with steam at 5 bar and 300°C. As a result of heat transfer, the temperature of the steam drops to 120°C. Determine the final pressure of steam in the cylinder and the fraction of steam that condenses.

3.35 A piston–cylinder assembly contains 1 kg superheated steam at 11 bar and 30 0°C. The steam is allowed to expand isentropically to a pressure of 1 bar. Making use of the Mollier diagram, determine the change in enthalpy of steam.

3.36 Compare the specific volume of steam at 100 kPa and 500°C as given by the ideal gas law, the generalized compressibiljty chart and the steam tables.

3.37 Compare the specific volume of steam at 50 MPa and 500°C as given by the ideal gas law, the generalized compressibility chart and the steam tables.

3.38 The van der Waals equation of state is cubic in volume. Develop an iterative scheme to estimate the value of v for given values of P and T.

3.39 A tank of 1 m^3 volume contains steam at 50 bar and 500°C. Determine the mass of steam using the ideal gas equation, van der Waals equation, compressibility chart and steam tables.

3.40 Estimate the temperature to which water at the bottom of a 300 m deep lake should be heated before it starts to boil.

3.41 It is desired to boil water in an industrial digestor at a temperature of 300°C. Making use of the steam tables, determine the pressure at which the digestor should be maintained.

3.42 Compare the change in the specific volume of water upon evaporation at 100 kPa, 2 MPa, and at the critical point.

3.43 A steam boiler has a total volume 3 m^3. The boiler initially contains 2 m^3 of saturated steam and 1 m^3 of saturated liquid at 3 MPa. Calculate the mass of vapor, mass of liquid, quality of the steam, the specific internal energy of the steam, and the specific enthalpy of the steam.

3.44 A rigid tank of volume of 1 m^3 contains saturated steam of quality 0.7 at 200°C. Energy is transferred as heat to the tank until the pressure rises to 3 MPa. Using the steam tables, determine the change in enthalpy of steam.

3.45 A piston–cylinder assembly contains one kg of saturated steam of quality 0.6 at 1 MPa pressure. Energy transfer as heat takes place at constant pressure till the temperature rises to 300°C. Using the steam tables, determine the change in volume and enthalpy of steam.

3.46 A rigid vessel of 10 litres capacity contains Freon-12 at 40°C and 0.6 MPa (v = 0.031 867 m^3/kg and h = 209.2007 kJ/kg). The vessel is cooled to 0°C by immersing it in melting ice. Determine the final conditions of Freon in the cylinder and the change in enthalpy of Freon-12.

3.47 A storage tank for propane has a bursting pressure of 12 MPa. Safety considerations dictate that the tank can be loaded with propane at a pressure not more than half the bursting pressure. If the volume of the tank is 1 m^3, determine the amount of propane that can be charged into the tank at 150°C. Make use of the compressibility chart to solve the problem.

3.48 Freon-12 for use in domestic refrigerators is usually sold in small cylinders having a volume of 3 litres. Determine the mass of Freon-12 contained in such a cylinder if the pressure is 10 MPa at 450 K.

3.49 A liquid nitrogen Dewar flask was filled to one half its volume with liquid nitrogen at its normal boiling point (77 K). The molar volume of liquid nitrogen at its normal boiling point is 34.7 cm^3/mol. The flask was accidentally closed and due to energy transfer as heat the temperature rose to 300 K. Determine the pressure developed by the nitrogen.

3.50 The gas cylinders are usually color coded to ensure that the industrial gases are filled in the corresponding cylinders only. By mistaken identity a nitrogen gas cylinder is filled with ethylene at 8.00 MPa and 300 K. The cylinder would contain 7 kg of nitrogen, when filled at the same conditions of pressure and temperature. Determine the amount of ethylene filled in the cylinder.

4 Zeroth Law of Thermodynamics

Learning Objectives

This chapter will enable the readers to:

- Understand the state postulate.
- State the Zeroth Law of Thermodynamics.
- Understand the ideal gas temperature scale.
- Know about the International Practical Temperature Scale (IPTS) and the primary fixed points used in the establishment of IPTS.

4.1 State Postulate

We know that a thermodynamic system has been defined as a fixed quantity of matter bounded by a closed surface, on which attention is focused for the purpose of thermodynamic analysis. How do we characterize a system without any ambiguity? For this purpose it is necessary to describe the system in terms of certain measurable quantities. Since every thermodynamic system contains some matter with energy in its various forms, the system can be completely described by specifying the following variables:

1. Composition of the matter in terms of mole numbers of each constituent,
2. Energy of the system,
3. Volume of the system, and
4. Measurable properties, like pressure and temperature.

By specifying these quantities, the state of the system is defined. Once the system is in a given state, it possesses a unique set of properties like pressure (P), temperature (T), density, etc. All the properties of a system cannot be varied independently since they are interrelated through expressions of the following type

$$f(P, v, T) = 0 \tag{4.1}$$

For example, the pressure, temperature and molar volume (v) of an ideal gas are related by the expression

$$Pv = RT \tag{4.2}$$

where R is a constant. Therefore, only two of the three variables P, v and T can be varied independently in this case. Now, the question that arises is that for a given thermodynamic system, how many variables can be varied independently?

Our task now is to determine the number of thermodynamic properties of a system that can be varied independently. It has been stated earlier that the thermodynamic properties are those variables which are relevant to energy and its transformation. Therefore, the number of independently variable thermodynamic properties are related to the number of ways in which the energy of a system can be varied.

Consider a compressible substance, like a gas as the system. The energy of the system can be increased by compressing the gas, that is, by doing (PdV) work on it. The energy can also be increased through heat interaction. If the system contains a magnetic substance, the energy of the system can be increased through magnetization and also by heating. If the system is an elastic rod, the energy of the system can be increased by subjecting it to a stress as well as heat interaction. We observe that the energy of the system can be changed by heat as well as work interactions. With every work mode, one independently variable property is associated. In the above examples, the independently variable properties are volume for the compression/expansion work mode, magnetization for the magnetization work mode and strain for the elastic solid extension work mode.

Let us consider now a certain amount of an ideal gas contained in a cylinder–piston assembly as our system. How many thermodynamic variables of this system can be varied independently? The cylinder can be maintained at constant temperature by immersing it in a constant temperature bath, and the gas can be compressed which decreases its volume. In this case the energy of the system is held constant while the volume is changed. Alternatively, the position of the piston can be fixed and the cylinder can be brought into contact with a hot body, thus transferring energy to the system through heat interaction. In this case the volume is held constant while the energy is changed. Therefore, energy and volume are two independently variable thermodynamic properties of a compressible gas.

Now let us see whether the energy and volume of a system containing a gas can be held constant, and some other thermodynamic variable like pressure changed independently. The pressure of the gas can be changed by (a) compressing the gas (b) heating the gas (c) vigorous stirring of the gas.

If the gas is compressed, the volume decreases, though the energy can be held constant. By heating the gas, the energy of the gas increases, though the volume can be held constant. By stirring the gas the energy increases, though the volume can be held constant. Whatever method one employs to change the pressure of the gas, it alters either the volume or the energy, but both of them cannot be held constant.

Based on this observation it is found that the number of variables that can be independently varied for a given thermodynamic system is equal to the number of reversible work modes in addition to the one, which accounts for the transfer of energy as heat . This can be elegantly stated by the state postulate which reads: *For a given thermodynamic system, the number of independently variable thermodynamic properties is equal to the number of reversible work modes plus one.*

If the given thermodynamic system is capable of exchanging energy as work with its surroundings through n reversible work modes, then the number of independently variable thermodynamic properties are $(n + 1)$. For a complete description of the thermodynamic state of a system, these many variables are to be specified. If we consider a simple compressible substance (containing only one component, or pure gas), the only reversible work mode is the compression/expansion or (Pdv), and the generalized displacement associated with the work mode is the volume (v). The state of such a system is completely described by the variables (u, v). Every other thermodynamic property of the system will have a unique value in that state. Therefore $T = T(u, v)$; $P = P(u, v)$ etc. It is possible to vary pressure and volume independently if we do not insist that the energy of the system should remain constant. Then the set of independent variables is (P, v). Likewise, it is possible to find a set of independently variable properties for any specified thermodynamic system.

4.2 Zeroth Law of Thermodynamics

Suppose two systems 1 and 2 (see Fig. 4.1) are brought into contact with each other through a diathermal wall. Let system (1) be hot and system (2) be cold. Because of interaction, the energies of both the systems, as well as their independent properties, undergo a change. The hot body becomes cold and the cold body becomes hot. After some time, the states of the two systems do not undergo any further change and they assume fixed values of all thermodynamic properties. These two systems are then said to be in a state of thermal equilibrium with each other. The two bodies which are in thermal equilibrium with each other have a common characteristic called temperature. Therefore, temperature is a property which has the same value for all the bodies in thermal equilibrium.

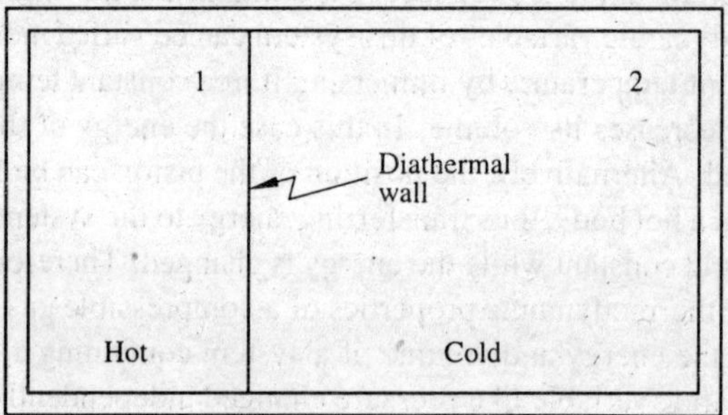

Fig. 4.1 *Two systems in diathermal contact*

Suppose we have three systems 1, 2 and 3 placed in an adiabatic enclosure as shown in Fig. 4.2(a). The systems 1 and 2 do not interact with each other, and they are separated by an adiabatic wall. But systems 1 and 2 interact separately with system 3 through a diathermal wall. Then system 1 is in thermal equilibrium with system 3 and system 2 is also in thermal equilibrium with system 3. By intuition we can say that though systems 1 and 2 are not interacting, they are in thermal equilibrium with each other. Suppose systems 1 and 2 are brought into contact with each other by replacing the adiabatic wall separating them by a diathermal wall as shown in Fig. 4.2(b). Further, they are isolated from system 3 by replacing the diathermal wall, separating systems 1 and 2 from 3, by an adiabatic wall. Then one observes no change in the state of systems 1 and 2. This fact is known as the zeroth law of thermodynamics which can be stated as *if each of the two given systems 1 and 2 are in thermal equilibrium with a third system 3, then the two systems 1 and 2 are in thermal equilibrium with each other.*

The above observation, which has been assumed to be true, was formulated by R.H. Fowler in 1931. The first and second laws of thermodynamics were formulated much before 1931. For a logical development of thermodynamics the above principle has to be stated before stating the first and second laws of thermodynamics. Therefore the above principle is called the zeroth law of thermodynamics. The zeroth law of thermodynamics is the basis for all temperature measurements.

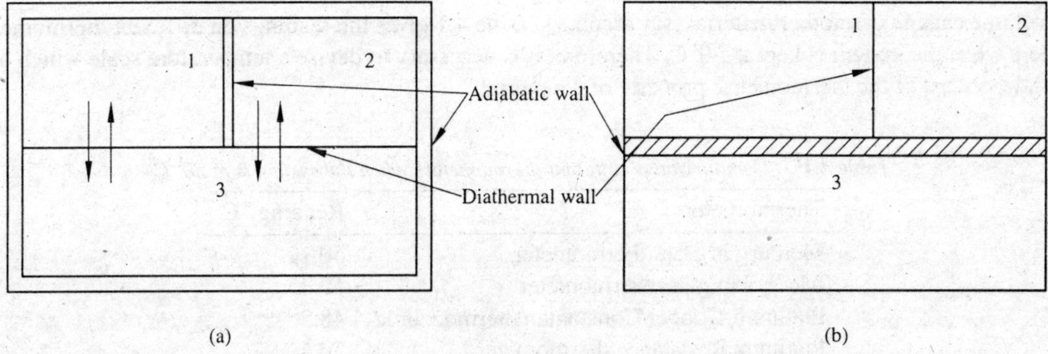

Fig. 4.2 *Sketch to illustrate the Zeroth law of thermodynamics*

4.3 Temperature Scale

Based on the zeroth law of thermodynamics, the temperature of a group of bodies can be compared by bringing a particular body (say a thermometer) into contact with each of them in turn and by observing changes in the properties of the particular body. To quantify the measurement, the instrument should have an easily observable and measurable characteristic called thermometric property. The length of a mercury column in a capillary tube, the resistance of a wire, the pressure of a gas in a closed vessel, the e.m.f generated in a junction of two dissimilar metal wires etc. are the commonly used thermometric properties. To assign numerical values to the thermal state of the system, it is necessary to establish a temperature scale on which the temperature of a system can be read. This requires the selection of a basic unit. For measurement of mass and length, which are extensive quantities, the choice of zero is quite easy because they are associated with no mass or length. Temperature being an intensive parameter, it requires the selection of an origin, or reference state. Therefore, the temperature scale is established by assigning numerical values to certain easily reproducible states. For this purpose, it is customary to use the following two fixed points:

(a) Ice point - the equilibrium temperature of ice with air- saturated water at standard atmospheric pressure which is assigned a value of 0°C.

(b) Steam point - the equilibrium temperature of pure water with its own vapor at standard atmospheric pressure, which is assigned a value of 100°C.

This scale is called the Celsius scale, named after Anders Celsius (1701–1744), the Swedish astronomer who devised it. (This scale was previously called the centigrade scale).

4.4 Ideal Gas Temperature Scale

It is known that all materials do not change their thermometric properties linearly with temperature. Therefore, different thermometers, made of different materials, show different temperatures except at the ice point and the steam point which are used for calibration purposes. This is so because different materials have different temperature - volume relationships so that a temperature scale based upon the volume change of one material (say mercury) does not agree at all points with a scale based on the volume change of another material (say alcohol). Table 4.1 gives the readings on different thermometers when the system is kept at 50°C. Therefore, it is necessary to devise a temperature scale which is independent of the thermometric property of a material.

Table 4.1 *Comparison of different thermometers when the system is at 50° C*

Thermometer	Reading °C
Mercury in glass thermometer	50.12
Alcohol in glass thermometer	50.45
Platinum-Copper/Constantan thermocouple	48.30
Platinum Resistance thermometer	50.4

A detailed discussion on the thermodynamic properties of fluids is presented in Chapter 3. We know that an ideal gas obeys the relation $Pv = RT$, where R is the universal gas constant (R = 8.314 J/mol K). This equation is only an approximation to the actual behavior of the gases. The behavior of all gases approaches the ideal gas limit at sufficiently low pressure (in the limit $P \to 0$). In particular, at low pressures, helium, hydrogen, nitrogen and oxygen behave like ideal gases. The ideal gas temperature scale is based on the observation that the temperature of a gas held at constant volume increases monotonically with pressure. If the gas pressure is made to approach zero, the gas behavior follows the relation

$$Pv = RT$$

Figure 4.3 shows a constant-volume gas thermometer. Helium, or some other gas, is enclosed in the glass bulb B. The bulb B is placed in a system whose temperature is to be measured. The mercury column is so adjusted, by raising or lowering the limb C which is connected to a flexible tubing, such that the level of mercury stands at the reference mark S. This ensures that the volume of the gas is held at a constant value. Let the pressure of the gas be read as P. Let a similar measurement be made when the gas bulb is maintained at the triple point of water. Let the pressure reading be denoted by P_{tp}. It has been internationally accepted to use a single fixed point temperature scale based on the triple point of water. The temperature of the triple point of water has been assigned a value of 273.16 K. The triple point occurs at a fixed temperature and pressure for a specified substance. Any variation in pressure or temperature will cause disappearance of at least one phase. The triple point of water, which can be easily reproduced in any laboratory, has been found to be superior compared to the ice point or the steam point for standardization purposes.

Since for an ideal gas at constant volume $T \propto P$

$$\frac{T}{T_{tp}} = \frac{P}{P_{tp}} \quad \text{or} \quad T = 273.16\frac{P}{P_{tp}} \tag{4.3}$$

where T_{tP} is the triple point temperature of water. Suppose a series of measurements with different amounts of gas in the gas bulb are made. The measured pressures at the triple point as well as at the system temperature change depending on the amount of gas in the bulb. A plot of the temperature T_{cal}, calculated from Eq. (4.3), as a function of the pressure at the triple point results in a curve as shown in Fig. 4.4. When these curves are extrapolated to zero pressure, all of them yield the same intercept. This behavior can be expected since all gases behave like ideal gas when their pressure approaches

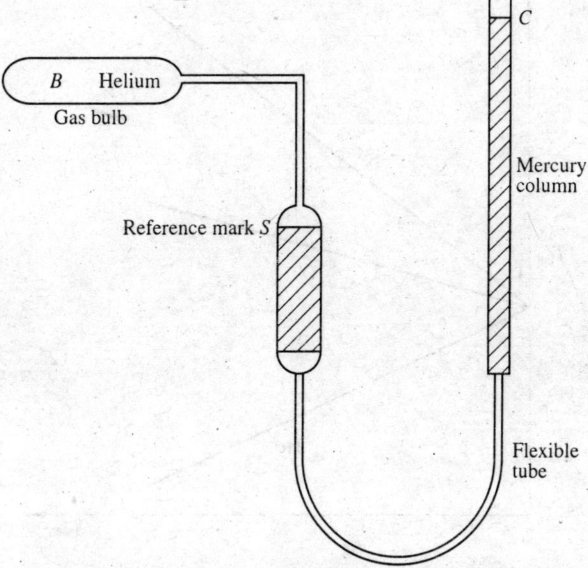

Fig. 4.3 *Schematic of a constant volume gas thermometer*

zero. The correct temperature of the system (calculated from Eq. (4.3) can be obtained only when the gas behaves like an ideal gas, and hence the value is to be calculated in the limit $P_{tP} \to 0$. Therefore,

$$T = 273.16 \lim_{P_{tp} \to 0} \left(\frac{P}{P_{tp}} \right) \tag{4.4}$$

This expression contains only positive quantities on the right hand side and hence, the ideal gas temperature is always a positive quantity.

4.5 International Practical Temperature Scale

The ideal gas thermometer can be used at temperatures where the gases do not liquefy. Moreover, great skill and experience is required to operate a gas thermometer because it requires many corrections when precise results are needed. Of all the gases, helium has the lowest boiling point 4.215 K at standard atmospheric pressure and this sets the limit to which the thermometer can be used. Therefore, a more convenient temperature scale is needed for practical temperature measurement. The International Committee on Weights and Measures have adopted the International Practical Temperature Scale in 1968 (IPTS-68) based on a number of easily reproducible fixed points. They have recognized the Kelvin Thermodynamic Temperature Scale (which will be discussed in Chapter 7) as the fundamental scale. The primary fixed points (expressed in degrees celsius) on which the IPTS-68 is based are given in Table 4.2.

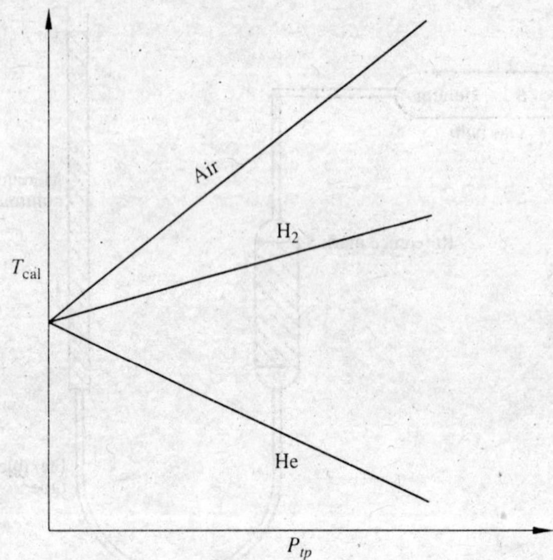

Fig. 4.4 *Plot of T_{cal} versus P_{tp} for a constant volume gas thermometer*

Table 4.2 *Primary Fixed points For International Practical Temperature Scale (1968)*

Primary fixed point	Temperature °C
Triple point of hydrogen	−259.34
Boiling point of hydrogen at 33.33 kPa pressure	−256.108
Normal boiling point of hydrogen	−252.870
Normal boiling point of neon	−246.048
Triple point of oxygen	−218.789
Normal boiling point of oxygen (oxygen point)	−182.962
Triple point of water	0.010
Normal freezing point of zinc	419.580
Normal freezing point of silver(silver point)	961.930
Normal freezing point of gold(gold point)	1064.430

4.6 Summary

A thermodynamic system can be described by specifying some of its properties like $P,\ v,\ T,\ u$ etc. All these properties cannot be varied independently since they are interrelated through expressions of the type $f(P, v, T) = 0$. The number of independently variable thermodynamic properties is related to the number of ways in which the energy of a system can be varied. This leads to the state postulate which states that for a given thermodynamic system, the number of independently variable thermodynamic properties is equal to the number of reversible work modes plus one.

Zeroth law of thermodynamics can be stated as: If each of the two given systems are in thermal equilibrium with a third system, then the two given systems are in thermal equilibrium with each other. This law is the basis for all temperature measurements.

Ideal gas temperature scale is a single fixed point temperature scale. This fixed point is the triple point of water (273.16 K). For practical temperature measurement the International Practical Temperature Scale (IPTS-68) based on a number of easily reproducible fixed points was devised.

4.7 Review Questions

1. What does the State Postulate say and what is the necessity of this postulate?
2. How many independent thermodynamic properties are to be specified for a complete description of the state of a simple compressible substance?
3. How many independent thermodynamic properties are needed to describe the state of an elastic and magnetic body?
4. Is it possible to describe the state of an ideal gas by specifying the variables u and T? If not explain why.
5. How does the state postulate differ from the Gibbs phase rule?
6. State the Zeroth law of thermodynamics.
7. What is an ice point and what is a steam point?
8. How many fixed points are used in establishing the Celsius temperature scale and what are those fixed points?
9. What is the triple point of water?
10. How many fixed points are used in establishing the ideal gas temperature scale and what are those fixed points?
11. Describe, with the help of a neat sketch, a constant volume gas thermometer and explain how it can be used to measure the temperature of a body.
12. Which principle is made use of in devising the ideal gas temperature scale?
13. What is the International Practical Temperature Scale (IPTS) and what is its necessity?
14. Name a few primary fixed points used in establishing IPTS.

5 First Law of Thermodynamics

Learning Objectives

This chapter will enable readers to:

- Know the genesis of the First Law of Thermodynamics.
- State the first law of thermodynamics and its consequences.
- Apply the first law of thermodynamics to analyze isometric, isothermal, isobaric, isentropic and polytropic processes

5.1 Historical Background

A study of the historical background of thermodynamics provides a good example of scientific methodology. Experimentation on thermometry led to the development of concepts like adiabatic walls, specific heats and latent heats. This consequently resulted in the development of the Caloric theory. This in turn suggested further experimentation and led to new concepts. The study of the historical background also reveals the pitfalls in the scientific analysis, like the fascinating attachment with a particular hypothesis (Caloric theory), which may result in overlooking many valid observations and facts. In addition, the study also reveals how difficult it is to put forward new ideas, which contradict the existing notions, and to convince the scientific community. No attempt will be made here to provide a detailed account of the chronological development of the laws of thermodynamics; but, a brief account of the historical background regarding the origin and development of classical thermodynamics is presented below.

The first law of thermodynamics is very well known as the law of conservation of energy. It was realized that there are two modes of energy transfer—work and heat. The concepts of work and heat do not present any difficulties to the scientists of the twenty-first century. Now it is easy to understand the relation between the two modes of energy transfer—heat and work, and hence the law of conservation of energy appears to be a natural consequence of the conversion of one form of energy into another.

Though the concept of work was clear to the scientists of the nineteenth century, the concept of heat caused some problems. Hueygens and Newton introduced the concept of energy which led to the proposition of the principle of conservation of energy for mechanical systems by Leibnitz in 1693. This principle states that in an isolated mechanical system the sum of the kinetic energy and the potential energy remains constant. A century and half later, Joule proposed the First Law of Thermodynamics which takes into account the internal energy also. The scientific community took more than 150 years to realize that heat is a form of energy transfer.

To understand the term heat it is essential to have a knowledge of thermometry. The early attempts in the understanding of heat and temperature can be attributed to Galileo (1564–1642) who tried to quantify the degree of 'hotness', and in the attempt invented the first thermometer in 1592. It was called a thermoscope which responded to changes in pressure as well as temperature. The scale was marked off at pleasure as 'the greatest Summer heat' and 'the severest winter cold'. Sanctorius, a professor of medicine at Padua University and a colleague of Galileo, made use of the thermoscope to detect fever. The men of medicine in those days were interested in developing such instruments. The thermoscope was a simple bulb and stem device filled with air. The development of glass blowing techniques made available narrow glass tubes and led to the development of liquid-in-glass thermometers. Jean Rey, a French doctor, used water as a thermometric fluid in 1631. Within ten years, alcohol replaced water and Grand Duke Ferdinand II of Tuscany developed the alcohol-in-glass thermometer in 1641. With the development of thermometers, quantification had become the main task of the then investigators and attempts were made to find the fixed points. In 1694 the freezing-point of water and the boiling-point of water were chosen as the two fixed points in developing a temperature scale.

As the thermometric experimentation advanced, it was observed that hot liquids cooled faster in metal containers than in wood containers. This observation led to the proposition of adiabatic walls and gave birth to the field of calorimetry. Fahrenheit (1668–1736) conducted a series of mixing experiments in which hot mercury was mixed with cold water and the temperature of mercury and water was measured before and after mixing. Joseph Black, a professor of chemistry at Glassgow University suggested the concept of specific heat in 1760, which was consistent with the mixing-experiments data for several fluids. Black was the first to introduce a clear distinction between the concepts of heat and temperature. These concepts led to the measurement of specific heats of several substances by several researchers. They repeatedly challenged Black's hypothesis. Consequently Black introduced the concepts of latent heats of vaporization and fusion to account for phase transitions, and also proposed that the specific heat of a substance varied with temperature. James Watt, Black's student, utilized the idea of latent heat of vaporization to improve the steam engine.

With the increased availability of the methods of measurements, attention was focused on finding out the nature of 'Heat'. The caloric theory was then proposed by Cleghorn in 1780. The caloric theory claimed that heat was an elastic fluid, which had no weight and was conserved. This fluid was later named as 'caloric' by Lavoisier in 1787. The caloric theory enjoyed the most popular support among the scientific community of that day. The researchers at that time mostly tried to find out whether or not the caloric had any weight. In 1785, George Fordyce, a British doctor, claimed to have detected an increase in the weight of frozen water by one part in 1,29,000 in support of the caloric theory.

Benjamin Thompson (1753–1814), also known as Count Rumford, founder of the Royal Society of London, was an exceptionally good philosopher and scientific investigator. While Rumford was engaged as a military superintendent at Munich, Germany, he noticed that boring operation raised the temperature of gun barrels. Based on this observation, he concluded that the heating was due to the conversion of mechanical work done on the barrels into heat. He also concluded that the caloric cannot be conserved and it cannot be a fluid, and provided convincing arguments against the caloric theory. He conducted a series of experiments (1787–1798) to check the results of Fordyce's experiments and the results were communicated to the Royal Society of London. Sir Humphry Davy (1799), an associate of Benjamin Thompson, supported his arguments through independent experimental observations. He maintained that the heat release was related to the work done. In spite of the convincing nature of these experiments, the scientific community continued to believe in the old caloric theory and this theory remained quite popular for another 50 years.

The doctors at that time were attempting to account for the heat released by the human body. Because of incorrect data regarding heats of combustion, they could not completely account for the heat released. They thought that the difference was due to the flow of blood through the veins. To them, Rumford's ideas were convincing. In 1840 Dr. Julius Robert Mayer thought of the mechanical equivalent of heat and tried to bring it to the attention of the physicists of the day. He communicated his first manuscript to the *Journal Annalen der physik*, in Germany. The well known editor, Poggendorf, refused to publish the paper. Later, in 1842 Liebig accepted the paper for publication in *Annalen der Chemie und pharmazie*. Mayer's papers were publicly ridiculed. Mayer did not conduct any experiments. He did not think that even an ingenious hypothesis needs an experimental verification. He had calculated the mechanical equivalent of heat based on the heat required to heat gases at constant volume and at constant pressure. His calculations were based on inaccurate specific heat data for gases and resulted in inaccurate value for the mechanical equivalent of heat.

James Prescott Joule (1818–1889), son of a rich brewer, had received no formal education in a school because of his bad health. He was tutored at home by the famous chemist John Dalton. At the age of twenty he carried out several experiments (1843–1850) in his own laboratory and published a series of five papers establishing the principle of conservation of energy. He also provided a more accurate value for the mechanical equivalent of heat. Joule was given the credit for establishing the law of conservation of energy based on strong experimental evidence. Joule's results appeared in 1851 when Mayer lost his children. Mayer could not succeed in establishing his claim of proposing the law of conservation of energy for the first time. In frustration, he tried to commit suicide by jumping from the third floor and in that act he broke his legs. He was considered as insane. Mayer received some credit in recognition of his work, due to the efforts of John Tyndall and Rudolph Clausius. In 1862, Mayer was elected as a member of the Royal Society of London.

The above account shows that the development of the first law of thermodynamics and the understanding of the phenomena of heat, temperature and energy took approximately 250 years.

5.2 First Law of Thermodynamics

Having gained some knowledge regarding the concepts of energy, work and heat, we are in a position to discuss the first law of thermodynamics. It was noticed that the principle of conservation of energy for mechanical systems, proposed by Leibnitz in 1693, dealt only with the kinetic energy and potential energy of a system and did not take into account the thermal effects. The first law of thermodynamics, proposed by Joule in 1851, known as the principle of conservation of energy, took into account all forms of energy. There was a lapse of about 150 years before it was realized that work was also a mode of energy interaction between a system and its surroundings. The laws of thermodynamics are not like theorems in mathematics, which can be proved by deduction or inductive logic. They are postulates based on experimental facts. Limited experiments were conducted and the results were interpreted in terms of energy, heat and work. The validity of these results was generalized. The generalization so arrived at were further applied to new situations and got refined over the years. So far, no experimental evidence is available to cast any doubt on the correctness of these laws. The first law of thermodynamics, one of such generalizations, which we are going to discuss now, deals with the conservation of energy of a system which undergoes a change of state as a result of work and heat interactions.

A series of experiments carried out by Joule between 1843 and 1848 form the basis for the first law of thermodynamics. Of the several experiments conducted by Joule, the paddle-wheel experiment is considered as the classic. In Joule's paddle-wheel experiment, [see Fig. 5.1], a known amount of water was taken in a rigid and insulated vessel. The vessel was provided with a paddle-wheel driven by a mass m. Work was done on the system (water in the vessel) by lowering the mass m, through a distance $(Z_1 - Z_2)$. The work done on the system was computed in terms of the change in the potential energy of the mass m. The thermometer, immersed in the water, was used to determine the change in the state of the system. The temperature of the water was found to increase after work had been performed on the system. Once the system had come to 'rest', it was brought into contact with a water bath and the system was allowed to come to the initial state. The amount of energy transferred as heat from the system to the bath was estimated in terms of the temperature rise of the bath. In this particular experiment, the system underwent a series of processes and was restored to the initial state. In other words, it went through a cyclic change. *A process, or a combination of processes, undergone by a system as a result of which the system is restored to the initial state, is called a cycle.*

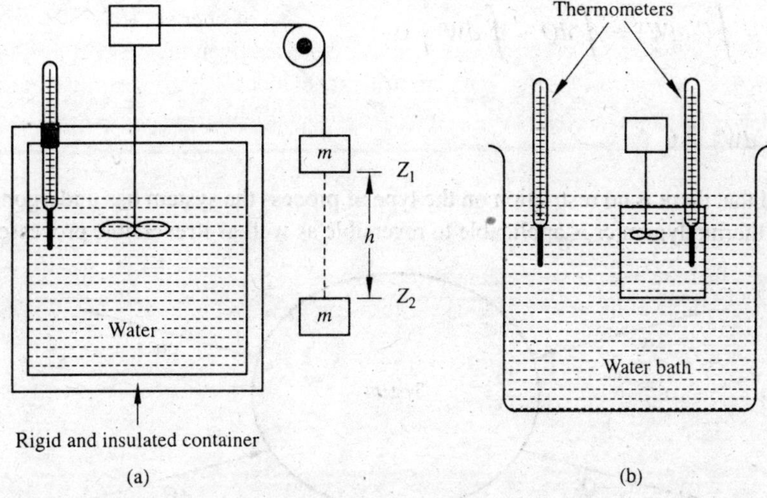

Fig. 5.1 *Schematic arrangement of Joule's paddle wheel experiment*

Joule carried out many experiments involving different types of work interactions. In some experiments, the paddle-wheel was replaced by an electrical resistor which was supplied with an electric current i, by using a reversible cell and the work performed on the system was equal to Eit, where E is the emf of the cell, and t is the time for which the current is passed through the resistor. As a result of these experiments, Joule found that the net work done on the system was always proportional to the net energy removed from the system as heat, irrespective of the type of work interaction, the rate at which work was done on the system, and the method employed for transferring the energy in the form of work into thermal energy.

That is,

$$\oint_{\text{(from system)}} dQ = \oint_{\text{(on system)}} dW \tag{5.1}$$

or

$$\oint_{\text{(from system)}} dQ - \oint_{\text{(on system)}} dW = 0 \tag{5.2}$$

where $\oint$ denotes the integral over a cycle. Therefore, the first law of thermodynamics can be stated as: *whenever a system undergoes a cyclic change, however complex the cycle may be, the algebraic sum of the work transfers is equal to the algebraic sum of energy transfers as heat.*

Sign convention

To avoid confusion as to whether the work is being done on the system by the surroundings, or by the system on the surroundings and whether energy is transferred as heat from the system, or to the system, we shall adopt the following sign convention with reference to Fig. 5.2. The work done by a system on its surroundings is treated as a positive quantity. Similarly, the energy transfer as heat to a system from its surroundings is also treated as a positive quantity. With this convention, Eq. (5.2) becomes

$$-\oint dQ - \oint (-dW) = \oint dQ - \oint dW = 0 \tag{5.3}$$

or

$$\oint (dQ - dW) = 0 \tag{5.4}$$

It may be noted that there is no restriction on the type of process the system has undergone. Therefore, the first law of thermodynamics is applicable to reversible as well as irreversible processes.

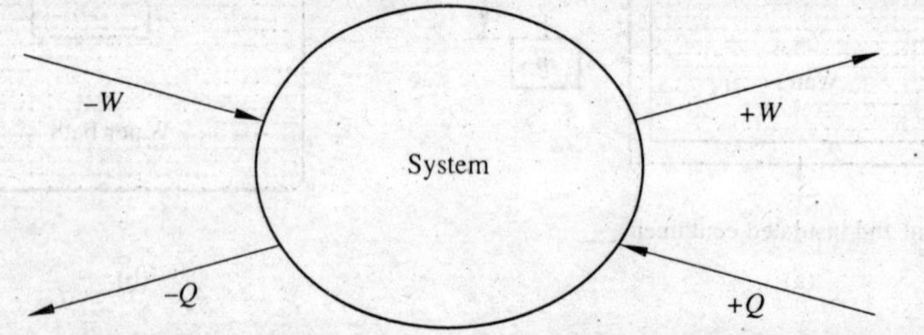

Fig. 5.2 *Sign convention for heat and work interactions*

5.3 Consequences of First Law of Thermodynamics

The first law of thermodynamics, as stated above, has a number of important consequences. A brief discussion of these is presented below.

5.3.1 Heat is a path function

Suppose a system is taken from an initial state 1 to state 2 by the path $1a2$ and it is restored to the initial state through the path $2b1$ (see Fig. 5.3). Then, the system has undergone a cyclic change. If the system is restored to the initial state from state 2 by following another path $2c1$, then $1a2c1$ is another cycle. Let us apply the first law to the two cycles $1a2b1$ and $1a2c1$.

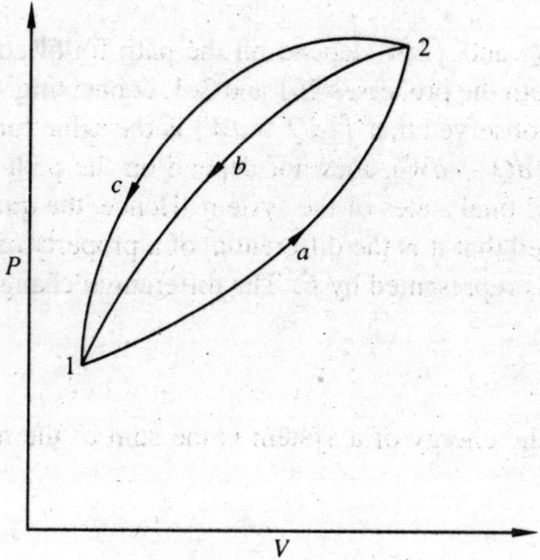

Fig. 5.3 *Reversible cycles on P–V diagram*

$$\int_{1a2} dQ + \int_{2b1} dQ - \int_{1a2} dW - \int_{2b1} dW = 0 \tag{5.5}$$

$$\int_{1a2} dQ + \int_{2c1} dQ - \int_{1a2} dW - \int_{2c1} dW = 0 \tag{5.6}$$

Subtracting Eq. (5.6) from Eq. (5.5), we get

$$\int_{2b1} dQ - \int_{2c1} dQ - \left(\int_{2b1} dW - \int_{2c1} dW\right) = 0 \tag{5.7}$$

Since work depends on the path, we have

$$\int_{2b1} dW - \int_{2c1} dW \neq 0 \tag{5.8}$$

Therefore

$$\int_{2b1} dQ \neq \int_{2c1} dQ \tag{5.9}$$

That is, energy transfer as heat is not a point function, neither is it a property of the system. Therefore, heat interaction is a path function and its differential is not exact.

5.3.2 Energy is a property of system

Quite often, we are concerned with the changes in the state of a system when it undergoes a process, rather than when it passes through a cycle. Moreover, our interest lies in evaluating the energy changes of the system while it interacts with its surroundings. By rearranging Eq. (5.7), we get

$$\int_{2b1} (dQ - dW) = \int_{2c1} (dQ - dW) \tag{5.10}$$

This shows that while $\int dQ$ and $\int dW$ depend on the path followed by the system, the quantity $\int (dQ - dW)$ is the same for both the processes $2b1$ and $2c1$, connecting states 2 and 1. By considering several other cycles, it can be observed that $\int (dQ - dW)$ is the same for all the processes connecting states 2 and 1. Therefore, $\int (dQ - dW)$ does not depend on the path followed by the system, but depends only on the initial and final states of the system. Hence, the quantity $(dQ - dW)$ is an exact differential. It can be concluded that it is the differential of a property of the system. This property is the energy of the system and is represented by E. The differential change in the energy of the system is given by

$$dE = dQ - dW \tag{5.11}$$

It was already observed that the energy of a system is the sum of the macroscopic and microscopic modes of energy, i.e.

$$E = KE + PE + U$$

and

$$dE = d(KE) + d(PE) + dU = dQ - dW \tag{5.12}$$

Therefore, whenever a system undergoes a change of state, energy may cross the boundary as either heat or as work, and the change in the energy of the system is equal to the total energy that crosses the boundary. The energy of the system may change from one form to the other, for example, kinetic energy may change into either potential energy or internal energy. Equation (5.12) simply gives us a procedure that accounts for the energy when the system interacts with the surroundings. The quantity $[d(KE)+d(PE)+dU]$ represents the net change in the energy of the system while dQ and dW represent the energy transfer across the boundary of the system in the form of heat and work, respectively. The net change in the energy of the system is exactly equal to the net energy transfer across the boundary of the system.

5.3.3 Energy of an isolated system is conserved

A system which does not exchange energy with the surroundings, in the form of either heat or work, is called an isolated system. During any process in such a system, $dQ = 0$, and $dW = 0$. The first law of thermodynamics then reduces to

$$dE = 0 \quad \text{or} \quad E_2 = E_1 \tag{5.13}$$

for a reversible or an irreversible process. Therefore, the energy of an isolated system remains constant. This fact is quite often referred to as the *principle of conservation of energy*.

5.3.4 Perpetual motion machine of the first kind is impossible

It has been already shown that the macroscopic modes of energy can be readily converted from one form to the other and work can be obtained. On the other hand, the microscopic modes of energy which are due to the molecular motion of the matter and which cannot be readily evaluated in terms of macroscopically measurable quantities, cannot be readily converted into work. They exist in 'disorganized' form and it is called internal energy. It has been stated earlier that thermodynamics originated as a result of man's endeavor to convert the disorganized form of energy (internal energy) into organized form of energy (work). One mechanism by which it is done is the heat engines which work on cyclic processes. Hence cyclic processes play an important role in thermodynamics. The first law of thermodynamics for a cycle is written as

$$\oint (dQ - dW) = 0 \quad \text{or} \quad \oint dQ = \oint dW$$

or

$$Q = W \tag{5.14}$$

A hypothetical device which would produce work continuously without absorbing any energy from its surroundings is called a Perpetual Motion Machine of the First Kind (PMMFK). A perpetual motion machine of the first kind must operate on a cycle to produce work continuously. If it does not operate on a cycle, its state would change continuously and it could not go on indefinitely. For such a device there cannot be any energy transfer in the form of heat from the surroundings; hence we can write $\oint dQ = 0$. Therefore, the work delivered by it is zero. Therefore, it is impossible to devise a perpetual motion machine of the first kind. A perpetual motion machine of the first kind is a device which violates the first law of thermodynamics. In spite of several attempts to create such a device, no device has yet been invented.

Let us consider an example where a certain amount of gas is held in an insulated piston–cylinder assembly at high pressure. The gas can be allowed to expand without requiring any energy interaction with the surroundings to extract some amount of work. The work is done by the system on the surroundings at the expense of the energy of the system. However, it cannot continuously deliver work, and at the same time the system does not undergo a cyclic change. Therefore, it is always possible to devise a machine which can deliver a limited amount of work at the expense of the energy of the system without restoring the system to its initial state. Such devices are not to be confused with the perpetual motion machine of the first kind.

5.4 First Law Analysis of Elementary Processes

The first law of thermodynamics can be applied to a system to evaluate the change in the energy of the system when it undergoes a change of state. The system, as defined earlier, refers to a definite quantity of matter which remains constant while the system undergoes a change of state. However, the system boundary may change in size, or shape. Quite often in engineering thermodynamics, the kinetic energy and potential energy of the system do not undergo any change. In such cases, the first law of thermodynamics reduces to

$$dU = dQ - dW \tag{5.15}$$

Most of the cycles or processes that are normally encountered in thermodynamic analysis of systems, can be identified as any one or a combination of some of the following elementary processes.

1. Constant volume, or isochoric process.
2. Constant pressure, or isobaric process.
3. Constant temperature, or isothermal process.
4. Reversible adiabatic, or isentropic process.
5. Polytropic process.
6. Constant internal energy process.

The first law analysis of the above elementary processes is presented below.

5.4.1 Constant volume process (isochoric)

Consider a gas confined in a rigid vessel of volume V as the system. Since the vessel has rigid walls, the expansion or (PdV), the work done by the system is zero. Let the system be brought into contact with a heat source so that it can reversibly exchange energy. Let the system change its state from 1 to 2 as shown in Fig. 5.4. If the process is carried out reversibly, the path of the process ($1 \longrightarrow 2$) can be traced on a P–V diagram as shown in Fig. 5.4. Applying the first law of thermodynamics, we obtain

$$dU = dQ \quad or \quad Q = U_2 - U_1 \tag{5.16}$$

Hence, the heat interaction is equal to the change in the internal energy of the system. If the temperature of the heat source is higher than that of the system, the process becomes irreversible and the path cannot be traced on the diagram though the final volume is identical with the initial volume of the system.

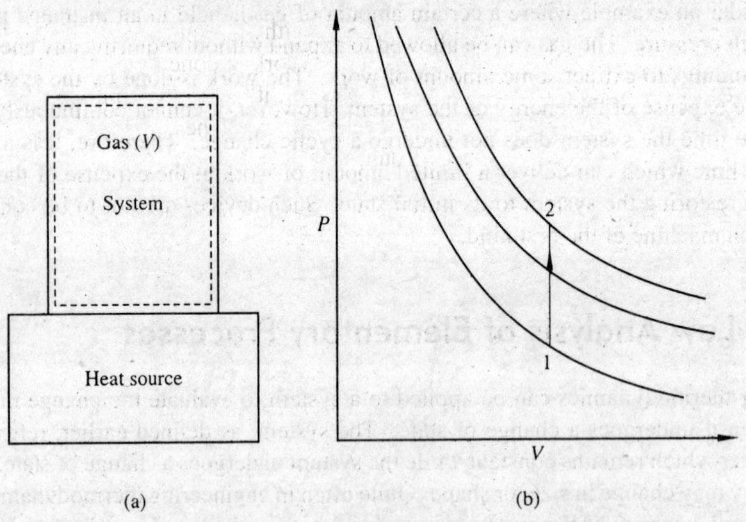

Fig. 5.4 *(a) Schematic arrangement for a reversible constant volume heating of a gas. (b) $P - v$ diagram for a constant volume process.*

Consider the constant volume adiabatic processes shown in Fig. 5.5. The figure depicts the change in the state of the system (liquid contained in the vessel) by performing paddle wheel work on the system. The paddle wheel is rotated by lowering a mass through a distance ($Z_1 - Z_2$) while the vessel is insulated. In Fig. 5.5(b), electrical work is performed on the system and the state of the system has changed from 1 to 2. Taking the liquid contained in the vessel as our system, there is no heat interaction across the boundary of the system ($Q = 0$). The processes are irreversible and hence cannot be represented on the P–V diagram since the intermediate states of the system are not equilibrium states. By discontinuing the process at intermediate stages, and allowing the system to reach a state of equilibrium, it is possible to realize some intermediate states. These states lie on the dotted line ($1 \longrightarrow 2$) shown in Fig. 5.5(c).

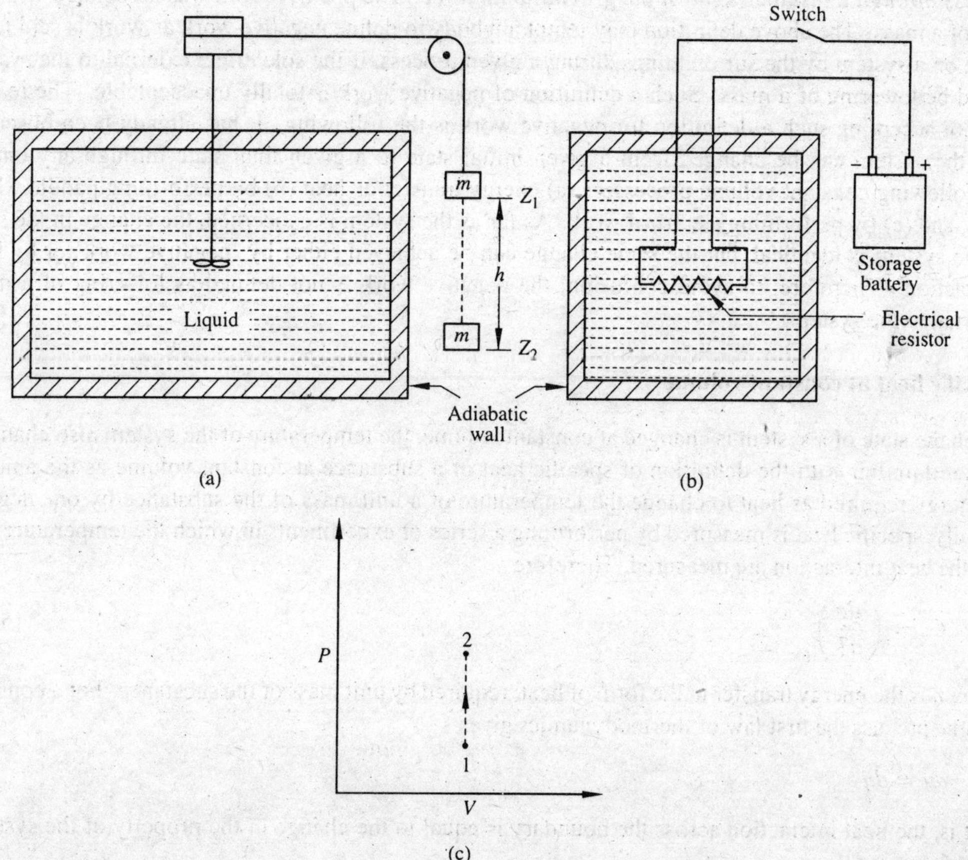

Fig. 5.5 *Irreversible constant volume process (a) involving paddle wheel work (b) involving electrical work and (c) P – v diagram*

Based on the above discussion the first law of thermodynamics gives the following relation for an irreversible adiabatic, constant volume process.

$$dU = -dW \quad \text{or} \quad -W = U_2 - U_1 \tag{5.17}$$

where $-W$ represents the work done on the system by the surroundings. The increase in the internal energy of the system is exactly equal to the work done. In this case the work done is negative. From the above observation it can be concluded that for irreversible adiabatic constant volume processes, the internal energy of the system can never be reduced and thereby made to perform work on the surroundings.

It can be observed that a system can be made to undergo a change from state 1 to state 2 either by work performed on the system (by lowering the mass through a distance) or by energy transfer as heat. It may be recalled that in the thermodynamic definition of work it was stated that *work is said to be that interaction whose sole effect, external to the system, could be viewed as the rising of a mass through a distance against the gravitational force.* The positive work was associated with the rise of a mass. The above definition may tempt anybody to define negative work as work is said to be done on a system by the surroundings during a given process, if the sole effect external to the system could be lowering of a mass. Such a definition of negative work is totally unacceptable. The reason for not accepting such a definition for negative work is the following. It has already been observed that the system can be changed from a given initial state to a given final state through any one of the following constant volume processes: (a) energy transfer as heat (b) by performing paddle wheel work and (c) by performing electrical work. As far as the system is concerned, the change in the state of the system is identical, but the same change can be achieved either by 'negative work' or by heat interaction. Therefore, to avoid confusion, the negative work is not defined as lowering of a mass external to the system.

Specific heat at constant volume

When the state of a system is changed at constant volume, the temperature of the system also changes. We are familiar with the definition of specific heat of a substance at constant volume as the amount of energy required as heat to change the temperature of a unit mass of the substance by one degree. Usually, specific heat is measured by performing a series of experiments in which the temperature rise and the heat interaction are measured. Therefore

$$C_v = \left(\frac{dq}{dT}\right)_v \tag{5.18}$$

where q is the energy transfer in the form of heat, required by unit mass of the substance. For a constant volume process the first law of thermodynamics gives

$$du = dq$$

That is, the heat interaction across the boundary is equal to the change in the property of the system. Therefore,

$$C_v = \left(\frac{dq}{dT}\right)_v = \left(\frac{\partial u}{\partial T}\right)_v \tag{5.19}$$

where u is the specific internal energy of the system. We know that the same change in the state can as well be achieved by performing work on the system. Hence it is appropriate to redefine C_v in terms of the energy change du of the substance as given by the second part of Eq. (5.19).

If C_v varies with temperature, one can use the mean specific heat at constant volume which is defined as

$$\bar{C}_v = \frac{\int_{T_1}^{T_2} C_v dT}{T_2 - T_1} \tag{5.20}$$

The total quantity of energy transferred during a constant volume process when the system changes from T_1 to T_2, can be easily estimated in terms of its specific heat at constant volume as

$$Q = m \int_{T_1}^{T_2} C_v dT = U_2 - U_1 \tag{5.21}$$

The units of the specific heat are kJ/kg K. Sometimes it is preferable to use the molar heat capacity (kJ/kmol K).

***Example* 5.1** A radiator of a heating system with a volume of 0.1 m^3 contains saturated steam at 0.2 MPa. The inlet and outlet valves of the radiator are closed. Due to the energy transfer as heat to the surroundings the pressure drops to 0.15 MPa. Determine:

(a) the temperature of steam in the final state
(b) the quality of steam in the final state and
(c) the masses of liquid and vapor in the final state.

Solution System: Steam contained in the radiator.

The process followed by the system is shown on a P–v diagram in Fig. 5.6 and on a $h - s$ diagram in Fig. 5.7.

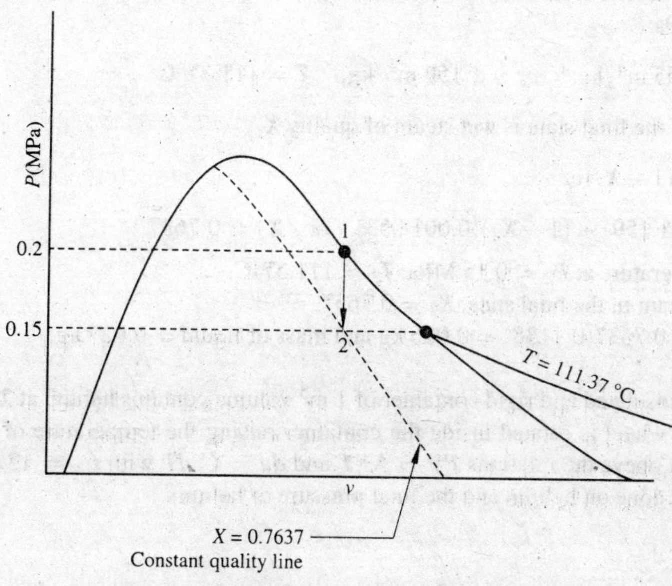

Fig. 5.6 *Sketch for Example 5.1 on a $P - v$ diagram*

Specific volume of the saturated steam at 0.2 MPa, $v_1 = 0.8854\ m^3/kg$
Volume of the radiator, $V = 0.1\ m^3$
Total mass of the steam, $m = \dfrac{V}{v_1} = \dfrac{0.1}{0.8854} = 0.113$ kg.

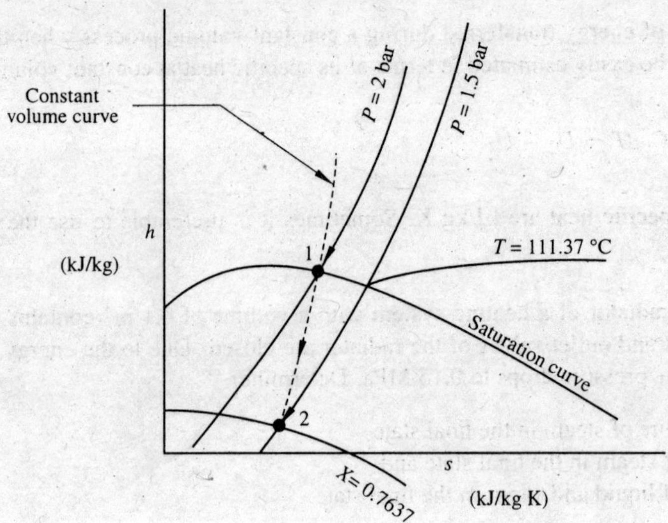

Fig. 5.7 *Sketch for Example 5.1 on a h – s diagram*

Since the quantity of steam and the volume of the radiator are constant, the steam undergoes a constant volume process. Therefore, v = constant; or, $v_1 = v_2$.

At $P_2 = 0.15$ MPa:

$$v_f = 0.001\,053 \text{ m}^3/\text{kg}; \quad v_g = 1.159 \text{ m}^3/\text{kg}; \quad T = 111.37°\text{C}$$

Since $v_f < v_2 < v_g$, the final state is wet steam of quality X_2.

$$v_2 = X_2 v_g + (1 - X_2) v_f$$

$$0.8854 = X_2(1.159) + (1 - X_2)(0.001\,053) \quad \text{or} \quad X_2 = 0.7637$$

The saturation temperature at $P_2 = 0.15$ MPa, $T_2 = 111.37°$C.

Quality of the steam in the final state, $X_2 = 0.7637$.

Mass of vapor = 0.7637(0.1138) = 0.086 kg and mass of liquid = 0.027 kg.

***Example* 5.2** An insulated and rigid container of 1 m^3 volume contains helium at 20°C and 100 kPa pressure. A paddle wheel is rotated inside the container raising the temperature of helium to 60°C. Assume that helium obeys the relations $PV = NRT$ and $du = C_v dT$ with $C_v = 12.4717$ kJ/kmol K. Determine the work done on helium and the final pressure of helium.

Solution

$$N = \frac{P_1 V_1}{RT_1} = \frac{100 \times 10^3 \times 1}{8.314 \times 10^3 \times 293.15} = 0.041 \text{ kmol}$$

$$\Delta u = C_v \Delta T$$

or

$$\Delta U = NC_v \Delta T = 0.041 \times 12.4717(40) = 20.4685 \text{ kJ}.$$

$$\Delta U = -W \quad \text{or} \quad W = -20.4685 \text{ kJ}$$

$$\frac{P_2}{P_1} = \frac{T_2}{T_1} \quad \text{or} \quad P_2 = \frac{P_1 T_2}{T_1} = \frac{100 \times 10^3 \times 313.15}{293.15} = 106.82 \text{ kPa}$$

5.4.2 Constant pressure (isobaric) process

Quite a few industrially important processes occur at constant pressure. Suppose a gas enclosed in a cylinder–piston assembly shown in Fig. 5.8(a, c, d) is considered as the system. The pressure of the gas is maintained at a constant value by loading the piston with a mass. Let us ignore the friction between the cylinder wall and piston. In Fig. 5.8(a) the cylinder is brought into contact with a heat source and energy transfer as heat takes place reversibly and the system is always maintained in a state of equilibrium. Then the work done by the system when it changes from the initial state (1) to the final state (2) (see Fig. 5.9) is given by

$$W = \int_1^2 P dV = P(V_2 - V_1) \tag{5.22}$$

Applying the first law of thermodynamics,

$$dU = dQ - dW = dQ - PdV = dQ - d(PV)$$

or

$$dQ = dU + d(PV) = d(U + PV)$$

or

$$dQ = dH \quad \text{or} \quad Q = H_2 - H_1 \tag{5.23}$$

As the quantity $(U + PV)$ occurs so frequently in thermodynamics, it has been given a separate name, enthalpy, denoted by H. Thus

$$H = U + PV \tag{5.24}$$

Therefore, in a constant pressure process, the heat interaction across the boundary is exactly equal to the change in the enthalpy (a property) of the system. The specific enthalpy h is defined as

$$h = H/m = u + Pv \tag{5.25}$$

and the molar enthalpy is $h = H/N$, where N is the mole number of the substance.

In Fig. 5.8(b) the system is a saturated liquid contained in the cylinder–piston assembly, where the piston is loaded with a mass. If the system is brought into contact with a constant temperature reservoir and energy transfer as heat takes place across the boundary of the system, the liquid starts vaporizing and follows the path $1 \longrightarrow 2$ as shown in Fig. 5.10. The process is reversible, because at any stage of the process, the direction of the process can be reversed by increasing the pressure of the system infinitesimally, by adding an infinitesimal mass on the piston. The vapor condenses and the directions of the work and heat interactions are reversed. The processes indicated in Fig. 5.8(a) and (b)

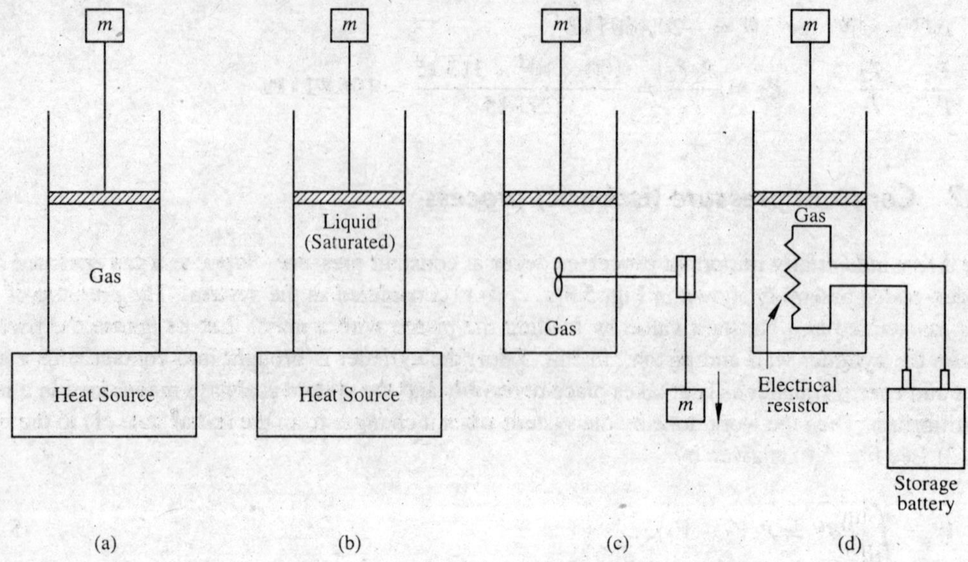

Fig. 5.8 *Schematic arrangement for constant pressure process (a) for heating of a gas (b) for phase change of liquid (c) involving paddle wheel work and (d) involving battery work*

are reversible and the path can be represented on a thermodynamic diagram. The processes can take place in either direction (from 1 to 2 or from 2 to 1)(see Fig. 5.9).

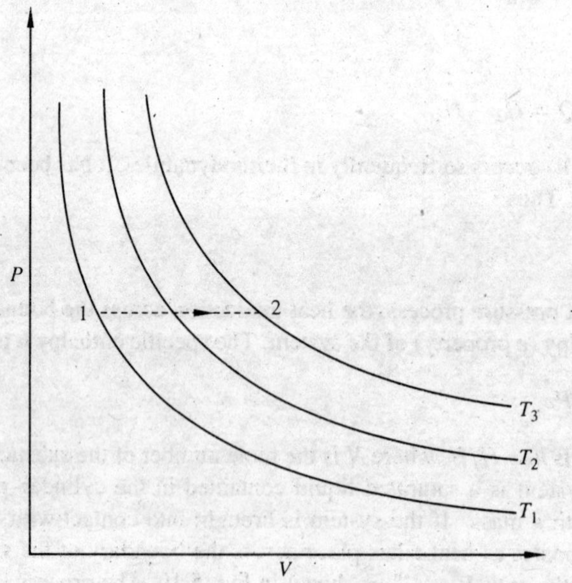

Fig. 5.9 *P − v diagram for an irreversible constant pressure process*

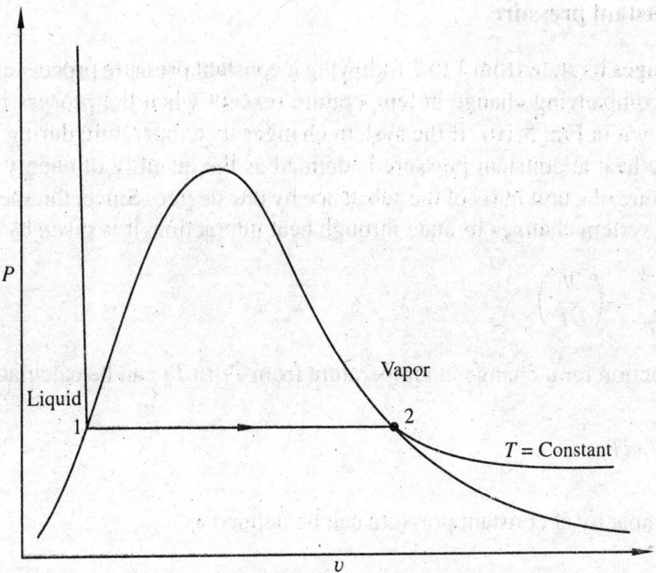

Fig. 5.10 *P–v diagram for a liquid showing vaporization process*

Now consider the constant pressure processes shown in Fig. 5.8(c) and (d) where a certain amount of gas (system) is enclosed in an adiabatic wall and work is done on the system. Since the wall is adiabatic

$$dQ = 0$$

Application of the first law of thermodynamics to the system gives

$$dU = dQ - dW = dQ - (PdV + dW_s) \tag{5.26}$$

Here (dW), the work interaction across the boundary, consists of the work (PdV) associated with the boundary motion, and $(-dW_s)$ the shaft work, or the paddle wheel work done by the surroundings on the system. Therefore

$$-dW_s = dU + PdV = dU + d(PV) = d(U + PV) = dH$$

or

$$-W_s = H_2 - H_1 \tag{5.27}$$

(W_s is always a negative quantity.)

Therefore, the increase in the enthalpy of the system is equal to the amount of shaft work done on the system. These processes (c and d of Fig. 5.8) are irreversible because it is not possible to decrease the energy of the system and to raise the mass. These processes can take place only in one direction 1 $\longrightarrow$ 2, but not in the other direction 2 $\longrightarrow$ 1 (see Fig. 5.9) and the path cannot be traced on a thermodynamic diagram.

Specific heat at constant pressure

When a system changes its state from 1 to 2 following a constant pressure process as shown in Fig. 5.9, there will be an accompanying change in temperature (except when the process is associated with a phase change as shown in Fig. 5.10). If the system changes its temperature during a constant pressure process, the specific heat at constant pressure is defined as the quantity of energy required as heat to change the temperature of a unit mass of the substance by one degree. Since, the specific heat is usually measured when the system changes its state through heat interaction, it is given by

$$C_P = \left(\frac{dq}{dT}\right)_P = \left(\frac{\partial h}{\partial T}\right)_P \tag{5.28}$$

The total heat interaction for a change in temperature from T_1 to T_2 can be calculated by

$$Q = m \int_{T_1}^{T_2} C_P dT \tag{5.29}$$

The molar heat capacity at constant pressure can be defined as

$$C_P = \left(\frac{\partial h}{\partial T}\right)_P \tag{5.30}$$

Example **5.3** One kg of saturated liquid water is contained in a cylinder–piston assembly shown in Fig. 5.11 at a pressure of 100 kPa. The cylinder is brought into contact with a thermal reservoir and energy transfer as heat takes place till the water is completely evaporated at constant pressure. Calculate the amount of heat interaction and the work done by the system.

Solution Read the following data at 1 bar from Appendix Table A4:
Enthalpy of saturated liquid $h_f = 417.54$ kJ/kg
Enthalpy of saturated vapor $h_g = 2675.4$ kJ/kg
Specific volume of saturated liquid $v_f = 0.001\,043\,4\ \text{m}^3/\text{kg}$
Specific volume of saturated vapor $v_g = 1.694\ \text{m}^3/\text{kg}$
Work done by the system = $W = \int PdV = P(v_g - v_f)$

$$W = 100 \times 10^3 \times (1.694 - 0.001\,043\,4) = 169.3 \text{ kJ}$$

For a constant pressure process $q = \Delta h = h_g - h_f = 2675.4 - 417.54 = 2257.86$ kJ. Therefore, 2257.86 kJ of energy is transferred as heat to the system and the system does 169.3 kJ of work.

Example **5.4** The density of water at 0°C is 999.8 kg/m^3, while that of ice at the same temperature is 916.23 kg/m^3. One kg of ice at 0°C is melted to water at 0°C at a pressure of 100 kPa increasing its energy by 334.98 kJ. Calculate the work done by the ice and the amount of energy transferred as heat during the melting process.

Solution

$$v_w = 1/999.8 = 1.0002 \times 10^{-3}\ \text{m}^3/\text{kg} \quad \text{and} \quad v_i = 1/916.23 = 1.091 \times 10^{-3}\ \text{m}^3/\text{kg}$$

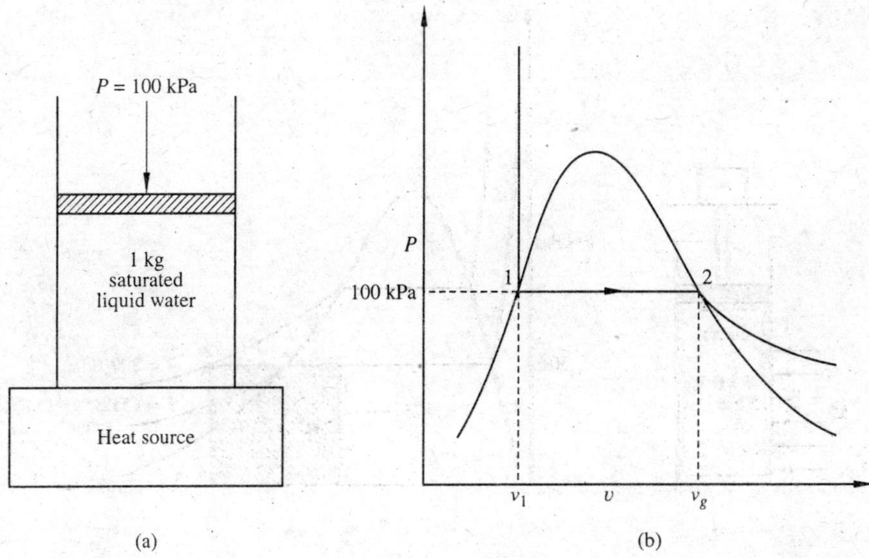

Fig. 5.11 *Sketch for Example 5.3 (a) Schematic arrangement to show vaporization of a liquid at constant pressure (b) Vaporization process on a $P - v$ diagram*

(The subscripts w and i stand for water and ice, respectively.)

Work done $= W = \int P dv = 100 \times 10^3 (v_w - v_i)$

$$W = 100 \times 10^3 (1.0002 - 1.091) \times 10^{-3} = -9.08 \text{ J}$$

$$\Delta u = q - W \quad \text{or} \quad q = \Delta u + W = 334.98 - 9.08 \times 10^{-3} = 334.97 \text{ kJ}$$

Therefore, 9.08 J work was done by the atmosphere on the ice and 334.97 kJ of energy is transferred as heat to ice during the melting process.

***Example* 5.5** One kg of saturated steam of quality 0.75 at 5 bar is contained in a piston–cylinder assembly. Energy transfer as heat takes place to the steam till its temperature attains a value of 200°C. During this process the pressure remains constant at all times. Calculate: (a) the work done by the steam, and (b) the energy transferred as heat.

Solution System: The steam contained in the cylinder–piston assembly.

The constant pressure process followed by the steam is shown on a $P - v$ diagram in Fig. 5.12 and on a $h - s$ diagram in Fig. 5.13.

Work done, $W = \int_1^2 P dv = P(v_2 - v_1)$

Application of the first law of thermodynamics to the above process gives:

$$\Delta u = u_2 - u_1 = q - W = q - P(v_2 - v_1)$$

or

$$h_2 - h_1 = q$$

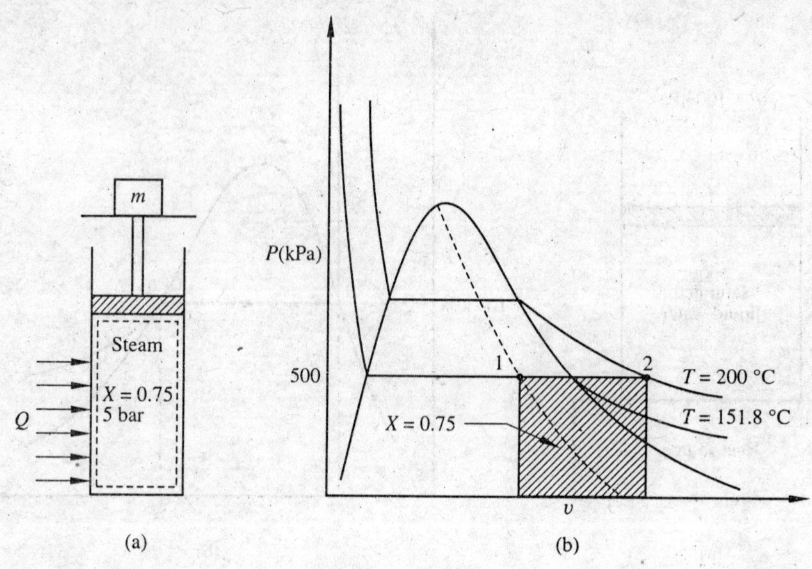

Fig. 5.12 *Sketch for Example 5.5 (a) System (b) P − v diagram. The shaded area represents the work done by the system*

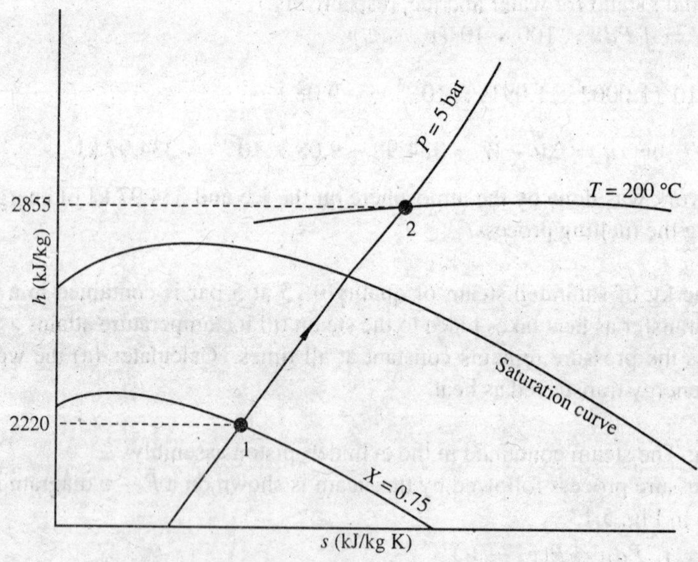

Fig. 5.13 *Sketch for Example 5.5 illustrating the use of Mollier diagram*

At 5 bar, $h_f = 640.12$ kJ/kg and $h_g = 2747.5$ kJ/kg. Initial enthalpy of steam, h_1 is given by

$$h_1 = X_1 h_g + (1 - X_1) h_f = 0.75 \times 2747.5 + 0.25 \times 640.12 = 2220.66 \text{ kJ/kg}$$

Final state of the steam: $P = 5$ bar; $T = 200°C$. From steam tables, we find that $h_2 = 2855.1$ kJ/kg.

$$q = h_2 - h_1 = 2855.1 - 2220.66 = 634.44 \text{ kJ/kg}$$

Energy transfer as heat, $q = 634.44$ kJ/kg.

Initial volume, $v_1 = X_1 v_g + (1 - X_1)v_f$

At 5 bar, $v_f = 0.001\,092\,8$ m^3/kg; $v_g = 0.3747$ m^3/kg

$$v_1 = 0.001\,092\,8 \times 0.25 + 0.75 \times 0.3747 = 0.2813 \text{ m}^3/\text{kg}$$

Final volume v_2 from superheated steam tables at $P = 5$ bar and $T = 200°C$ is $v_2 = 0.425$ m^3/kg. Therefore,

$$W = P(v_2 - v_1) = 5 \times 10^5(0.425 - 0.2813) = 72.35 \text{ kJ/kg.}$$

5.5 Constant Temperature (Isothermal) Process

Consider a gas enclosed in a cylinder–piston assembly shown in Fig. 5.14(a) as the system. The system is allowed to undergo an expansion process while in contact with a constant temperature bath. During the expansion process, the opposing force is continuously reduced such that the system is in equilibrium at all times. Since the process is reversible, the path followed by the system can be traced on a thermodynamic diagram as shown in Fig. 5.14(b). The shaded area in Fig. 5.14(b) represents the work done by the system while reaching state 2 from state 1.

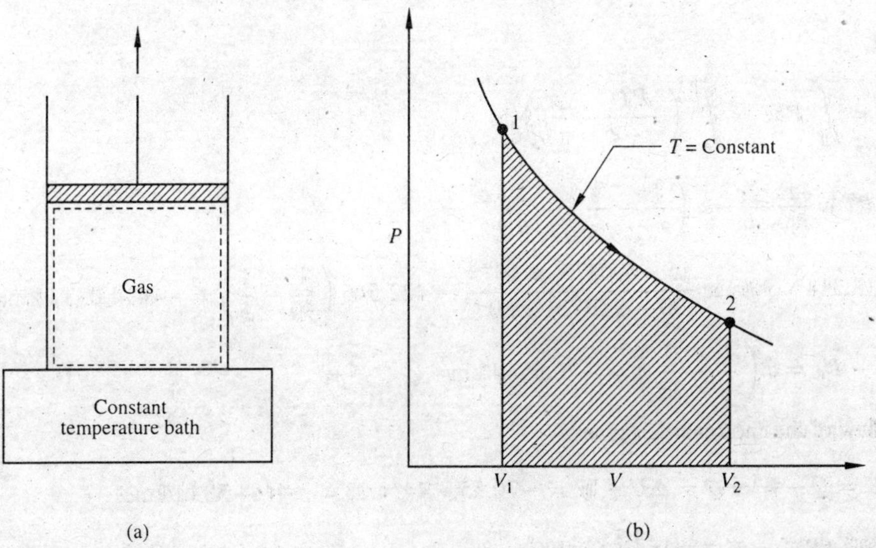

Fig. 5.14 *(a) Schematic arrangement to show isothermal process (b) P–v diagram for a gas undergoing isothermal process*

Applying the first law of thermodynamics for the path 1 ⟶ 2, we get

$$dU = dQ - dW = dQ - PdV \tag{5.31}$$

It is not possible to calculate the work done and the heat interaction, unless the relationships between the thermodynamic properties and independent variables of the system are known.

Suppose the gas under consideration is an ideal gas and we would like to estimate the heat and work interactions. For an ideal gas the desired property relations are $Pv = RT$ and $u = u(T)$ only. Then

$$du = C_v dT \tag{5.32}$$

Since the temperature is held constant, $du = 0$. Therefore,

$$q = W = \int_1^2 P dv = \int \frac{RT dv}{v} = RT \ln \frac{v_2}{v_1} \tag{5.33}$$

In this case we find that the internal energy of the system does not change and the energy transfer as heat to the system is exactly equal to the work done by the system.

***Example* 5.6** It is desired to compress isothermally one kmol of ammonia from the initial state 30 m³/kmol and 300 K to the final state 5 m³/kmol and 300 K. It is known that ammonia obeys the relations

$$\left(P + \frac{a}{v^2}\right)(v - b) = RT \quad \text{and} \quad du = C_v dT + \left(\frac{a}{v^2}\right) dv$$

where, $a = 422.546 \times 10^{-3}$ Pa(m³/mol)²; $b = 0.037$ m³/kmol and $C_v = 27.2376$ kJ/kmol K. Calculate the heat and work interactions associated with the compression process.

Solution

$$W = \int_1^2 P dv = \int_1^2 \left(\frac{RT}{v - b} - \frac{a}{v^2}\right) dv$$

$$= RT \ln \frac{v_2 - b}{v_1 - b} - a\left(\frac{1}{v_1} - \frac{1}{v_2}\right)$$

$$= (8.314 \times 300) \ln \frac{(5 - 0.0373) \times 10^{-3}}{(30 - 0.0373) \times 10^{-3}} - 422.546\left(\frac{1}{30} - \frac{1}{5}\right) = -4414.03 \text{ kJ/kmol}$$

$$U_2 - U_1 = a\left(\frac{1}{v_1} - \frac{1}{v_2}\right) = -70.55 \text{ kJ/kmol}$$

The first law of thermodynamics gives:

$$\Delta U = Q - W \text{ or } Q = \Delta U + W = -70.55 - 4414.03 = -4484.58 \text{ kJ/kmol}$$

Hence, work done $W = -4414.03$ kJ/kmol

Energy transferred as heat, $Q = -4484.58$ kJ/kmol

Change in the internal energy $= -70.55$ kJ/kmol

When a substance undergoes a transition from liquid to vapor, or from solid to liquid, the pressure as well as the temperature of the substance remain constant. Therefore, such transitions are both isobaric and isothermal processes. The examples dealing with such transitions have been already discussed under constant pressure process.

5.6 Adiabatic Processes

5.6.1 Irreversible adiabatic processes

A process in which there is no energy transfer as heat across the boundaries of the system, that is, when the system is enclosed by an insulated wall, is called adiabatic. For an adiabatic process:

$$Q = 0$$

Application of the first law of thermodynamics to an adiabatic process gives:

$$dU = -dW \quad \text{or} \quad \Delta U = -W \tag{5.34}$$

Some of the irreversible adiabatic processes are shown in Fig. 5.15. In Fig. 5.15(a) the system is enclosed by a rigid, adiabatic wall and paddle-wheel work is performed on the system. In Fig. 5.15(b) the wall enclosing the system is again adiabatic, but here electrical work is performed on the system.

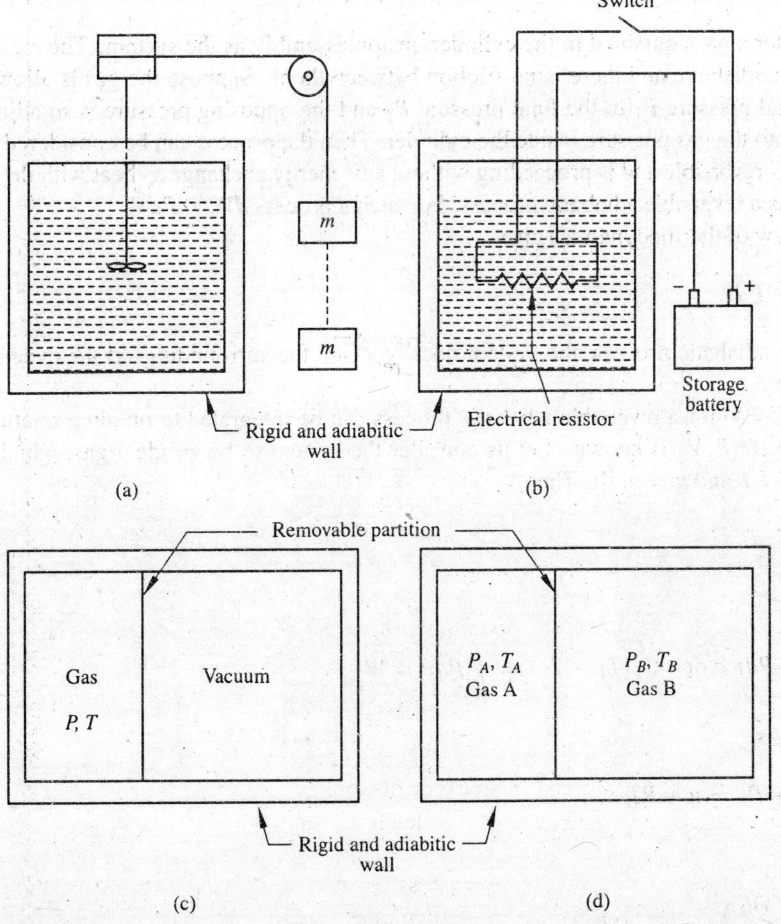

Fig. 5.15 *Schematic arrangement of irreversible adiabatic processes involving (a) Paddle wheel work (b) Storage battery work (c) Free expansion of a gas (d) Mixing of gases at different temperatures and pressures*

In Fig. 5.15(c) the system is enclosed by a rigid and adiabatic wall. The system is divided into two compartments by a removable partition. One compartment contains a gas at a pressure P and temperature T, and the other compartment is evacuated. When the partition is removed, the gas occupies the whole volume and the process occurs without exchange of energy as heat with the surroundings. In Fig. 5.15(d), one compartment contains a gas A at pressure P_A and temperature T_A, whereas the second compartment contains a gas B (which may be the same gas as in compartment A or different gas) at pressure P_B and temperature T_B. When the partition is removed, the gases mix. For such adiabatic processes, the first law gives:

$$\Delta U = -W$$

For processes shown in Fig. 5.15(a) and (b), W is negative and for processes shown in Fig. 5.15(c) and (d) $W = 0$.

5.6.2 Reversible adiabatic (isentropic) process

Let us consider a gas contained in the cylinder–piston assembly as the system. The cylinder wall and the piston are adiabatic and there is no friction between them. Suppose the gas is allowed to expand from the initial pressure P_1 to the final pressure P_2 and the opposing pressure is so adjusted that it is always equal to the gas pressure inside the cylinder. Then the process can be considered as reversible. The process is reversible and is proceeding without any energy exchange as heat with the surroundings and hence it is a reversible adiabatic process. For such a process $dW = PdV$.

The first law of thermodynamics gives

$$dU = -PdV \tag{5.35}$$

That is, in an adiabatic process, the system does work on the surroundings at the expense of its own internal energy.

Equation (5.35) for a reversible adiabatic process can be integrated to obtain a relation between P and V if $U = U(P, V)$ is known. Let us consider the system to be an ideal gas, which satisfies the relation $Pv = RT$ and $u = u(T)$. Then,

$$du = C_v dT$$

Therefore,

$$du = -Pdv \quad \text{or} \quad C_v(T_1 - T_2) = \int Pdv = W \tag{5.36}$$

For an ideal gas:

$$h = u + Pv = u + RT \tag{5.37}$$

Therefore,

$$C_P = \left(\frac{\partial h}{\partial T}\right)_P = C_v + R \tag{5.38}$$

The ratio of specific heats γ is given by

$$\gamma = \frac{C_P}{C_v} = 1 + \frac{R}{C_v} \tag{5.39}$$

or

$$C_v = \frac{R}{\gamma - 1} \tag{5.40}$$

Then

$$W = C_v(T_1 - T_2) = \frac{R(T_1 - T_2)}{\gamma - 1} = \frac{P_1 v_1 - P_2 v_2}{\gamma - 1} \tag{5.41}$$

Therefore, when an ideal gas expands reversibly and adiabatically from the initial state (P_1, v_1, T_1) to the final state (P_2 , v_2, T_2), the work done per mole of the gas is given by Eq. (5.41).

Now let us find the path followed by the system in reaching the final state starting from the initial state. We have seen that for an ideal gas:

$$C_v dT - -P dv = \frac{-RT}{v} dv \tag{5.42}$$

or

$$\frac{dT}{T} = -\frac{R dv}{C_v v} = -(\gamma - 1)\frac{dv}{v} \tag{5.43}$$

or

$$\ln\frac{T_2}{T_1} = -(\gamma - 1)\ln\frac{v_2}{v_1} \tag{5.44}$$

$$\frac{T_2}{T_1} = \left(\frac{v_1}{v_2}\right)^{\gamma - 1} \tag{5.45}$$

Since $Pv = RT$

$$\frac{T_2}{T_1} = \frac{P_2 v_2}{P_1 v_1} \tag{5.46}$$

Substituting Eq. (5.46) in Eq. (5.45), we obtain

$$\frac{T_2}{T_1} = \left(\frac{P_2}{P_1}\right)^{\frac{\gamma - 1}{\gamma}} \tag{5.47}$$

Substituting for T_2/T_1 in Eq. (5.45) we get

$$\frac{P_2}{P_1} = \left(\frac{v_1}{v_2}\right)^{\gamma} \tag{5.48}$$

$$Pv^{\gamma} = \text{Constant} \tag{5.49}$$

The reversible adiabatic path can be traced on a diagram by making use of the relations (5.45), (5.47) and (5.48).

***Example* 5.7** At the start of the compression stroke in an Otto engine the fuel–air mixture is at a pressure of 100 kPa and at a temperature of 300 K. The fuel–air mixture is compressed to a pressure of 2 MPa by an adiabatic and reversible process. Assuming the fuel–air mixture to behave as an ideal gas with $\gamma = 1.4$, calculate the final temperature, the compression ratio which is defined as the ratio of the initial volume to the final volume, and the work done per mole of the fuel–air mixture.

Solution

$$T_2 = T_1 \left(\frac{P_2}{P_1}\right)^{\frac{\gamma-1}{\gamma}} = 300 \left(\frac{2 \times 10^6}{10^5}\right)^{\frac{0.4}{1.4}} = 706.06 \text{ K}$$

Compression ratio is given by

$$= \frac{V_1}{V_2} = \frac{P_2 T_1}{P_1 T_2} = \frac{2 \times 10^6 \times 300}{10^5 \times 706.06} = 8.4979$$

$$W = \frac{R(T_1 - T_2)}{\gamma - 1} = \frac{8.314(300 - 706.06)}{0.4} = -8.44 \text{ kJ/mol}$$

***Example* 5.8** One kg of superheated steam at 3 MPa and 500°C contained in a cylinder–piston assembly, is allowed to expand reversibly and adiabatically to a pressure of 100 kPa. Determine the final condition of the steam and the work done.

Solution System: The steam contained in the cylinder–piston assembly.

As specified in the problem, steam undergoes a reversible adiabatic process. For a reversible adiabatic (isentropic) process, $s_2 = s_1$. (It is shown in Chapter 7 that a reversible adiabatic process is an isentropic process). Application of the first law of thermodynamics gives:

$$\Delta u = -W \quad \text{or} \quad W = u_1 - u_2$$

Initial entropy of steam, $s_1 = 7.2345$ kJ/kg K
Initial internal energy of steam u_1 is given by

$$u_1 = 3456.2 \times 10^3 - 3 \times 10^6 \times 0.116\,08 = 3107.96 \text{ kJ/kg}$$

But $s_1 = s_2 = X_2 s_g + (1 - X_2) s_f$
At 100 kPa; $s_f = 1.3027$ kJ/kg K; $s_g = 7.3598$ kJ/kg K.
Therefore,

$$7.2345 = 7.3598 X_2 + (1 - X_2)1.3027 \quad \text{or} \quad X_2 = 0.979$$

$$h_2 = h_g X_2 + (1 - X_2) h_f = 2675.4 \times 0.979 + (1 - 0.979)417.54 = 2627.99 \text{ kJ/kg.}$$

$$v_2 = X_2 v_g + (1 - X_2) v_f = 0.979 \times 1.694 + (1 - 0.979) \times 0.001\,043\,4 = 1.6584 \text{ m}^3\text{/kg}$$

$$u_2 = h_2 - P_2v_2 = 2627.99 \times 10^3 - 100 \times 10^3 \times 1.6584 = 2462.15 \text{ kJ/kg}$$

Therefore, work done W is given by

$$W = u_1 - u_2 = 3107.96 - 2462.15 = 645.81 \text{ kJ/kg}$$

***Example* 5.9** Ten moles of an ideal gas with $\gamma = 1.4$, is compressed reversibly and adiabatically from 100 kPa and 27°C to 1 MPa. Determine the work done on the gas, the change in the internal energy and the final temperature of the gas.

Solution For a reversible-adiabatic process

$$\Delta U = -W \quad \text{or} \quad U_2 - U_1 = -W$$

The work done per mole is given by

$$= C_v(T_1 - T_2) = \frac{R(T_1 - T_2)}{\gamma - 1}$$

$$\frac{T_2}{T_1} = \left(\frac{P_2}{P_1}\right)^{\frac{\gamma-1}{\gamma}} = (10)^{0.2857} = 1.9306$$

Final temperature, $T_2 = 300(1.9306) = 579.2$ K
Therefore,

$$W = \frac{8.314(300 - 579.2)}{0.4} = -5803.2 \text{ J/mol}$$

and total work done = −58.032 kJ
Increase in the internal energy of the gas = 58.032 kJ

5.7 Polytropic Process

We have seen that when an ideal gas undergoes a reversible and adiabatic process, the path followed by the system is given by:

$$Pv^{\gamma} = \text{constant}$$

and the work done,

$$W = \frac{P_1v_1 - P_2v_2}{\gamma - 1}$$

These types of relations are very convenient in engineering calculations and can also be extended to other processes. When a gas undergoes a reversible process in which energy is transferred as heat, the process frequently occurs in such a manner that a plot of $\ln P$ versus $\ln v$ yields a straight line. For such a process

$$Pv^n = constant \qquad (5.50)$$

where n is the index of expansion. Such processes are called polytropic processes. One can verify that for such a reversible polytropic process the work done is given by:

$$W = \frac{P_1 v_1 - P_2 v_2}{n - 1} \tag{5.51}$$

By applying the first law of thermodynamics to a reversible polytropic process, we get

$$u_2 - u_1 = q - \frac{P_1 v_1 - P_2 v_2}{n - 1} \tag{5.52}$$

If the index of expansion n assumes a value of zero, then the polytropic process reduces to a constant-pressure process. On the other hand, if n assumes a value of one, the process reduces to an isothermal process. The values of n for some familiar processes are given below.

$n = 0$, constant pressure process.
$n = \infty$, constant volume process.
$n = 1$, constant temperature process.
$n = \gamma$, reversible adiabatic process.

The polytropic processes commonly encountered in practice are shown on the P–v diagram in Fig. 5.16. Starting from the initial state the lower right quadrant shows the expansion processes and the upper left quadrant shows the compression processes.

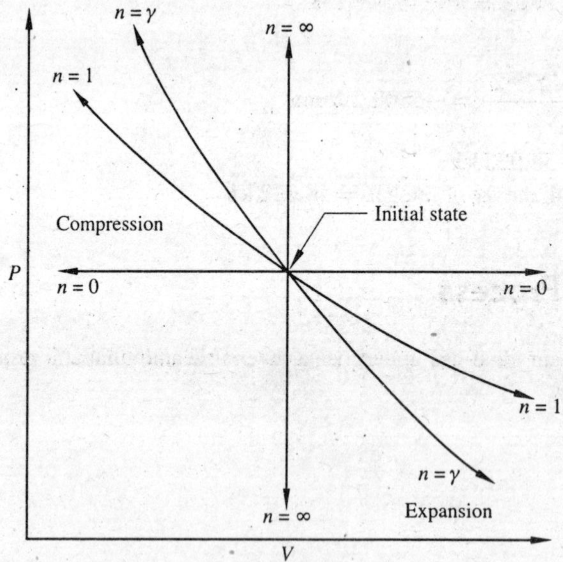

Fig. 5.16 *Polytropic processes on P–v diagram*

***Example* 5.10** At the start of compression in a diesel engine, air is at a pressure of 100 kPa and at a temperature of 27°C. The air is compressed till the final volume is (1/16) of the original volume. The compression process in the diesel engine can be expected to follow the polytropic process, $Pv^{1.25} =$

constant. Determine (a) the pressure, and the temperature of the gas at the end of the compression. (b) the work done on the gas and (c) the energy transfer as heat per mole of air.

Solution (a) $P_1 v_1^{1.25} = P_2 v_2^{1.25}$ or,

$$P_2 = P_1 \left(\frac{v_1}{v_2}\right)^{1.25} = 10^5 (16)^{1.25} = 3.2 \text{ MPa}$$

The air may be treated as an ideal gas. Hence

$$T_2 = \frac{T_1 P_2 v_2}{P_1 v_1} = \frac{3.2 \times 10^6 \times 300}{10^5 \times 16} = 600 \text{ K}$$

At the end of compression the temperature of the gas = 600 K. Pressure of the gas = 3.2 MPa

(b) For a polytropic process

$$W = \frac{P_1 v_1 - P_2 v_2}{n-1} = \frac{R(T_1 - T_2)}{n-1} = \frac{8.314(300 - 600)}{0.25} = -9.9768 \text{ kJ/mol}$$

Work done on the gas = 9.9768 kJ/mol

(c) The first law of thermodynamics gives

$$q = (u_2 - u_1) + W = C_v(T_2 - T_1) + W$$

$$q = \frac{R(T_2 - T_1)}{\gamma - 1} + W$$

$$q = \frac{8.314(600 - 300)}{0.4} - 9.9768 \times 10^3 = -3.7413 \text{ kJ/mol}$$

Therefore, energy transfer as heat $= q = -3.7413$ kJ/mol

5.8 Constant Internal Energy Processes

5.8.1 Irreversible constant internal energy process

A process in which $\Delta U = 0$ is called a constant internal energy process. This can be achieved either when $Q = 0$, and $W = 0$, or when $Q = W$, i.e., when the work done is exactly equal to the energy transfer as heat. Suppose a rigid and insulated tank is divided into two compartments by a partition. One compartment contains a gas at a given pressure and temperature and the other compartment is evacuated. When the partition is removed, the gas expands and occupies the whole volume. Clearly this process is irreversible and the path followed by the system cannot be traced on a thermodynamic diagram. However, one can imagine placing a few removable partitions in the second compartment as shown in Fig. 5.17 and removing the partitions individually, one by one, allowing the gas to reach a state of equilibrium at intermediate stages and obtaining the intermediate states of the system. Consider the gas as the system. Since the system is enclosed by a rigid and an adiabatic wall, the work done, as well as the energy transfer as heat is zero ($W = 0$; $Q = 0$).

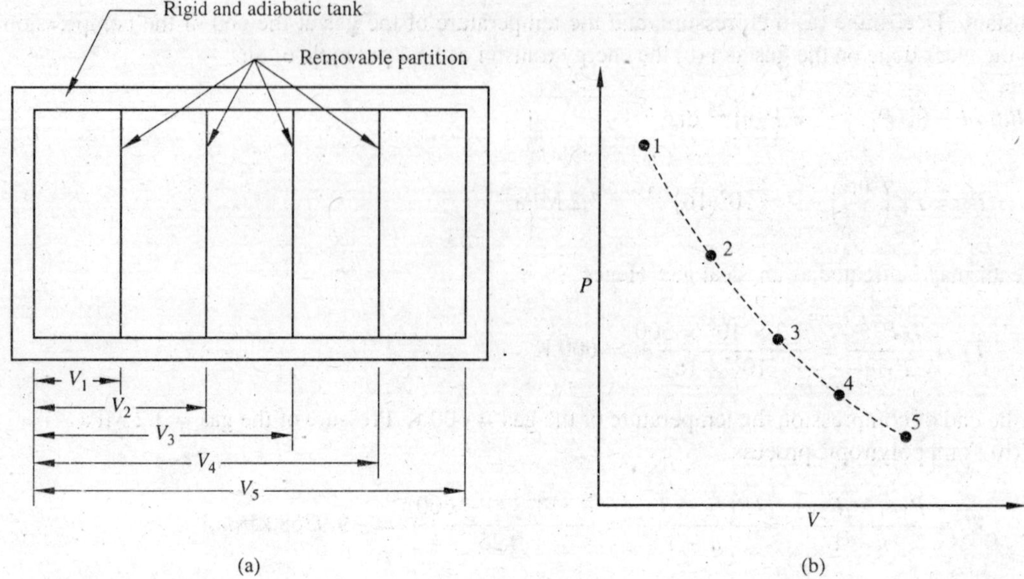

Fig. 5.17 *(a) Schematic arrangement for an irreversible free expansion of a gas (b) P–v diagram for irreversible free expansion of a gas*

Another example of an irreversible constant internal energy process is the mixing of fluids at different states in a rigid and adiabatic container.

5.8.2 Reversible constant internal energy process

It is possible to conceive of a reversible constant internal energy process in which the energy transfer as heat is exactly balanced by the work done by the system. Consider a gas contained in a cylinder–piston assembly (see Fig. 5.18) as the system. The system is brought into contact with a heat source and energy transfer as heat takes place reversibly across the boundary as the gas undergoes expansion. The external pressure can be continuously reduced such that the system is in equilibrium at all times. In this case, the work is done by the system on the surroundings ($W > 0$) while the energy transfer as heat takes place from the surroundings to the system ($Q > 0$). If the work and heat interactions are such that $Q = W$, then the energy of the system does not undergo any change. Now for an ideal gas, we know that $u = u(T)$ only. When the internal energy remains constant, the temperature of the gas also remains constant. Therefore, an isothermal process is also a constant internal energy process for an ideal gas.

In a reversible, constant internal-energy process it was noticed that all the energy flowing into the system as heat is completely converted into work. However, one should not come to the conclusion that it is possible for complete conversion of the energy transferred as heat into mechanical work continuously.

In a reversible, constant internal-energy process the state of the system continuously changes and reaches a particular state beyond which work cannot be obtained. It is not possible to continuously

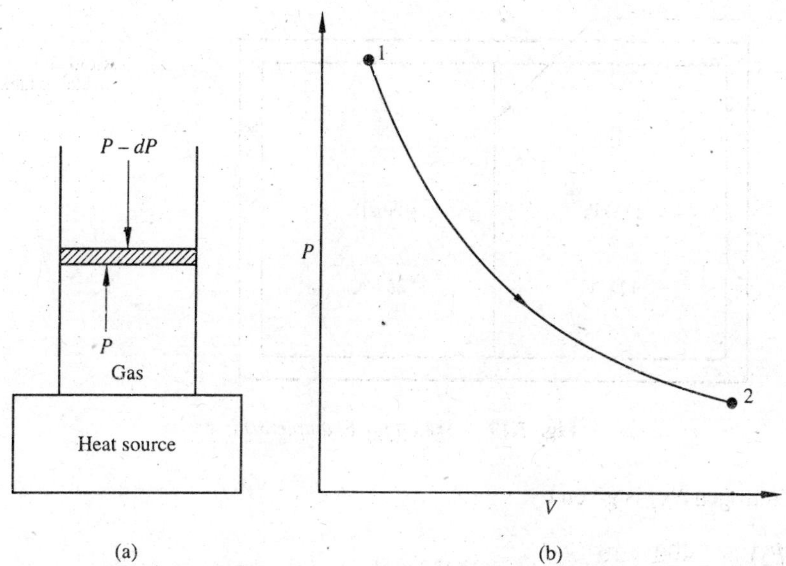

Fig. 5.18 *(a) Schematic arrangement for reversible constant internal energy process (b) P – v diagram for a gas undergoing constant internal energy process*

convert the internal energy into work unless the system is periodically restored to the initial state, that is, unless the system undergoes a cyclic process. When the process is reversed, work has to be done on the system ($W < 0$) to restore it to the original state. To maintain constant internal energy, energy should flow as heat from the system to the surroundings such that $Q = W$. Therefore, in a cycle, the net work done is zero. Then, is it possible for the continuous conversion of energy transferred as heat into mechanical work? What are the limitations on such conversion? These questions will be answered by the second law of thermodynamics which will be discussed in Chapter 7.

***Example* 5.11** A rigid and insulated container of 2 m^3 capacity is divided into two equal compartments by a membrane as shown in Fig. 5.19. One compartment contains helium at 200 kPa and 127°C, while the second compartment contains nitrogen at 400 kPa and 227°C. The membrane is punctured and the gases are allowed to mix. Determine the temperature and pressure after equilibrium has been established. Consider helium and nitrogen to be ideal gases with C_v (He) $= 3R/2$ and $C_v(N_2) = 5R/2$, respectively.

Solution

Considering the gases contained in both the compartments as the system, $W = 0$ and $Q = 0$. Therefore $\Delta U = 0$.

Amount of helium N_{He} is given by

$$N_{\text{He}} = \frac{P_1 V_1}{RT_1} = \frac{200 \times 10^3 \times 1}{8.314 \times 400} = 60.1395 \text{ mol}$$

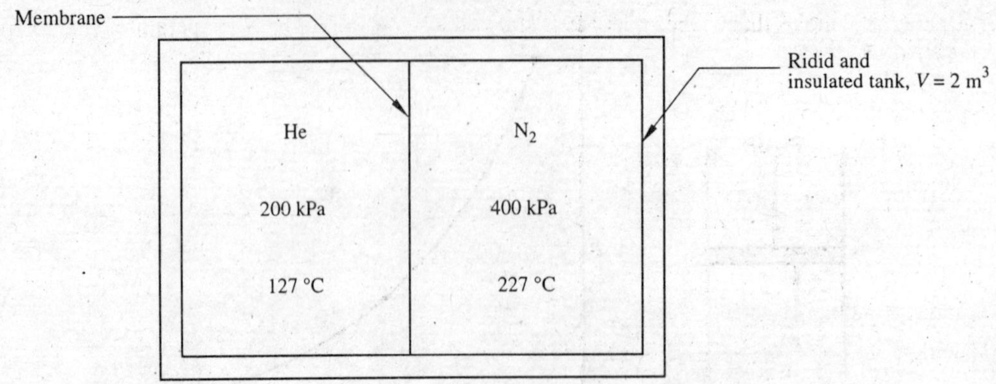

Fig. 5.19 *Sketch for Example 5.11*

Amount of nitrogen N_{N_2} is given by

$$N_{N_2} = \frac{P_2 V_2}{RT_2} = \frac{400 \times 10^3 \times 1}{8.314 \times 500} = 96.2232 \text{ mol}$$

Let T_f be the final temperature after equilibrium has been established. Then

$$\{NC_v(T_f - 400)\}_{He} + \{NC_v(T_f - 500)\}_{N_2} = 0$$

or

$$\frac{R\{60.1395(T_f - 400)3 + 96.2232(T_f - 500)5\}}{2} = 0$$

or $T_f = 472.73$ K

The final pressure of the mixture can be obtained by applying the equation of state:

$$P_f V_f = (N_{He} + N_{N_2})RT_f$$

$$2P_f = (60.1395 + 96.2232)8.314 \times 472.73 \quad \text{or} \quad P_f = 307.27 \text{ kPa}$$

***Example* 5.12** A rigid and insulated tank of 1 m^3 capacity is divided into two compartments. One compartment of volume 0.05 m^3 contains water at 170°C and 1 MPa. The second compartment is evacuated. The partition separating the compartments is punctured and the water fills the whole tank. Determine the final temperature and pressure of the water in the tank.

Solution Since the tank is rigid and insulated; $Q = 0$; $W = 0$. Application of the first law of thermodynamics to the system (water in the tank) gives

$$\Delta U = Q - W = 0 \quad \text{or} \quad U_2 = U_1$$

or, $u_2 = u_1$ since the total mass remains constant.

At 1 MPa, the saturation temperature is 179.88°C. Therefore, the water is in a subcooled state though close to the saturation condition. We know that for a compressed liquid at low pressures, the properties are the same as that of the saturated liquid at the same condition of temperature. Hence, for water at 170°C:

$$v_1 = 0.001\ 114\ 5\ \text{m}^3/\text{kg}.$$

$$\text{Mass of liquid} = \frac{0.05}{0.001\ 114\ 5} = 44.88\ \text{kg}.$$

$$u_1 = h_1 - P_1 v_1 = 719.12 - 0.792 \times 10^3 \times 0.001\ 114\ 5 = 718.24\ \text{kJ/kg}$$

Final volume of the tank $= 1\ \text{m}^3$

Final specific volume, $v_2 = \dfrac{1}{44.88} = 0.022\ 282\ \text{m}^3/\text{kg}.$

$$u_2 = u_1 = 718.24\ \text{kJ/kg}.$$

In the final state, u_2 and v_2 are known. Therefore, it should be possible to determine the final state (that is the pressure and temperature) of the water. Assume the final pressure to be equal to 4.5 bar. Then at 4.5 bar;

$$T = 147,73°\text{C}; \quad v_f = 1.088 \times 10^{-3}\ \text{m}^3/\text{kg}; \quad h_f = 622.4\ \text{kJ/kg}$$

$$v_g = 0.418\ 45\ \text{m}^3/\text{kg}; \quad h_g = 2742.55\ \text{kJ/kg}$$

$$v_2 = X_2 v_g + (1 - X_2) v_f$$

$$0.022\ 282 = X_2(0.418\ 45) + (1 - X_2) \times 1.088 \times 10^{-3}$$

or $X_2 = 0.0508$

$$h_2 = 0.0508 \times 2742.55 + (1 - 0.0508)622.4 = 730.1\ \text{kJ/kg}$$

$$u_2 = h_2 - P_2 v_2 = 730.1 \times 10^3 - 4.5 \times 10^5 \times 0.022\ 282 = 720.07\ \text{kJ/kg}$$

The calculated value of u_2, with the assumed P_2 is close to the value of u_1. Therefore, the final state corresponds to wet steam of quality 0.0508 at a pressure of 4.5 bar and the temperature is 147.73°C.

***Example* 5.13** It is proposed to compress air (ideal gas) reversibly from an initial state of 100 kPa and 27°C to a final state of 500 kPa and 27°C. Compare the work required for the following processes. (a) Heating at constant volume followed by cooling at constant pressure. (b) Isothermal compression. (c) Adiabatic compression followed by cooling at constant volume.

For air $C_v = 20.93$ J/mol K and $C_P = 29.302$ J/mol K.

Solution (a) Constant volume heating followed by cooling at constant pressure: The path of the process is $1a2$ as shown in Fig. 5.20

Process $1 - a$: Constant volume

$$W = \int P dV = 0$$

$$\Delta u = u_a - u_1 = q - W = C_v(T_a - T_1)$$

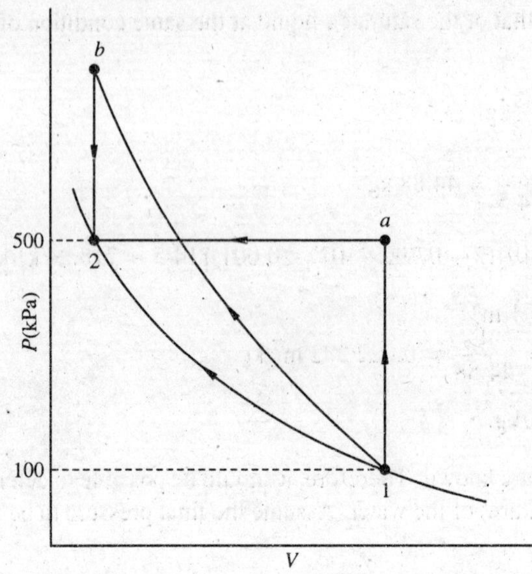

Fig. 5.20 *Sketch for Example 5.13*

$$\frac{P_1 V_1}{T_1} = \frac{P_a V_a}{T_a} \quad \text{or} \quad T_a = \frac{P_a V_a T_1}{P_1 V_1} = \frac{P_a T_1}{P_1} = \frac{500 \times 10^3 \times 300}{100 \times 10^3} = 1500 \text{ K}$$

(Since $V_a = V_1$). Since $u = u(T)$ for an ideal gas, $du = C_v dT$, $dh = C_P dT$.

$$u_a - u_1 = 20.93(1500 - 300) = 25.116 \text{ kJ/mol} = q_{1a}$$

Process $a - 2$: Constant pressure
For a constant pressure process it was shown that $dq = dh = C_P dT$.

$$q_{a2} = h_2 - h_a = C_P(T_2 - T_a) = 29.302(300 - 1500) = -35.162 \text{ kJ/mol}$$

$$u_2 - u_a = C_v(T_2 - T_a) = 20.93(300 - 1500) = -25.116 \text{ kJ/mol}$$

$$\Delta u = q - W \quad or \quad W = q - \Delta u = -35.162 - (-25.116) = -10.046 \text{ kJ/mol}$$

For the complete process $1 - a - 2$: $W = 0 - 10.046 = -10.046$ kJ/mol

(b) Isothermal compression: The path of the process is 1-2 as shown in Fig. 5.20.
Work done

$$W = \int_1^2 P dv = \int \frac{RT dv}{v} = RT \ln \frac{v_2}{v_1} = RT \ln \frac{P_1}{P_2}$$

$$W = 8.314(300) \ln(1/5) = -4.0143 \text{ kJ/mol}$$

(c) Adiabatic compression followed by constant volume cooling: The path of the process is $1 - b - 2$ as shown in Fig. 5.20. Process $1 - b$ is a reversible adiabatic process. For a reversible adiabatic process (for an ideal gas) it was shown that

$$T v^{\gamma - 1} = \text{constant}$$

$$\frac{T_b}{T_1} = \left(\frac{v_1}{v_b}\right)^{\gamma-1}$$

But $v_b = v_2$ and $T_1 = T_2$. For an ideal gas,

$$\frac{P_1 v_1}{T_1} = \frac{P_2 v_2}{T_2} \quad \text{hence} \quad \frac{v_1}{v_2} = \frac{P_2}{P_1}$$

Therefore,

$$T_b = T_1\left(\frac{v_1}{v_2}\right)^{\gamma-1} = T_1\left(\frac{P_2}{P_1}\right)^{\gamma-1} = 300(5)^{0.4} = 571 \text{ K}$$

For the adiabatic process: $q = 0$ and $\Delta u = -W$.
Hence $u_b - u_1 = C_v(T_b - T_1) = 20.93(571 - 300) = 5.672$ kJ/mol $= -W$
Process $b - 2$ is cooling at constant volume. $W = 0$
Total work done, $W = -5.672 + 0 = -5.672$ kJ/mol

***Example* 5.14** A cylinder contains 0.1 kg of saturated steam at 2 bar as shown in Fig. 5.21. The cylinder is now brought into contact with a thermal reservoir and energy transfer as heat takes place causing the piston to rise. The spring force is proportional to the distance moved from the normal position with a spring constant of 250 kN/m. The cross-sectional area of the piston is 0.05 m^2. (a) Determine the steam temperature in the cylinder when its pressure reaches 4 bar, and (b) calculate the energy transfer as heat.

Solution System: Steam enclosed in the cylinder

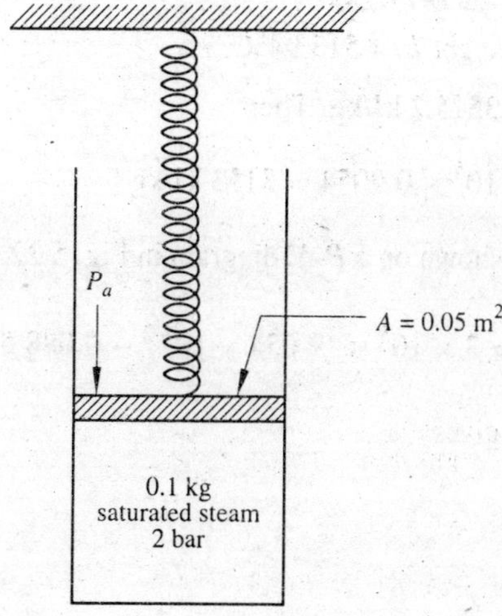

Fig. 5.21 *Sketch for Example 5.14*

Initially the cylinder contains saturated steam at 2 bar.
$P_1 = 2$ bar; $v_g = 0.8854$ m^3/kg

A force balance across the piston gives

$$P_1 = P_a + \frac{K(x_1 - x_0)}{A} = P_a + \frac{K(V_1 - V_0)}{A^2}$$

where P_1 = initial steam pressure
P_2 = final steam pressure
K = spring constant
A = cross-sectional area of piston
V_0 = volume corresponding to unstretched spring
V_1 = volume of steam when $P = P_1$

$$2 \times 10^5 = 1 \times 10^5 + \frac{250 \times 10^3(0.088\,54 - V_0)}{(0.05)^2}$$

or, $V_0 = 8.754 \times 10^{-2}$ m^3

$$P_2 = P_a + \frac{K(V_2 - V_0)}{A^2}$$

$$4 \times 10^5 = 1 \times 10^5 = \frac{250 \times 10^3(V_2 - 8.754 \times 10^{-2})}{(0.05)^2}$$

or $V_2 = 9.054 \times 10^{-2}$ m^3.

Therefore, $v_2 = 0.9054$ m^3/kg

At 4 bar and $T = 500$°C; $v = 0.8892$ m^3/kg

At 4 bar and at 600°C ; $v = 1.0054$ m^3/kg

Therefore, by interpolation, we get $T = 513.94$°C.

At 4 bar and 513.94°C: $h_2 = 3515.2$ kJ/kg. Then

$$u_2 = 3515.2 \times 10^3 - 4 \times 10^5 \times 0.9054 = 3153 \text{ kJ/kg}$$

The work done by the system is shown on a P–V diagram in Fig. 5.22. We know that

$$W = \frac{P_1 + P_2}{2}(V_2 - V_1) = 3 \times 10^5 \times (9.054 \times 10^{-2} - 0.088\,54) = 0.6 \text{ kJ}$$

The application of first law now gives:

$$\Delta U = Q - W$$

or

$$Q = \Delta U + W = (315.3 - 252.92) + 0.6 = 62.98 \text{ kJ}$$

(Note that $u_1 = 2706.3 \times 10^3 - 2 \times 10^5 \times 0.8854 = 2529.2$ kJ/kg)

Therefore, $Q = 62.98$ kJ.

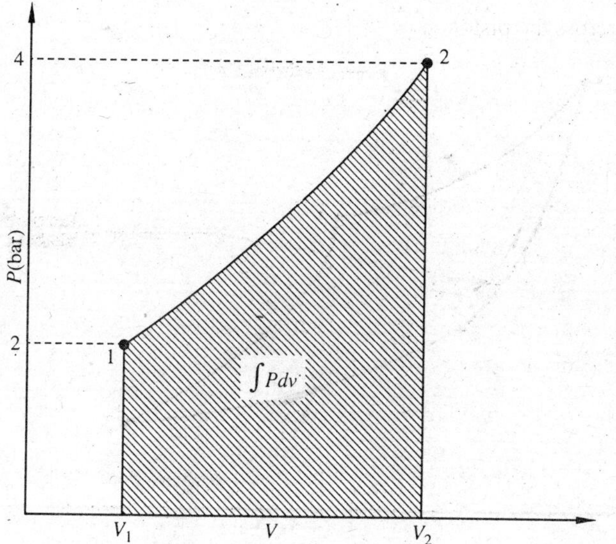

Fig. 5.22 *P–V diagram showing the work done by steam of Example 5.14*

***Example* 5.15** A simple engine uses an ideal gas as the working fluid in a cylinder–piston assembly. The gas is first expanded reversibly and isothermally from state 1 state 2 followed by a reversible adiabatic expansion to state 3. Then it is compressed reversibly and isothermally to state 4 followed by a reversible adiabatic compression to restore the system to state 1 as shown in Fig. 5.23. Derive expressions for the energy transfer as heat and work for each process in terms of the temperatures and volumes at each state. Derive the expressions for the net work delivered during a cycle and the energy conversion efficiency, defined as the net work output/energy input as heat, for the cycle.

Solution Process 1–2 is isothermal. Hence, $\Delta u = C_v \Delta T = 0$ for an ideal gas, or $u_2 - u_1 = 0$

Applying the first law of thermodynamics, we get:

$$u_2 - u_1 = 0 = q_{12} - W_{12}$$

$$q_{12} = W_{12} = \int_1^2 Pdv = RT_1 \int_1^2 \frac{dv}{v} = RT_1 \ln \frac{v_2}{v_1}$$

Process 2–3 is reversible and adiabatic: $q = 0$. Applying the first law of thermodynamics we get,

$$\Delta u = u_3 - u_2 = C_v(T_2 - T_1) = q - W_{23}$$

Therefore, $W_{23} = C_v(T_1 - T_2)$

Process 3–4 is reversible isothermal. Hence,

$$\Delta u = u_4 - u_3 = 0 = q_{34} - W_{34}$$

or

$$q_{34} = W_{34} = \int_3^4 Pdv = RT_2 \int_3^4 \frac{dv}{v} = RT_2 \ln \frac{v_4}{v_3}$$

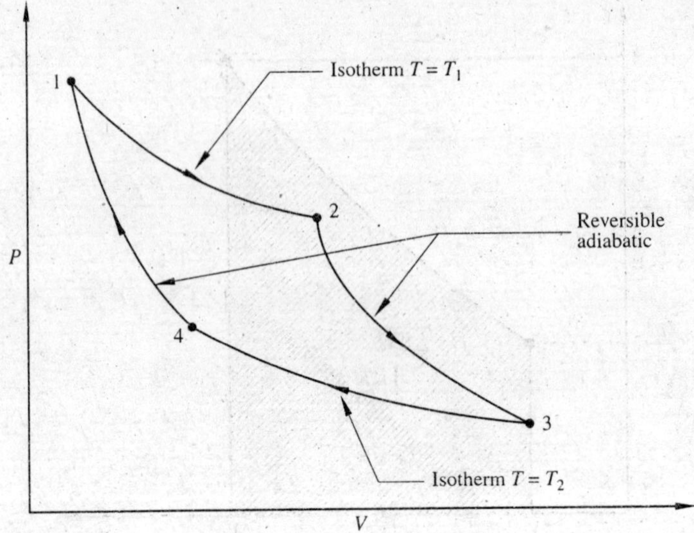

Fig. 5.23 *Sketch for Example 5.15*

Process 4–1 is reversible adiabatic; $q_{41} = 0$ and

$$u_1 - u_4 = C_v(T_1 - T_2) = q_{41} - W_{41}$$

or $W_{41} = C_v(T_2 - T_1)$

Since processes 2–3 and 4–1 are reversible adiabatic,

$$T_1 v_2^{\gamma-1} = T_2 v_3^{\gamma-1} \quad \text{and} \quad T_2 v_4^{\gamma-1} = T_1 v_1^{\gamma-1}$$

$$\frac{v_3}{v_2} = \left(\frac{T_1}{T_2}\right)^{\frac{1}{\gamma-1}} \quad \text{and} \quad \frac{v_4}{v_1} = \left(\frac{T_1}{T_2}\right)^{\frac{1}{\gamma-1}}$$

Therefore

$$\frac{v_3}{v_2} = \frac{v_4}{v_1} \quad \text{or} \quad \frac{v_2}{v_1} = \frac{v_3}{v_4}$$

Net work done

$$= RT_1 \ln \frac{v_2}{v_1} + C_v(T_1 - T_2) + RT_2 \ln \frac{v_4}{v_3} + C_v(T_2 - T_1)$$

$$= RT_1 \ln \frac{v_2}{v_1} - RT_2 \ln \frac{v_3}{v_4} = R(T_1 - T_2) \ln \frac{v_2}{v_1}$$

Energy conversion efficiency = Net work/q_{12}

$$\frac{R(T_1 - T_2) \ln \frac{v_2}{v_1}}{RT_1 \ln \frac{v_2}{v_1}} = 1 - \frac{T_2}{T_1}$$

5.9 Summary

A study of the genesis of the first law of thermodynamics provides a good example of scientific methodology. Experimentation on thermometry led to the development of concepts like specific heat and latent heat. This consequently resulted in the proposition of caloric theory. This in turn suggested further experimentation and led to the development of the first law of thermodynamics.

A cycle is a process or a combination of processes undergone by a system as a result of which the system is restored to its initial state.

The first law of thermodynamics can be stated as: Whenever a system undergoes a cyclic change, the algebraic sum of the work interactions is equal to the algebraic sum of the energy transfers as heat. That is, $\oint dW = \oint dQ$.

The first law of thermodynamics leads to the following consequences:

(a) Heat is a path function.
(b) Energy is a property of a system.
(c) Energy of an isolated system is conserved.
(d) A perpetual motion machine of the first kind is impossible.

Application of the first law of thermodynamics for a system which undergoes a change of state yields $dE = dQ - dW$. If the kinetic energy and potential energy of the system do not undergo any change, the first law reduces to:

$$dU = dQ - dW$$

Application of the first law of thermodynamics to some elementary processes yields the following results:

- For a constant volume process involving expansion/compression work: $dU = dQ$.
- For a constant volume adiabatic process involving other modes of work : $dU = -dW$.
- For a constant pressure process: $dH = dQ$.
- For a constant temperature process: $dU = dQ - dW$.
- If an ideal gas undergoes an isothermal process: $q = w = RT \ln \frac{v_2}{v_1}$.
- For an adiabatic process: $dU = -dW$ and if an ideal gas undergoes a reversible adiabatic process

$$\Delta u = w = \frac{R(T_1 - T_2)}{\gamma - 1} = \frac{P_1 v_1 - P_2 v_2}{\gamma - 1}$$

5.10 Review Questions

1. State the first law of thermodynamics.
2. Is the first law of thermodynamics applicable to irreversible processes also? Explain.
3. What are the consequences of the first law of thermodynamics?
4. Prove that heat is a path function.
5. State the principle of conservation of energy.
6. What is a Perpetual Motion Machine of First Kind (PMMFK)? Is it possible to devise a PMMFK? If not explain why.

7. Define the specific heats in terms of property changes.
8. What is an adiabatic process and how does it differ from isentropic process?
9. What is the work done by an ideal gas when it expands reversibly and adiabatically from (P_1, v_1) to (P_2, v_2)?
10. What is the relation between P and v for an ideal gas which undergoes a reversible and adiabatic process?
11. What is the relation between T and P for an ideal gas, if it undergoes an isentropic process?
12. What is a polytropic process?

5.11 Problems

5.1 A system is taken from state 1 to state 2 along the path 1*a*2. During this process, 100 kJ of energy is added to system and 25 kJ of work was done by the system. If the system is taken from state 1 to state 2 by a different path 1*b*2, it does 50 kJ work on the surroundings. How much energy is transferred as heat during the path 1*b*2? If the system is restored to state 1 from state 2 by an adiabatic path, how much work is done by the system?

5.2 A system is taken through several processes as a result of which the final state of the system is identical with the initial state. The work done and energy transferred as heat during the processes are given below.

Process	W (kJ)	Q (kJ)	ΔU (kJ)
1–2	500	200	–
2–3	300	–	400
3–4	–200	–	–
4–1	0	50	–

Complete the table and determine the net work done by the system.

5.3 A system receives 200 kJ of energy as heat at constant volume. Then it is cooled at constant pressure. During the constant pressure process 50 kJ of work was done on the system while it rejects 70 kJ of energy as heat. Suppose the system is restored to the initial state by an adiabatic process, how much work will be done by the system?

5.4 Water is flowing with negligible velocity on the top of a hill, the height of which is 2500 m above the surface of the earth. The water falls on to the surface of the earth. Determine the change in the energy, kinetic energy, potential energy and internal energy of the water corresponding to the following conditions: (a) when the water is about to hit the surface of the earth, (b) after the water has come to rest on the surface of the earth.

5.5 A 1 m^3 tank with rigid walls contains steam at 200 kPa with a quality of 0.8. If 540 kJ of energy is added as heat to the steam, determine the final state of the steam.

5.6 A sealed and rigid tank of 0.5 m^3 capacity is filled with saturated steam at 500 kPa. The tank is left in the open and after some time the temperature of the steam is found to be 143.62°C. Determine the final state of the steam and the amount of heat interaction.

5.7 A domestic pressure-cooker of 8 litres capacity is initially filled with one kg of water at 30°C. Then it is heated on a stove. The weight placed on the steam vent releases the steam if the pressure inside the cooker exceeds 200 kPa. Determine the conditions of the steam inside the cooker just before the weight is about to release the steam. Also calculate the amount of energy transferred as heat to the steam until the weight is about to release the steam.

5.8 In a particular plant a storage tank of 10 m^3 capacity is initially filled with 4000 kg of saturated Freon-12 at 20°C. Owing to poor insulation the temperature of the contents of the tank rises to 40°C. Determine the final state of Freon-12 in the tank and the energy transferred as heat to the Freon-12 in the tank.

5.9 A sealed and rigid tank of 50 litres capacity is initially half filled with saturated liquid Freon-12 at 0°C and the remaining space is filled with saturated vapor. Owing to the energy transfer as heat from the ambient the temperature of the Freon-12 rises to 30°C. Determine the pressure of Freon-12 in the tank and the energy transferred as heat.

5.10 A sealed and rigid tank of 50 litres capacity is filled with saturated Freon-12 of quality 0.3 at −10°C. The tank is left in the vicinity of a furnace. The bursting pressure of the tank is 1.2 MPa. Determine the energy that can be transferred as heat to the Freon-12 in the tank when the tank is about to burst.

Extract from Superheated Freon-12 Tables at $P = 1.2$ MPa

T (°C)	v (m^3/kg)	h (kJ/kg)
150	0.022 402	282.5288
160	0.023 089	290.0215
170	0.023 767	297.5333

5.11 An insulated rigid container of 1 m^3 volume contains air at a pressure of 500 kPa and 100°C. A paddle wheel is rotated inside the container raising the temperature of the air by 50°C. If the air obeys the relations $PV = NRT$ and $du = C_v dT$ with $C_v = 20.7889$ kJ/kmol K, calculate the amount of work done and the final pressure of the air.

5.12 At a certain instant one kg of Freon-12, of quality 0.2 at −30°C is contained in the evaporator of a refrigeration system. The inlet and outlet valves of the evaporator are closed. As a result of energy transfer in the form of heat to the Freon-12, its temperature rises to 0°C. Determine the final state of the Freon-12 and the energy transferred as heat.

5.13 One kg of saturated liquid Freon-12 at −30°C is contained in an insulated piston–cylinder assembly. A paddle wheel is inserted inside the cylinder and it is rotated for 30 min with a 40 W motor. Determine the final state of Freon-12 in the cylinder.

5.14 One kg of saturated liquid Freon-12 is contained in a cylinder–piston assembly at a temperature of 30°C. Energy is transferred as heat at constant pressure till the Freon-12 is completely converted into saturated vapor. Calculate the work done by Freon-12 and the energy transferred as heat.

5.15 One kg of saturated steam at 100 kPa is held in a cylinder piston assembly. Energy in the form of heat is added to the steam at constant pressure till the volume is doubled. Determine the amount of heat interaction and the final temperature of the steam.

5.16 An adiabatic cylinder–piston assembly contains one kg of saturated steam at 100 kPa with a quality of 0.8. 300 kJ of paddle wheel work is performed on the steam, while the pressure is held constant. Determine the final state of the steam.

5.17 A cylinder–piston assembly contains one mole of nitrogen (ideal gas) at 300 K. Nitrogen is allowed to expand at constant temperature until the pressure is one half of its initial value. Determine the work done and the heat interaction.

5.18 One kg of superheated steam at 30 bar and 350°C contained in a cylinder–piston assembly is allowed to expand reversibly and adiabatically (isentropic) to a pressure of 10 kPa. Determine the final conditions of the steam and the work done.

5.19 One kg of Freon-12 at 0.4 MPa and 30°C contained in a cylinder is allowed to expand isentropically to a pressure of 100 kPa. Determine the final conditions of Freon and the work done.

5.20 Ten moles of an ideal gas with $\gamma = 1.4$ is held in an insulated piston–cylinder assembly at 2 bar and 300 K. A paddle wheel is inserted inside the gas space and rotated while maintaining the gas at constant pressure, until the volume of the gas is doubled. Calculate the work done on the gas by the paddle wheel and the work done by the gas.

5.21 One kmol of an ideal gas with $\gamma = 1.4$ is compressed reversibly and adiabatically from 1 bar and 300 K to 5 bar and then it is cooled at constant pressure to the original temperature. The gas is then restored to the initial pressure of 1 bar by following a reversible and isothermal process. Calculate the heat and work interactions for each of the processes. Also determine the net work done by the gas and the net energy transferred as heat to the gas.

5.22 An ideal gas is heated at constant volume until its temperature is doubled and then cooled at constant pressure until it is returned to the original temperature. Finally the gas is allowed to expand at constant temperature until it is returned to the original volume. Derive an expression to determine the net work done by the gas.

5.23 An insulated and rigid tank of volume 2 m^3 is divided into two equal compartments by a partition. One compartment contains an ideal gas at 100 kPa and 300 K, while the second compartment is evacuated. When the partition is removed the gas occupies the entire tank. Determine the final temperature and pressure of the gas.

5.24 A cylinder–piston assembly contains one mole of air at 300 K and 100 kPa. The air is compressed reversibly until the volume is one fourth of its initial value. During the compression process the relation between pressure and volume is given by $Pv^{1.25} = \text{Const}$. Determine the work done and heat interaction. C_v for air is 20.789 J/mol K.

5.25 A cylinder of 0.5 m in diameter with a frictionless piston contains saturated steam at 500 kPa. The piston is 0.5 m above the base of the cylinder. The piston is held in position by means of latches and energy is added as heat until the pressure of the steam is 1 MPa. Determine the final temperature of the steam and the energy transferred as heat. Now, the piston is released and the energy is added as heat at constant pressure until the steam temperature is 700°C. Determine the amount of energy added as heat and the final position of the piston.

5.26 A spherical balloon contains one kg of saturated steam at 1 bar. Energy is transferred as heat to the steam until the pressure reaches 1.5 bar. The pressure of steam inside the balloon is always proportional to the volume of the balloon. Determine the final temperature of steam and the energy transferred as heat to the steam.

5.27 A spherical balloon has a diameter of 25 cm and contains air at a pressure of 150 kPa. The diameter of the balloon increases to 30 cm in a certain process. During this process the pressure of gas inside the balloon is proportional to the diameter of the balloon. Calculate the work done by the air inside the balloon during this process.

5.28 State whether the following relations are true or false for (i) any process (ii) any reversible process (iii) any cycle (iv) any reversible cycle?
(a) $dQ - dW = 0$ (b) $dU = dQ - dW$; (c) $dU = dQ - PdV$

5.29 An insulated and rigid container of volume 1 m^3 contains a capsule of volume 0.1 m^3. The capsule contains an ideal gas at 1 MPa and 500 K. If the capsule breaks, determine the conditions of the gas inside the container.

5.30 It is desired to compress one kmol of air (ideal gas with $\gamma = 1.4$) from the initial state 1 bar and 300 K to the final state 10 bar and 300 K. Calculate the work required for the following processes: (a) Isothermal compression (b) Cooling at constant pressure followed by heating at constant volume (c) Adiabatic compression followed by cooling at constant volume.

5.31 A simple engine uses an ideal gas with $\gamma = 1.4$ as the working fluid in a piston–cylinder assembly. The gas is first heated at constant volume from 1 bar and 300 K to 5 bar, then it is allowed to expand isothermally to a pressure of 1 bar. Finally the gas is cooled at constant pressure until its temperature is reduced to 300 K. Calculate the efficiency (defined as the ratio of the net work done to the energy absorbed as heat) of the cycle.

5.32 A piston–cylinder assembly contains 10 kg liquid water at 30°C and 1 bar. Energy is transferred as heat to the water at constant pressure till the water is completely converted into saturated vapor. Determine the energy transferred as heat and the work done.

5.33 A cylinder contains ten moles of an ideal gas ($\gamma = 1.4$) at 100 kPa and 300 K as shown in Fig. 5.24. At this stage the spring just touches the piston but exerts no force on it. Energy is transferred as heat to the gas causing the piston to move. During this process the force exerted by the spring is proportional to the displacement, with a spring constant of 50 kN/m. The cross-sectional area of the piston is 0.05 m^2. Determine (a) the temperature of the gas when the pressure reaches 500 kPa, and (b) the energy transferred as heat.

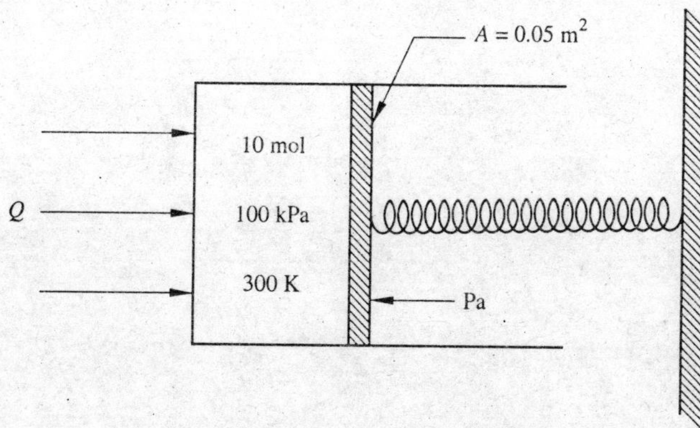

Fig. 5.24 *Sketch for Problem 5.33*

5.34 Repeat the above problem if the cylinder is insulated and the energy is added as work by rotating a paddle wheel inside the gas space instead of the energy transfer as heat. Determine the work done by the gas, the work done by the paddle wheel and the final temperature of the gas.

5.35 A cylinder contains one mole of an ideal gas ($\gamma = 1.4$) at 300 K and 100 kPa as shown in Fig. 5.25. At this stage the spring is at its natural length and exerts no force on the piston. The cylinder is heated causing the piston to move. The spring force is proportional to the distance moved from the normal position with a spring constant of 40 kN/m. The cross-sectional area of the piston is 0.04 m^2. Determine (a) the temperature of the gas when its pressure is 400 kPa, (b) the work done by the gas, and (c) the energy transfer as heat.

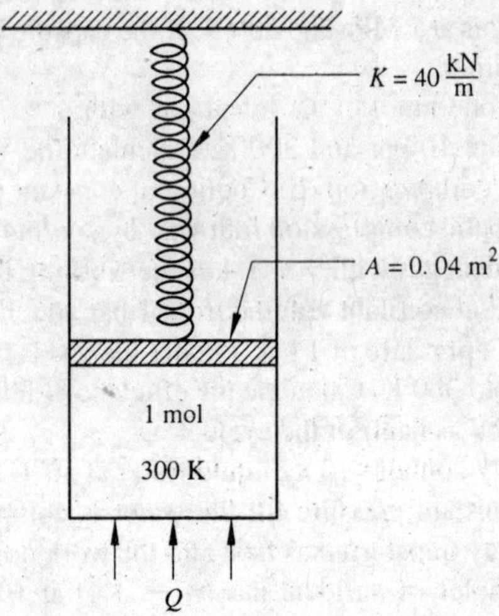

Fig. 5.25 *Sketch for Problem 5.35*

6 First Law Analysis of Processes

Learning Objectives

This chapter will enable the readers to:

- Understand control mass analysis.
- Define control volume and to compare the control-mass and control-volume.
- Derive a general expression for the first law of thermodynamics for a control-volume.
- Analyze steady state flow processes.
- Define Joule–Thompson coefficient, inversion point and inversion curve.
- Know about throttling calorimeter and combined separating and throttling calorimeter.
- Analyze transient flow processes.
- Know about various types of gaseous fuels.
- Define stoichiometric or theoretical air and excess air.
- Calculate the stoichiometric air requirement for combustion of a fuel.
- Calculate the composition of combustion products.
- Understand the Orsat analysis of flue gases.
- Calculate the air-fuel ratio from a knowledge of the Orsat analysis of flue gases.
- Define a formation reaction.
- Understand the standard states used in chemical thermodynamics.
- Know about the standard heat of formation, the standard heat of combustion and the standard heat of a reaction.
- Calculate the standard heat of a reaction from a knowledge of standard heats of formation and also from standard heats of combustion.
- Estimate the standard heat of reaction as a function of temperature.
- Calculate the adiabatic flame temperature.

The first law of thermodynamics was discussed in Chapter 5 and it was applied for analyzing some processes. In essence, the first law of thermodynamics relates the energy transfer as heat and work to the overall change in the energy of the system. Such an analysis is known as the energy analysis. Basically there are two types of energy analysis, namely control mass and control volume analysis.

6.1 Control Mass Analysis

In the control-mass analysis, the system, or the control mass on which the attention is focused, must be clearly identified. Suppose a certain quantity of liquid contained in an insulated containers is heated by immersing into it an electrical resistor which is supplied with a steady flow of electrical current.

If the liquid is considered as the system the energy interaction is heat. On the other hand, the energy interaction is work if the combination of the liquid and the electrical resistor is considered as the system. Therefore, the energy interactions depend on the choice of the control mass. Hence the analyst should specify clearly the control mass under consideration. The steps involved in the control mass analysis are:

1. Choose the control mass carefully. The choice of the control mass is arbitrary and dictated by convenience.
2. Represent the process on a thermodynamic diagram (like P–v or h–s). This is helpful in identifying the states of the system and analyzing the process.
3. Follow the energy interactions carefully while applying the principle of conservation of energy.
4. Incorporate the information regarding the P–v–T behavior of substances in the form of an equation of state, charts or tables, while relating the energy changes of the system in the final solution of the problem.

The control mass analysis is usually used for analyzing nonflow processes and it is illustrated through several examples in Chapter 5.

6.2 Control Volume Analysis

In the first law analysis of non-flow processes, attention is focused on a definite quantity of matter termed as a thermodynamic system. As the system interacts with the surroundings, the heat and work interactions across the boundary are related to the change of energy of the matter within the boundary. Quite often, a practicing engineer in industry encounters equipment such as turbines, compressors, pumps, nozzles, diffusers, heat exchangers, reactors, etc., through which material flows in and out continuously . In the analysis of such equipment, it is quite difficult to follow the course of each element of matter and to evaluate the energy changes and the heat and work interactions across the boundary. Moreover, the properties of the fluid are not uniform throughout the system; they change from point to point within the system. Such systems are usually called continuous systems.

6.2.1 Continuous system

The description of continuous systems require the specification of spatial variation of properties. In other words, a continuous system cannot be described by specifying only a few properties like P, v and T, because these properties change from point to point in the region. The description of a continuous system requires the specification of several variables in terms of dependent parameters (like the pressure distribution with length in a turbine, temperature distribution with length in a nozzle, etc.).

Consider a small portion of volume, dV, around a point i in the center of a continuous system as shown in Fig. 6.1. Imagine that this small elemental volume is isolated from the rest of the continuous system. Then this elemental volume reaches a state of equilibrium and in that state of equilibrium it is possible to specify definite values of P, T, etc. However, the values of P, T, etc. differ from element to element. In the limit as $dV \rightarrow 0$, the element tends to a point and the point values or local values are specified as

$$T_i = \lim_{dV \to 0} T \quad and \quad P_i = \lim_{dV \to 0} P \tag{6.1}$$

The state of the continuous system can now be described by specifying functions of the type $T(x, y, z)$ or $P(x, y, z)$, etc.

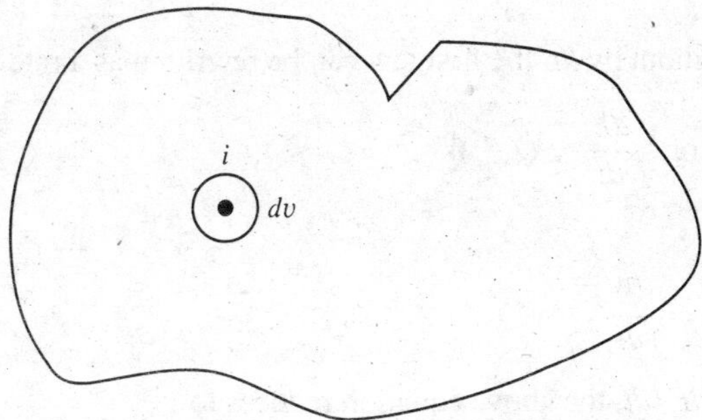

Fig. 6.1 *Continuous system*

It is assumed that the relations between the local thermodynamic properties of a continuous system are identical to the relations between the same properties in a homogeneous system at equilibrium. This is usually called the *principle of local equilibrium*. This principle is true when the gradients are not very large. The local equilibrium principle has been used in the study of heat transfer, fluid mechanics and many other branches of science and has led to results which are in good agreement with the experimental observations. This principle has, therefore, been widely accepted as true.

Since mass and energy are extensive properties, they are additive over the constituent subsystems. Mass and energy of a continuous system can be evaluated by integrating the local values over the entire volume of the continuous system. Any extensive property of the continuous system can be evaluated in a similar manner.

If $\rho(x, y, z, t)$ is the density distribution in the continuous system, then the mass of the continuous system at time t is given by:

$$m = \int_V \rho(x, y, z, t) dV \tag{6.2}$$

where the integration is taken over the entire volume. Similarly, if $e(x, y, z, t)$ is the specific energy distribution, then the energy of the continuous system E at time t is given by:

$$E = \int_V \rho(x, y, z, t) e(x, y, z, t) dV \tag{6.3}$$

6.2.2 First law of thermodynamics for a continuous system

Consider a continuous system which undergoes a change of state during a given process. Let the continuous system be in state 1 at time t and after a differential time dt let it be in the state 2. During this time, the change in the energy of the continuous system is:

$$dE = \frac{d(\int \rho e dV)}{dt} dt \tag{6.4}$$

As seen earlier in Chapter 5 the first law of thermodynamics is given by

$$dE = dQ - dW \tag{5.11}$$

Dividing Eq. (5.11) throughout by dt, the first law can be rewritten as a rate equation. Thus

$$\frac{dE}{dt} = \frac{dQ}{dt} - \frac{dW}{dt} \quad \text{or} \quad \frac{dE}{dt} = \dot{Q} - \dot{W} \tag{6.5}$$

where

$$\dot{Q} = \frac{dQ}{dt} \quad \text{and} \quad \dot{W} = \frac{dW}{dt}$$

Substituting Eq. (6.4) for dE/dt, the above equation reduces to

$$\frac{d}{dt} \int_V \rho e dV = \dot{Q} - \dot{W} \tag{6.6}$$

Now let us consider a device like a turbine which is supplied with high pressure steam. The steam expands in the turbine and emerges at a lower pressure; but with a velocity greater than with which it enters the turbine. The turbine delivers a certain amount of shaft work. During this process energy may be transferred from the turbine casing to the environment. In analyzing this flow process, let us focus our attention on the turbine and draw an imaginary envelope enclosing the turbine (see Fig. 6.2). This envelope passes through the inlet pipe and the exhaust pipe of the turbine at points i and e. This imaginary envelope enclosing the device is called the *control surface*. The space bounded by the envelope is termed as the *control volume*. Therefore, the *control volume is any defined region in space through which matter flows in and out continuously, and heat and work interactions are present across the boundary*. The space bounded by the control surface may change in shape and size, and may be moving through space. At this stage, it is important to clearly note the difference between the system and the control volume.

System

- Refers to a definite quantity of matter on which attention is focused. The matter is enclosed by a boundary.
- The system boundary need not be constant in shape and volume and may undergo configuration change.
- Matter does not cross the system boundary.
- Heat and work interactions are present across the system boundary.

Control volume

- Refers to a definite volume on which attention is focused. The control volume is enclosed by a boundary.

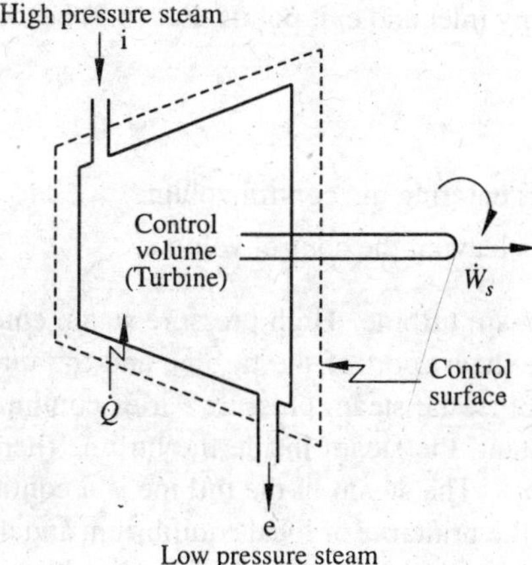

Fig. 6.2 *Schematic representation of a control-volume*

- The control volume also need not be constant in volume and shape and may undergo configuration changes. The control volume may also move through space. Most often, we deal with control volumes that are of fixed shape and size and are fixed in the reference frame.
- Matter continuously flows in and out of the control volume.
- Heat and work interactions are present across the control surface.

6.2.3 First law of thermodynamics for a control volume

Let us consider first the law of conservation of mass as applied to the control volume. During a time interval dt, let the mass dm_i enter the control volume and the mass dm_e leave the control volume. Let $m(t)$ represent the mass inside the control volume at time t, and $m(t+dt)$ represent the mass inside the control volume at time $(t+dt)$. The law of conservation of mass then gives:

$$m(t+dt) = m(t) + dm_i - dm_e$$

or

$$m(t+dt) - m(t) = dm_i - dm_e \tag{6.7}$$

The left-hand side of the above equation represents the change in mass within the control volume during the time interval dt. The right-hand side of the equation represents the net mass flow into the control volume during the time interval dt. Dividing Eq. (6.7) by dt, we get

$$\frac{\{m(t+dt) - m(t)\}}{dt} = \frac{dm_i}{dt} - \frac{dm_e}{dt} \tag{6.8}$$

In the limit as $dt \to 0$, we obtain

$$\frac{dm}{dt} = \dot{m}_i - \dot{m}_e \tag{6.9}$$

If the control volume has many inlet and exit points, Eq. (6.9) becomes:

$$\frac{dm}{dt} = \sum \dot{m}_i - \sum \dot{m}_e \tag{6.10}$$

where $\dot{m}_i$ = rate of mass entering the control volume,

$\dot{m}_e$ = rate of mass leaving the control volume.

Consider the example of a steam turbine. High pressure steam enters the inlet port of the turbine, low pressure steam leaves the exhaust port of the turbine and the turbine delivers some shaft work. As steam flows through the turbine, the steam pressure varies continuously from the inlet port to the exhaust port. At any given instant, the steam inside the turbine (here the control volume) is not at uniform pressure and temperature. The steam in the turbine is a continuous system. In analyzing the flow processes, we make use of the principle of local equilibrium and the average changes of energy of the continuous system can be related to the energy transferred and the work done. Let $\dot{m}_i$ represent the rate of mass entering the control volume at inlet i and $\dot{m}_e$ the rate of mass leaving the control volume at exit e (see Fig. 6.3). To apply the first law of thermodynamics, which is applicable to a definite quantity of matter, we should choose the system first. That is, we should identify a definite quantity of matter which remains constant as the matter flows. For this purpose, let the boundary of the system include all matter inside the control volume and that which is about to enter the control volume during the differential time interval dt.

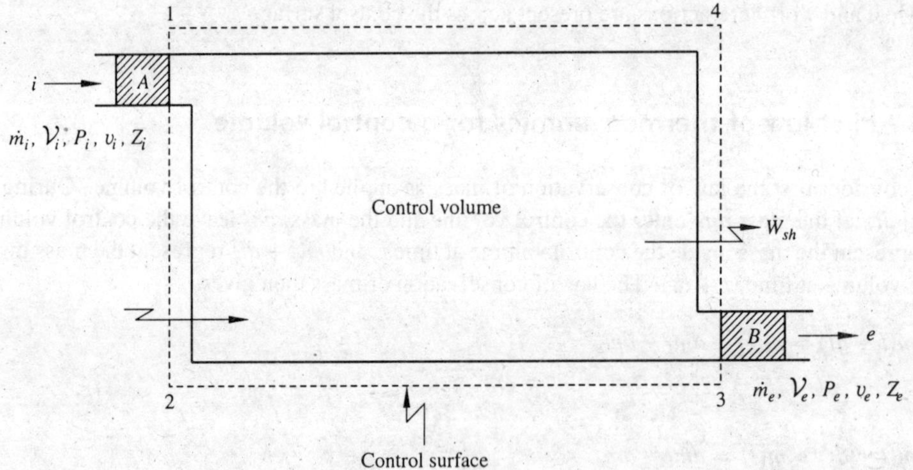

Fig. 6.3 *Control-volume for first law analysis of flow processes*

At time t, the system is defined as the mass contained in the control volume and the mass in region A which is about to enter the control volume in a differential time dt. At time $(t + dt)$, the system is defined as the mass contained in the control volume and the mass in region B. Therefore, during the differential time dt, the system configuration undergoes a change. The continuous flow can be considered as a succession of these processes applied to every element of matter that enters the control volume. During this differential time dt, the mass occupying the region A enters the control volume and some mass leaves the control volume to occupy region B.

Let

$m(t)$ = total mass inside the control volume at time t.
$m(t+dt)$ = total mass inside the control volume at time $(t+dt)$.
$e(t) = (u + \mathcal{V}^2/2 + gZ)$ = specific energy of matter inside the control volume at time t.
$e(t+dt)$ = specific energy of matter inside the control volume at time $(t+dt)$.
P_i and P_e = pressure at the inlet and exit ports, respectively.
$\mathcal{V}_i$ and $\mathcal{V}_e$ = flow velocity at the inlet and exit ports, respectively.
v_i and v_e = specific volume at the inlet and exit ports, respectively.
e_i and e_e = specific energy of the material at inlet and exit ports, respectively.
$\dot{Q}$ = rate of energy flow as heat into the control volume.
$\dot{W}_s$ = rate of shaft work done by the control volume.

Then, mass contained in region A which enters the control volume during the time interval $dt = \dot{m}_i dt$. Mass contained in region B which leaves the control volume during the time interval $dt = \dot{m}_e dt$. From mass balance, we can write

$$m(t) + \dot{m}_i dt = m(t+dt) + \dot{m}_e dt \tag{6.11}$$

To accommodate this, the mass inside the control volume has to be compressed such that its volume decreases by the amount $(\dot{m}_i v_i dt)$. This is accomplished by the pressure P_i acting on the material entering the control volume.

Therefore, the work done = $-P_i v_i \dot{m}_i dt$. Since the mass $\dot{m}_e dt$ has to leave the control volume at the exit port, the work done (by similar reasoning) = $P_e v_e \dot{m}_e dt$

Energy of the system at time $t = m(t)e(t) + \dot{m}_i e_i dt = E(t) + \dot{m}_i e_i dt$

Energy of the system at time $(t+dt) = m(t+dt)e(t+dt) + \dot{m}_e e_e dt = E(t+dt) + \dot{m}_e e_e dt$

During the time interval dt:

Energy transferred as heat to the system = $\dot{Q}dt$

Shaft work done by the system = $\dot{W}_s dt$

For heat and work interactions, the sign convention employed is the same as that used for the system. Energy flow as heat into the control volume and the shaft work delivered by the control volume are taken as positive. By applying the first law of thermodynamics, we get

$$E_2 - E_1 = Q - W \tag{6.12}$$

$$(E(t+dt) + \dot{m}_e e_e dt) - (E(t) + \dot{m}_i e_i dt) = \dot{Q}dt - \dot{W}_s dt - (\dot{m}_e P_e v_e - \dot{m}_i P_i v_i)dt \tag{6.13}$$

$$\dot{m}_e(e_e + P_e v_e) - \dot{m}_i(e_i + P_i v_i) = \dot{Q} - \dot{W}_s - \frac{E(t+dt) - E(t)}{dt} \tag{6.14}$$

In the limit as $dt \longrightarrow 0$,

$$\dot{m}_e\left(h_e + \frac{\mathcal{V}_e^2}{2} + gZ_e\right) - \dot{m}_i\left(h_i + \frac{\mathcal{V}_i^2}{2} + gZ_i\right) = \dot{Q} - \dot{W}_s - \frac{dE}{dt} \tag{6.15}$$

where,

$$h_e = u_e + P_e v_e \tag{6.16}$$

$$h_i = u_i + P_i v_i \tag{6.17}$$

Z_e and Z_i = Elevation of the exit and the inlet ports above the datum plane. The expression (6.15) can be rearranged as:

$$\frac{dE}{dt} = \left(\dot{m}_i(h_i + \frac{V_i^2}{2} + gZ_i) + \dot{Q}\right) - \left(\dot{m}_e(h_e + \frac{V_e^2}{2} + gZ_e) + \dot{W}_s\right) \tag{6.18}$$

or, stated in words: Rate of accumulation of energy = Rate of inflow of energy − Rate of outflow of energy.

6.2.4 Steady state flow processes

Many of the engineering devices like turbines, compressors, pumps, nozzles, condensers, heat exchangers, etc. operate at steady state conditions. An analysis of steady state flow processes is useful in designing and evaluating the performance of such devices. The charging and discharging of cylinders are unsteady state or transient processes. The analysis of such transient processes is dealt with in a later section. The general expression developed above for the control volume can be simplified if the steady state flow conditions are assumed. These are

(a) The mass flows into the control volume at a constant rate and leaves the control volume at the same rate.

$$\dot{m}_i = \dot{m}_e = \dot{m} \tag{6.19}$$

Therefore, there is no accumulation of mass inside the control volume.

$$\frac{d}{dt}\int_V \rho dV = \frac{dm}{dt} = 0 \tag{6.20}$$

(b) The state of the matter at the inlet, exit and at any given point inside the control volume does not change with respect to time. Therefore

$$\frac{d}{dt}\int_V \rho e dV = \frac{dE}{dt} = 0 \tag{6.21}$$

(c) The rate of energy transfers as heat and work across the control surface are constant.

$$\dot{Q} = \text{Constant} \quad \text{and} \quad \dot{W}_s = \text{Constant} \tag{6.22}$$

With these assumptions, the first law for the control volume reduces to:

$$\left(h_e + \frac{V_e^2}{2} + gZ_e\right) - \left(h_i + \frac{V_i^2}{2} + gZ_i\right) = \frac{\dot{Q} - \dot{W}_s}{\dot{m}} \tag{6.23}$$

The control volume analysis for steady flow processes is illustrated below through several examples.

***Example* 6.1** The turbine shown in Fig. 6.4, operating under steady state conditions receives 3,600 kg of steam per hour. The superheated steam at 0.2 MPa and 400°C enters the turbine at a velocity of

80 m/s. The inlet to the turbine is at an elevation of 10 m above the ground level. The steam leaves the turbine at 0.01 MPa with 5% moisture content and a velocity of 150 m/s. The exit port of the turbine is at an elevation of 3 m above the ground level. Energy losses as heat from the turbine to the surroundings are estimated to be 36 MJ/h. Estimate the power output of the turbine.

Solution Turbine operates under steady state flow conditions. Therefore,

$$\dot{m}_i = \dot{m}_e = \dot{m} = 3600 \text{ kg/h} = 1 \text{ kg/s}$$

$$\dot{Q} = -36 \text{ MJ/h} = -10 \text{ kJ/s}$$

$$h_i = 3276.6 \text{ kJ/kg} \quad \text{and} \quad \mathcal{V}_i = 80 \text{ m/s}; \quad Z_i = 10 \text{ m}$$

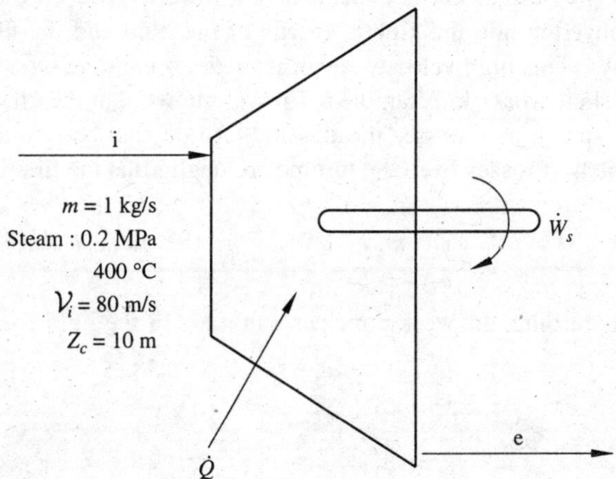

Fig. 6.4 *Sketch for Example 6.1*

Exit steam is at 0.01 MPa with 5% moisture content.

$$h_e = h_g X_e + (1 - X_e) h_f = 2584.8 \times 0.95 + 0.05 \times 191.83 = 2465.15 \text{ kJ/kg}$$

The first law of thermodynamics for the steady flow process is:

$$\{h_e + \mathcal{V}_e^2/2 + gZ_e\} - \{h_i + \mathcal{V}_i^2/2 + gZ_i\} = \frac{\dot{Q} - \dot{W}_s}{\dot{m}}$$

Substituting the values of h, $\mathcal{V}$, Z, $\dot{Q}$ and $\dot{m}$, we get

$$\left(2465.15 \times 10^3 + \frac{(150)^2}{2} + 9.81(3)\right) - \left(3276.7 \times 10^3 + \frac{(80)^2}{2} + 98.1\right) = -10 \times 10^3 - \dot{W}_s$$

$$(2465.15 - 3276.6)10^3 + \frac{(150)^2 - (80)^2}{2} + 9.81(3 - 10)$$

$$= -811.55 \times 10^3 + 8.05 \times 10^3 + 0.06867 \times 10^3$$

$$= -10 \times 10^3 - \dot{W}_s$$

or $\dot{W}_s = 793.43$ kJ/s $= 793.43$ kW

The kinetic and potential energy changes are negligible (8.05 kJ and 0.068 kJ, respectively) compared to the change in enthalpy of the fluid. The omission of the kinetic and potential energy terms introduces an error of only about 1%. Therefore, the changes in the kinetic and potential energy in many of the flow processes can be ignored without introducing much error.

6.3 Applications of Steady State Flow Processes

6.3.1 Turbine

In a steam turbine, high pressure steam is expanded to a lower pressure in a set of nozzles, where the thermal energy is converted into the kinetic energy of the fluid and the fluid emerges out of the nozzles at a high velocity . This high velocity steam is made to impinge on a set of blades fixed to a rotor which delivers the shaft work. In Example 6.1 it was shown that the changes in the kinetic and potential energies of the steam, as it passes through the turbine, can be ignored without introducing much error. Further, if the heat losses from the turbine are negligible, the first law for this steady flow reduces to

$$h_e - h_i = -\frac{\dot{W}_s}{\dot{m}} \tag{6.24}$$

Therefore, in an adiabatic turbine, the work done per unit mass of the fluid is equal to the decrease in the enthalpy of the fluid.

6.3.2 Compressor

A compressor can be considered as a turbine operating in the reverse. Fluid enters the compressor at a low pressure and emerges at a higher pressure. If the changes in the kinetic energy and the potential energy of the fluid are ignored, and the energy losses as heat are negligible, then the first law for this flow process reduces to

$$h_e - h_i = -\frac{\dot{W}_s}{\dot{m}} \tag{6.25}$$

In a compressor $\dot{W}_s$ is negative, because work is being done on the control volume and hence $h_e > h_i$. The work done on the compressor per unit mass of the fluid is equal to the increase in the enthalpy of the fluid.

6.3.3 Nozzle

A nozzle, shown in Fig. 6.5, is primarily used to increase the flow velocity. Since its objective is to increase the kinetic energy of the fluid, the change in the kinetic energy cannot be ignored. Neglecting the potential energy change, for an adiabatic nozzle the first law expression for a steady flow process reduces to

$$h_e + \frac{\mathcal{V}_e^2}{2} = h_i + \frac{\mathcal{V}_i^2}{2}$$

or

$$\mathcal{V}_e^2 - \mathcal{V}_i^2 = 2(h_i - h_e) \tag{6.26}$$

If the inlet velocity is negligible, that is $\mathcal{V}_i \sim 0$,

$$\mathcal{V}_e = \sqrt{2(h_i - h_e)} \tag{6.27}$$

If an ideal gas is flowing through the nozzle, the exit velocity $\mathcal{V}_e$ can be expressed in terms of the inlet and outlet pressures and temperatures by making use of the relations:

$$Pv = RT \quad \text{and} \quad dh = C_P dT \tag{6.28}$$

Therefore,

$$\mathcal{V}_e^2 = 2C_P(T_i - T_e) = 2C_P T_i \left(1 - \frac{T_e}{T_i}\right) \tag{6.29}$$

For an adiabatic expansion,

$$\frac{T_e}{T_i} = \left(\frac{P_e}{P_i}\right)^{\frac{\gamma-1}{\gamma}} \tag{6.30}$$

Hence,

$$\mathcal{V}_e = \sqrt{2C_P T_i \left\{1 - \left(\frac{P_e}{P_i}\right)^{\frac{\gamma-1}{\gamma}}\right\}} \tag{6.31}$$

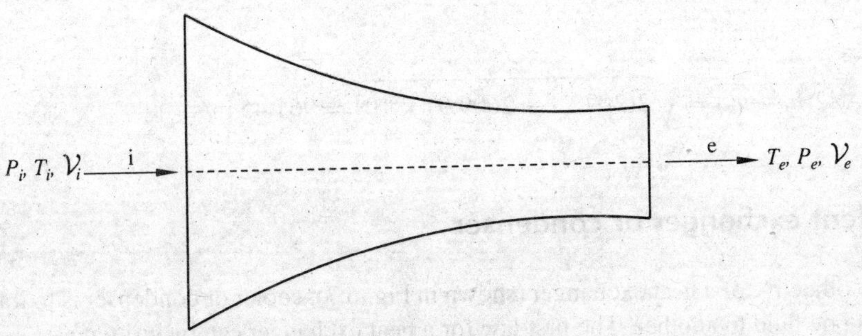

Fig. 6.5 *Schematic representation of a nozzle*

6.3.4 Diffuser

A diffuser, shown in Fig. 6.6, can be thought of as a nozzle in which the direction of flow is reversed. For an adiabatic diffuser, $\dot{Q}$ and $\dot{W}_s$ are zero and the first law of thermodynamics reduces to

$$h_e + \frac{\mathcal{V}_e^2}{2} = h_i + \frac{\mathcal{V}_i^2}{2}$$

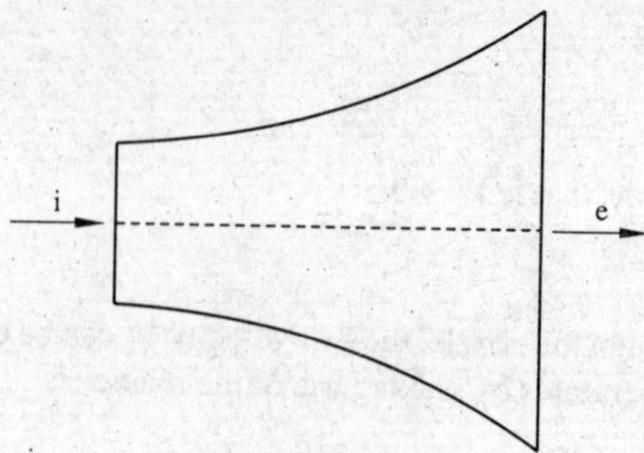

Fig. 6.6 *Schematic representation of a diffuser*

$$\mathcal{V}_e^2 - \mathcal{V}_i^2 = 2(h_i - h_e)$$

***Example* 6.2** Superheated steam at 2 bar and 300°C enters a horizontal adiabatic nozzle with negligible velocity and leaves as saturated vapor at 0.2 bar. Determine the exit velocity of steam.

Solution For saturated steam at 0.2 bar:

$$h_e = 2609.9 \text{ kJ/kg}$$

For steam at 2 bar and 300°C:

$$h_i = 3072.1 \text{ kJ/kg}$$

$$h_e + \mathcal{V}_e^2/2 = h_i + \mathcal{V}_i^2/2$$

Therefore,

$$\mathcal{V}_e = \sqrt{2(h_i - h_e)} = \sqrt{2(3072.1 - 2609.9) \times 10^3} = 961.45 \text{ m/s}$$

6.3.5 Heat exchanger or condenser

The primary objective of a heat exchanger (shown in Fig. 6.7), cooler or condenser is to transfer energy as heat from one fluid to another. The first law for a heat exchanger can be written as:

$$\sum \dot{m}_e(h_e + \mathcal{V}_e^2/2 + gZ_e) - \sum \dot{m}_i(h_i + \mathcal{V}_i^2/2 + gZ_i) = \dot{Q} - \dot{W}_s \tag{6.32}$$

No shaft work is done in a heat exchanger, $\dot{W}_s = 0$; energy losses as heat are neglected, $\dot{Q} = 0$; and changes in KE and PE can be ignored in a heat exchanger $\Delta(\mathcal{V}^2/2) = 0$, $\Delta(gZ) = 0$. Hence,

$$\sum \dot{m}_e h_e = \sum \dot{m}_i h_i \tag{6.33}$$

or

$$\dot{m}_{1e}h_{1e} + \dot{m}_{2e}h_{2e} = \dot{m}_{1i}h_{1i} + \dot{m}_{2i}h_{2i} \tag{6.34}$$

Further

$$\dot{m}_{1e} = \dot{m}_{1i} = \dot{m}_1 \quad \text{and} \quad \dot{m}_{2e} = \dot{m}_{2i} = \dot{m}_2 \tag{6.35}$$

or

$$\dot{m}_1(h_e - h_i)_1 = \dot{m}_2(h_i - h_e)_2 \tag{6.36}$$

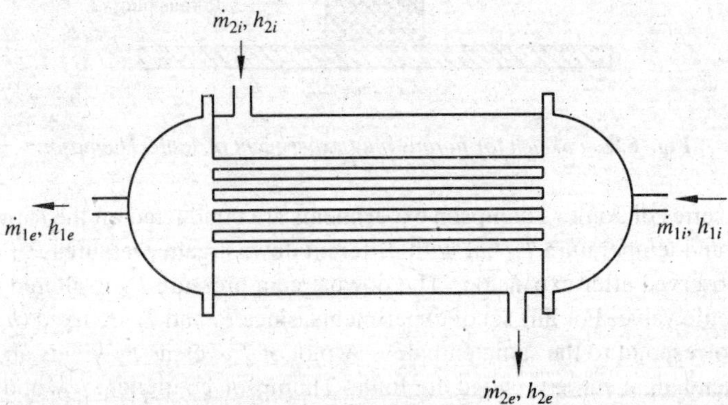

Fig. 6.7 *Schematic diagram of a heat exchanger*

6.4 Throttling Process

Consider the porous plug experiment conducted by Joule and Thompson (Lord Kelvin) in 1853. The porous plug experiment was designed to measure the resulting temperature change when a fluid flows steadily through a porous plug which is inserted in a thermally insulated, horizontal pipe. The apparatus used by Joule and Thompson is shown in Fig. 6.8. A gas at pressure P_i and temperature T_i flows continuously through a porous plug in a tube and emerges into the space which is maintained at a constant pressure P_e. The device is thermally insulated and kept horizontal. The device is operated for a time long enough to attain steady state and the temperature T_e of the emerging gas is measured. Consider the dotted portion in Fig. 6.8 as the control volume and apply the first law of thermodynamics to this steady flow process. Since the pipe is thermally insulated and also does not exchange work, $\dot{Q} = 0$, $\dot{W}_s = 0$. Neglecting the changes in the kinetic and potential energy terms, the first law expression for the flow process reduces to

$$h_i = h_e \tag{6.37}$$

Therefore, whenever a fluid expands from a region of high pressure to a region of low pressure through a porous plug, partially opened valve or some obstruction without exchanging any energy as heat and work with the surroundings (neglecting the changes in KE and PE), the enthalpy of the fluid remains constant, and the fluid is said to have undergone a throttling process.

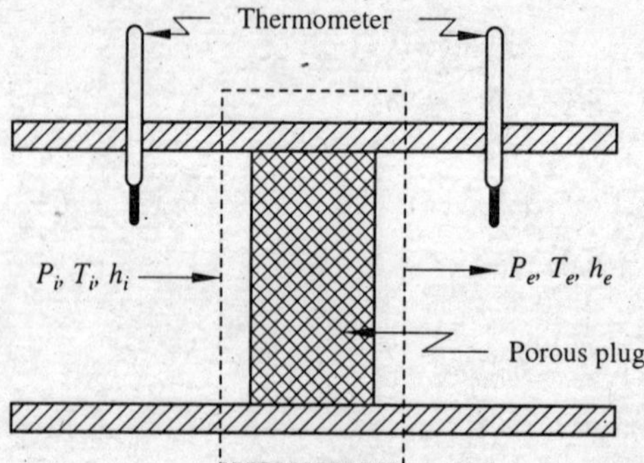

Fig. 6.8 *Sketch for porous plug experiment of Joule-Thompson*

Suppose that a series of Joule–Thompson experiments are conducted on the same gas with a given initial pressure P_i and temperature T_i, but with different downstream pressures P_e; different exit temperatures T_e are observed after expansion. The downstream pressure P_e is altered by controlling the opening of the throttle valve. For this set of experiments, since P_i and T_i are fixed (h_i is fixed), the final set of P_e and T_e correspond to the same enthalpy. A plot of T_e versus P_e yields an isenthalpic curve. The slope of the isenthalpic curve is called the Joule–Thompson coefficient μ_{JT} and is given by:

$$\mu_{JT} = \left(\frac{\partial T}{\partial P}\right)_h \tag{6.38}$$

If experiments are conducted with different P_i and T_i, different isenthalpic curves can be obtained. A family of isenthalpic curves is shown in Fig. 6.9 which is typical of all real gases. The point at which $\mu_{JT} = 0$ is called the *inversion point.* The locus of all inversion points is the *inversion curve.* In the region to the left of the inversion curve $\mu_{JT} > 0$. In the throttling process the downstream pressure P_e is always less than the upstream pressure P_i. Therefore, whenever a real gas, the state of which lies in the region to the left of the inversion curve, is subjected to throttling, the temnerature of the gas decreases. To the right of the inversion curve, $\mu_{JT} < 0$. If the initial state of the gas lies in the region to the right of the inversion curve, the temperature of the gas increases upon throttling.

For almost all gases at ordinary temperatures and pressures, $\mu_{JT} > 0$ and the maximum inversion temperature is above room temperature. The exceptions are hydrogen, helium and neon. For hydrogen the maximum inversion temperature is 200 K and for helium the maximum inversion temperature is 24 K. If hydrogen is throttled at room temperature, the temperature of the gas increases. A leak from the high pressure hydrogen cylinders, or supply mains, may result in the production of high temperatures leading to favorable conditions for an explosion. A knowledge of the throttling process and inversion curve for hydrogen is essential in the safe handling of high pressure cylinders. To produce cooling by throttling, the initial temperature of hydrogen should be reduced below 200 K. This is usually accomplished by cooling with liquid nitrogen. Similarly, in the production of liquid helium by throttling, the initial temperature of helium should be below 24 K. Hence, it is usually cooled with liquid hydrogen prior to throttling. A knowledge of the inversion curve for real gases is of considerable importance in the design of refrigerators and liquefaction equipment.

Suppose an ideal gas is throttled. Since the throttling process is isenthalpic, and for an ideal gas the enthalpy is a function of temperature only, the temperature of the ideal gas does not change. Hence $\mu_{JT} = 0$ for an ideal gas.

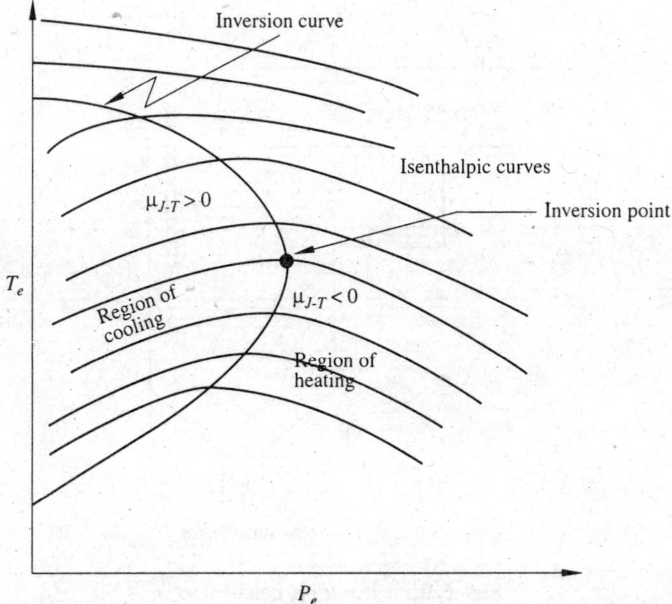

Fig. 6.9 *Inversion and isenthalpic curves of a real gas*

6.5 Applications of Throttling

6.5.1 Throttling calorimeter

A throttling calorimeter, shown in Fig. 6.10, is a device which is used to measure the quality of wet steam. This calorimeter is based on the principle of Joule–Thompson expansion. Part of the wet steam flowing through the mains is throttled to a low pressure, usually atmospheric pressure, in a throttling calorimeter. While the steam is still in the calorimeter, the steam becomes superheated and the temperature of the steam in the calorimeter is measured. Knowing the pressure (usually atmospheric) and temperature of the steam in the calorimeter, its enthalpy can be determined. Since throttling is an isenthalpic process, the enthalpy of steam before throttling is the same as the enthalpy of steam after throttling. Knowing the enthalpy and the pressure of the steam in the mains, its quality can be determined by making use of the relation:

$$h_e = h_i = Xh_g + (1 - X)h_f = h_f + Xh_{fg} \tag{6.39}$$

where, X is the quality of the steam in the main, and h_g and h_f are the enthalpies of saturated vapor and saturated liquid respectively at the pressure P_i.

***Example* 6.3** A throttling calorimeter is used to measure the quality of wet steam in a pipe carrying steam at 10 bar. The pressure and temperature in the calorimeter are measured as 1 bar and 110°C. Determine the quality of steam in the mains.

Solution For steam at 10 bar:

$$h_f = 762.61 \text{ kJ/kg} \quad \text{and} \quad h_g = 2776.2 \text{ kJ/kg}$$

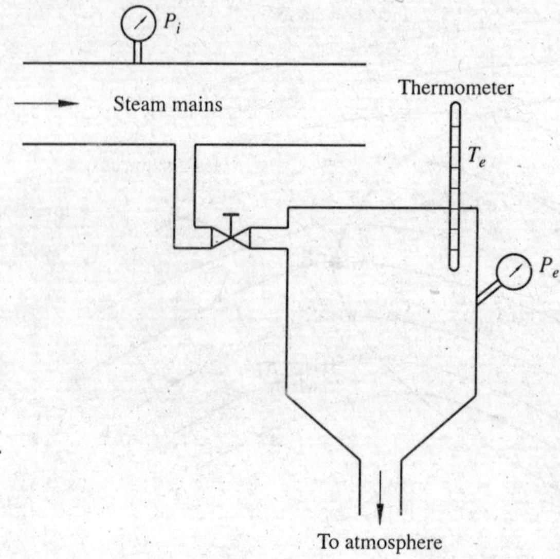

Fig. 6.10 *Throttling calorimeter*

For steam at 1 bar and 110°C:

$$h_e = 2696.12 \text{ kJ/kg}$$

Throttling process is an isenthalpic process. Therefore,

$$h_i = h_e = 2696.12 \text{ kJ/kg}.$$

$$2696.12 = h_g X + (1 - X)h_f = 2776.2X + (1 - X)762.61$$

or, $X = 0.96$

6.5.2 Refrigeration

A simple refrigeration cycle is shown in Fig. 6.11. In a refrigerator, the working fluid (called refrigerant) leaving the evaporator as a mixture of liquid and vapor at state 1 undergoes an adiabatic compression in a compressor and leaves as saturated vapor at state 2. This compressed refrigerant enters a condenser, where it exchanges energy with the ambient atmosphere, and leaves as saturated liquid at high pressure at state 3. Then the liquid refrigerant at high pressure undergoes throttling and leaves as a mixture of liquid and vapor at low pressure at state 4. During throttling, the liquid vaporizes partially and its temperature decreases . The cooled refrigerant then passes through an evaporator, where it absorbs energy from the body to be cooled, and leaves at state 1. Finally, the hot refrigerant vapor goes to the compressor, thus completing a cycle. Applying the first law of thermodynamics to each of the units (control volume), the following relations are obtained.
Compressor (adiabatic):

$$h_2 - h_1 = -\frac{\dot{W}_s}{\dot{m}} \tag{6.40}$$

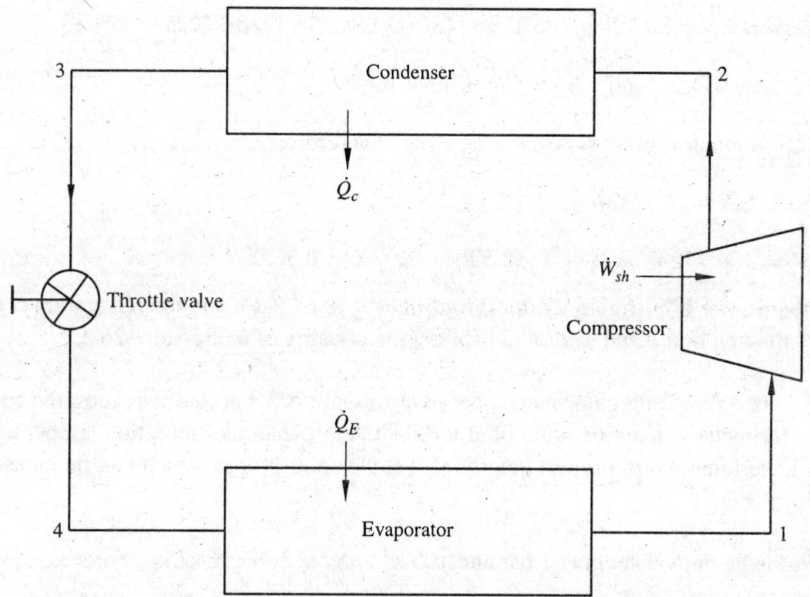

Fig. 6.11 *Schematic diagram of a simple refrigeration cycle*

Condenser (constant pressure cooling):

$$h_3 - h_2 = -\frac{\dot{Q}_c}{\dot{m}} \tag{6.41}$$

Throttling valve :

$$h_4 - h_3 = 0 \tag{6.42}$$

Evaporator :

$$h_1 - h_4 = \frac{\dot{Q}_E}{\dot{m}} \tag{6.43}$$

The Coefficient Of Performance (COP) of the refrigerator is defined as the ratio of the energy absorbed at the evaporator to the energy required to perform the task. That is,

$$\text{COP}_\text{R} = \frac{h_1 - h_4}{h_2 - h_1} \tag{6.44}$$

***Example* 6.4** In a vapor compression refrigerator, Freon-12 leaves the compressor as saturated vapor at 40°C which enters a condenser. The refrigerant leaves the condenser as saturated liquid at 40°C. This liquid refrigerant is throttled to a low pressure. The low pressure refrigerant is −26°C. Determine the pressure to which the refrigerant is throttled and the state of the refrigerant.

Solution Enthalpy of saturated liquid Freon-12 at 40 °C = h_3 = 75.1134 kJ/kg

Saturation pressure of Freon-12 at −26°C = 118.643 kPa. For Freon-12 at −26°C:

$$h_f = 12.5201 \text{ kJ/kg} \quad \text{and} \quad h_g = 176.1723 \text{ kJ/kg}$$

Throttling is an isenthalpic process. Hence, $h_3 = h_4$. Therefore,

$$h_3 = h_4 = h_g X + (1 - X)h_f$$

$$75.1134 = 176.1723X + (1 - X)12.5201 \quad \text{or} \quad X = 0.3825$$

Therefore, the pressure of refrigerant after throttling = 118.643 kPa and the refrigerant is available as a mixture of saturated liquid and saturated vapor, with a quality of 0.3825 at −26°C.

Example 6.5 In a throttling calorimeter, the exit pressure is 1 bar and it is required for the steam to leave the calorimeter at a temperature of at least 5°C more than the saturation temperature for easy observation. Determine the minimum quality of wet steam at 20 bar which can be measured with a throttling calorimeter.

Solution Enthalpy of exit steam at 1 bar and 105°C = h_e = 2686.16 kJ/kg. For steam at 20 bar:

$$h_f = 908.59 \text{ kJ/kg} \quad \text{and} \quad h_g = 2797.2 \text{ kJ/kg}$$

Since throttling is an isenthalpic process, $h_i = h_e$ = 2686.16 kJ/kg. Or

$$2686.16 = h_g X + (1 - X)h_f = 2797.2X + (1 - X)908.59$$

Or $X = 0.9412$. Therefore, the minimum quality, which can be measured is 0.9412.

In a throttling calorimeter, the steam is throttled to a low pressure so that it leaves the calorimeter as superheated steam. Usually, the exit pressure of a throttling calorimeter is the atmospheric pressure (approximately 1 bar). The degree of superheat of the exit steam has to be at least 5°C for convenient measurement. Hence, if the exit steam is at 1 bar and 105°C, the minimum quality of wet steam at 20 bar which can be measured is 0.9412 (see Example 6.5). The minimum quality of wet steam which can be measured with the help of a simple throttling calorimeter, depends on the pressure of the wet steam as shown in Fig. 6.12.

If the quality of wet steam is less than the minimum that can be measured with a simple throttling calorimeter, then a combined separating and throttling calorimeter, shown in Fig. 6.13, is used. In the combined separating and throttling calorimeter, wet steam from the steam main is fed to a separator, where part of the moisture separates out due to sudden change in the direction of flow and the separated condensate is drained out. The separator unit operates at constant pressure. After the removal of the condensate, the quality of wet steam leaving the separator increases. This steam then enters the throttling calorimeter, where it is throttled to yield superheated steam at 1 bar. The process followed by the steam is shown in Fig. 6.14, on a Mollier diagram.

Let

m_1 = mass of wet steam that enters the combined separating and throttling calorimeter at state 1.

m_2 = mass of wet steam that enters the throttling calorimeter at state 2.

m_c = mass of condensate that leaves the separator unit at c.

X_1 = quality of wet steam in the steam mains.

X_2 = quality of wet steam that enters the throttling calorimeter.

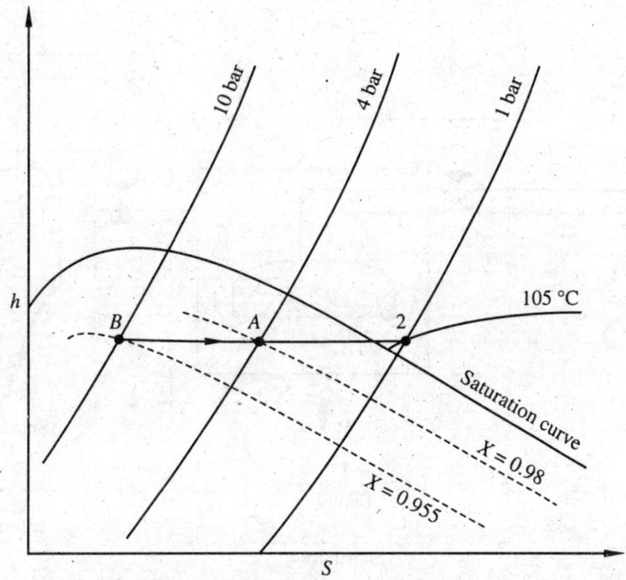

Fig. 6.12 *Mollier diagram indicating the pressure dependence of minimum quality of steam which can be measured with a simple throttling calorimeter*

From material balance, we get the mass of steam throttled m_2 as

$$m_2 = m_1 - m_c$$

$$X_2 = \frac{\text{mass of vapor}}{\text{total mass of wet steam fed to throttling calorimeter}}$$

$$X_2 = \frac{\text{mass of vapor}}{m_2} = \frac{\text{mass of vapor}}{m_1 - m_c}$$

Or mass of vapor = $(m_1 - m_c)X_2$ and the quality X_1 is given by:

$$X_1 = \frac{(m_1 - m_c)X_2}{m_1} = X_2\left(1 - \frac{m_c}{m_1}\right) \tag{6.45}$$

***Example* 6.6** Wet steam at 30 bar is fed to a combined separating and throttling calorimeter. The steam leaves the throttling calorimeter at 1 bar and 105°C and 30 g of condensate is collected for every kg of wet steam fed to the combined unit. Determine the quality of wet steam.

Solution See Fig. 6.13 for a sketch of the combined separating and throttling calorimeter. Steam at 1 bar and 105°C :

$$h_3 = 2686.16 \text{ kJ/kg}$$

Steam at 30 bar :

$$h_f = 1008.4 \text{ kJ/kg} \quad \text{and} \quad h_g = 2802.3 \text{ kJ/kg}$$

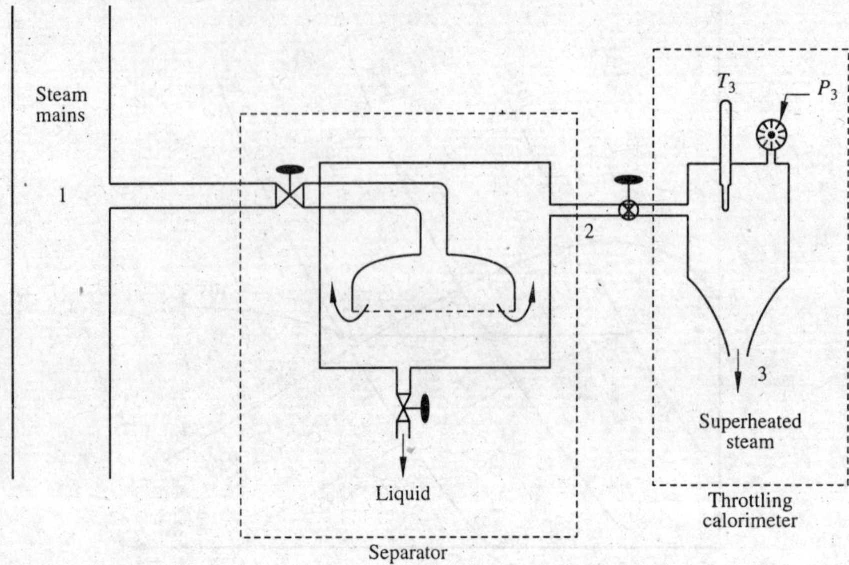

Fig. 6.13 *Schematic of a combined separating and throttling calorimeter*

$$h_3 = h_2 = h_g X_2 + (1 - X_2) h_f$$

$$2686.16 = 2802.3 X_2 + (1 - X_2) 1008.4 \quad \text{or} \quad X_2 = 0.935$$

We know that quality X_1 of wet steam is given by:

$$X_1 = X_2 \left(1 - \frac{m_c}{m_1}\right) = 0.935(1 - 0.03) = 0.907$$

6.5.3 Liquefaction of gases

Throttling process is an isenthalpic process and hence when an ideal gas is throttled, its temperature does not change. However, when a real gas is throttled the temperature of the gas either decreases or increases depending on whether the Joule–Thompson coefficient is positive or negative. When a real gas is throttled in the region where the Joule–Thompson coefficient is positive, there will be a reduction in temperature and hence a throttling process can be employed to produce temperatures low enough for liquefaction of gases. Linde made use of the throttling process in the liquefaction of gases. A schematic diagram of the Linde process is shown in Fig. 6.15. The gas is compressed in a multistage compressor with intercooling to a high pressure. The high pressure gas is cooled in a counter-current heat exchanger with cold gas coming out of the separator. The cooled high pressure gas is then throttled to a low pressure. Because of the throttling process some of the gas is liquefied. The gas is separated from the liquid and the cold gas is recirculated through a counter-current heat exchanger in which the incoming gas is cooled.

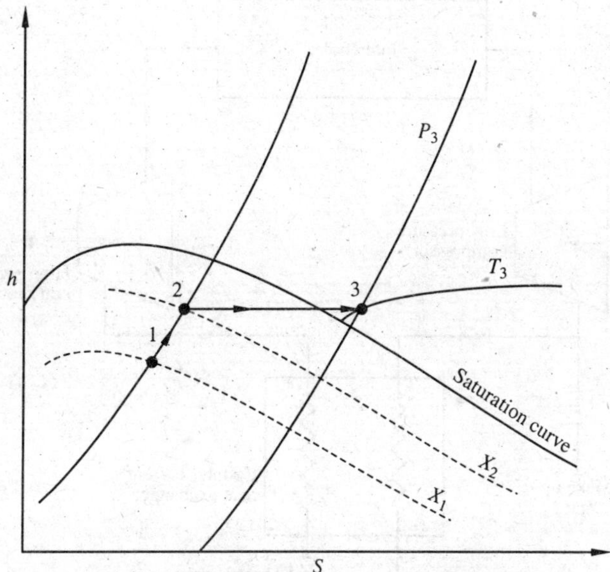

Fig. 6.14 *Mollier diagram showing the process in a combined separating and throttling calorimeter*

6.6 Transient Flow Processes

So far in the discussion on the first law for flow processes, emphasis was placed on steady state flow conditions. It was assumed that the properties of the fluid at any given location are invariant with respect to time, and the fluid flows in and out of the control volume at a constant rate. Many processes of interest to the engineer involve unsteady flow which cannot be analyzed with the above assumptions. While charging and discharging a gas cylinder, the properties of the fluid inside the cylinder vary with time. The mass flow rates into and out of the cylinder are not the same and there is an accumulation of mass inside the cylinder. Such flow processes are transient and the analysis of these transient flow processes is illustrated below with the help of a specific example.

6.6.1 Charging of a cylinder

Consider the case of charging a cylinder from the supply mains. With the passage of time, the mass and state of the gas inside the cylinder change continuously. This problem is solved by the control mass analysis first, and later by control volume analysis.

Control mass analysis

In the control mass analysis, the system is selected such that it contains a definite quantity of matter throughout the charging process. Accordingly, the system boundary is selected as shown by the dotted line in Fig. 6.16. This boundary is variable in size. In the final state, the boundary of the system is the same as the boundary of the cylinder itself. During the process of filling, the system boundary

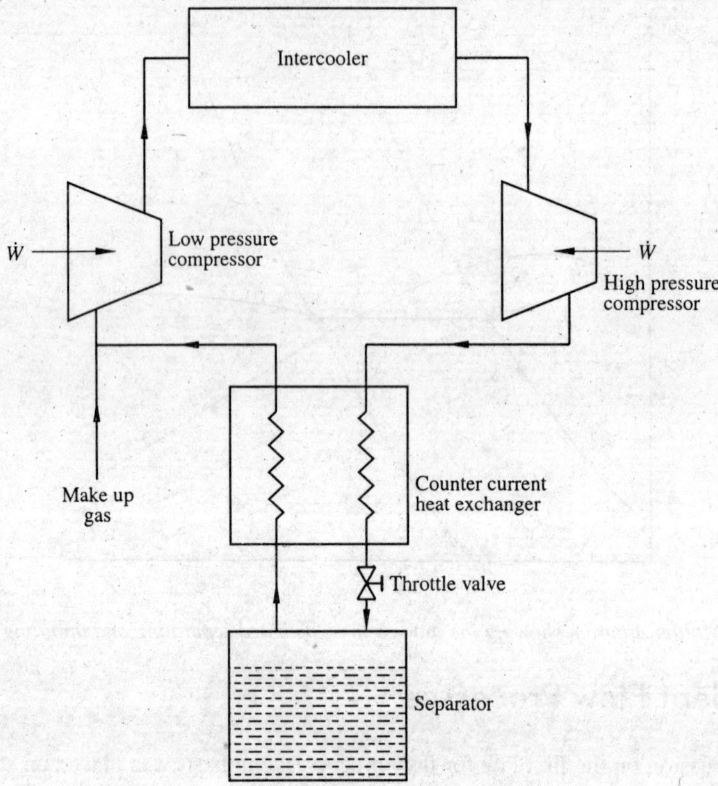

Fig. 6.15 *Schematic diagram for liquefaction of gases*

containing the total mass collapses continuously, and at the end of the charging operation it coincides with the cylinder walls.

Let

m_0 = mass of gas in the cylinder at the start

m_f = mass of gas in the cylinder at the end of the charging operation

P = pressure of gas in the supply mains

v = specific volume of gas in the supply mains

u_0 = specific internal energy of the gas in the cylinder at the start

u_f = specific internal energy of the gas in the cylinder at the end of charging operation

u = specific internal energy of gas in the supply mains

$\mathcal{V}$ = Velocity of gas in the supply mains, and

Q = amount of energy transferred as heat into the system.

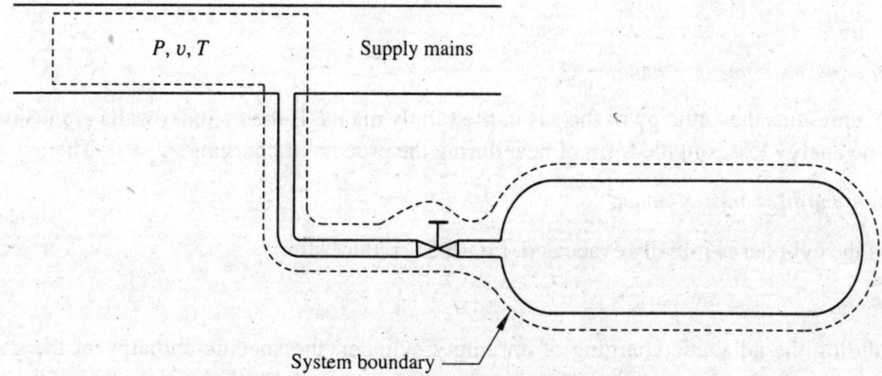

Fig. 6.16 *Control mass for charging a cylinder*

The mass of gas in the supply mains which enters the cylinder during the charging operation is given by:

$$m_f - m_0 \tag{6.46}$$

Energy of the system before the start of charging process equals the sum of the energy of the gas inside the cylinder before charging operation and energy of the gas in the supply main before the charging operation. That is,

$$E_0 = m_0 u_0 + (m_f - m_0)(u + \mathcal{V}^2/2) \tag{6.47}$$

(where the potential energy term is ignored).

Energy of the system at the end of the charging operation $= E_f$

$$E_f = m_f u_f \tag{6.48}$$

For the mass $(m_f - m_0)$ to enter the cylinder, the system boundary is to be collapsed reducing its volume by $(m_f - m_0)v$. This is accomplished by the pressure P acting on this mass in the supply mains. Therefore, work is being done on the system in forcing the gas into the cylinder.

Work done in forcing the gas into the cylinder is given by:

$$W = -(m_f - m_0)Pv \tag{6.49}$$

Application of the first law of thermodynamics to the system gives

$$E_f - E_0 = Q - W \tag{6.50}$$

$$m_f u_f - \left(m_0 u_0 + (m_f - m_0)\left(u + \frac{\mathcal{V}^2}{2} \right) \right) = Q + (m_f - m_0)Pv \tag{6.51}$$

Neglecting the kinetic energy change (which is a legitimate assumption for the process of charging of a cylinder), Eq. (6.51) reduces to:

$$m_f u_f - m_0 u_0 = Q + (m_f - m_0)(u + Pv) = Q + (m_f - m_0)h$$

or

$$(m_f - m_0)h_i = m_f u_f - m_0 u_0 - Q \tag{6.52}$$

where, h_i represents the enthalpy of the gas in the supply mains. If the cylinder walls are insulated and there are no energy losses in the form of heat during the process of charging, $Q = 0$. Then

$$(m_f - m_0)h_i = m_f u_f - m_0 u_0 \tag{6.53}$$

Further, if the cylinder is initially evacuated, Eq. (6.53) reduces to

$$h_i = u_f \tag{6.54}$$

That is, during the adiabatic charging of an empty cylinder, the specific enthalpy of the gas in the supply mains is equal to the specific internal energy of the gas in the cylinder at the end of the charging operation.

If the gas is treated as ideal ($h = C_P T$ and $u_f = C_v T_f$), then Eq. (6.54) gives

$$T_f = \gamma T \tag{6.55}$$

where T_f is the final temperature of the gas in the cylinder and γ is the ratio of specific heats.

Control volume analysis

Consider the cylinder as the control volume as shown in Fig. 6.17. The first law expression for a control volume is given by:

$$\dot{m}_e \left(h_e + \frac{\mathcal{V}_e^2}{2} + gZ_e \right) - \dot{m}_i \left(h_i + \frac{\mathcal{V}_i^2}{2} + gZ_i \right) = \dot{Q} - \dot{W}_s - \frac{dE}{dt}$$

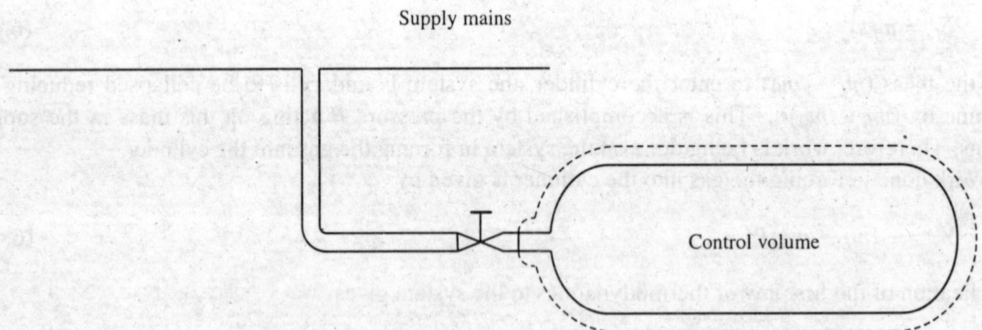

Fig. 6.17 *Control volume for charging of a cylinder*

While charging a cylinder from the supply mains, no gas leaves the control volume, and $\dot{m}_e = 0$. No shaft work is done; $\dot{W}_s = 0$. Let us also ignore the changes in the kinetic energy and potential energy of the gas during the charging operation. Then Eq. (6.15) reduces to

$$-m_i h_i = \dot{Q} - \frac{dE}{dt} \tag{6.56}$$

or

$$-\dot{m}_i h_i = \dot{Q} - \frac{d(mu)}{dt} \tag{6.57}$$

where m denotes the mass of gas in the control volume and u denotes the specific internal energy of the gas inside the control volume at time t. It is assumed that the state of the gas at any instant of time inside the control volume is uniform, though the state of the gas varies with time. That is, the fact that the system within the cylinder is a continuous system is ignored, and the gas is treated as if it is homogeneous. The properties of this homogeneous system are taken to be the average values of the real continuous system. It is also assumed that the state of the gas in the supply mains is constant. Expression (6.57) can be integrated from the start of the charging operation ($t = 0$) to the end of the charging operation ($t = t$), to obtain

$$-h_i \int_0^t \dot{m}_i dt = \int_0^t \dot{Q} dt - \int_0^t \frac{d(mu)}{dt} \tag{6.58}$$

The principle of conservation of mass for the charging process reduces to:

$$\frac{dm}{dt} = \dot{m}_i \tag{6.59}$$

Integrating Eq. (6.59) over time t gives the change of mass in the control volume during the charging operation. That is,

$$\int_0^t \frac{dm}{dt} dt = \int_0^t \dot{m}_i dt = m_f - m_0 \tag{6.60}$$

In equation (6.58)

$$\int_0^t \frac{d(mu)}{dt} = m_f u_f - m_0 u_0 \tag{6.61}$$

and

$$\int_0^t \dot{Q} dt = Q \tag{6.62}$$

Substituting these values, Eq. (6.58) reduces to:

$$(m_f - m_0) h_i = m_f u_f - m_0 u_0 - Q \tag{6.63}$$

Equation (6.63) is the same as Eq. (6.52) that was obtained by a control mass analysis.

Example 6.7 Air at 50 bar and 300 K is flowing through a pipeline. An evacuated and insulated cylinder of volume 0.1 m^3 is connected to the pipeline through a valve. The valve is opened and the cylinder is filled with air till the pressure in the cylinder reaches 50 bar and then the valve is closed. Assuming that air behaves like an ideal gas with $\gamma = 1.4$, determine the temperature of air in the cylinder at the end of filling operation and the mass of air that is filled in the cylinder.

Solution We know that the temperature of an ideal gas at the end of charging an evacuated tank is given by

$$T_f = \gamma T = 1.4 \times 300 = 420 \text{ K}$$

Moles of air that enters the cylinder, N is given by

$$N = \frac{PV}{RT} = \frac{50 \times 10^5 \times 0.1}{8.314 \times 420} = 143.19$$

Molar mass of air = 28.97 kg/kmol

Mass of air that is filled in the cylinder $= 143.19 \times 28.97 \times 10^{-3} = 4.15$ kg

Example 6.8 Steam at a pressure of 20 bar and 500°C is flowing in a pipe. An evacuated tank is connected to this pipe through a valve. The valve is opened and the tank is filled with steam until the pressure is 20 bar, and then the valve is closed. The process takes place adiabatically and the kinetic energy and potential energy changes can be assumed to be negligible. Determine the temperature of the steam in the tank at the end of the filling operation.

Solution The first law for a transient flow process is given by the expression

$$(m_f - m_0)h_i = m_f u_f - m_0 u_0$$

When the tank is initially empty, $m_0 = 0$. Therefore,

$$m_f h_i = m_f u_f \quad \text{or} \quad h_i = u_f$$

From the superheated steam tables, the enthalpy of steam at 20 bar and 500°C is, $h_i = 3467.3$ kJ/kg.

Hence, the final state of steam is: $P = 20$ bar ; $u_f = 3467.3$ kJ/kg.

From a knowledge of P and u, the state of the steam can be determined. Let us calculate u at 600°C and 700°C. From superheated steam tables, we read:

$P = 20$ bar; $T = 600$°C: $h = 3689.2$ kJ/kg and $v = 0.1995$ m^3/kg.

$$u = 3689.2 \times 10^3 - 20 \times 10^5 \times 0.1995 = 3290.2 \text{ kJ/kg.}$$

$P = 20$ bar; $T = 700$°C : $h = 3916.5$ kJ/kg and $v = 0.2232$ m^3/kg.

$$u = 3916.5 \times 10^3 - 20 \times 10^5 \times 0.2232 = 3470.1 \text{ kJ/kg}$$

By interpolation, we find $u = 3467.3$ kJ/kg corresponding to $T = 698.4$°C.

Example 6.9 A 2 m^3 tank with perfectly insulated walls contains saturated steam at a pressure of 1.0 MPa. This tank is connected through a valve to a steam line through which flows superheated steam at a pressure of 4 MPa and 400°C . The valve is opened and steam is admitted rapidly into the tank until the pressure in the tank is 4.0 MPa. Estimate the mass of steam that enters the tank.

Solution For transient filling of a tank, the first law of thermodynamics gives

$$(m_f - m_0)h_i = m_f u_f - m_0 u_0 \tag{A}$$

For saturated steam at 1.0 MPa: $v_0 = 0.1943$ m^3/kg and

$$u_0 = 2776.2 - 0.1943 \times 10^3 = 2581.9 \text{ kJ/kg}$$

Initial mass of steam in the tank

$$= m_0 = \frac{V}{v_0} = \frac{2}{0.1943} = 10.293 \text{ kg}$$

For superheated steam at 4 MPa and 400°C; $h_i = 3215.7$ kJ/kg. The volume of the tank = 2 m^3. Therefore,

$$m_f v_f = 2 \tag{B}$$

Equations (A) and (B) can be solved to determine m_f. Equation (A) can be rearranged as

$$m_f(h_i - u_f) = m_0(h_i - u_0)$$

Therefore,

$$m_f(3215.7 - u_f) = 10.293(3215.7 - 2581.9) = 6523.7 \text{ kJ} \tag{C}$$

Equations. (B) and (C) can be solved by a trial and error procedure. Assuming the final temperature of steam T_f to be equal to 425°C, from steam tables,

$$h_f = 3273.03 \text{ kJ/kg}, \quad v_f = 0.0766 \text{ m}^3\text{/kg}$$

$$u_f = 3273.03 - 4 \times 10^3(0.0766) = 2966.63 \text{ kJ/kg}$$

Substituting in Eq. (B), we get

$$m_f = \frac{2}{0.0766} = 26.11 \text{ kg.}$$

Substituting the above value for m_f in equation (C), we get

$$m_f(3215.7 - u_f) = 26.11(3215.7 - 2966.63) = 6503.22 \text{ kJ}$$

The assumed value for T_f gives 6503.22 kJ for the left-hand side of Eq. (C), which is close to the value 6523.7 kJ on the right-hand side of Eq. (C). Therefore, the final temperature = 425°C. Then the mass of steam that enters the tank is given by:

$$m_f - m_i = 26.11 - 10.293 = 15.817 \text{ kg.}$$

***Example* 6.10** An ideal gas ($C_P = 7R/2$ and $C_v = 5R/2$) is flowing in a pipe line at 10 bar and 600 K. An insulated tank containing a piston and spring is connected to the line as shown in Fig. 6.18. Initially the tank is empty and the spring is in its natural position and exerts no force on the piston. When the valve is opened, the gas enters the tank and the force exerted by the spring is proportional to the distance moved. Determine the temperature of the gas inside the tank, when the pressure reaches a value of 10 bar.

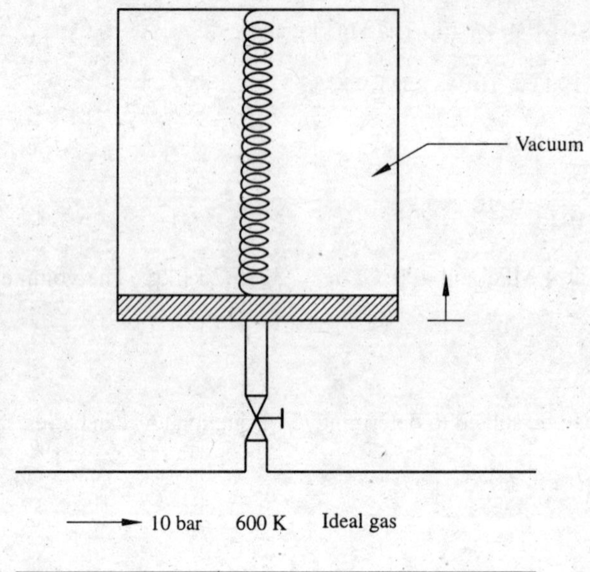

Fig. 6.18 *Sketch for Example 6.10. Control volume is the region below the piston*

Solution Control volume: The space below the piston.

In all the previous problems, the control volume was selected as a definite region in space. This problem is different from the others because the control volume size is not constant. Moreover, in place of the shaft work, the work done by the control volume is "spring work".

The force F_s exerted by the spring is given by

$$F_s = Kx = \frac{KV}{A}$$

where A is the cross-sectional area of piston and K is the spring constant,

$$P = \frac{F}{A} = \frac{KV}{A^2}$$

$$W = \int P dV = \int \frac{KV dV}{A^2} = \frac{KV_f^2}{2A^2} = \frac{P_f V_f}{2}$$

The first law of thermodynamics for any control volume is:

$$\dot{m}_e \left(h_e + \frac{\mathcal{V}_e^2}{2} + gZ_e \right) - \dot{m}_i \left(h_i + \frac{\mathcal{V}_i^2}{2} + gZ_i \right) = \dot{Q} - \dot{W}_s - \frac{dE}{dt}$$

For the problem at hand:

(a) Ignore the kinetic energy and potential energy terms.
(b) $\dot{Q} = 0$; since the tank is insulated.
(c) $\dot{m}_e = 0$, because the gas is not escaping from the control volume. Then,

$$-\dot{m}_i h_i = -\dot{W}_s - \frac{dE}{dt}$$

$$h_i \int \dot{m}_i dt = W + E_f - E_i$$

or

$$h_i m_f = \frac{P_f V_f}{2} + m_f u_f$$

$m_i = 0$ and $E_i = 0$ (since the tank is initially empty)

This can also be written as

$$h_i N_f = \frac{N_f P_f v_f}{2} + N_f u_f \tag{A}$$

where, N_f is the mole number of the ideal gas in the tank at the end of the filling operation. h_i is the molar enthalpy of the gas in the line and u_f is the molar internal energy of the gas inside the tank at the end of the filling operation.

Equation (A) can be written as:

$$C_P T = \frac{RT_f}{2} + C_v T_f$$

$$\frac{7RT}{2} = \frac{RT_f}{2} + \frac{5RT_f}{2} = \frac{6RT_f}{2}$$

Therefore,

$$T_f = \frac{7T}{6} = \frac{7 \times 600}{6} = 700\ K$$

Alternative choice of control volume

Consider the tank with the piston and spring as the control volume. The first law expression for the control volume is:

$$\dot{m}_e \left(h_e + \frac{V_e^2}{2} + gZ_e \right) - \dot{m}_i \left(h_i + \frac{V_i^2}{2} + gZ_i \right) = \dot{Q} - \dot{W}_s - \frac{dE}{dt}$$

There is no shaft work delivered by the control volume ($\dot{W}_s = 0$). Hence the first law expression reduces to:

$$h_i \int \dot{m}_i dt = m_f h_i = E_f$$

The term E_f consists of the internal energy of the gas ($m_f u_f$) and the energy stored in the spring ($P_f V_f/2$). Therefore,

$$m_f h_i = m_f u_f + \frac{P_f V_f}{2}$$

This relation is identical with the expression obtained in the first case. Therefore, the choice of the control volume by the analyst depends on convenience—the result, however, remains the same.

6.6.2 Discharging of a cylinder

Consider a cylinder containing a gas at high pressure. If the valve is opened, the gas inside the cylinder rushes out into the surroundings. During the process of discharging a tank, the rate at which the gas escapes from the cylinder, its condition and the condition of the gas inside the cylinder vary with time. The solution of this transient flow problem is illustrated below.

Control volume analysis

Consider the cylinder shown in Fig. 6.19 as the control volume.

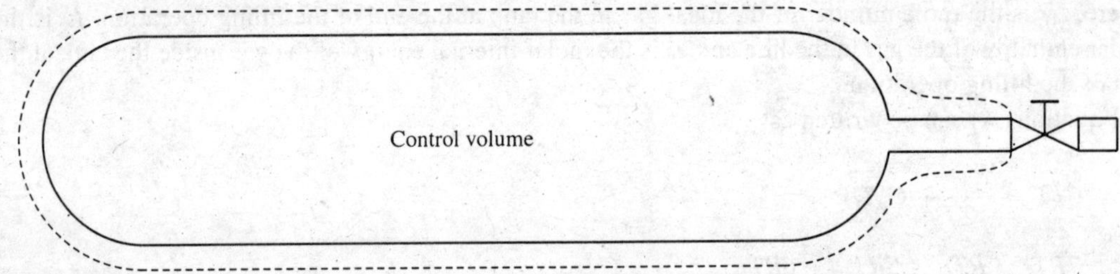

Fig. 6.19 *Control volume for discharging of a cylinder*

Let

P_0 = pressure of gas inside the cylinder at time zero.

T_0 = temperature of the gas in the cylinder at time zero.

u_0 = specific internal energy of the gas in the cylinder at time zero

$P_f,\ T_f,\ u_f$ denote the corresponding quantities of the gas left in the cylinder at the end of the discharging operation.

$\mathcal{V}$ = Velocity of the escaping gas.

The general first law expression for any control volume is given by

$$\dot{m}_e\left(h_e + \frac{\mathcal{V}_e^2}{2} + gZ_e\right) - \dot{m}_i\left(h_i + \frac{\mathcal{V}_i^2}{2} + gZ_i\right) = \dot{Q} - \dot{W}_s - \frac{dE}{dt}$$

In the discharging of a cylinder, no material flows into the cylinder ($\dot{m}_i = 0$), no shaft work is done ($\dot{W}_s = 0$), and the potential energy changes can be ignored. Then

$$\dot{m}_e(h_e + \mathcal{V}_e^2/2) = \dot{Q} - \frac{dE}{dt} \tag{6.64}$$

For the discharging process, the conservation of mass gives:

$$\frac{dm}{dt} = -\dot{m}_e \tag{6.65}$$

Integrating the above equation over time t gives:

$$\int_0^t \frac{dm}{dt}dt = -\int_0^t \dot{m}_e dt = m_f - m_0 \tag{6.66}$$

Integrate Eq. (6.64) over time t, the time required for discharging the cylinder to obtain:

$$\int_0^t \dot{m}_e \left(h_e + \frac{\mathcal{V}_e^2}{2}\right) dt = \int_0^t \dot{Q}dt - \int_0^t \frac{dE}{dt}dt \tag{6.67}$$

or

$$\int_0^t \dot{m}_e \left(h_e + \frac{\mathcal{V}_e^2}{2}\right) dt = (m_0 - m_f)\left(h + \frac{\mathcal{V}^2}{2}\right) \tag{6.68}$$

In evaluating this, it is assumed that the gas escapes with a uniform velocity $\mathcal{V}$ and specific enthalpy h.

$$\int_0^t \dot{Q}dt = Q$$

$$\int_0^t \frac{dE}{dt}dt = \int_0^t \frac{d(mu)}{dt}dt = m_f u_f - m_0 u_0 \tag{6.69}$$

Substituting Eqs. (6.62), (6.68) and (6.69) in Eq. (6.67), we get

$$(m_0 - m_f)(h + \mathcal{V}^2/2) = Q + m_0 u_0 - m_f u_f \tag{6.70}$$

Control mass analysis

In the control mass analysis, attention is focused on a definite quantity of matter enclosed by a boundary. Hence, the total mass must remain constant at all times. Accordingly a boundary surface is selected such that the total quantity of matter is contained within it at all times. Therefore, draw an imaginary envelope enclosing the cylinder and the gas that is escaping from the cylinder as shown in Fig. 6.20. Initial energy of the system is given by

$$E_0 = m_0 u_0 \tag{6.71}$$

The final energy of the system is equal to the sum of the energy of the gas left in the tank and the energy of the gas that has escaped from the tank.

Final energy of the system, E_f is given by:

$$E_f = m_f u_f + (m_0 - m_f)\left(u + \frac{\mathcal{V}^2}{2}\right) \tag{6.72}$$

where, u is the specific internal energy of the gas which has left the tank. Here also, it is assumed that the gas escapes with a uniform velocity $\mathcal{V}$, specific internal energy u, and specific volume v. Since the gas escapes from the tank, the change in the volume of the system is associated with the gas outside the tank. External to the tank, the volume of the gas changes from zero to $(m_0 - m_f)v$.

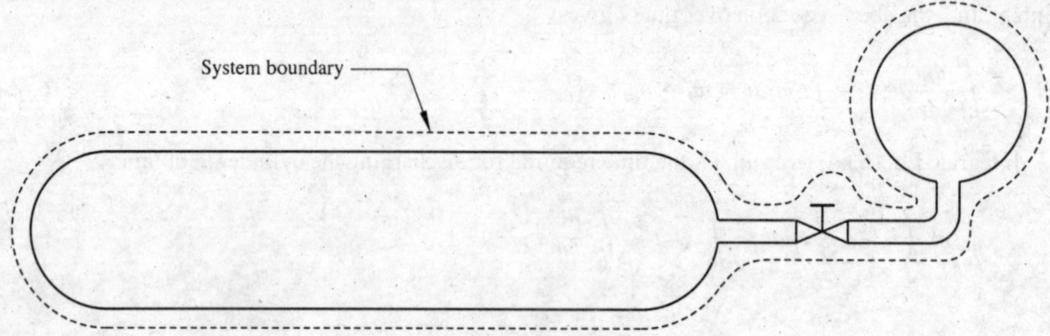

Fig. 6.20 *Control mass for discharging of a cylinder*

Work done W by the gas is given by:

$$W = P\{(m_0 - m_f)v - 0\} \tag{6.73}$$

where P is the pressure at which the gas escapes from the cylinder.

The first law for a control mass is given by

$$E_f - E_0 = Q - W$$

Substituting for E_0, E_f and W from Eqs. (6.71)–(6.73) in the above equation, we get

$$m_f u_f + (m_0 - m_f)\left(u + \frac{V^2}{2}\right) - m_0 u_0 = Q - Pv(m_0 - m_f)$$

or

$$(m_0 - m_f)\left(h + \frac{V^2}{2}\right) = Q + m_0 u_0 - m_f u_f \tag{6.74}$$

This expression is identical to the expression derived with the help of the control volume analysis.

6.7 Case of an Ideal Gas

It is shown below that when an ideal gas escapes from an insulated cylinder, the gas that remains in the tank can be considered to have undergone a reversible adiabatic expansion.

Consider a small amount of gas (confined in an imaginary envelope) in the cylinder as the system. When the gas from the cylinder escapes, the boundary of the system changes continuously and eventually occupies the entire volume of the cylinder. That is, the gas which is left in the cylinder does not exchange energy in the form of heat with the surroundings. It undergoes an adiabatic expansion. To prove that the gas has undergone a reversible adiabatic expansion, let us choose the control volume as shown by the dotted line in Fig. 6.21.

The general expression for the control volume is given by:

$$\dot{m}_e\left(h_e + \frac{V_e^2}{2} + gZ_e\right) - \dot{m}_i\left(h_i + \frac{V_i^2}{2} + gZ_i\right) = \dot{Q} - \dot{W}_s - \frac{dE}{dt} \tag{6.15}$$

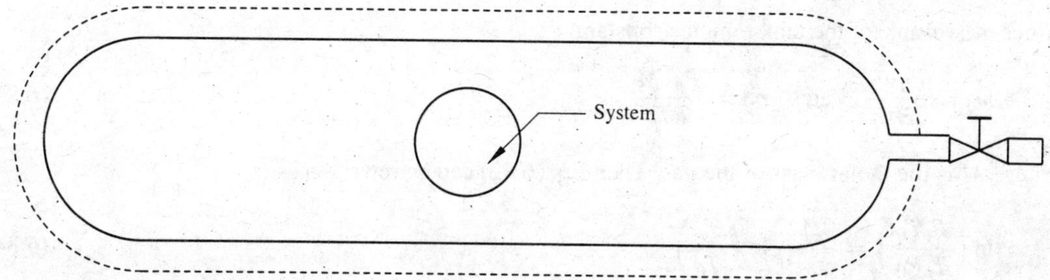

Fig. 6.21 *Control volume for adiabatic expansion of an ideal gas*

In this particular case of tank discharge $\dot{Q} = 0$, since the tank is insulated. As no mass flows into the control volume, $\dot{m}_i = 0$. There is no shaft work ($\dot{W}_s = 0$) delivered by the control volume. Ignore the changes in the kinetic and the potential energy terms. Under these conditions Eq. (6.15) reduces to:

$$\dot{m}_e h_e = -\frac{dE}{dt} = -\frac{d(mu)}{dt} \tag{6.75}$$

From the principle of conservation of mass we get,

$$\dot{m}_e = -\frac{dm}{dt}$$

Therefore,

$$-h_e\frac{dm}{dt} = -\frac{d(mu)}{dt} \tag{6.76}$$

For an ideal gas:

$$h = C_P T \quad \text{and} \quad u = C_v T \tag{6.77}$$

Then Eq. (6.76) can be rewritten as:

$$C_P T\frac{dm}{dt} = \frac{d(mC_v T)}{dt} \tag{6.78}$$

(It is assumed that the temperature of the gas which escapes is the same as that of the gas that remains in the tank at any instant of time.)

$$T\gamma\frac{dm}{dt} = m\frac{dT}{dt} + T\frac{dm}{dt} \tag{6.79}$$

$$(\gamma - 1)T\frac{dm}{dt} = m\frac{dT}{dt} \tag{6.80}$$

$$\frac{(\gamma - 1)dm}{mdt} = \frac{dT}{Tdt} \tag{6.81}$$

Integrating over time t, we get:

$$(\gamma - 1)\ln\left(\frac{m_f}{m_0}\right) = \ln\left(\frac{T_f}{T_0}\right) \tag{6.82}$$

Since the volume of the tank remains constant:

$$m_0 = \frac{P_0 VM}{RT_0} \quad \text{and} \quad m_f = \frac{P_f VM}{RT_f} \tag{6.83}$$

where, M is the molar mass of the gas. Then Eq. (6.82) can be rewritten as:

$$\ln\left(\frac{P_f T_0}{T_f P_0}\right) = \frac{1}{\gamma - 1} \ln\left(\frac{T_f}{T_0}\right) \tag{6.84}$$

or

$$\ln\left(\frac{P_f}{P_0}\right) = \frac{\gamma}{\gamma - 1} \ln\left(\frac{T_f}{T_0}\right) \tag{6.85}$$

or

$$\frac{T_f}{T_0} = \left(\frac{P_f}{P_0}\right)^{\frac{\gamma-1}{\gamma}} \tag{6.86}$$

Equation (6.86) is identical with Eq. (5.47), which is valid for an ideal gas undergoing reversible adiabatic process. Therefore, the process of discharge is reversible as far as the change in the state of the gas that remains in the tank is concerned. Hence, the gas that remains in the tank can be considered to have undergone a reversible adiabatic expansion from the initial pressure P_0 to the final pressure P_f.

***Example* 6.11** An insulated gas cylinder of volume 0.1 m^3 contains air (an ideal gas) at 50 bar and 300 K. The valve of the cylinder is opened allowing the air to escape till the air pressure in the cylinder reduces to 30 bar. Determine the temperature of air left in the cylinder and the mass of air that has escaped from the cylinder.

Solution We know that when an ideal gas discharges from an insulated tank, the gas remaining in the tank can be considered to have undergone a reversible adiabatic expansion from P_0, T_0 to P_f and T_f. That is,

$$\frac{T_f}{T_0} = \left(\frac{P_f}{P_0}\right)^{\frac{\gamma-1}{\gamma}}$$

$$\frac{T_f}{300} = \left(\frac{30}{50}\right)^{\frac{0.4}{1.4}} = 0.864$$

or

$$T_f = 0.864 \times 300 = 259.2 \text{ K}$$

Initial mass of air in cylinder,

$$m_0 = MN_0 = \frac{MP_0 V}{RT_0}$$

Final mass of air in cylinder,

$$m_f = MN_f = \frac{MP_f V}{RT_f}$$

where, N_0 = initial moles of air in cylinder

N_f = final moles of air in cylinder

M = molar mass of air

V = volume of cylinder

Mass of air that escapes from the cylinder is given by

$$m_0 - m_f = \frac{MV}{R}\left(\frac{P_0}{T_0} - \frac{P_f}{T_f}\right)$$

$$= \frac{28.97 \times 10^{-3} \times 0.1}{8.314}\left(\frac{50 \times 10^5}{300} - \frac{30 \times 10^5}{259.2}\right) = 1.775 \text{ kg}$$

***Example* 6.12** The insulated tank shown in Fig. 6.22 contains 1000 kg of wet steam at 30 bar. Initially the tank contains 50 percent liquid and 50 percent vapor by volume. 100 kg of moisture-free vapor is slowly withdrawn from the top of the tank. Estimate the final pressure in the tank.

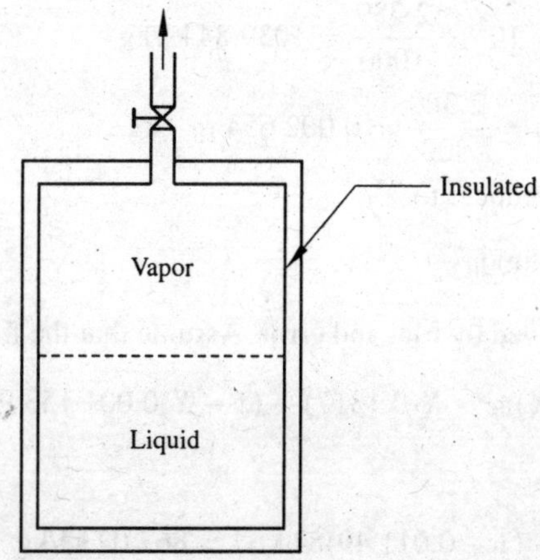

Fig. 6.22 *Sketch for Example 6.12*

Solution Choose the tank as the control volume . While saturated vapor is withdrawn from the tank, the enthalpy of the leaving vapor varies continuously with time. If the change in the enthalpy of the escaping steam is small, it can be assumed that the enthalpy of the escaping steam is the average of the enthalpies of the escaping steam at time $t = 0$ and at time $t = t$. Applying the control volume equation for transient flow (neglecting the velocity of the escaping steam) we get:

$$(m_1 - m_2)h_e = m_1 u_1 - m_2 u_2 \qquad \text{(A)}$$

where, h_e = average enthalpy of the escaping steam,

$m_1,\ m_2$ = initial and final mass of steam in the tank,

$u_1,\ u_2$ = initial and final specific internal energy of steam in the tank.

Let V = Volume of the tank.

$$\text{Volume of liquid} = \frac{V}{2} = m_f v_f = m_f(0.001\,216\,3)$$

$$\text{Volume of vapor} = \frac{V}{2} = m_g v_g = m_g(0.066\,63)$$

$$m_f + m_g = 1000 = \frac{V}{2}\left(\frac{1}{0.001\,216\,3} + \frac{1}{0.066\,63}\right)$$

Hence, $V = 2.389\ \text{m}^3$

Mass of liquid $= m_f = 982.07$ kg, and mass of vapor $= m_g = 17.93$ kg

$$m_1 = m_f + m_g = 1000\ \text{kg}$$

$$m_2 = 900\ \text{kg}$$

$$X_1 = \frac{17.93}{1000} = 0.017\,93$$

$$h_1 = 2802.3 \times 0.017\,93 + (1 - 0.017\,93) \times 1008.4 = 1040.56\ \text{kJ/kg}$$

$$u_1 = 1040.56 \times 10^3 - 3 \times 10^5 \times \frac{2.389}{1000} = 1039.84\ \text{kJ/kg}$$

$$\text{Final specific volume} = v_2 = \frac{2.389}{900} = 0.002\,654\ \text{m}^3/\text{kg} \qquad \text{(B)}$$

Substituting the values, Eq. (A) reduces to

$$100h_e = 1.03984 \times 10^6 - 900u_2 \qquad \text{(C)}$$

Equations (B) and (C) can be solved by trial and error. Assume that the final pressure = 15 bar. Then,

$$0.002\,654 = Xv_g + (1 - X)v_f = X(0.1317) + (1 - X)0.001\,153\,9$$

Therefore, $X = 0.011\,49$

$$h_2 = 2789.8 \times 0.011\,49 + (1 - 0.011\,49)844.67 = 867.02\ \text{kJ/kg}$$

$$u_2 = 867.02 \times 10^3 - 15 \times 10^5 \times 0.002\,654 = 863.04\ \text{kJ/kg}$$

or

$$h_e = \frac{h_{g1} + h_{g2}}{2} = \frac{2802.3 + 2789.8}{2} = 2796.05\ \text{kJ/kg.}$$

Substituting in (C), we obtain LHS = 279605 kJ and RHS = 263104 kJ

Now, let us assume the final pressure = 14 bar. Then

$$h_e = \frac{2802.3 + 2787.8}{2} = 2795.05\ \text{kJ/kg}$$

LHS of Eq. (C) = 279505 kJ

$$0.002\,654 = 0.1407X + (1 - X)0.001\,148\,9 \quad \text{or} \quad X = 0.0108$$

$$h_2 = 2787.8 \times 0.0108 + (1 - 0.0108)830.08 = 851.22\ \text{kJ/kg}$$

$$u_2 = 851.22 \times 10^3 - 14 \times 10^5 \times 0.002\,654 = 847.5\ \text{kJ/kg}$$

RHS of Eq. (C) = $1.03984 \times 10^6 - 900 \times 847.5 = 277090$ kJ

Since the values of LHS and RHS of Eq. (C) are not in agreement, let us revise the guess pressure as $P = 13.8$ bar. At 13.8 bar:

$$h_g = 2787.32\ \text{kJ/kg} \quad v_f = 1.1478 \times 10^{-3}\ \text{m}^3/\text{kg}$$

$$h_f = 827\ \text{kJ/kg} \quad v_g = 0.1428\ \text{m}^3/\text{kg}$$

$$h_e = \frac{2802.3 + 2787.32}{2} = 2794.81\ \text{kJ/kg}$$

LHS of Eq. (C) = 279 481 kJ

$$0.002\,654 = 0.1428X + (1 - X)1.1478 \times 10^{-3} \quad \text{or} \quad X = 0.0106$$

$$h_2 = 2787.32 \times 0.0106 + (1 - 0.0106)827 = 847.78\ \text{kJ/kg}$$

$$u_2 = 847.78 \times 10^3 - 13.8 \times 10^5 \times 0.002\,654 = 844.12\ \text{kJ/kg}$$

RHS of Eq. (C) = $1.03984 \times 10^6 - 900 \times 844.12 = 280\,132$ kJ. The LHS and RHS of Eq. (C) differ by about 0.2%. Let us say that the guess pressure satisfies Eq. (C). Therefore, the final pressure is 13.8 bar.

6.8 Chemically Reacting Systems

The applications of the first law of thermodynamics to chemically reacting systems is discussed in this section. Since the combustion of fuels is of great significance in many power producing devices, let us study the combustion of fuels first. So far in our discussion attention was focused on the thermodynamic analysis of nonreactive systems or systems of fixed composition. Situations do arise where an engineer or a scientist has to analyze chemically reactive systems since a large number of industries convert raw materials into products of greater value through chemical reactions. The synthesis of ammonia from nitrogen and hydrogen, and the production of methanol from carbon monoxide and hydrogen are some of the examples of industrially important reactions. Therefore, it is necessary to understand the problems associated with the design and operation of chemical reactors.

6.8.1 Fuels and combustion

Most of the fuels contain carbon and hydrogen and they are available as solid, liquid or gas. The proximate analysis of coal gives the mass per cent of volatile matter, water vapor, carbon, ash, etc. in the fuel. The chemical analysis known as the *ultimate analysis* of solid fuels is usually given in terms of the mass per cent of each constituent element. A typical solid fuel is coal the major constituent of which is carbon. In addition to carbon, coal contains hydrogen and small amounts of sulphur and nitrogen.

Most liquid fuels are a mixture of several hydrocarbons. For example, petroleum oil is a complex mixture of a large number of hydrocarbons and is obtained from crude oil. The commonly used liquid fuels are gasoline, kerosene, diesel oil and fuel oil. In dealing with the combustion of fuels, usually a liquid fuel is treated as a single hydrocarbon with an empirical formula C_xH_y even though it is a mixture of several hydrocarbons. For example, gasoline is treated as octane C_8H_{18} and diesel oil is treated as dodecane, $C_{12}H_{26}$.

The widely used gaseous fuels are natural gas (mainly methane), coal gas or coke oven gas (primarily a mixture of methane and hydrogen), water gas (essentially a mixture of CO and H_2), and producer gas (a mixture of N_2, CO and H_2). Usually, the analysis of the gaseous fuels is given in terms of mole per cent or volume per cent of the constituent chemical species.

The combustion of fuel involves the oxidation of the constituents of a fuel and the process can be represented by a chemical reaction. In combustion process, the mass of each element remains the same even though the reactants disappear and new products are formed. Consider the combustion of methane in presence of oxygen. When complete combustion is achieved all the carbon atoms in the fuel are converted into CO_2 and all the hydrogen atoms are converted into H_2O. Then the combustion reaction of methane can be represented by the chemical reaction.

$$CH_4 + 2O_2 \longrightarrow CO_2 + 2H_2O \tag{6.87}$$

Equation (6.87) shows that 2 moles of O_2 are required for the complete combustion of one mole CH_4. That is, for complete combustion, a minimum of two moles of O_2 are required for every mole of CH_4.

In most of the combustion processes, the required oxygen is not supplied as pure oxygen but is supplied as air. Air contains 21 mol per cent O_2 and 79 mol per cent N_2 (ignoring the presence of small quantities of argon, CO_2 and others). When air is used as a source of oxygen, equation (6.87) can be rewritten as

$$CH_4 + 2O_2 + 2\left(\frac{0.79}{0.21}\right)N_2 \longrightarrow CO_2 + 2H_2O + 2\left(\frac{0.79}{0.21}\right)N_2$$

or,

$$CH_4 + 2O_2 + 2(3.7619)N_2 \longrightarrow CO_2 + 2H_2O + 7.524N_2 \tag{6.88}$$

The minimum amount of air which supplies the required amount of oxygen for complete combustion of a fuel is called the stoichiometric or theoretical air. Usually, the reaction does not proceed to completion unless the amount of air supplied is more than the theoretical amount. The amount of air actually supplied is expressed in terms of per cent theoretical air. Sometimes the amount of air actually supplied is expressed in terms of per cent excess air. The excess air denotes the amount of air supplied over and above the theoretical air requirement. Thus, 120 per cent theoretical air means the actual amount of air supplied is 1.2 times the theoretical air, or 20 per cent excess air. An important parameter in the study of combustion of fuels is the air–fuel ratio, the ratio of the mass of the air to the mass of the fuel. The air–fuel ratio is also expressed on a molal basis. It is not desirable to supply more air than what is theoretically required as it lowers the temperature of the combustion products. However, the reaction does not go to completion unless excess air is supplied. For several combustion processes 20 per cent excess air is usually considered as desirable.

***Example* 6.13** Determine, (a) the stoichiometric or theoretical air–fuel ratio for the combustion of C_5H_{12} and (b) the composition of the combustion products when C_5H_{12} is burned with 150 per cent theoretical air.

Solution

(a)

$$C_5H_{12} + 8O_2 + 8(3.7619)N_2 \longrightarrow 5CO_2 + 6H_2O + 30.095N_2$$

Air–fuel ratio on mole basis

$$= \frac{(8 + 30.095)}{1} = 38.095 \, \frac{\text{mol air}}{\text{mol fuel}}$$

Molar mass of C_5H_{12} = 72; Molar mass of air = 28.97

$$\text{Air–fuel ratio} = \frac{38.095 \times 28.97}{72} = 15.328 \, \frac{\text{kg air}}{\text{kg fuel}}$$

(b)

$$C_5H_{12} + (1.5 \times 8)O_2 + (1.5 \times 8 \times 3.7619)N_2 \longrightarrow 5CO_2 + 6H_2O + 45.14N_2 + 4O_2$$

Total mole number of the product = 5 + 6 + 45.14 + 4 = 60.14.

The composition of the products expressed in mole fractions is as follows:

$$CO_2 = \frac{5}{60.14} = 0.0832 \quad H_2O = \frac{6}{60.14} = 0.0998$$

$$N_2 = \frac{45.14}{60.14} = 0.7506 \quad O_2 = \frac{4}{60.14} = 0.0664$$

***Example* 6.14** A particular fuel gas has the following composition expressed in mole fractions.

$CO = 0.26$; $H_2 = 0.12$; $CO_2 = 0.07$ and $N_2 = 0.55$

Calculate (a) theoretical air–fuel ratio on a molal basis (b) if the fuel gas is burned with 20 per cent excess air, calculate the product composition assuming complete combustion of the fuel.

Solution (a) Consider 1 mol of fuel gas. The amount of oxygen required for the complete combustion of the fuel is estimated from the following chemical reactions.

$$0.26CO + 0.13O_2 \longrightarrow 0.26CO_2$$

$$0.12H_2 + 0.06O_2 \longrightarrow 0.12H_2O$$

O_2 required = 0.13 + 0.06 = 0.19 mol. The complete combustion of the fuel is represented by the equation

$$(0.26CO + 0.12H_2 + 0.07CO_2 + 0.55N_2) + (0.19O_2 + 0.19 \times 3.7619N_2)$$

$$\longrightarrow 0.33CO_2 + 0.12H_2O + 1.2648N_2$$

$$\text{Theoretical air–fuel ratio} = \frac{0.19 + 0.19 \times 3.7619}{1} = 0.9048\ \frac{\text{mole air}}{\text{mole fuel}}$$

(b) If 20 per cent excess air is used, then the air–fuel ratio is 1.2. The combustion equation for 20 per cent excess air is

$$0.26CO + 0.12H_2 + 0.07CO_2 + 0.55N_2 + 1.2 \times 0.19O_2 + 1.2 \times 0.19 \times 3.7619N_2$$

$$\longrightarrow 0.33CO_2 + 0.12H_2O + 0.038O_2 + 1.408N_2$$

The product analysis is

Product	N	$y_i = \frac{N_i}{\sum N_i}$
CO_2	0.33	0.174
H_2O	0.12	0.063
O_2	0.038	0.020
N_2	1.408	0.743
	$\sum N_i = 1.896$	

The actual amount of air supplied in a combustion process can be determined from a knowledge of the composition of the combustion products. The combustion products can be easily analyzed with an Orsat apparatus. In the Orsat analysis, a known volume of the products of combustion is bubbled successively through several reagent solutions—20 per cent solution of potassium hydroxide which absorbs CO_2, an alkaline solution of pyrogallic acid which absorbs O_2, an acidic solution of cuprous chloride which absorbs CO—and the volume of the unabsorbed gas is measured each time. The decrease in volume represents the partial volume of the component which has been absorbed. The composition of the gas is then estimated with a knowledge of the partial volumes of the constituents. The analysis is carried out at a temperature which is usually lower than the dew point temperature of the combustion products, and hence the water vapor is not present in the combustion products analyzed by the Orsat apparatus. Thus the Orsat analysis gives the analysis on dry basis. The combustion products contain CO_2, CO, O_2 and N_2 only, and hence the unabsorbed gas is treated as N_2. The method of estimating the actual air–fuel ratio from Orsat analysis of a flue gas is illustrated in the following example.

***Example* 6.15** Pentane, C_5H_{12} is burned with air and the resulting flue gas has the following volumetric analysis on dry basis:

$$CO_2 = 9.27\%,\ \ CO = 2.31\%,\ \ O_2 = 4.86\% \text{ and } N_2 = 83.56\%$$

Determine the actual air–fuel ratio for the combustion process and the excess air.

Solution Consider 100 mol of dry combustion products as the basis for calculation. The combustion equation can be written as:

$$xC_5H_{12} + yO_2 + 3.7619yN_2 \longrightarrow 9.27CO_2 + 2.31CO + zH_2O + 4.86O_2 + 83.56N_2$$

Nitrogen mass balance:

$$3.7619y = 83.56 \text{ or } y = 22.21$$

Carbon balance:

$$5x = 9.27 + 2.31 \text{ or } x = 2.32$$

Hydrogen balance:

$$12x = 2z \text{ or } z = 13.92$$

Actual air–fuel ratio is given by:

$$= \frac{22.21 + 83.56}{2.32} = 45.591 \frac{\text{mol air}}{\text{mol fuel}}$$

Theoretical air–fuel ratio = 38.095 mole air/mole fuel (see Example 6.13)

$$\text{Percent theoretical air} = \frac{45.591}{38.095} \times 100 = 119.68$$

That is 19.68 per cent excess air is used for combustion.

***Example* 6.16** A hydrocarbon fuel of unknown composition is burned with air. The volumetric analysis on dry basis of the resulting flue gas is 9.27% CO_2, 2.31% CO, 4.86% O_2 and 83.56% N_2. Determine, (a) the composition of the fuel on mass basis. (b) the air–fuel ratio on mass basis, and (c) the percent excess or deficit air.

Solution (a) Basis: 100 mol of dry combustion products. The combustion equation is

$$C_xH_y + zO_2 + 3.7619zN_2 \longrightarrow 9.27CO_2 + 2.31CO + aH_2O + 4.86O_2 + 83.56N_2$$

Nitrogen balance:

$$3.7619z = 83.56 \text{ or } z = 22.21$$

Carbon balance :

$$x = 9.27 + 2.31 = 11.58$$

Oxygen balance:

$$z = 22.21 = 9.27 + \frac{2.31}{2} + \frac{a}{2} + 4.86 \text{ or } a = 13.85$$

Hydrogen balance:

$$y = 2a \text{ or } y = 27.7$$

The composition (overall) of the fuel on mass basis is

$$\text{Carbon} = \frac{11.58 \times 12}{11.58 \times 12 + 27.7} = 83.38\%$$

$$\text{Hydrogen} = \frac{27.7}{11.58 \times 12 + 27.7} = 16.62\%$$

(b) Air–fuel ratio

$$= \frac{(22.21 + 83.56) \times 28.97}{11.58 \times 12 + 27.7} = 18.386\ \frac{\text{kg air}}{\text{kg fuel}}$$

(c) The combustion equation can be written as

$$C_{11.58}H_{27.7} + 25.43O_2 + 25.43 \times 3.7619N_2 \longrightarrow 11.58CO_2 + 13.85H_2O + 95.665N_2$$

$$\text{Theoretical air–fuel ratio} = \frac{(25.43 + 95.665) \times 28.97}{11.58 \times 12 + 27.7}$$

$$= 21.05\ \frac{\text{kg air}}{\text{kg fuel}}$$

$$\text{Per cent theoretical air} = \frac{18.386}{21.05} = 87.34\%$$

Percent deficit air $= 12.66\%$

6.8.2 The standard enthalpy (heat) of reaction

Chemical reactions are generally accompanied by the evolution or absorption of energy in the form of heat because the energy associated with the products is different from the energy associated with the reactants. The difference in energy arises because of the difference in the structures of the product molecules and the reactants. Consider the following reaction as an example,

$$H_2 + Cl_2 \longrightarrow 2HCl$$

The reaction involves the breaking of H – H and Cl – Cl bonds, and the formation of H – Cl bonds. In a diatomic molecule if the atoms are separated and at rest, the energy possessed by the molecule is equal to zero. As the atoms come close to each other, the attraction force between the nuclei increases. If the atoms are very close they repel each other. The potential energy possessed by the molecule due to attraction and repulsion forces appears as shown in Fig. 6.23. The potential energy of the molecule reaches a minimum value at r_e where r_e is the equilibrium internuclear distance. This state corresponds to the equilibrium state of the molecule. The depth of the potential well, D_e, represents the energy of the chemical bond. A diatomic molecule has one vibrational degree of freedom and the energy of the vibrational mode is given by

$$\varepsilon_n = \left(n + \frac{1}{2}\right) h\nu \quad n = 0, 1, 2, \ldots$$

where ν is the vibrational frequency of the molecule. Suppose a diatomic molecule is held at a temperature of zero kelvin, it still possesses the zero point vibrational energy which is equal to $\frac{1}{2}h\nu$. Therefore, the energy required to dissociate the molecule, that is to break the chemical bond is given by

$$D_0 = D_e - \frac{1}{2}h\nu$$

where D_o is the dissociation energy.

The dissociation energies of H_2, Cl_2 and HCl are 429.78; 239.3 and 427.47 kJ/mol, respectively. Suppose the reaction between H_2 and Cl_2 is carried out at a temperature of zero kelvin, 429.78 kJ energy is required to break H – H bonds in one mol of H_2 and 239.3 kJ energy is required to break the Cl – Cl bonds in one mol Cl_2 and 854.94 kJ energy is liberated while two mol HCl is formed. That is, the net amount of energy liberated as the reaction proceeds is (854.94 – 429.78 – 239.3) = 185.86 kJ. Since the products possess less energy than the reactants, the energy change associated with the chemical reaction is –185.86 kJ, or –92.93 kJ/mol of HCl formed. That is, for every mol of HCl formed 92.93 kJ energy is liberated as the chemical reaction proceeds, and this reaction is termed exothermic.

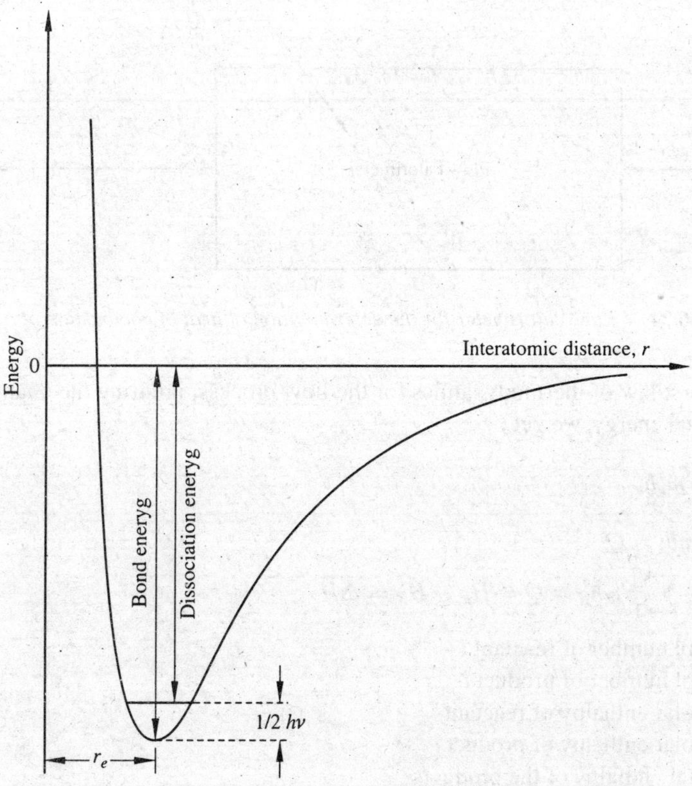

Fig. 6.23 *Potential energy diagram for a diatomic molecule*

The energy change associated with the above reaction at 298.15 K is experimentally found to be –184.62 kJ. The difference arises due to the fact that at 298.15 K all the reactant and product molecules are not in the ground state and they are distributed among the permissible vibrational energy levels according to the Maxwell–Boltzmann distribution law.

The amount of energy liberated or consumed during a chemical reaction depends upon the temperatures of the reactants and products. There are a vast number of chemical reactions and these reactions can be carried out in many different ways. Therefore, it is impossible to compile the data regarding the energy changes associated with chemical reactions carried out under different conditions. Hence, it is necessary to devise a method by which the energy changes associated with chemical reactions can be estimated from a limited amount of data which is based on a few reactions carried out in a standard way. In order to have a consistent basis for the treatment of energy changes associated with chemical reactions, it is convenient to measure the energy changes when the products have been restored to the temperature of the reactants. The energy change associated with the combustion reactions is usually measured with a flow calorimeter described below.

Consider the steady flow of hydrogen and oxygen through a flow-calorimeter (control volume) shown in Fig. 6.24. The flow calorimeter is surrounded by a water jacket through which cooling water flows. Suppose the reactants enter at 298.15 K and 1 bar and the products leave the control volume at 298.15 K and 1 bar.

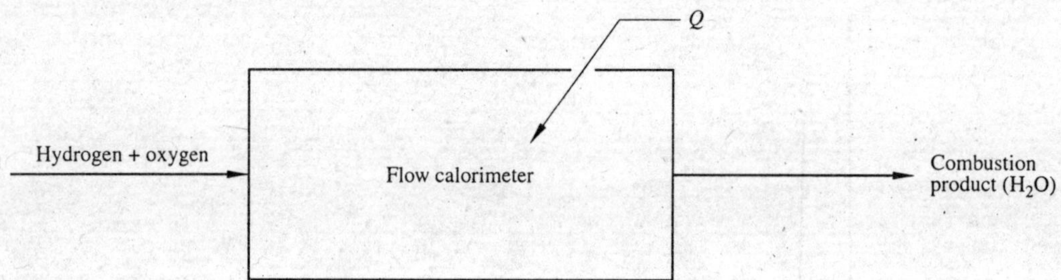

Fig. 6.24 *Flow calorimeter for the determination of heat of combustion of a fuel*

Applying the first law of thermodynamics for the flow process, ignoring the changes in the kinetic energy and potential energy, we get

$$\dot{m}_i h_i + \dot{Q} = \dot{m}_e h_e \tag{6.89}$$

$$\text{or } \dot{m}_e h_e - \dot{m}_i h_i = \dot{Q}$$

$$\sum N_{iP} h_P - \sum N_{iR} h_R = Q = H_P - H_R = \Delta H$$

where, N_{iR} = mol number of reactant i
N_{iP} = mol number of product i
h_{iR} = molar enthalpy of reactant i
h_{iP} = molar enthalpy of product i
H_P = total enthalpy of the products
H_R = total enthalpy of the reactants
Q = amount of energy transferred as heat to the surroundings.

Thus we find that the energy liberated by the combustion process is equal to Q which can be measured.

For purposes of compiling data it is necessary to define a standard heat of reaction. ΔH is a change in the property which does not depend upon the path but depends only upon the states of the reactants and products. Hence, it is necessary to specify the state of the reactants as well as products. In the thermodynamic analysis of systems of constant composition the choice of the reference state can be arbitrarily made and it does not affect the energy change of a system associated with a given process. For example, in the thermodynamic analysis of processes, tables of thermodynamic properties for steam and Freon-12 are used. In the steam tables, the saturated liquid water at its triple point is selected as the reference state and its internal energy is assigned a value of zero. In the Freon-12 and ammonia tables, saturated liquid at −40°C is selected as the reference state. Such a choice of the reference state is inadequate to analyze chemically reactive systems since the reactant molecules disappear while the product molecules form as the reaction proceeds. Though the reactant and product molecules are not conserved, atoms constituting these species are conserved. Hence, it is necessary to think of a chemically reactive system as a system of elements constituting the reactants and products. Then it is appropriate to consider a system of these atoms which are separated from each other and at rest as the reference state, to which the energy is assigned a zero value. However, such a choice of the reference state is not very convenient as it involves a large number of calculations, which take into account the distribution of molecules among the energy levels. For convenience in the analysis of reactive systems, the elements in their standard states are chosen as reference states and their enthalpy assigned a value of zero.

In chemical thermodynamics the following standard states are used. For gases, the standard state is taken as pure gas in the ideal gas state at 1 bar pressure. For liquids and solids, the standard state is taken as pure liquid or solid at 1 bar pressure. Carbon is found in the form of graphite and diamond. Whenever a substance is found in allotropic forms, its most stable form is selected as the standard state. Since graphite is more stable than diamond, the standard state for carbon is taken as graphite. The choice of the temperature for the standard state is entirely arbitrary. For convenience of tabulation of thermodynamic properties, the standard state temperature is selected as 298.15 K.

Whenever heat of reaction is specified for a particular chemical reaction, it applies for the stoichiometric coefficients indicated in the reaction. Usually, the standard heat of a reaction is represented by ΔH^0_{298} where the superscript zero indicates that the reactants and products are in their standard states and the subscript 298 indicates the temperature (298.15 K) at which the reaction is carried out. If each of the stoichiometric coefficients in the reaction equation is multiplied by a factor K, then the heat of reaction is also multiplied by the same constant K. For example, the reaction between H_2 and Cl_2 may be written as:

$$H_2(g) + Cl_2(g) \longrightarrow 2HCl(g) \qquad \Delta H^0_{298} = -184.62\text{ kJ}$$

or

$$\frac{1}{2}H_2(g) + \frac{1}{2}Cl_2(g) \longrightarrow HCl(g) \qquad \Delta H^0_{298} = -92.31\text{ kJ}$$

6.8.3 The standard enthalpy of formation

The standard heat of any reaction can be estimated and tabulated in principle. However, in view of the large number of chemical reactions it is impossible to do so. The standard heat of a reaction can be estimated from a knowledge of the standard heats of formation of the reactants and products involved in the reaction. Consider the reactions,

$$\frac{1}{2}H_2(g) + \frac{1}{2}Cl_2(g) \longrightarrow HCl(g)$$

$$C(s) + O_2(g) \longrightarrow CO_2(g)$$

$$\frac{3}{2}H_2(g) + \frac{1}{2}N_2(g) \longrightarrow NH_3(g)$$

where C(s) indicates that carbon is in the solid state, i.e., graphite form. The above reactions indicates that one mole of single compound (HCl, CO_2, NH_3) is formed from its constituent elements. Such reactions are called formation reactions. In other words, *a formation reaction is a reaction which forms one mol of a single compound from the constituent elements.* Thus, the formation reaction for ethanol can be written as

$$2C(s) + 3H_2(g) + \frac{1}{2}O_2(g) \longrightarrow C_2H_5OH(g)$$

Ethanol can also be manufactured by the gas phase hydration of ethylene according to the reaction,

$$C_2H_4(g) + H_2O(g) \longrightarrow C_2H_5OH(g)$$

However, the above reaction is not a formation reaction since ethanol is formed from some other compounds and not from its constituent elements. *The energy change associated with the formation reaction when the elements and the compound are in their standard states is called the standard enthalpy (heat) of formation.* Usually, the standard heat of formation is represented by ΔH^0_{f298} where f denotes that the reaction is a formation reaction and the superscript zero denotes that it is a standard value. The subscript 298 indicates that the standard state temperature is approximately 298 K (to be exact 298.15 K). The standard heats of formation of some common compounds are presented in Appendix H.

Suppose it is required to determine the standard heat of reaction at 298 K for the industrially important water gas shift reaction.

$$CO_2(g) + H_2(g) \longrightarrow CO(g) + H_2O(g) \tag{6.90}$$

Since enthalpy is a state function, its change does not depend upon the process by which the reactants are converted into products, but depends only upon the states of the reactants and products. Therefore, it is possible to replace reaction (6.90) by the following set of reactions the sum of which is equivalent to reaction (6.90)

$$CO_2(g) \longrightarrow C(s) + O_2(g) \qquad \Delta H^0_{298} = 393.8 \text{ kJ}$$

$$C(s) + \frac{1}{2}O_2(g) \longrightarrow CO(g) \qquad \Delta H^0_{298} = -110.6 \text{ kJ}$$

$$H_2 + \frac{1}{2}O_2(g) \longrightarrow H_2O(g) \qquad \Delta H^0_{298} = -242 \text{ kJ}$$

$$CO_2(g) + H_2(g) \longrightarrow CO(g) + H_2O(g) \qquad \Delta H^0_{298} = 41.2 \text{ kJ}$$

The first of the above reactions is the negative of the formation reaction for CO_2 while the other reactions represent the formation reactions of CO and H_2O respectively. The desired reaction (6.90) is expressed as the difference between the sum of the formation reactions of products and the sum of the formation reactions of reactants. Similarly, any chemical reaction can be expressed as an algebraic sum of the formation reactions of compounds involved in the reactions. The change in the enthalpy or the standard heat of reaction, ΔH^0_{298} is then given by the algebraic sum of the standard heats of formation of the compounds involved in the reaction. The standard heat of reaction ΔH^0_{298} for the water gas shift reaction is given by

$$\begin{aligned}\Delta H^0_{298} &= \Delta H^0_{298}\{CO(g)\} + \Delta H^0_{298}\{H_2O(g)\} - \Delta H^0_{298}\{CO_2(g)\}\\ &= -242 - 110.6 - (-393.8) = 41.2 \text{ kJ}\end{aligned}$$

In general, the standard heat of reaction ΔH^0_{298} can be estimated by the relation,

$$\Delta H^0_{298} = \sum_P \nu_P \Delta H^0_{298}(P) - \sum_R \nu_R \Delta H^0_{298}(R) \tag{6.91}$$

where, ν_p, ν_R = stoichiometric coefficients associated with the product (P) and reactant (R), respectively. The first term on the right-hand side of equation (6.91) is summed over all the product P, while the second term is summed over all the reactants R.

Now let us consider a general method of representing a chemical reaction. The combustion of methane in the presence of oxygen occurs according to the reaction

$$CH_4 + 2O_2 \longrightarrow CO_2 + 2H_2O$$

This can be rewritten as

$$CO_2 + 2H_2O - CH_4 - 2O_2 = 0 \tag{6.92}$$

The stoichiometric coefficients of the chemical species involved in the reaction (6.92) are $\nu_{CH_4} = -1$; $\nu_{O_2} = -2$; $\nu_{CO_2} = 1$ and $\nu_{H_2O} = 2$, where a positive sign is assigned to the stoichiometric coefficients associated with the products and a negative sign is assigned to the stoichiometric coefficients associated with the reactants since the reactants are consumed and products are formed as the chemical reaction proceeds. With this sign convention for the stoichiometric coefficients of the chemical species involved in the reaction, reaction (6.92) can be expressed as

$$\sum \nu_i A_i = 0 \tag{6.93}$$

where, A_i denotes the chemical species i. Adopting the same sign convention, the standard heat of reaction for (6.92) can be expressed as

$$\Delta H^0_{298} = \sum_i \nu_i \Delta H^0_{298}(A_i) \tag{6.94}$$

***Example* 6.17** Calculate the standard heat ΔH^0_{298} for the following reaction

$$C_5H_{12}(g) + 8O_2(g) \longrightarrow 5CO_2(g) + 6H_2O(g)$$

Solution The standard heats of formation ΔH^0_{f298} of the chemical species involved in the reaction are:

$$CO_2(g): \quad -393.8\ \text{kJ}: \quad H_2O: \quad -242\ \text{kJ}$$

$$C_5H_{12}(g): \quad -146.5\ \text{kJ}; \quad O_2(g): \quad 0\ \text{kJ}$$

The following combination of the formation reactions gives the desired reaction

$$C_5H_{12}(g) \longrightarrow 5C(s) + 6H_2(g) \qquad \Delta H^0_{298} = 146.5\ \text{kJ}$$

$$5\{C(s) + O_2(g) \longrightarrow CO_2(g)\} \qquad \Delta H^0_{298} = 5(-393.8)\ \text{kJ}$$

$$6\{H_2(g) + \frac{1}{2}O_2(g) \longrightarrow H_2O(g)\} \qquad \Delta H^0_{298} = 6(-242)\ \text{kJ}$$

$$C_5H_{12}(g) + 8O_2(g) \longrightarrow 5CO_2(g) + 6H_2O(g) \qquad \Delta H^0_{298} = -3274.5\ \text{kJ}$$

Alternatively, the standard heat of reaction ΔH^0_{298} can be estimated directly by applying the relation (6.94) as shown below.

$$\begin{aligned}\Delta H^0_{298} &= 5\left[\Delta H^0_{f298}\left(CO_2(g)\right)\right] + 6\left[\Delta H^0_{f298}\left(H_2O(g)\right)\right] - \Delta H^0_{f298}\left(C_5H_{12}(g)\right)\\ &= 5(-393.8) + 6(-242) - (-146.5) = -3274.5\ \text{kJ}\end{aligned}$$

***Example* 6.18** Estimate the standard heat ΔH^0_{298} of the following reaction:

$$C_5H_{12}(g) + 8O_2(g) \longrightarrow 5CO_2(g) + 6H_2O(l)$$

Solution From steam tables, we read the following values at 25°C:

$h_g = 2547.3$ kJ/kg $\quad h_f = 104.77$ kJ/kg

Therefore

$$H_2O(l) \longrightarrow H_2O(g) \quad \Delta H^0_{298} = 2547.3 - 104.77 = 2442.53\ \text{kJ/kg} = 43.966\ \text{kJ/mol}$$

In Example 6.17, the standard heat ΔH^0_{298} of the reaction

$$C_5H_{12}(g) + 8CO_2(g) \longrightarrow 5CO_2(g) + 6H_2O(g)$$

is determined as $\Delta H^0_{298} = -3274.5$ kJ. To determine ΔH^0_{298} for the reaction in which H_2O is present in the liquid phase, the following physical process involving the phase change from vapor to liquid can be added.

$$6\,(H_2O(g) \longrightarrow H_2O(l)) \qquad \Delta H^0_{298} = 6(-43.966)\ \text{kJ}$$

Hence the required value is

$$= \Delta H^0_{298} = -3274.5 + 6(-43.966) = -3538.296\ \text{kJ}$$

6.8.4 Standard enthalpy of combustion

The formation reaction for n-pentane is given

$$5C(s) + 6H_2(g) \longrightarrow C_5H_{12}(g)$$

This reaction cannot be carried out in practice to determine the standard heat of formation of n-pentane. However, this reaction can be seen as a sum of the following reactions:

$$\begin{array}{rl}
5C(s) + 5O_2(g) \longrightarrow 5CO_2(g) & \Delta H^0_{298} = -1969 \text{ kJ} \\
6H_2(g) + 3O_2(g) \longrightarrow 6H_2O(g) & \Delta H^0_{298} = -1452 \text{ kJ} \\
5CO_2(g) + 6H_2O(g) \longrightarrow C_5H_{12}(g) + 8O_2(g) & \Delta H^0_{298} = 3274.5 \text{ kJ} \\
\hline
5C(s) + 6H_2(g) \longrightarrow C_5H_{12}(g) & \Delta H^0_{298} = -146.5 \text{ kJ}
\end{array}$$

Thus, the standard heat of formation of n-pentane is -146.5 kJ. In the above set of reactions, the first two reactions are the formation reactions which can be carried out. The third reaction is the opposite of the combustion of C_5H_{12}. A *combustion reaction is defined as a reaction between the element or compound and oxygen to form specified combustion products.* For hydrocarbons, the combustion products are CO_2 and H_2O, but water may be present either as gas or liquid. The combustion reaction for n-pentane is given by

$$C_5H_{12}(g) + 8O_2(g) \longrightarrow 5CO_2(g) + 6H_2O(g) \quad \Delta H^0_{298} = -3274.5 \text{ kJ}$$

The enthalpy change associated with a combustion reaction when the reactants and products are in their respective standard states is called the standard enthalpy (heat) of combustion. The standard heats of combustion of several compounds are presented in Appendix I. Frequently, the term 'heating value' is used. This represents the amount of energy that can be transferred as heat from the reactor to the surroundings and is equal to $-\Delta H^0_c$. If the water in the products is in the liquid state, the standard heat of combustion has a higher value and it is usually referred as 'gross heating value' or 'higher heating value'. On the other hand, if the water in the combustion products is in the gaseous state, the standard heat of combustion has a lower value and is usually termed as 'net heating value' or 'lower heating value'. The standard heat of a reaction can also be estimated from a knowledge of the standard heats of combustion of the chemical species involved in the reaction.

***Example* 6.19** Estimate the gross heating value and net heating value of pentane if the reactants and products are at 25°C.

Solution The combustion reaction is given by

$$C_5H_{12}(g) + 8O_2(g) \longrightarrow 5CO_2(g) + 6H_2O(g)$$

The standard heat of the above reaction is estimated as $\Delta H^0_{298} = -3274.5$ kJ per mol of pentane. Therefore,

$$\text{Net heating value} = \text{Lower heating value} = -\Delta H^0_c = 3274.5 \text{ kJ}$$

If water in the products is in the liquid state, the combustion reaction is given by

$$C_5H_{12}(g) + 8O_2(g) \longrightarrow 5CO_2(g) + 6H_2O(l)$$

The standard heat of this reaction is $\Delta H^0_{298} = -3538.32$ kJ per mol of pentane (see Example 6.18). Gross heating value or higher heating value, $-\Delta H^0_c = 3538.296$ kJ.

6.8.5 Effect of temperature on the standard heat of reaction

The standard heats of reaction are enthalpy changes between the products and reactants in their standard states. The standard states of gases, liquids and solids are chosen at a pressure of 1 bar and at the reaction temperature. If a chemical reaction is carried out at a temperature which is different from 298 K, the standard heat of reaction depends on the temperature. Since the enthalpy change does not depend on the path by which a system changes from the initial state to the final state, the change in the enthalpy can be estimated by devising a path which is mathematically convenient to deal with. Suppose a reaction is carried out at temperature T which is above 298.15 K. The standard heat of reaction at temperature T can be estimated by the following three-step procedure.

1. The reactants in their standard states are cooled at constant pressure (1 bar) from T to 298.15 K. The change in the enthalpy associated with this process is given by

$$\Delta H^0_R = \sum_R \nu_i \int_{298}^{T} C^0_P dT$$

where the summation is over all the reactants, and C^0_P denotes the molar heat capacity of reactant i in the ideal gas state. Note the ν_i is negative for reactants.
2. The reaction is allowed to proceed at 298.15 K. The change in the enthalpy associated with this process is given by ΔH^0_{298}.
3. The products in their standard states are heated form 298.15 K to T. The change in the enthalpy associated with this process is given by

$$\Delta H^0_P = \sum_P \nu_j \int_{298}^{T} C^0_P dT$$

The desired enthalpy change or the standard heat of reaction at temperature T is then given by

$$\Delta H^0_T = \Delta H^0_R + \Delta H^0_P + \Delta H^0_{298}$$

$$= \sum_R \nu_i \int_{298}^{T} C^0_P dT + \sum_P \nu_j \int_{298}^{T} C^0_P dT + \Delta H^0_{298}$$

$$\text{or } \Delta H^0_T = \Delta H^0_{298} + \sum_i \nu_i \int_{298}^{T} C^0_P dT = \Delta H^0_{298} + \int_{298}^{T} \Delta C^0_P dT \tag{6.95}$$

where the summation is over all the reactants and products and

$$\Delta C^0_P = \sum \nu_i C^0_{Pi} \tag{6.96}$$

The method of calculation is illustrated in the following example.

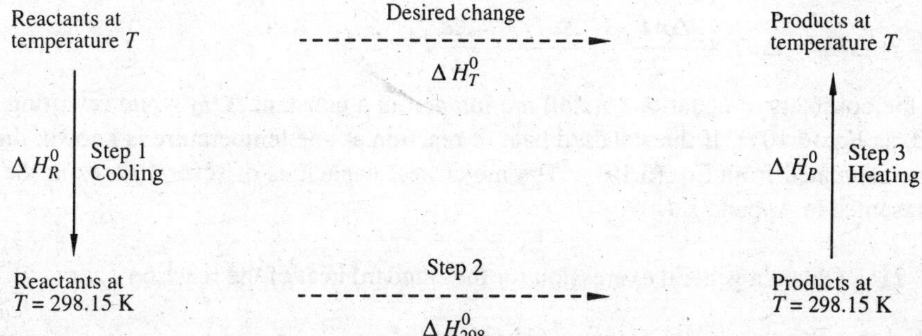

Fig. 6.25 *Three-step process for estimating the standard enthalpy change of a reaction at temperature T*

***Example* 6.20** Calculate the standard heat of the reaction

$$C_5H_{12}(g) + 8O_2(g) \longrightarrow 5CO_2(g) + 6H_2O(g)$$

at 800 K. The average molar heat capacities in the ideal gas state are C_5H_{12}: 247 J/mol K; O_2: 33.62 J/mol K; CO_2: 52.32 J/mol K and H_2O: 38.49 J/mol K; respectively.

Solution

$$\Delta H_{800}^0 = \Delta H_{298}^0 + \int_{298}^{800} \Delta C_P^0 dT$$

$$\Delta C_P^0 = \sum \nu_i C_{Pi}^0 = 5(52.32) + 6(38.49) - 247 - 8(33.62) = -23.42$$

$$\Delta H_{800}^0 = -3274.5 - 23.42 \times 10^{-3}(800 - 298) = -3286.26 \text{ kJ}$$

(Note that $\Delta H_{298}^0 = -3274.5$ kJ from Example 6.17.)

If the molar heat capacities in the ideal gas state for the reactants and products are expressed as a function of temperature by equations of the form

$$C_P^0 = a + bT + cT^2 + dT^3 \tag{6.97}$$

An analytical expression for ΔH_T^0 as a function of temperature can be obtained as shown below.

$$\Delta C_P^0 = \Delta a + (\Delta b)T + (\Delta c)T^2 + (\Delta d)T^3 \tag{6.98}$$

where

$$\Delta a = \sum \nu_i a_i; \quad \Delta b = \sum \nu_i b_i; \quad \Delta c = \sum \nu_i c_i \quad \text{and} \quad \Delta d = \sum \nu_i d_i \tag{6.99}$$

Substituting Eq. (6.98) in (6.95), we get

$$\Delta H_T^0 = \Delta H_{298}^0 + \int_{298}^{T} \left[\Delta a + (\Delta b)T + (\Delta c)T^2 + (\Delta d)T^3\right] dT$$

$$\Delta H_T^0 = \Delta H_{298}^0 + \Delta a(T - 298) + \frac{\Delta b}{2}(T^2 - 298^2) + \frac{\Delta c}{3}(T^3 - 298^3) + \frac{\Delta d}{4}(T^4 - 298^4) \tag{6.100}$$

$$\Delta H_T^0 = \Delta H_0 + \Delta aT + \frac{\Delta bT^2}{2} + \frac{\Delta cT^3}{3} + \frac{\Delta d}{4}T^4 \tag{6.101}$$

where, all the constants of equation (6.100) are lumped in a constant ΔH_0 while rewriting the equation (6.100) as Eq. (6.101). If the standard heat of reaction at one temperature is known, the value of ΔH_0 can be estimated from Eq. (6.101). The molar heat capacities of several gases in the ideal gas state are presented in Appendix *J*.

***Example* 6.21** Obtain a general expression for the standard heat of the reaction

$$C_5H_{12}(g) + 8O_2(g) \longrightarrow 5CO_2(g) + 6H_2O(g)$$

as a function of temperature. Use the molar heat capacities of the reactants and products presented in Appendix J.

Solution

$$\Delta a = 5(19.8) + 6(32.24) - (-3.626) - 8(28.11) = 63.934$$

$$\Delta b = [5(7.334) + 6(0.1924) - 48.73 - 8(-3.68 \times 10^{-4})] \times 10^{-2} = -10.91 \times 10^{-2}$$

$$\Delta c = [5(-5.602) + 6(1.055) - (-25.80) - 8(1.746)] \times 10^{-5} = -9.848 \times 10^{-5}$$

$$\Delta d = [5(1.715) + 6(-0.3596) - 5.305 - 8(-1.065)] \times 10^{-8} = 9.632 \times 10^{-8}$$

$$\Delta H_{298}^0 = -3274.5 \text{ kJ (see Example 6.17)}$$

Substituting $T = 298$ in equation (6.101), we get

$$\Delta H_{298}^0 = -3274.5 \times 10^3 = \Delta H_0 + 63.934(298) - \frac{10.91 \times 10^{-2}}{2}(298)^2 + \frac{-9.848 \times 10^5}{3}(298)^3 + \frac{9.632 \times 10^{-8}}{4}(298)^4$$

or $\Delta H^0 = -3288.03$ kJ. Therefore,

$$\Delta H_T^0 = -3288.03 \times 10^3 + 63.934T - \frac{10.91 \times 10^{-2}}{2}T^2 - \frac{-9.848 \times 10^{-5}}{3}T^3 + \frac{9.632 \times 10^{-8}}{4}T^4$$

where ΔH_T^0 is in J and T is in K.

6.8.6 Adiabatic flame temperature

Suppose a mixture of fuel and oxygen enter an adiabatic steady flow burner and the fuel is completely burned into products. Since the burner is not exchanging energy in the form of heat with the surroundings, the products leave the burner at high temperature. It is necessary to know the maximum temperature that can be attained by a burning a fuel because the maximum permissible temperature of the combustion chamber is governed by metallurgical considerations. The maximum temperature can be achieved only when the fuel is burned with the stoichiometric amount of oxygen. The use of excess oxygen, or air as a source of oxygen, reduces the maximum temperature attained by the flame.

In addition to this, incomplete combustion and the dissociation of the combustion products also reduce the flame temperature. The flame temperature can also be controlled by adjusting the air–fuel ratio. It is possible to estimate the maximum flame temperature of the air–fuel mixture assuming complete combustion and neglecting the dissociation of combustion products. This maximum flame temperature is called the adiabatic flame temperature. The method of estimating the adiabatic flame temperature is illustrated below by an example.

***Example* 6.22** Estimate the adiabatic flame temperature that can be reached by the combustion of n-pentane with 25 percent excess air. Both the fuel and air enter the burner at 25°C. The average molar heat capacities of $CO_2(g)$, $H_2O(g)$, $O_2(g)$ and $N_2(g)$ are 62.75 J/mol K; 52.96 J/mol K; 38.67 J/mol K and 37.13 J/mol K; respectively. Assume complete combustion.

Solution The first law of thermodynamics for a steady flow process, ignoring the changes in the kinetic energy and potential energy, is given by

$$H_e - H_i = Q = 0 \quad \text{or} \quad \Delta H = 0$$

since the burner is adiabatic ($Q = 0$) and no shaft work ($W_s = 0$) is delivered in the process.

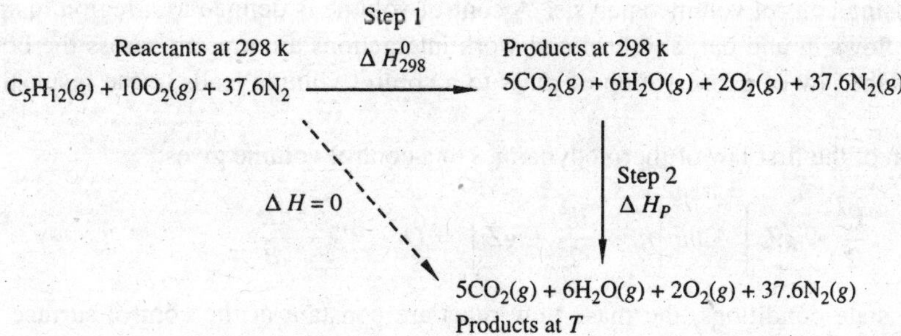

Fig. 6.26 *Two-step process for calculating the adiabatic flame temperature*

The process that occurs in the reactor can be treated as consisting of two steps. First the reactants are converted into products at 298 K. The energy change associated with this step is equal to ΔH^0_{298}. In the second step, the products are raised to temperature T. The energy change associated with this step is equal to ΔH_P and is given by

$$\Delta H_P = \sum_P N_i C^0_{Pi}(T - 298)$$

The energy transferred as heat to the surroundings is Q which is equal to the overall change in the enthalpy. That is,

$$Q = \Delta H^0_{298} + \Delta H_P = 0 \tag{A}$$

$$\Delta H^0_{298} = -3274.5 \text{ kJ (see Example 6.17)}$$

$$\Delta H_P = (5 \times 62.75 + 6 \times 52.96 + 2 \times 38.67 + 37.619 \times 37.13) \times 10^{-3}(T - 298)$$

Hence, Eq. (A) gives

$$-3274.5 + (5 \times 62.75 + 6 \times 52.96 + 2 \times 38.67 + 37.619 \times 37.13) \times 10^{-3}(T - 298) = 0$$

or $T = 1853$ K.

In the above example the molar heat capacities of the products are assumed to be temperature independent. If the heat capacities of the products are expressed by the relations of the form (6.97), the adiabatic flame temperature can be determined by solving the following equation for T.

$$\Delta H^0_{298} + \Delta H_P = \Delta H^0_{298} + \sum_P \int_{298}^{T} N_i C^0_{Pi} dT = 0$$

where the summation is over all the species present in the product.

6.9 Summary

The term control mass is used for a thermodynamic system to emphasize the fact that the quantity of matter remains constant while the system exchanges energy with its surroundings. Flow processes can be analyzed using control volume analysis. A control volume is defined as a region in space through which matter flows in and out, and heat and work interactions are present across the boundary. The application of the law of conservation of mass to a control volume leads to the result that $dm/dt = \dot{m}_i - \dot{m}_e$.

Application of the first law of thermodynamics to a control volume gives:

$$\dot{m}_e \left[h_e + \frac{\mathcal{V}_e^2}{2} + gZ_e\right] - \dot{m}_i \left[h_i + \frac{\mathcal{V}_i^2}{2} + gZ_i\right] = \dot{Q} - \dot{W}_s - \frac{dE}{dt}$$

Under steady state conditions, the mass flow rates are constant at the control surface and there is no accumulation of energy within the control volume. For steady state flow processes, the first law expression reduces to:

$$\left(h_e + \frac{\mathcal{V}_e^2}{2} + gZ_e\right) - \left(h_i + \frac{\mathcal{V}_i^2}{2} + gZ_i\right) = \left(\frac{\dot{Q} - \dot{W}_s}{\dot{m}}\right)$$

Application of the first law for an adiabatic turbine/compressor gives:

$$h_e - h_i = -\frac{\dot{W}_s}{\dot{m}}$$

The exit velocity of an ideal gas leaving an adiabatic nozzle (when the inlet velocity of the gas is ignored) is given by

$$\mathcal{V}_e = \sqrt{2C_P T_i \left[1 - \left(\frac{P_e}{P_i}\right)^{\frac{\gamma-1}{\gamma}}\right]}$$

A throttling process is shown to be an isenthalpic process. The Joule–Thompson coefficient is defined as $\mu_{JT} = (\partial T/\partial P)_h$. The temperature of the fluid either decreases or increases upon throttling depending on whether $\mu_{JT} > 0$ or $\mu_{JT} < 0$. The point at which $\mu_{JT} = 0$ is called the inversion point. The locus of all inversion points is the inversion curve. The importance of a knowledge of inversion curves for fluids is discussed. Throttling calorimeter, a device based on the principle of Joule–Thompson expansion, is used to measure the quality of wet steam. The combined separating and throttling calorimeter is also discussed.

The cases of charging a cylinder and discharging a cylinder are considered. In these cases the mass and the state of the fluid inside the control volume changes with time, and hence they are called transient flow processes. It is demonstrated that these problems can be solved either through control mass analysis or control volume analysis. In the case of adiabatic charging of a cylinder, the first law reduces to:

$$(m_f - m_0)\, h_i = m_f u_f - m_0 u_0$$

In the case of discharging of a cylinder, the first law of thermodynamics reduces to:

$$(m_0 - m_f)\left(h + \frac{V^2}{2}\right) = Q + m_0 u_0 - m_f u_f$$

Through control volume analysis it is shown that when an ideal gas escapes from an insulated tank, the gas that remains in the tank can be considered to have undergone a reversible adiabatic expansion.

Most of the fuels contain carbon and hydrogen and are available as solid, liquid or gas. The chemical analysis of solid fuels is usually given in terms of the mass percent of each constituent element. Most of the liquid fuels are a mixture of several hydrocarbons. A liquid fuel is usually treated as a single hydrocarbon while dealing with the combustion of fuels. For example, gasoline is treated as pure octane (C_8H_{18}). The analysis of the gaseous fuels is given in terms of mole percent (or volume percent) of the constituent chemical species.

The combustion of a fuel involves the oxidation of the constituents and the process can be represented by a chemical reaction.

In most of the combustion processes the required oxygen is not supplied as pure oxygen but in the form of air. The combustion of a hydrocarbon fuel is considered to be complete if all the carbon and hydrogen contained in the fuel are converted into carbon dioxide and water, respectively. The minimum amount of air required for the complete combustion of a fuel is called the stoichiometric or theoretical air. Usually, the combustion process does not go to completion unless the amount of air supplied is more than the theoretical air. The amount of air actually used in a combustion process is expressed in terms of percent theoretical air or percent excess air. The amount of air supplied over and above the theoretical requirement is called excess air.

An important parameter in the study of combustion of fuels is the air–fuel ratio, the ratio of the mass (moles) of air to the mass (moles) of the fuel. The higher the air–fuel ratio, the lower is the temperature of the combustion products and hence the use of high air–fuel ratio is undesirable.

The combustion products can be analyzed with an Orsat apparatus. Orsat analysis gives the composition of the combustion products on dry basis. The actual air–fuel ratio used in a combustion process can be calculated from a knowledge of the fuel composition and the result of Orsat analysis of the combustion products.

Chemical reactions are usually accompanied by the evolution or absorption of energy in the form of heat because the energy of the products is different from the energy of the reactants. This difference in energy arises because of the structural difference between the products and reactants. The amount of energy liberated or absorbed during a chemical reaction depends upon the temperature of the reactants and the products. The energy change associated with a chemical reaction is called the heat of reaction and it can be measured with the help of a flow calorimeter.

While reporting the heat of a chemical reaction, it is necessary to specify the state of the reactants and products. For chemically reacting systems, the following standard states are used. The pure component gas in its ideal gas state at 1 bar is selected as the standard state for a gaseous species. For liquids and solids, the standard state is taken as the pure liquid or solid species at 1 bar. Whenever a substance is found in its allotropic forms, the most stable form is chosen as the standard state. Carbon (graphite) is the standard state for solid carbon. The choice of temperature for the standard state is entirely arbitrary. For convenience in the tabulation of data, 298.15 K is selected as the standard state temperature. If the reactants and products are in their standard states, the energy change associated with the reaction is called the standard heat of reaction. The standard heat of a reaction is denoted by ΔH^0_{298}. The standard heat of a reaction refers to the stoichiometric coefficients shown in the reaction.

A formation reaction is a chemical reaction which represents the formation of one mole of compound from the constituent elements. The standard heat of formation which is denoted by ΔH^0_{f298}. The standard heat of formation of all the elements in their standard states have been assigned a zero value. The standard heat of a reaction can be estimated from a knowledge of the standard heats of formation of the products and reactants. If a chemical reaction can be expressed as

$$\sum \nu_i A_i = 0$$

where, A_i denotes the chemical species, i and ν_i is the stoichiometric coefficient of species i. By convention ν_i is positive for products and negative for reactants. Then the standard heat of reaction is given by

$$\Delta H^0_{298} = \sum_i \nu_i \Delta H^0_{f298}(A_i)$$

A combustion reaction is defined as a reaction between an element or compound and oxygen to form specified combustion products. For hydrocarbons the combustion products are CO_2 and H_2O, but water may be present either as gas or liquid. The heat of a combustion reaction when the reactants and products are in their respective standard states is called the standard heat of combustion. The negative of the standard heat of combustion is frequently called the heating value as it represents the amount of energy that can be transferred as heat from the reactor to the surroundings. If water in the combustion products is in liquid state, the heating value has a higher value and is called as gross heating value. On the other hand, if water in the combustion products is in the gaseous state, the heating value has a lower value and is termed as net heating value. The standard heat of a reaction can also be determined from a knowledge of the standard heats of combustion of the chemical species participating in the reaction.

The standard heat of a reaction at any temperature other than 298 K can be evaluated by making use of the relation:

$$\Delta H^0_T = \Delta H^0_{298} + \int_{298}^{T} \Delta C^0_P dT$$

where, ΔH^0_T is the standard heat of reaction at temperature T and

$$\Delta C^0_P = \sum_i \nu_i C^0_{Pi}$$

If the molar heat capacities at constant pressure of the species are expressed as

$$C_P^0 = a + bT + cT^2 + dT^{-3}$$

the standard heat of reaction ΔH_T^0 is given by:

$$\Delta H_T^0 = \Delta H_0 + \Delta aT + \frac{\Delta bT^2}{2} + \frac{\Delta cT^3}{3} + \frac{\Delta d}{4}T^4$$

where, ΔH_0 is a constant, the value of which can be determined from a knowledge of ΔH^0 at any one temperature, and

$$\Delta a = \sum \nu_i a_i; \quad \Delta b = \sum \nu_i b_i; \quad \Delta c = \sum \nu_i c_i \quad \text{and} \quad \Delta d = \sum \nu_i d_i$$

If a fuel is burned in an adiabatic reactor, the products leave the reactor at a higher temperature called the adiabatic flame temperature. The adiabatic flame temperature depends on the air–fuel ratio.

6.10 Review Questions

1. Distinguish between control mass analysis and control volume analysis.
2. What is a control volume?
3. What is a continuous system and how can one specify the state of a continuous system?
4. What does the principle of local equilibrium state?
5. Compare the characteristics of control mass and control volume.
6. What is a control surface?
7. What is meant by steady state steady flow process?
8. What is a throttling process?
9. What is an inversion point and what is an inversion curve?
10. Sketch a few isenthalpic curves on T versus P diagram and show the inversion curve and the zones of cooling and heating.
11. Define the Joule–Thompson (or Joule–Kelvin) coefficient.
12. What will be the value of Joule–Thompson coefficient for an ideal gas and why should it have that value?
13. Describe a simple throttling calorimeter and explain how it can be used to measure the quality of wet steam.
14. With the help of a neat sketch, describe the combined separating and throttling calorimeter and explain how it is used to measure the quality of wet steam.
15. What is a transient flow process?
16. Is it possible to analyze a transient flow process following control mass analysis? If so, derive a relation for analyzing the process of charging of a gas cylinder.
17. Apply the control mass approach and derive a relation to analyze the process of discharging a tank.
18. Show that if an ideal gas discharges from an insulated cylinder, the state of the gas remaining in the cylinder can be estimated by making use of the reversible adiabatic relations for an ideal gas.
19. Suppose an evacuated gas cylinder is connected to a high pressure line carrying an ideal gas at temperature T and the cylinder is adiabatically filled. What will be the temperature of the gas in the cylinder at the end of filling operation?

20. Name a few gaseous fuels and specify their constituent species.
21. Define a combustion reaction.
22. What is meant by stoichiometric air or theoretical air?
23. What is meant by excess air and what is the necessity of using excess air in the combustion of a fuel?
24. Describe the method of analysis followed in the Orsat analysis.
25. Explain the difference between dissociation energy and bond energy of a diatomic molecule with the help of a potential energy diagram.
26. Describe the method of determining the heat of combustion of a fuel with the help of a flow calorimeter.
27. What are the standard states which are commonly used in chemical thermodynamics?
28. What is the necessity of using standard states in chemical thermodynamics?
29. What is a formation reaction?
30. What is meant by standard heat of formation?
31. How does one determine the standard heat of a reaction from a knowledge of the standard heats of formation of the constituent species?
32. What is meant by standard heat of combustion?
33. What is higher heating value or gross heating value?
34. What is the difference between gross heating value and net heating value?
35. How does one estimate the standard heat of a reaction from a knowledge of the standard heats of combustion of the reactants and products?
36. What is an adiabatic flame temperature?
37. What factors influence the adiabatic flame temperature?
38. Does the addition of an inert gas change the standard heat of a reaction? Explain.
39. Does the addition of an inert gas change the adiabatic flame temperature? Explain.
40. What is the necessity of a knowledge of adiabatic flame temperature?

6.11 Problems

6.1 A particular tank has one inlet and one exit. Mass enters the tank at a rate of 2 kg/s. At the exit, the mass leaves the tank at the rate $\dot{m}_e = 2(1 - e^{-0.02t})$, where $\dot{m}_e$ is in kg/s and t is in s. If the tank initially contains 50 kg of mass, determine the amount of mass in the tank at $t = 200$ s.

6.2 Air at 10 bar and 600 K enters a device operating under steady state conditions at a rate of 2 kg/s. The air enters the device with a velocity of 2 m/s. The air leaves the device at 2 bar and 400 K with a velocity of 20 m/s. If the device is well insulated, determine the power output of the device.

6.3 A turbine in a steam power plant operating under steady state conditions receives one kg/s of superheated steam at 3 MPa and 350°C with a velocity of 50 m/s at an elevation of 2 m above the ground level. The steam leaves the turbine at 10 kPa with a quality of 0.95 at an elevation of 5 m above the ground level. The exit velocity of the steam is 120 m/s. The energy losses as heat from the turbine are estimated at 5 kJ/s. Estimate the power output of the turbine. How much error will be introduced if the kinetic energy and the potential energy terms are ignored?

6.4 In an air-conditioning plant, saturated Freon-12 at $-20°C$ with a quality of 0.8 enters an adiabatic compressor and leaves as saturated vapor at 40°C. If the flow rate of Freon through the compressor is one kg/s, estimate the power input to the compressor.

6.5 Consider the steady-state operation of a compressor. The fluid enters the compressor at P_1 and v_1 and leaves the compressor at P_2 and v_2. Show that the work done on the compressor is given by $W = -\int_1^2 v dP$. Sketch the P–v diagram to represent the work done on the compressor and compare it with the work done if the compression is carried out in a piston–cylinder assembly under non-flow conditions.

6.6 Air enters a well insulated compressor operating at steady state at 1 bar and 300 K, and exits at 6 bar and 440 K. If the flow rate of air through the compressor is 3 kg/s, calculate the power consumption of the compressor.

6.7 In a steam power plant, saturated liquid water at 10 kPa enters a feed pump at the rate of one kg/s. The feed pump delivers the water to the boiler at a pressure of 3 MPa. Assuming that the pump is adiabatic, estimate the power input to the pump.

6.8 Steam enters an adiabatic nozzle operating at steady state at 4 bar and 200°C with negligible velocity, and exits at 2 bar with a velocity of 300 m/s. Determine the temperature of the steam leaving the nozzle.

6.9 Air (ideal gas with $\gamma = 1.4$) expands reversibly and adiabatically through a nozzle from inlet conditions of 3 bar and 300°C to an exit pressure of 1 bar. Determine the exit velocity and the mass flow rate if the exit area is 5×10^{-4} m^2.

6.10 A nozzle receives 5 kg/s of steam at 3 bar and 300°C and discharges it at 1 bar and 200°C. The inlet velocity of the steam is negligible and the energy loss, to the surroundings, in the form of heat is estimated to be 50 kJ/s. Determine the discharge velocity of the steam.

6.11 A throttling calorimeter is used to measure the quality of steam in a pipe where the pressure is 10 bar. The conditions in the calorimeter are 1 bar and 120°C. Determine the quality of steam in the pipeline.

6.12 When wet steam at 10 bar is throttled to 1 bar, the temperature of the exit steam is 110°C. Determine the quality of wet steam.

6.13 In a commercial refrigerator, the refrigerant (saturated liquid Freon-12 at 40°C) is throttled to a low temperature of $-26°C$. Determine the quality of Freon-12 after throttling.

6.14 Saturated vapor Freon-12 at 5 bar enters a steady flow adiabatic device with a velocity of 2 m/s. The Freon-12 leaves the device as saturated vapor at 1 bar. Calculate the exit velocity.

6.15 Air enters an adiabatic nozzle at 3 bar and 200°C with a velocity of 10 m/s and leaves the nozzle at 2 bar and 150°C. Calculate the exit velocity of the air.

6.16 Air at 3 bar and 200°C enters a reversible and adiabatic nozzle with negligible velocity and leaves at 2 bar. Calculate the exit velocity of air.

6.17 In a small gas turbine operating under steady state conditions, air enters at 3 bar and 400 K. The turbine discharges air at 1 bar and 320 K. Ignoring the changes in kinetic energy, estimate the flow rate of air, if the power output of the turbine is 15 kW.

6.18 Steam at 20 bar and 220°C enters a throttling valve and exhausts at 5 bar. Determine the temperature and specific volume of the exhaust steam.

6.19 A 1 m^3 tank contains air (ideal gas with $\gamma = 1.4$) at 100 kPa and 300 K. Attached to this tank is a line in which air is flowing at 1 MPa and 600 K. The air from this line enters the tank until the pressure reaches 1 MPa. If there is no energy transfer as heat from the tank, calculate the final temperature of the air in the tank, and the quantity of air that enters the tank.

6.20 An insulated and rigid tank of volume 10 m^3 is connected to a steam line through which steam flows at 20 bar and 400°C. The tank is initially evacuated. The steam is allowed to flow into the tank until the pressure inside is 20 bar. Calculate the final temperature of steam in the tank and the mass of steam that flows into the tank.

6.21 A well insulated, rigid tank of volume 5 m^3 initially contains steam at 2 bar and 200°C. The tank is connected to a line carrying steam at 10 bar and 400°C. If 10 kg steam is allowed to flow into the tank, determine the final pressure and temperature of steam in the tank.

6.22 A tank of volume 1 m^3 initially contains steam at 50 bar and 400°C. Steam is slowly withdrawn from the tank until the pressure drops to 20 bar. Energy is transferred to the steam in the tank to maintain its temperature at 400°C while the steam is withdrawn. Determine the amount of energy transferred to the steam.

6.23 An insulated tank of 1 m^3 capacity contains saturated steam at 100 kPa. Attached to this tank is a line in which steam is flowing at 1 MPa and 200°C . The steam from this line enters the tank until the pressure reaches 1 MPa. Calculate the mass of steam that enters the tank.

6.24 A tank of 0.2 m^3 capacity contains ethane (non-ideal gas with $\gamma = 1.19$) at 15 bar and 306 K. Attached to this tank is a line in which ethane is flowing at 14 MPa and 33°C . The gas from this line enters the tank until the pressure reaches 14 MPa. (a) If the tank is filled slowly so that the final temperature of the ethane is 33°C, how much ethane enters the tank? (b) If the tank is filled rapidly so that the process is essentially adiabatic, how much ethane is added to the tank?

6.25 An insulated cylinder of volume 0.12 m^3 contains nitrogen at 1.5 bar and 27°C. The cylinder is connected to a line carrying nitrogen at 10 bar and 27°C through a valve. The valve is opened and held open till the nitrogen pressure in the cylinder reaches 10 bar. Assuming that nitrogen is an ideal gas, determine the final temperature and mass of nitrogen in the cylinder at the end of the filling operation.

6.26 A tank of volume 3 m^3 contains saturated water vapor at 2 bar. Attached to this tank is a line in which steam at 10 bar and 300°C flows. Steam from this line enters the tank until the pressure is 10 bar. If there is no heat transfer from the tank, calculate the mass of steam that enters the tank.

6.27 An evacuated, insulated tank containing a piston and spring is connected to a line carrying steam at 2 MPa and 300°C as shown in Fig. 6.27. Initially, the spring is in its natural position and exerts no force on the piston. When the valve is opened, the steam enters the tank, and the force exerted by the spring is proportional to the displacement from its natural position. Determine the state of steam in the tank when the pressure inside the tank is 2 MPa.

6.28 An insulated tank of 1 m^3 volume contains saturated steam at 2 MPa. Only vapor is withdrawn from the tank by opening a valve at the top of the tank. If the final pressure of steam in the tank is 1 MPa, estimate the amount of steam withdrawn from the tank.

6.29 An insulated cylinder of 0.05 m^3 volume contains Freon-12 at 0.4 MPa at 30°C. The valve of the cylinder is opened and the Freon is allowed to escape as vapor from the cylinder until the pressure is reduced to 0.3 MPa. Determine the amount of Freon that escapes from the cylinder. Properties of superheated Freon-12 are given below.

P (MPa)	T (°C)	v (m^3/kg)	h (kJ/kg)
0.4	30	0.047 870	205.5454
0.3	10	0.059 901	194.2576
0.3	20	0.062 637	200.6529

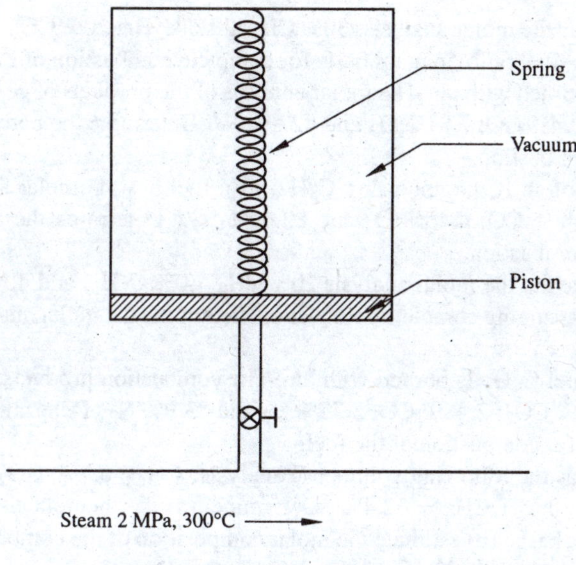

Fig. 6.27 *Sketch for Problem 6.27*

6.30 At a certain instant, a boiler of 1 m^3 volume contains 50% (by volume) of saturated vapor and 50% (by volume) of saturated liquid at 3 MPa. A valve at the bottom of the boiler is opened and 100 kg of liquid is withdrawn. Determine the final quality of steam in the boiler.

6.31 An insulated tank of 1 m^3 volume contains an ideal gas ($\gamma = 1.4$) at 1 MPa and 600 K. A valve at the top of the tank is opened and the gas pressure is reduced to 200 kPa. Determine the number of moles of the gas that escapes from the tank. Note that the temperature of the gas leaving the tank changes continuously.

6.32 A rigid tank of volume 0.2 m^3 contains a mixture of saturated liquid water and saturated water vapor at 120°C. The tank initially contains 0.05 m^3 of liquid and 0.15 m^3 of vapor. A valve on the top of the tank is opened while energy is transferred as heat to the tank, allowing only vapor to escape till the pressure drops to 1 bar and at that instant; the temperature of the contents of the tank is 100°C. Calculate the mass which leaves the tank and the quantity of energy transferred.

6.33 A rigid and insulated tank having a volume of 0.5 m^3 contains nitrogen at 20 bar and 400 K. A valve on the tank is opened allowing the nitrogen pressure to drop to 4 bar. Assuming that nitrogen behaves as an ideal gas, estimate the mass of nitrogen discharged from the tank.

6.34 A rigid vessel of volume 1 m^3 contains steam at 20 bar and 300°C. A valve on the tank is opened allowing the steam to escape until the pressure reduces to 2 bar and the temperature of the tank is maintained constant while the steam discharges from the tank. Calculate the energy transferred as heat.

6.35 The molar composition of a particular natural gas is as follows: $CH_4 = 67.6\%$; $C_2H_6 = 31.3\%$; $N_2 = 1.1\%$. Calculate, (a) the theoretical air–fuel ratio on molar basis and on mass basis, (b) Estimate the molar composition of the combustion products if the fuel is burned with 150 percent theoretical air. Assume complete combustion of the fuel.

6.36 A gaseous fuel has the molar analysis: 30% CH_4; 20% C_2H_6; 20% CO and 30% N_2. Determine the theoretical air–fuel ratio on mass basis for complete combustion of the fuel gas.

6.37 Octane vapor is burned with air. The molar analysis of the products of combustion on a dry basis is 8.45% CO_2; 0.94% CO; 7.81% O_2 and 82.8% N_2. Determine the percent excess air, on mass basis, used for combustion.

6.38 The exhaust gas of an IC engine using C_8H_{18} as a fuel has the molar analysis on a dry basis: 10.38% CO_2; 1.48% CO; 4.45% O_2 and 83.69% N_2. Determine the air–fuel ratio (on mass basis), used for combustion.

6.39 A gaseous fuel having the molar analysis 20% CH_4; 40% C_2H_6; and 40% C_3H_8 is burned with 40% excess air. Assuming complete combustion, determine the molar analysis of the combustion products.

6.40 A hydrocarbon fuel C_xH_y is burned with air. The combustion products has the molar analysis on dry basis: 9.9% CO_2; 2.48% CO; 3.72% O_2 and 83.9% N_2. Determine the air–fuel ratio, on mass basis, used for combustion of the fuel.

6.41 Coke-oven gas has the following volumetric analysis: CO = 6.7%, CO_2 = 2.2%, H_2 = 54.0% N_2 = 4.1% O_2 = 0.6%; CH_4 = 32.4%. Determine, (a) the theoretical air–fuel ratio on molar basis and on mass basis; (b) estimate the molar composition of the combustion products (assuming complete combustion) if 20 percent excess air is used.

6.42 Gasoline may be treated as a single hydrocarbon octane. Determine, (a) theoretical air–fuel ratio on a mass basis, (b) to ensure complete combustion it is desirable to introduce 20 percent excess air along with the fuel vapor into the cylinder of car engine. Assuming complete combustion determine the composition of the exhaust gases.

6.43 Diesel oil may be treated as pure dodecane, $C_{12}H_{26}$ for combustion calculations. Estimate the theoretical air–fuel ratio on mass basis. Determine the molar composition of the combustion products if diesel oil is burned with 50 percent excess air.

6.44 Gasoline (octane) is burned with air in an internal combustion engine and the resulting exhaust gas has the following molar composition on dry basis: 8.83% CO_2; 1.26% CO; 6.94% O_2; 82.97% N_2. Determine the actual air–fuel ratio on mole basis and the percent theoretical air.

6.45 The exhaust gas of a diesel engine has the following molar composition on dry basis: 9.98% CO_2; 2% CO; 4.69% O_2; 83.33% N_2. Determine the actual air–fuel ratio on mass basis and the percent excess air.

6.46 A hydrocarbon fuel is burned with air and the molar composition of the resulting combustion products on dry basis is 7.26% CO_2: 2.42% CO; 7.5% O_2; 82.82% N_2. Determine the composition of the fuel on mass basis and the percent theoretical air used in the combustion process.

6.47 A gaseous fuel having volumetric analysis of 70% CH_4; 20% C_2H_6 and 10% C_3H_8 is burned with 30 percent excess air. Assuming complete combustion, determine the air–fuel ratio (on molar basis) used and the volumetric analysis of the combustion products, on dry basis.

6.48 The Orsat analysis of the combustion products of a hydrocarbon fuel is 13.6% CO_2; 0.8% O_2; 0.8% CO and 84.8% N_2. Determine the composition of the fuel on mass basis and the air–fuel ratio (on mass basis) used for combustion.

6.49 In a particular combustion chamber methane is burned in the presence of air and the Orsat analysis of the flue gases gives CO_2 = 8%; CO = 1.7% and O_2 = 4.32%. Determine the air–fuel ratio on mole basis and the percent air used for combustion.

6.50 When propylene (C_3H_6) is burned with 20 percent excess air, the products of combustion shows 4.3% of O_2 on dry basis. Determine the percentage of carbon in the fuel which is converted into CO instead of CO_2.

6.51 In a particular combustion chamber a mixture of ethane and air enters at 25°C and 1 bar. The products of combustion leave at 1100 K. For the combustion of ethane 15% excess air is used. Assuming complete combustion determine the energy lost (in kJ/kmol of fuel) as heat to the surroundings.

6.52 When ethylene is burned with 10% excess air, 96% of carbon in the fuel is converted into CO_2 while the rest is converted into CO. Ethylene and air enter the burner at 298 K and 1 bar. If the products of combustion leave the burner at 1500 K and 1 bar, determine the energy lost, in kJ/kmol of fuel, as heat to the surroundings.

6.53 Calculate the standard heat of reaction at 298 K for the following chemical reactions

$$4NH_3(g) + 5O_2(g) \longrightarrow 4NO(g) + 6H_2O(g)$$

$$H_2S(g) + \frac{3}{2}O_2(g) \longrightarrow H_2O(g) + SO_2(g)$$

$$C_2H_2(g) + \frac{5}{2}O_2(g) \longrightarrow 2CO_2(g) + H_2O(g)$$

6.54 Estimate the standard heat of reaction at 298 K for the reactions of Problem 6.53 if H_2O in the products is in the liquid state.

6.55 Determine the enthalpy change of the following reaction

$$4H_2O_2(l) + CH_4(g) \longrightarrow 6H_2O(g) + CO_2(g)$$

at 25°C and 1 bar. The enthalpy of formation of $H_2O_2(l)$ at 298 K is −187.58 kJ/mol.

6.56 Calculate the standard heat at 25°C of the following reactions from the standard heat of combustion data

$$C_5H_{12}(g) + 8O_2(g) \longrightarrow 5CO_2(g) + 6H_2O(l)$$

$$CH_4(g) + H_2O(g) \longrightarrow CO(g) + 3H_2(g)$$

$$CO(g) + H_2O(g) \longrightarrow CO_2(g) + H_2(g)$$

6.57 An internal combustion engine uses octane as fuel. The air and fuel vapor mixture enter the engine at 25°C and 0.1 MPa and the engine uses 120 percent theoretical air. Supposing 75% of the fuel's carbon is converted into CO_2 and the rest is converted to CO, and the combustion products leave the engine at 800 K, calculate the amount of energy transferred as heat to the engine (per kg of fuel).

6.58 Calculate the standard heat of reaction at 1000°C for the following reactions

$$CH_4(g) + H_2O(g) \longrightarrow CO(g) + 3H_2(g)$$

$$CO(g) + H_2O(g) \longrightarrow CO_2(g) + H_2(g)$$

6.59 Develop a general expression for the standard heat of reaction

$$CH_4(g) + 2O_2(g) \longrightarrow CO_2(g) + 2H_2O(g)$$

as a function of temperature.

6.60 Estimate the maximum flame temperature that can be attained when methane is burned with theoretical air. Both the fuel and air enter the burner at 298 K. Assume complete combustion. The average molar heat capacities of CO_2, H_2O, O_2 and N_2 are 62.75 J/mol K; 52.96 J/mol K; 38.67 J/mol and 37.13 J/mol K, respectively.

6.61 Determine the adiabatic flame temperature for complete combustion of octane (C_8H_{18}) gas with 200% of theoretical air in a burner operating at steady state. The reactants enter the burner at 25°C and 1 bar. Assume complete combustion. Assume that C_P^0 is given by $C_P^0 = a + bT$ with the values of the constants a and b as given in Appendix J.

6.62 Ethanol (C_2H_5OH) vapor is burned with 100% excess air in a steady-state, steady flow combustion chamber. The fuel and air enter the combustion chamber at 25°C and 1 bar. Assuming complete combustion calculate the adiabatic flame temperature. Assume that C_P^0 is given by

$$C_P^0 = a + bT$$

with the values of the constants a and b as given in Appendix J.

6.63 Gaseous propane is burned with 300% theoretical air. The fuel and air enter the burner at 25°C and 1 bar. Assuming complete combustion, estimate the adiabatic flame temperature. Make use of the fact that $C_P^0 = a + bT$ with the values of a and b as given in Appendix J.

6.64 Propane is burned with 100% excess air in an adiabatic burner. Both the fuel and air enter the burner at 25°C and 1 bar. Estimate the maximum temperature that can be attained in the flame. The molar heat capacities of the product gases are as follows:

$$C_P^0(CO_2) = 19.8 + 7.344 \times 10^{-2}T$$

$$C_P^0(H_2O) = 32.24 + 1.924 \times 10^{-3}T$$

$$C_P^0(O_2) = 28.11 - 3.68 \times 10^{-6}T$$

$$C_P^0(N_2) = 31.15 - 1.357 \times 10^{-2}T$$

where C_P^0 is in J/mol K and T is in K.

6.65 Carbon monoxide is burned with 100% excess air in an adiabatic combustion chamber. Both the fuel and air enter the combustion chamber at 25°C and 1 bar. Assuming complete combustion, estimate the maximum flame temperature that can be attained. Make use of the molar heat capacity data given in Problem 6.64

7 Second Law of Thermodynamics

Learning Objectives

This chapter will enable the readers to:

- Know the limitations of the First Law of Thermodynamics.
- Define source, sink, heat engine, heat pump and refrigerator.
- State the Kelvin–Planck statement and Clausius statement of the second law of thermodynamics.
- Prove the equivalence of the Kelvin–Planck and Clausius statements of the second law of thermodynamics.
- Know about the Perpetual Motion Machine of the Second Kind.
- State the criterion of irreversibility based on the second law of thermodynamics.
- Judge whether a given process is irreversible or not.
- Understand Carnot cycle.
- Estimate the thermal efficiency of a Carnot heat engine.
- State and prove the Carnot theorems.
- Understand the thermodynamic or absolute temperature scale.
- State the Clausius inequality.
- Understand entropy.
- Calculate the entropy change of a system.
- Understand the principle of entropy increase.
- Understand the second law analysis of a control volume.
- Define available energy and unavailable energy.
- Estimate the loss in available energy in a given process.
- Understand the meaning of *dead state*.
- Define availability for nonflow and for steady flow processes.
- Calculate the irreversibility associated with a given process.

7.1 First Law of Thermodynamics: Limitations

Consider the spontaneous process of water flowing downhill. The potential energy of the water at the top of the hill gets converted into kinetic energy of the flowing water. The first law of thermodynamics tells us that the decrease in the potential energy of water is equal to the increase in its kinetic energy while the water is falling. Let us now imagine the reverse of the process, that is, water flowing uphill instead of downhill. Applying the first law of thermodynamics, we find that the law would be equally well satisfied if the water was to flow uphill by transferring its kinetic energy into potential energy.

Thus we see that the first law of thermodynamics is satisfied regardless of the direction of the process. However, in practice we never see water flowing uphill on its own.

Similarly, if a running car is stopped by applying brakes, kinetic energy of the car is converted into the internal energy of the brakeshoe and the wheels. According to the first law, the increase in the internal energy of the brakeshoe and the wheels is equal to the decrease in the kinetic energy of the car. The first law would be well satisfied if the brakeshoe and the wheels were to cool off to the original temperature by transferring the internal energy causing the car to run. However, we never see such an occurrence happen in nature.

We can consider several other spontaneous processes also and observe that all such spontaneous processes proceed in one direction only. These processes never proceed in the reverse direction on their own; though, such a feasibility is not ruled out by the first law of thermodynamics. In other words, the first law is only a principle of conservation of energy and does not specify the direction of the process. There is nothing in the first law to deny the possibility of a spontaneous process reversing itself. All these examples indicate that there exists a directional law which determines the direction in which a spontaneous process will take place. This law is called the second law of thermodynamics.

In the thermodynamic analysis of a given process, we are also concerned with the following two important aspects of the process in addition to a knowledge of the direction of the process:

1. What energy changes are associated with the given process?
2. Does the process occur spontaneously? If so, to what extent? What is the point beyond which no further change can occur?

Let us consider a specific example—the synthesis of ammonia from N_2 and H_2 at 500°C and 30 MPa pressure. The synthesis reaction can be represented as

$$N_2 + 3H_2 \longrightarrow 2NH_3$$

To carry out the above synthesis reaction, the reactants N_2 and H_2, which are available at low pressure (100 kPa), are compressed to a high pressure. The reactants are raised to the reaction temperature before they are fed to the reactor. We would like to know how much energy is needed to bring about the desired change of state of the reactants? If the reaction occurs from the given initial state of the reactants to the given final state of products, what are the energy changes associated with the process? In the production of ammonia, the reactor effluent is cooled. Ammonia is condensed and separated. The unreacted gases are recycled. What energy changes are associated with the cooling of the reactor effluent and the recycling process? All these questions are associated with the first aspect. They can be answered by applying the first law of thermodynamics to the various steps of the synthesis process.

Before initiating an expensive pilot plant study, or commercial scale production, it is necessary to know the answers to the following questions. Is the contemplated reaction feasible under the given conditions of temperature and pressure? in other words, does the reaction proceed starting with the given state of reactants to the given state of products? If so to what extent? Is complete conversion of N_2 and H_2 to NH_3 possible? Or, does the reaction proceed to a certain extent only? If so, to what extent? All these questions are related to the second aspect. They cannot be answered by the application of the first law of thermodynamics alone.

As a second example, let us consider the operation of a heat engine. In a heat engine we deal with the continuous conversion of internal energy into work. The engine receives a certain amount of energy as heat from a source. A part of it is converted into work and the rest of it is rejected to a sink at a temperature lower than that of the source. Is complete conversion of energy into work possible in a heat engine? The efficiency of a heat engine is defined as the ratio of the net work done to the energy absorbed. The above question then gets reduced to: Is it possible to achieve an efficiency of one in a heat engine? If not, what is the limit on the efficiency of a heat engine? These questions cannot be answered by the application of the first law of thermodynamics alone. Therefore, the first law is not adequate for answering questions relating to the second aspect. To answer these questions, it is essential to have some other principle of wide generality—the second law of thermodynamics.

7.2 Heat Engine

Before discussing the second law of thermodynamics in detail, it is necessary to define a few terms which are used quite often.

A *thermal reservoir* is a large system from which a finite quantity of energy can be extracted as heat or to which a finite quantity of energy can be added as heat without changing its temperature. A *source* is a thermal reservoir at high temperature from which energy in the form of heat is extracted by a heat engine. A *sink* is a thermal reservoir at lower temperature to which energy in the form of heat is rejected by the engine. The *working fluid* or the *working medium* is a substance which receives energy from the source and which rejects energy to the sink while undergoing a cyclic change in a heat engine.

A heat engine is an energy conversion device. It is a cyclically operating device and its primary objective is the conversion of energy absorbed as heat into work. In a heat engine, the working fluid receives energy as heat from a high temperature source, delivers some net work and rejects the rest of the energy as heat to a low temperature sink. The working fluid returns periodically to the initial state thus executing a cycle. A thermal power plant shown in Fig. 7.1 can be viewed as a heat engine in which the working fluid (water) undergoes the following changes:

1. The working fluid (water) enters the boiler where it receives a certain amount of energy (Q_1) as heat from the hot combustion products of coal or fuel oil, and is converted into superheated steam.
2. The superheated steam enters a turbine, where it undergoes an expansion performing the shaft work (W_T) and emerges as low-pressure steam.
3. The low-pressure steam enters a condenser where it exchanges energy as heat at constant pressure with the cooling water and emerges as the condensate. During this process, it rejects a certain amount of energy (Q_2) as heat to the cooling water.
4. The low-pressure condensate from the condenser enters the pump. Work (W_p, part of the work delivered by the turbine) is done on the pump to compress the condensate to the boiler pressure and return it to the boiler. Thus, the working fluid completes a cycle.

An index of performance of the heat engine is expressed by the *thermal efficiency* η, which is defined as the ratio of the net work done to the energy absorbed as heat. If the system boundary is drawn as shown by the dotted line, in Fig. 7.1 the heat and work interactions across the boundary are given by

$$\begin{aligned}\text{Work done by the system} &= W_T - W_P \\ \text{Energy absorbed as heat by the system} &= Q_1 \\ \text{Energy rejected as heat by the system} &= Q_2\end{aligned}$$

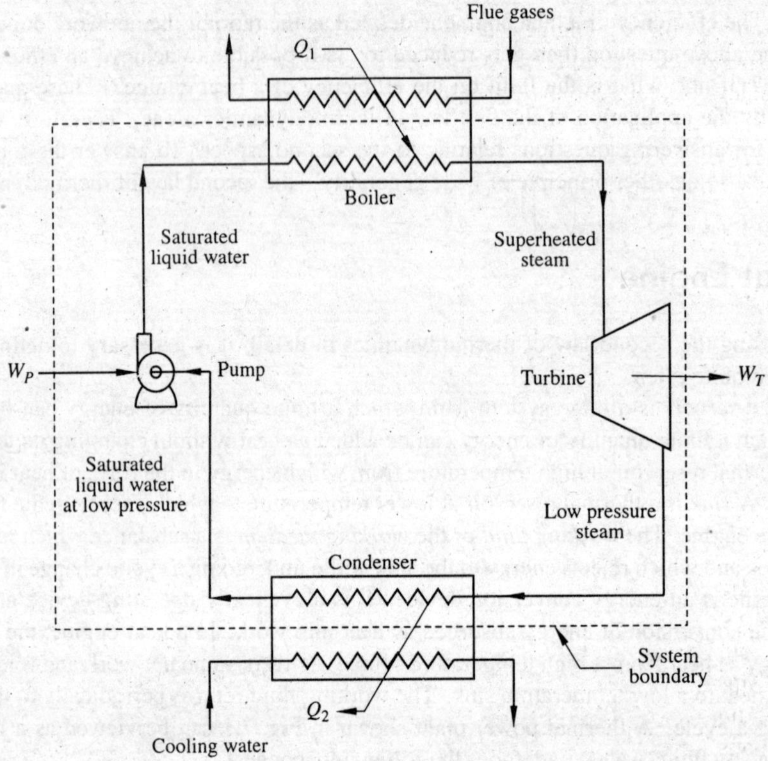

Fig. 7.1 *Schematic representation of a thermal power plant*

According to the first law of thermodynamics, the heat and work interactions are related by the equation

$$\oint dQ = \oint dW$$

or

$$Q_1 - Q_2 = W_T - W_P \tag{7.1}$$

The heat interaction Q_2 is negative as per the sign convention, because energy is transferred as heat from the system to the surroundings. For convenience, let the quantity of energy absorbed and rejected as heat by the system be represented as Q_1 and Q_2, respectively. That is, in expressing these quantities only the magnitude is taken into account but not the sign. Then the thermal efficiency of a heat engine can be expressed as:

$$\eta = \frac{\text{Energy absorbed as heat} - \text{Energy rejected as heat}}{\text{Energy absorbed as heat}}$$

$$\eta = \frac{\text{Net work done}}{\text{Energy absorbed as heat}} = \frac{W_T - W_P}{Q_1} = \frac{Q_1 - Q_2}{Q_1} \tag{7.2}$$

A second example of a heat engine is an automobile engine shown in Fig. 7.2 (idealized as a cylinder fitted with a frictionless piston). The working fluid (a certain amount of gas) is contained in the cylinder at all times. The working fluid undergoes the following processes and executes a cycle.

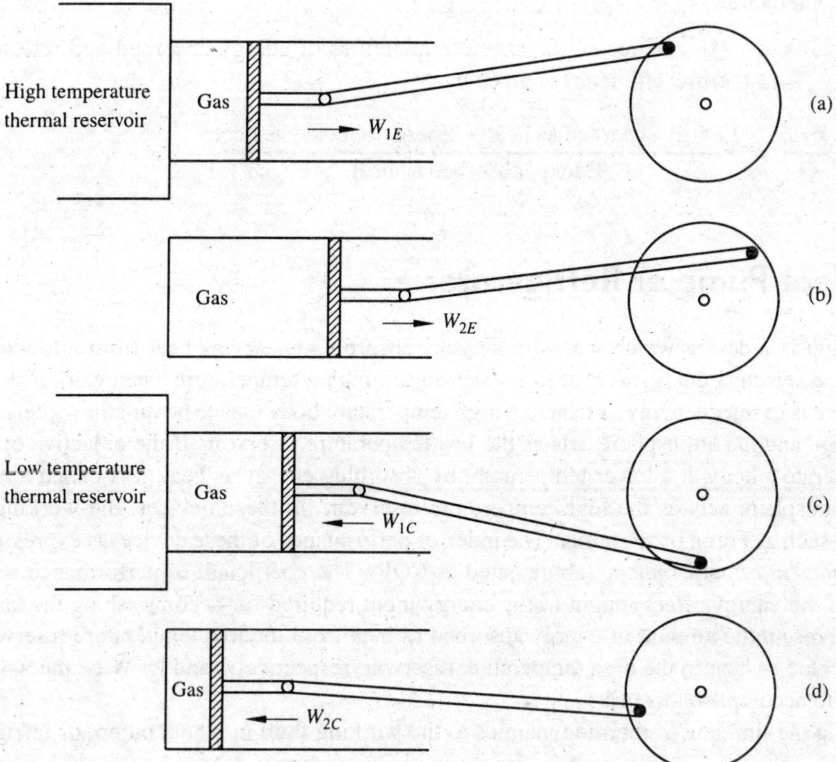

Fig. 7.2 *Schematic representation of processes in an automobile engine*

1. In the first process of the cycle, the cylinder is brought into contact with a thermal reservoir at high temperature (source). The working fluid absorbs a certain amount of energy as heat (Q_1), and the gas expands. The work done by the gas in this process is W_{1E}.
2. During the second process of the cycle, the cylinder is isolated and the working fluid is allowed to expand adiabatically. During this process the work done is W_{2E} and there is no heat interaction. The temperature of the gas at the end of this process becomes much less than the temperature of the source.
3. In the third process of the cycle, the cylinder is brought into contact with a thermal reservoir at lower temperature (sink). The gas is compressed while it exchanges energy as heat with the sink. During this process, the energy rejected as heat by the gas is Q_2 and the work done on the gas is W_{1C}.
4. The above process is terminated at a point such that further compression of the gas under adiabatic conditions brings the gas to the initial state, thus completing the cycle. During this process, there is no heat interaction and the work done on the gas is W_{2C}.

Applying the first law of thermodynamics to the cyclic device, we get

$$Q_1 - Q_2 = W_{1E} + W_{2E} - (W_{1C} + W_{2C}) = W_{\text{Net}} \tag{7.3}$$

$$\text{Heat interaction} = Q_1 - Q_2 = | Q_1 | - | Q_2 | \tag{7.4}$$

where $| Q_1 |$ and $| Q_2 |$ represent the absolute quantities of energy absorbed and rejected as heat, respectively. Then thermal efficiency is given by

$$\eta = \frac{W_{Net}}{Q_1} = \frac{\text{Energy absorbed as heat} - \text{Energy rejected as heat}}{\text{Energy absorbed as heat}} \tag{7.5}$$

7.3 Heat Pump or Refrigerator

A heat pump is a device which works in a cycle, absorbing energy as heat from a low temperature reservoir and rejecting energy as heat to a reservoir at a higher temperature when work is done on it. If the objective is to reject energy as heat to a high temperature body (space heating in winter), it is called a *heat pump*, and the atmosphere acts as the low temperature reservoir. If the objective of the device is to refrigerate a body at a lower temperature by absorbing energy as heat, it is called a *refrigerator* and the atmosphere acts as the high temperature reservoir. In these devices, the working fluid is a refrigerant such as Freon or ammonia. The index of performance of these devices is expressed in terms of the *coefficient of performance* (abbreviated as COP). The coefficient of performance is defined as the ratio of the energy effect sought to the energy input required for accomplishing the task. Let Q_L and Q_H represent the amount of energy absorbed as heat from the low temperature reservoir and the energy rejected as heat to the high temperature reservoir, respectively; and let W be the work done on the device to accomplish this task.

Applying the first law of thermodynamics to the working fluid in a heat pump, or refrigerator, we get

$$Q_H - Q_L = W \tag{7.6}$$

Therefore,

$$(COP)_R = \frac{Q_L}{W} = \frac{Q_L}{Q_H - Q_L} \tag{7.7}$$

and

$$(COP)_{HP} = \frac{Q_H}{W} = \frac{Q_H}{Q_H - Q_L} \tag{7.8}$$

A close look at the various heat engines reveals that they have a few common characteristics. They are:

1. Energy is absorbed as heat from a high temperature reservoir.
2. Energy is rejected as heat to a low temperature reservoir.
3. Some net work is done on the surroundings.

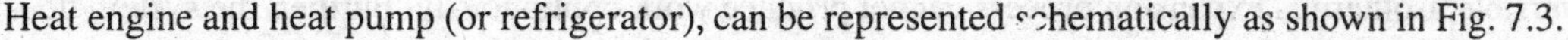

Heat engine and heat pump (or refrigerator), can be represented schematically as shown in Fig. 7.3.

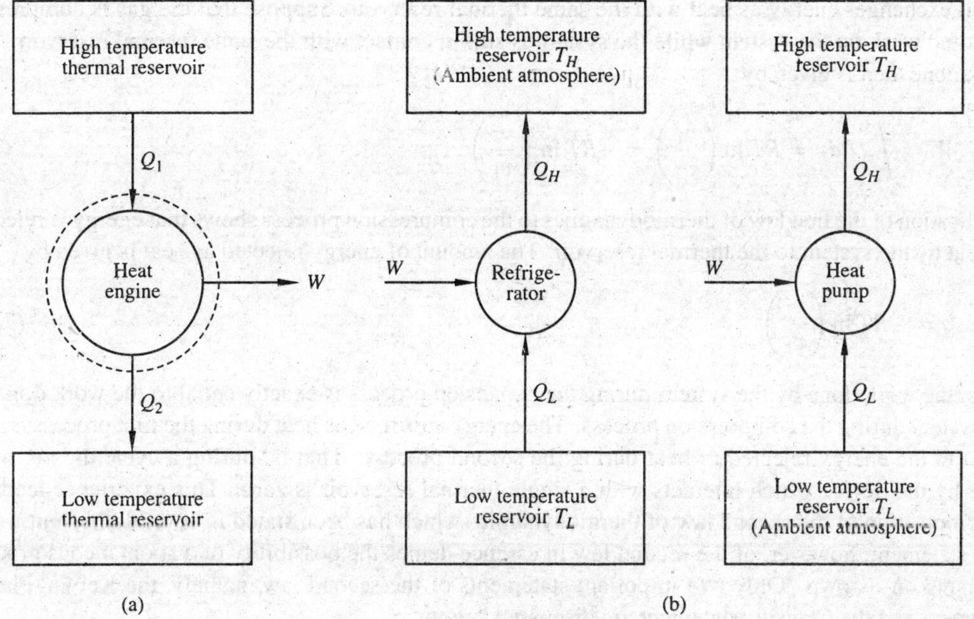

Fig. 7.3 *Schematic diagram of (a) a heat engine and (b) a heat pump or refrigerator*

7.4 Second Law of Thermodynamics

Let us consider as the system one mole of an ideal gas enclosed in a cylinder fitted with a frictionless piston. Let the cylinder be brought into contact with a thermal reservoir the temperature of which is infinitesimally greater than that of the gas inside the cylinder. When the gas undergoes a reversible expansion under isothermal conditions, the system does work on the surroundings. The work done during this expansion process is given by

$$W = \int_1^2 Pdv = \int_1^2 RT\frac{dv}{v} = RT\ln\left(\frac{v_2}{v_1}\right)$$

Since the gas is an ideal gas and the expansion is isothermal, the change in the internal energy of the gas $\Delta u = 0$. The application of the first law of thermodynamics to the isothermal expansion process gives

$$q = W = RT\ln\left(\frac{v_2}{v_1}\right)$$

During the reversible isothermal expansion of an ideal gas, all the energy absorbed as heat by the system is converted completely into work. Based on this, one should not immediately come to the conclusion that the energy absorbed as heat can be converted completely into work continuously. The reversible, isothermal expansion process cannot produce work continuously. Further, the system does not undergo a cyclic change. Let the cycle be completed by restoring the system to its initial state such that it exchanges energy as heat with the same thermal reservoir. Suppose that the gas is compressed by doing work on the system while the system is still in contact with the same thermal reservoir. The work done then is given by

$$W = \int_2^1 P dv = RT \ln\left(\frac{v_1}{v_2}\right) = -RT \ln\left(\frac{v_2}{v_1}\right) \tag{7.9}$$

Application of the first law of thermodynamics to the compression process shows that energy is rejected as heat by the system to the thermal reservoir. The amount of energy rejected as heat is given by

$$q = -RT \ln\left(\frac{v_2}{v_1}\right) \tag{7.10}$$

Thus, the work done by the system during the expansion process is exactly equal to the work done on the system during the compression process. The energy absorbed as heat during the first process is also equal to the energy rejected as heat during the second process. That is, during a cycle the net work done by the device which interacts with a single thermal reservoir is zero. This experience leads to the proposition of the second law of thermodynamics which has been stated in several different ways. Any statement, however, of the second law in essence denies the possibility of a spontaneous process reversing on its own. Only two important statements of the second law, namely, the Kelvin–Planck statement and the Clausius statement are discussed below.

7.4.1 Kelvin–Planck statement

The Kelvin–Planck statement of the second law of thermodynamics reads *it is impossible to construct a cyclically operating device such that it produces no other effect than the absorption of energy as heat from a single thermal reservoir and performs an equivalent amount of work.*

The Kelvin–Planck statement tells us that it is impossible to construct an engine which, while operating cyclically, absorbs a certain amount of energy as heat from a high temperature reservoir (source) and converts all of it into work. The only option then is that the engine converts part of the energy it receives as heat into work, and rejects the rest to another thermal reservoir the temperature of which is less than the temperature of the source. Then, at least two thermal reservoirs, one of high temperature (source) from which the working fluid receives energy as heat, and another of low temperature (sink) to which the working fluid rejects energy as heat, are needed for a heat engine. Once the heat engine rejects part of the energy it receives, its efficiency becomes less than one. Thus, the Kelvin–Planck statement further implies that no heat engine can have a thermal efficiency of one. This does not violate the first law of thermodynamics either.

Consider an ideal gas enclosed in a frictionless piston–cylinder assembly as a system. During the isothermal expansion, $\Delta U = 0$ for an ideal gas. Then $Q = W$. This implies that all the energy transferred as heat to the system is converted completely into work. Does this imply that it is possible to construct a device which exchanges energy as heat with only one thermal reservoir and converts all the energy absorbed as heat into work in violation of the Kelvin–Planck statement? In fact, the example does not violate the Kelvin–Planck statement, because the device is not working cyclically; that is, the gas in the cylinder-piston assembly has not returned to the initial state. Moreover, the system does not give a continuous supply of work. In other words, during a given process (for example, the one cited above), it may be possible to convert the energy absorbed as heat completely into work when the system exchanges energy with only one thermal reservoir. However, when considered over a cycle it is impossible for any device to have a thermal efficiency of one.

7.4.2 Perpetual motion machine of the second kind

The first law of thermodynamics does not stipulate any restriction on the thermal efficiency of a heat engine. However, the second law restricts the thermal efficiency of a heat engine to less than one. It stipulates that some portion of the energy absorbed as heat from a source must always be rejected to a low temperature sink. Wilhelm Ostwald introduced the concept of Perpetual Motion Machine of the Second Kind, that is, of a device which could perform work solely by absorbing energy as heat from a single thermal reservoir (see Fig. 7.4). Clearly, such a device does not violate the first law of thermodynamics because it would perform work at the expense of the internal energy of a body. A Perpetual Motion Machine of the Second Kind (PMMSK) is a hypothetical device which, working cyclically, receives energy as heat from a single thermal reservoir and delivers an equivalent amount of work.

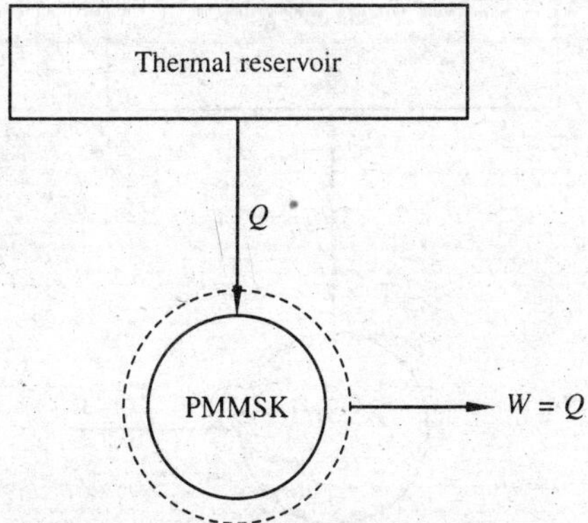

Fig. 7.4 *Perpetual motion machine of the second kind*

Several attempts have been made in the past to devise a perpetual motion machine of the second kind and all of them have been unsuccessful. The Kelvin–Planck statement of the second law tells us that it is impossible to construct a perpetual motion machine of the second kind.

7.4.3 Clausius statement

The Clausius statement of the second law of thermodynamics reads *"it is impossible to construct a device that, working cyclically, will produce no other effect than the transfer of energy in the form of heat from a low temperature body to a high temperature body"*.

This statement tells us that it is impossible for any device, unaided by an external agency, to transfer energy as heat from a cooler body to a hotter body. The device for transferring energy as heat from a cooler body to a hotter body (like a refrigerator or a heat pump), has to receive some external energy. In other words, energy in the form of heat cannot pass spontaneously from a body at a lower temperature to a body at a higher temperature. Whereas, the flow of energy as heat from a body of higher temperature to a body of lower temperature is a spontaneous process. The Clausius statement of the second law of thermodynamics denies the possibility of self-reversal of such a spontaneous process. That is, the second law specifies the direction of a spontaneous process.

Consider the device shown in Fig. 7.5. The Clausius statement implies that it is not possible to operate the device when $W = 0$. When $W = 0$, $(COP)_R = \infty$ and $(COP)_{HP} = \infty$. That is, it is not possible to construct a refrigerator, or a heat pump whose COP is infinity. Consider a domestic refrigerator as an example. This device extracts energy as heat from the material to be cooled, and transfers it to the surroundings. This does not violate the Clausius–statement, for the device is externally aided, that is, the refrigerator is supplied with electrical power. Such a device, not only transfers the energy as heat from a low temperature body to a high temperature body, but also leaves other effects (receives energy from an external source) on the surroundings other than the two thermal reservoirs between which it is working.

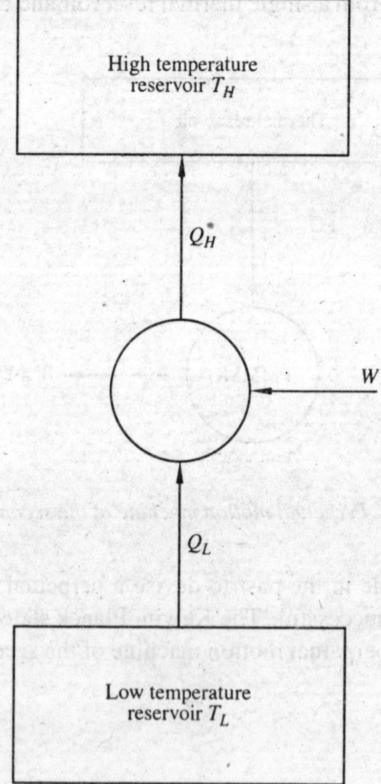

Fig. 7.5 *Schematic diagram of a refrigerator or heat pump*

7.5 Equivalence of Kelvin–Planck and Clausius Statements

The second law of thermodynamics—for that matter, all the laws of thermodynamics—cannot be proved like a mathematical theorem. It is an axiom based on experimental evidence. Every experiment that has been conducted so far to disprove the validity of the second law has been unsuccessful.

The two statements of the second law have been discussed in Section 7.4. At first it appears as if the two statements are different altogether. However, it can be shown that the Kelvin–Planck statement and the Clausius statement are, in fact, equivalent. To show that they are equivalent, it is enough if we prove that the Kelvin–Planck proposition implies the Clausius proposition and *vice-versa.* Alternatively, their equivalence can be proved by contradiction, that is, the violation of one statement should imply the violation of the other statement. In proving that the Kelvin–Planck statement and the Clausius statement are equivalent, the second method is adopted below.

To prove that violation of the Kelvin–Planck statement leads to violation of the Clausius statement, let us suppose that the Kelvin–Planck statement is incorrect. That is, it is possible to construct a device (1) which, working cyclically, absorbs energy as heat (Q_1) from a source at temperature T_H and performs an equivalent amount of work $W (W = Q_1)$. Let us consider another cyclically working device (2) which absorbs energy as heat (Q_L) from a reservoir at temperature T_L and delivers energy as heat (Q_H) to a high temperature reservoir at temperature T_H ($T_H > T_L$). To accomplish this, work W is done on the device. The device (2) is such that it does not violate any of the two statements. Then, according to the first law of thermodynamics

$$Q_H = Q_L + W \tag{7.11}$$

Now let us consider the combination of these two devices such that the work delivered by device (1) is used by device (2) as shown in Fig. 7.6.

Then

$$W = Q_1 \tag{7.12}$$

and

$$Q_H = Q_L + W = Q_L + Q_1 \tag{7.13}$$

This combined device (which is not aided by any external agency), working cyclically, is not producing any effect other than the transfer of energy as heat (Q_L) from the low temperature reservoir to the high temperature reservoir. This is in violation of the Clausius statement. Hence, the violation of the Kelvin–Planck statement leads to violation of the Clausius statement.

To prove that violation of the Clausius statement leads to violation of the Kelvin–Planck statement, let us assume that the Clausius statement is incorrect. That is, it is possible to construct a device (1) such that it transfers energy as heat (Q) from a body at a lower temperature to a body at a higher temperature unaided by any external agency as shown in Fig. 7.7. Consider another device (2) which receives energy as heat (Q_1) from a body at high temperature, delivers work (W) and rejects energy as heat (Q) to the body at a low temperature. This device (2) does not violate any of the above two statements.

Application of the first law of thermodynamics to device (2) gives

$$Q_1 - Q = W \tag{7.14}$$

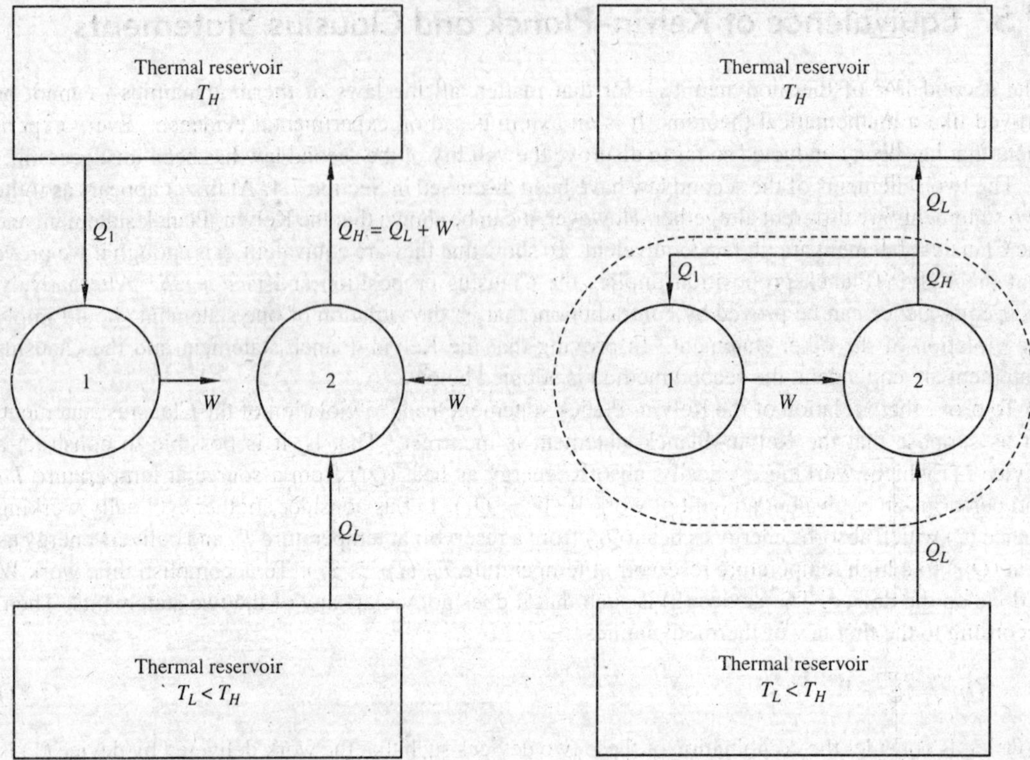

Fig. 7.6 *Illustration of violation of Kelvin–Planck statement leads to violation of Clausius statement*

Now consider the combination of devices (1) and (2) as a single device (see Fig. 7.7). This combined device, working cyclically, absorbs energy ($Q_1 - Q$) as heat from the thermal reservoir at temperature T_H and delivers work ($W = Q_1 - Q$), leaving the thermal reservoir at temperature T_L unaffected. That is, the resulting device is a perpetual motion machine of the second kind, which is in violation of the Kelvin–Planck statement. Thus, violation of the Clausius statement leads to violation of the Kelvin–Planck statement. Therefore, the Kelvin–Planck statement and the Clausius statement are equivalent.

7.6 Reversible and Irreversible Processes

It has been shown in Section 5.3 that the first law of thermodynamics leads to certain consequences. Similarly, the second law of thermodynamics, as stated above, also leads to certain important consequences. Before discussing the consequences of the second law, it is important to understand reversible and irreversible processes. An engineer is most interested in reversible processes since these reversible processes yield the maximum work in engines and require minimum work in devices such as refrigerators, compressors, pumps etc. Therefore, an engineer should be able to recognize the irreversible processes, and try to reduce irreversibility in any given process. Processes can be classified as reversible processes and irreversible processes as follows.

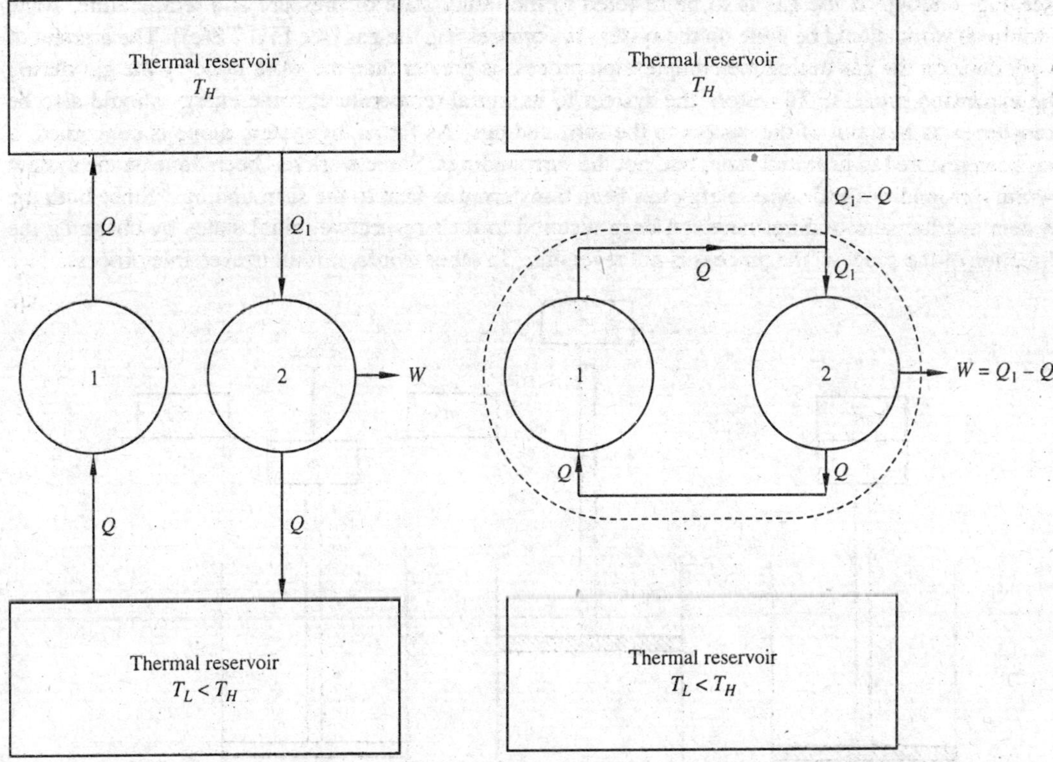

Fig. 7.7 *Illustration of the fact that violation of Clausius statement leads to violation of Kelvin–Planck statement*

A process is said to be reversible if both the system and the surroundings can be restored to their respective initial states by reversing the direction of the process. Otherwise, the process is said to be irreversible.

A system can, however, always be restored to its initial state; but this does not imply that the process is reversible. Only if the surroundings are also simultaneously restored to their initial states, the process becomes reversible. In other words, a reversible process should not leave a permanent change in both the system and its surroundings. The changes should be completely erased by reversing the direction of the process. Let us consider a few specific examples to distinguish between reversible and irreversible processes.

7.6.1 Expansion/compression with finite pressure difference

Consider a certain amount of gas enclosed in an adiabatic cylinder fitted with a frictionless piston as the system. Suppose the piston is loaded with blocks that exactly balance the gas pressure. Let the piston be almost at the bottom of the cylinder initially as shown in Fig. 7.8(a). If one block is shifted on to a platform which is at the same level as the block, then the piston suddenly moves up. Some work is done during this expansion process since the remaining block has been raised over a certain distance [see Fig. 7.8(b)]. If the gas is to be restored to the initial state of pressure and temperature, some additional work should be done on the system in compressing the gas [see Fig. 7.8(c)]. The amount of work done on the gas during the compression process is greater than the work done by the gas during the expansion process. To restore the system to its initial temperature, some energy should also be transferred as heat out of the system to the surroundings. As far as the system alone is concerned, it has been restored to its initial state; but, not the surroundings. Some work has been done on the system by the surroundings and some energy has been transferred as heat to the surroundings. Since both the system and the surroundings have not been restored to their respective initial states by changing the direction of the process, the process is not reversible. In other words, it is an irreversible process.

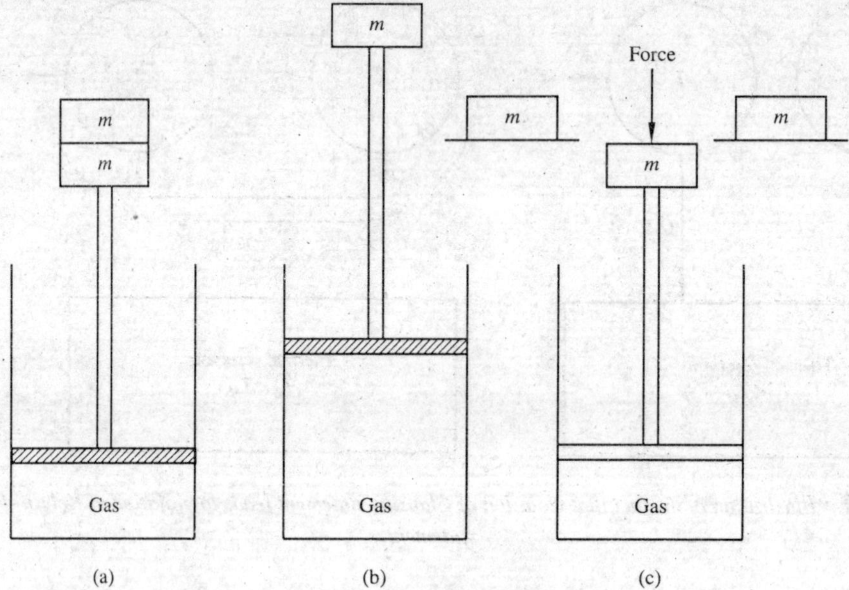

Fig. 7.8 *Expansion with finite pressure difference: Irreversible process*

7.6.2 Expansion/compression with infinitesimal pressure difference

Expansion or compression of a gas enclosed in a cylinder fitted with a frictionless piston can, however, be made reversible in the following way. Suppose the piston is loaded with a large number of small weights (in the limit, the piston is loaded with a heap of fine sand). Suppose the sand particles are removed one after another allowing the expansion process to proceed slowly. At each instant the external pressure is infinitesimally different from the gas pressure. This process is shown in Fig. 7.9. The system reaches a state of equilibrium at every successive step with its unique properties. The state of the system can be described by specifying its properties and the path followed by the system can be traced on a thermodynamic diagram.

The compression process can be carried out by restoring the sand particles, one after another, on to the piston from the platforms which are at the same level as the sand particles. No additional work is involved in compressing the gas. That is, the process can be carried out in either direction. By reversing

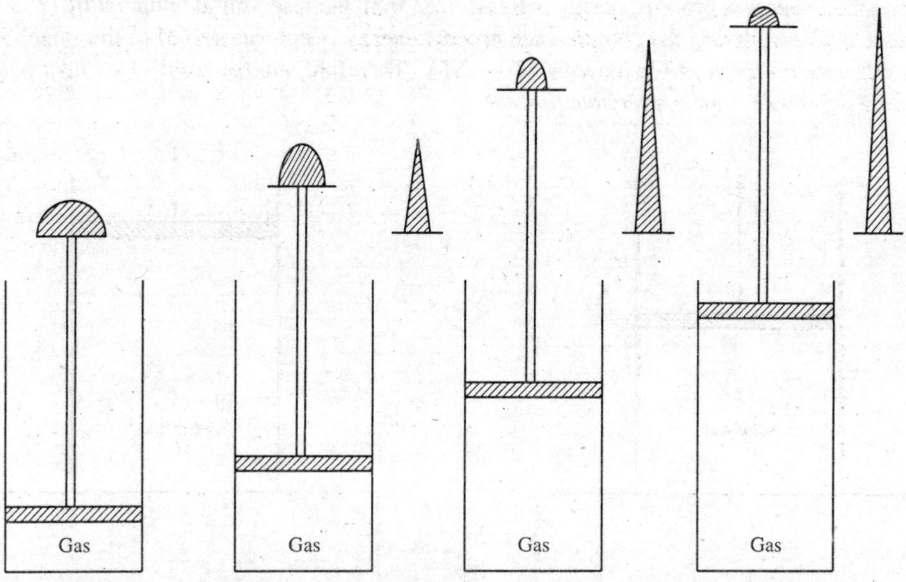

Fig. 7.9 *Expansion with infinitesimal pressure difference: Reversible process*

the direction of the process at any instant, both the system and the surroundings can be restored to their respective initial states. Such a process is called a reversible process. Necessarily, a reversible process is quasi-static, but the converse need not be true. Thus we see that expansion or compression with finite pressure differences is an irreversible process. Expansion or compression with infinitesimal pressure differences (in the limit $dP = 0$) is a reversible process.

7.6.3 Heat transfer with finite temperature difference

Let us consider another example of an irreversible process. Suppose a certain amount of an ideal gas enclosed in a cylinder fitted with a frictionless piston shown in Fig. 7.10 is taken as the system. Let the temperature of the gas be T, and let it undergo an isothermal expansion while in contact with a thermal reservoir. For an isothermal expansion of an ideal gas, $dU = 0$ and $dQ = dW$. Then, energy should flow as heat into the system from the surroundings during the expansion process. For the energy to flow, it should be brought into contact with a thermal reservoir the temperature of which is greater than the temperature of the gas. Let $(T+\Delta T)$ be the temperature of the thermal reservoir. Now, let us reverse the direction of the process. During an isothermal compression, work is done on the system to restore it to its original volume. The work done on the system during compression is equal in magnitude to the work done by the system during expansion. Also, according to the first law, energy should flow as heat from the system to the surroundings. The heat interaction during the compression process is equal in magnitude to the heat interaction during the expansion process. For the energy to flow as heat out of the system, let it be brought into contact with a thermal reservoir the temperature $(T - \Delta T)$ of which is less than the system temperature. The system has now been restored to the initial state by reversing the direction of the process; but the surroundings have not reached their initial states. It should be noted that during the expansion process, energy is transferred from the reservoir at temperature $(T + \Delta T)$ to the system; whereas, during the compression process energy is not transferred to the same reservoir but to a different reservoir at temperature $(T - \Delta T)$. Therefore, *energy transfer as heat with finite temperature difference is an irreversible process.*

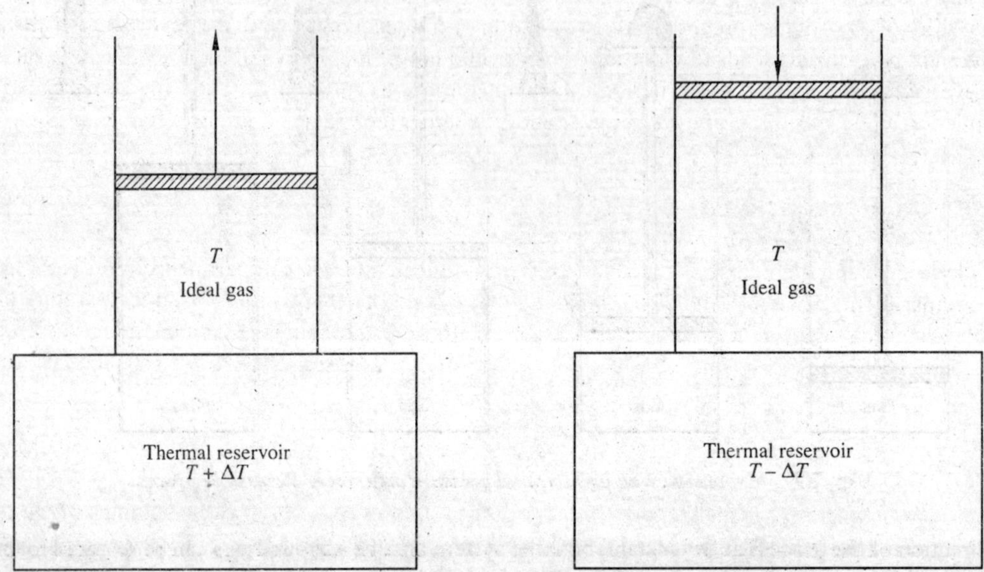

Fig. 7.10 *Heat transfer with finite temperature difference: Irreversible process*

7.6.4 Heat transfer with infinitesimal temperature difference

Suppose the above system shown in Fig. 7.10 is now brought into contact with a thermal reservoir the temperature of which is equal to the system temperature. If the system undergoes an isothermal expansion, energy flows as heat from the reservoir to the system. The work done by the system is equal to the heat interaction. If the direction of the process is reversed, the energy rejected as heat to the reservoir during the compression process is equal to the work done on the system. The work and heat interactions between the system and the surroundings during the compression process are exactly balanced by the work and heat interactions during the expansion process. During both the expansion and compression processes, the system exchanges energy as heat with the same reservoir. That is, by reversing the direction of the process, both the system and the surroundings are restored to their respective initial states. Hence, *energy transfer as heat with no temperature difference is a reversible process.*

7.7 Criterion for Irreversible Processes

It has been stated in Section 7.4 that the second law of thermodynamics stipulates that certain processes (spontaneous processes) proceed in one direction only and these processes are irreversible. In other words, the second law distinguishes between reversible and irreversible processes. *If a process can proceed in either direction without violating the second law of thermodynamics, it is a reversible process.* On the other hand, *if a process can proceed in one direction only and the reversal of the direction leads to violation of the second law, the process is called an irreversible process.* Therefore, the criterion that can be used to classify processes as reversible or irreversible is to see whether the assumption of reversibility of a particular process leads to violation of the second law of thermodynamics. If the assumption of reversibility leads to violation of the second law of thermodynamics, the process is said to be irreversible. By applying this criterion irreversible processes can be identified. The above criterion is applied to a few specific examples below to judge whether the processes are irreversible or not.

7.7.1 Free expansion of a gas

Consider a rigid and thermally insulated tank which is divided into two compartments by a diaphragm. One compartment contains N moles of an ideal gas at some known pressure (P) and temperature (T), and the second compartment is completely evacuated. If the diaphragm is punctured, the gas expands and occupies the entire tank. Since the tank is rigid and thermally insulated, $W = Q = 0$, and $\Delta U = 0$. Then

$$\Delta U = NC_v(T_f - T) = 0 \tag{7.15}$$

Hence, $T_f = T$ where, T_f is the final temperature of the gas after expansion. If the volumes of the two compartments are equal, the final pressure $P_f = P/2$.

The above expansion process is clearly irreversible. No work is done during free expansion. To restore the system to its initial state, work is to be done on the system by the surroundings and the amount of work to be done is given by

$$W = \int PdV = NRT \ln \frac{1}{2} = -NRT \ln 2 \tag{7.16}$$

It is also necessary to transfer energy as heat by an amount $Q = (-NRT \ln 2)$ from the system to the surroundings. Though the system is restored to its initial state, the surroundings are not, and hence the process is irreversible.

Now let us apply the criterion stated in the beginning of this Section to judge whether the process can be classified as irreversible or not. Let us assume that the process is reversible, that is the gas flows through the punctured diaphragm and occupies only one compartment, leaving the second compartment empty as shown in Fig. 7.11. Then the gas can be expanded adiabatically till the pressure on both sides of the diaphragm is the same. Since some work is done during the expansion process, the internal energy of the gas decreases. Then the tank (after removing part of the insulation) can be brought into contact with a thermal reservoir and energy can be transferred as heat from the reservoir to the tank, raising the internal energy of the gas to its original value. Thus, the cycle can continue. The net result of these processes is that energy is being extracted as heat from a single thermal reservoir and work is being done. This is in violation of the Kelvin–Planck statement of the second law. Hence *free expansion is an irreversible process.*

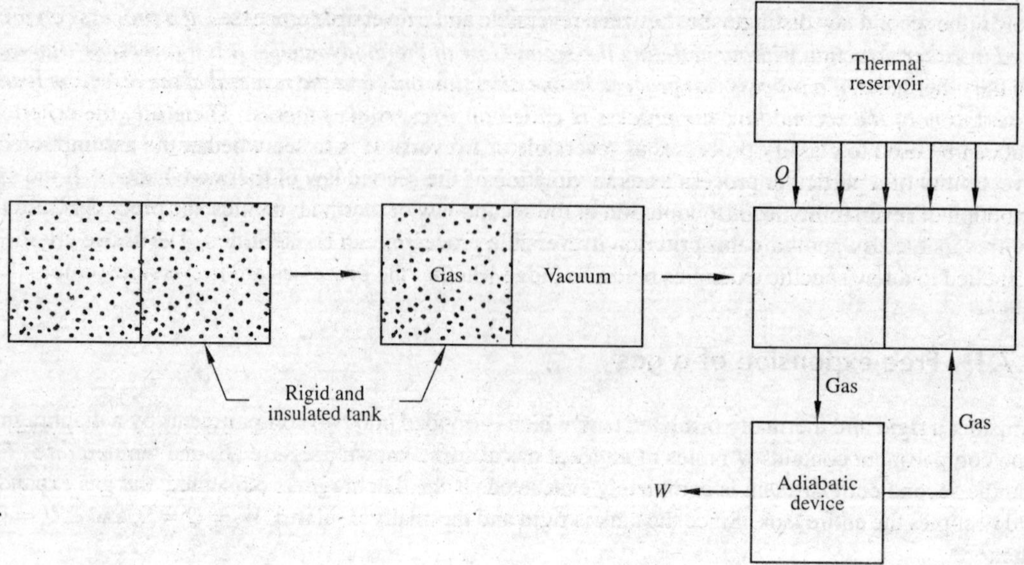

Fig. 7.11 *Free expansion of a gas: Irreversible process*

7.7.2 Mixing of matter at different states

Let us consider a rigid and thermally insulated tank divided into two compartments by a partition as shown in Fig. 7.12. One compartment contains a hot fluid and the other a cold fluid. If the partition is removed, the fluids mix and attain an equilibrium temperature. Let us now assume that the process is reversible, that is, the mixed fluid at a common equilibrium temperature would separate into the original hot and cold fluids. This clearly involves the transfer of energy as heat from a lower temperature body to a higher temperature body unaided by any external agency. This is certainly in violation of the Clausius statement of the second law of thermodynamics, and hence the process is irreversible.

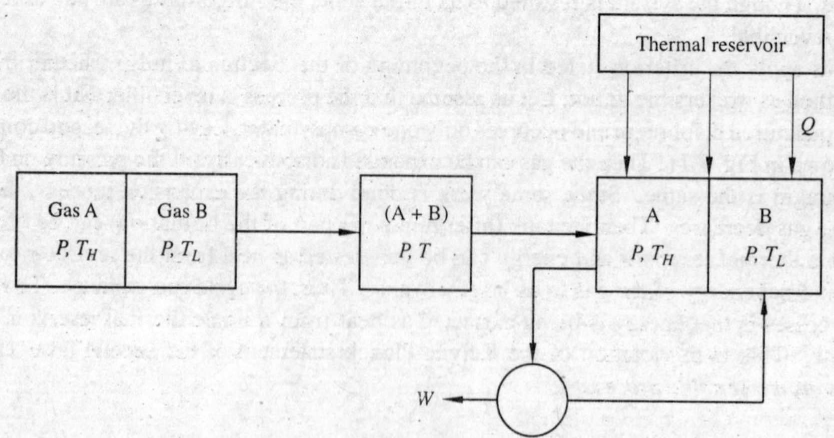

Fig. 7.12 *Mixing of matter at different states: Irreversible process*

Now let us apply the criterion for irreversibility to judge whether the process is reversible or not. Assume that the process is reversible. Then, the fluid at equilibrium temperature separates and part of it is available as a hot fluid in one compartment and the rest is available as a cold fluid in the other compartment. Then the hot fluid and the cold fluid can be used as source and sink, respectively to run a reversible heat engine. Since the source and sink are finite amounts of fluid, their temperatures change continuously. The heat engine runs till the source and sink attain a common temperature and in the process some net work is done. Then, energy can be transferred as heat from a thermal reservoir to the fluid and it can be raised to its equilibrium temperature. The fluid at equilibrium temperature separates into hot and cold fluids and the cycle can continue. The result of these processes is that energy is extracted from a single thermal reservoir and work is done resulting in a perpetual motion machine of the second kind leading to violation of the Kelvin–Planck statement of the second law of thermodynamics. Hence *mixing of matter at different states is an irreversible process.*

The criterion stated earlier in Section 7.7 has been applied to identify irreversible processes. The processes in which the causes of irreversibility are not present are reversible processes. All naturally occurring processes are spontaneous or irreversible. In practice, it is very difficult to carry out a process under perfectly reversible conditions and, therefore, reversible process is only an idealization. It can be approached to a large extent by carrying out the process at a very slow rate. The processes which can be idealized as reversible are:

1. Motion without friction.
2. Expansion or compression with no pressure difference between the system and surroundings.
3. Energy transfer as heat with no temperature difference across the system boundary.

The examples of irreversible processes are:

1. Motion with friction.
2. Free expansion.
3. Expansion or compression with finite pressure difference.
4. Energy transfer as heat with finite temperature difference.
5. Mixing of matter at different states.
6. Mixing of different gases.
7. Spontaneous chemical reactions.

The characteristics of irreversible processes are:

1. An irreversible process can be carried out in one direction only.
2. An irreversible process occurs at a finite rate.
3. During an irreversible process, the system is not in equilibrium.
4. An irreversible process cannot be reversed without causing permanent changes in the surroundings.

7.8 Carnot Cycle

Reversible processes can be combined to form a cycle called a reversible cycle. Since all the processes constituting the cycle are reversible, the cycle can be reversed. When the cycle is reversed, the heat and work interactions for the reversed cycle are the same in magnitude to the heat and work interactions for the original cycle. However, the direction of the interactions is reversed compared to the direction of the original cycle.

French military engineer, Nicholas Sadi Carnot (1796–1832), was the first to introduce the idea of a reversible cycle which is named after him. From the second law of thermodynamics, it has been observed that the efficiency of a heat engine cannot be equal to unity. If the efficiency of a heat engine is less than unity, what is the maximum efficiency of a heat engine? This can be answered by considering the Carnot cycle. The second law stipulates that it is impossible to construct a perpetual motion machine of the second kind, that is, a device which interacts with only one thermal reservoir. Therefore, at least two thermal reservoirs are needed for a device, namely, a high temperature reservoir from which the system absorbs energy and a low temperature reservoir to which the system rejects energy as heat. The Carnot cycle uses only two thermal reservoirs. The working fluid in the Carnot cycle receives energy as heat reversibly from the high temperature thermal reservoir and rejects energy as heat reversibly to the low temperature thermal reservoir. That is, the working fluid undergoes a cyclic change in which two of the processes are reversible and isothermal. By connecting the two isothermal processes with two reversible adiabatic processes, the cycle can be completed. The Carnot cycle consists of two reversible, isothermal processes and two reversible, adiabatic processes.

7.8.1 Carnot engine

Now let us consider the operation of a hypothetical engine which employs the Carnot cycle. The Carnot engine consists of a cylinder–piston assembly in which a certain amount of gas (working fluid) is enclosed. Let the cycle be started with the gas in state 1 (see Fig. 7.13).

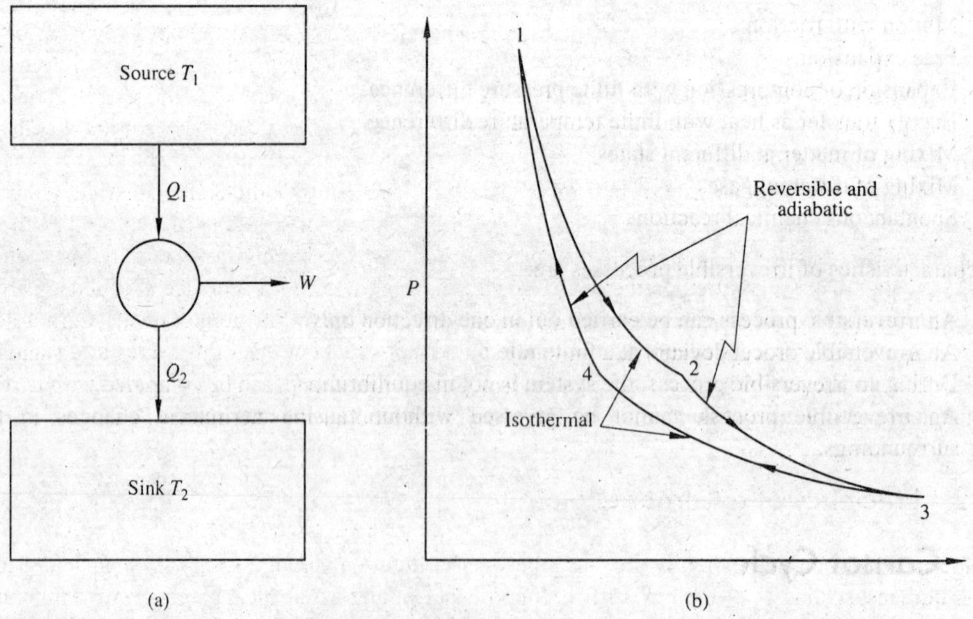

Fig. 7.13 *(a) Schematic representation of a Carnot engine. (b) P–v diagram of a Carnot engine*

1. In the first process the cylinder head is brought into contact with a source, the temperature (T_1) of which is equal to the temperature of the gas inside the cylinder. The gas expands reversibly and isothermally. During this process, the system absorbs energy as heat (Q_1) from the source. The system changes its state from 1 to 2 as shown on the $P - V$ diagram in Fig. 7.13(b).
2. In the second process the cylinder head is insulated and the gas is allowed to expand further till its temperature is equal to the sink temperature T_2. The system reaches state 3.
3. In the third process, the cylinder head is brought into contact with the sink and the gas is compressed reversibly and isothermally till it reaches state 4. During this process the system rejects energy as heat (Q_2) to the sink.
4. In the next process, the gas is compressed reversibly and adiabatically till it reaches the initial state 1, thus completing the cycle.

Since the system has undergone a cyclic change, application of the first law of thermodynamics to the system yields:

$$\oint dQ - \oint dW = 0$$

$$\oint dQ = Q_1 - Q_2 = \oint dW \tag{7.17}$$

and the thermal efficiency is given by:

$$\eta = \frac{W}{Q_1} = \frac{Q_1 - Q_2}{Q_1} = 1 - \frac{Q_2}{Q_1} \tag{7.18}$$

The Carnot engine is a hypothetical device and it is not practical to construct one. It consists of all reversible processes and the engine works at a very slow rate converting the thermal energy into work. The size of the engine will be large for a reasonable amount of work output. Nevertheless, the concept of Carnot engine serves as a yard stick to measure the performance of any other device; for, it consists of only reversible processes.

Since the Carnot cycle consists of reversible processes, one can hypothetically construct a device which executes the cycle in the reverse order. That is, the device absorbs energy as heat (Q_L) from the low temperature reservoir at T_L and delivers energy as heat (Q_H) to the high temperature reservoir at T_H when work (W) is performed on the device as shown in Fig. 7.5. Such a device is called a Carnot refrigerator or heat pump, depending upon the objective. The Carnot refrigerator or the Carnot heat pump, is also a hypothetical device and it is impossible to construct one. But this serves as a standard of perfection and the performance of any other refrigerator or heat pump, can be compared with it. The coefficient of performance of the Carnot refrigerator or heat pump is the maximum.

7.8.2 Efficiency of a Carnot engine

Consider a Carnot engine which is operating between a high temperature reservoir at T_1 and a low temperature reservoir at T_2 [see Fig. 7.13(b)]. Suppose the engine uses an ideal gas as the working fluid. To evaluate the thermal efficiency of the engine, a knowledge of the work done and the heat interaction during all the processes is necessary. The work done during a given process can be evaluated from the relation $W = \int Pdv$. To evaluate $\int Pdv$, a knowledge of the path and the nature of the working fluid is necessary. 1–2 is an isothermal expansion process. The first law of thermodynamics for process 1–2 gives:

$$q_1 = W = \int_1^2 P dv = RT_1 \ln \frac{v_2}{v_1}$$

2–3 and 4–1 are reversible adiabatic processes and hence there is no heat interaction in these processes. 3–4 is a reversible isothermal compression process. The first law of thermodynamics for process 3–4 gives

$$q_2 = W = \int_3^4 P dv = RT_2 \ln \frac{v_4}{v_3} \tag{7.19}$$

Thermal efficiency η = Net work/Energy absorbed as heat

We know that $\oint dq = \oint dw$. Therefore,

$$\eta = \frac{RT_1 \ln \frac{v_2}{v_1} + RT_2 \ln \frac{v_4}{v_3}}{RT_1 \ln \frac{v_2}{v_1}} = 1 + \frac{RT_2 \ln \frac{v_4}{v_3}}{RT_1 \ln \frac{v_2}{v_1}} \tag{7.20}$$

The efficiency of the Carnot engine as expressed above is in terms of the volumes at state 1 to 4. To simplify this relation further, we can make use of the fact that states 2 and 3 are connected by a reversible adiabatic path, and so are the states 4 and 1. Therefore,

$$T_1 v_2^{\gamma-1} = T_2 v_3^{\gamma-1} \tag{7.21}$$

$$T_2 v_4^{\gamma-1} = T_1 v_1^{\gamma-1} \tag{7.22}$$

$$\text{or} \quad \frac{v_2}{v_3} = \left(\frac{T_2}{T_1}\right)^{\frac{1}{\gamma-1}} \tag{7.23}$$

$$\text{and} \quad \frac{v_1}{v_4} = \left(\frac{T_2}{T_1}\right)^{\frac{1}{\gamma-1}} \tag{7.24}$$

Hence

$$\frac{v_2}{v_3} = \frac{v_1}{v_4} \quad \text{or} \quad \frac{v_2}{v_1} = \frac{v_3}{v_4} \tag{7.25}$$

Substituting Eq. (7.25) in Eq. (7.20), we obtain

$$\eta = 1 - \frac{RT_2 \ln \frac{v_3}{v_4}}{RT_1 \ln \frac{v_2}{v_1}} = 1 - \frac{T_2}{T_1} \tag{7.26}$$

7.9 Carnot Theorems

The following two consequences of the second law of thermodynamics were proposed by Carnot regarding the efficiency of a Carnot engine and are known as Carnot theorems. The proof of these propositions is by contradiction. If an assumption contrary to the proposition leads to violation of the second law, then the assumption is incorrect; thus, proving that the proposition is correct. The two Carnot theorems are presented below.

7.9.1 Carnot theorem 1

No heat engine operating between the two given thermal reservoirs, each of which is maintained at a constant temperature, can be more efficient than a reversible engine operating between the same two thermal reservoirs.

To prove this, let us assume that the Carnot theorem 1 is incorrect. That is, one can construct an engine I (let us say an irreversible engine) which operates between the same thermal reservoirs as those of the reversible engine such that the efficiency of the irreversible engine is greater than that of the reversible engine. That is by assumption:

$$\eta_I > \eta_R \tag{7.27}$$

Let the irreversible engine I absorb energy as heat (Q_1) from the high temperature reservoir at T_1. It does work W_I and rejects energy as heat (Q) to the low temperature reservoir at T_2. Then the work done by the engine is given by

$$W_I = Q_1 - Q$$

and

$$\eta_I = \frac{W_I}{Q_1} = 1 - \frac{Q}{Q_1} \tag{7.28}$$

Consider a reversible engine R which works between the same two thermal reservoirs. It performs work W_R by absorbing energy as heat (Q_1) from the source at T_1 and rejecting energy as heat (Q_2) to the sink at T_2. Then

$$W_R = Q_1 - Q_2$$

or

$$\eta_R = \frac{W_R}{Q_1} = 1 - \frac{Q_2}{Q_1} \tag{7.29}$$

Since $\eta_I > \eta_R$ we get,

$$1 - \frac{Q}{Q_1} > 1 - \frac{Q_2}{Q_1}$$

Hence

$$Q < Q_2 \tag{7.30}$$

and

$$W_I(= Q_1 - Q) > W_R(= Q_1 - Q_2)$$

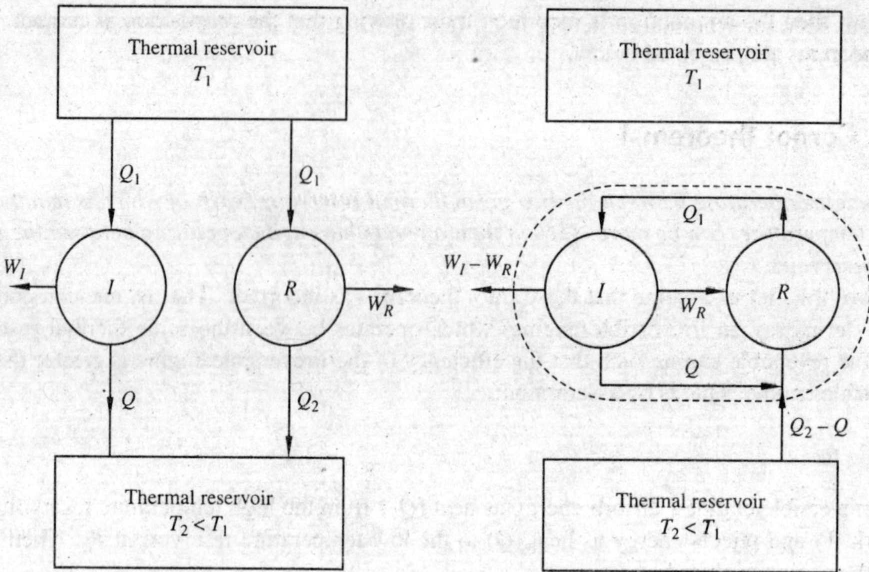

Fig. 7.14 *Illustration to prove the Carnot theorem 1*

Therefore, for the same amount of energy intake, the irreversible engine I does more work than the reversible engine R. Since the engine R is reversible, it can be made to execute the cycle in the reverse order. Since W_I is greater than W_R, the engine R can be made to execute the cycle in the reverse order utilizing part of the work done by the engine I. The combination of the two devices is as shown in Fig. 7.14.

The net work delivered by the combined device is given by

$$W_I - W_R = (Q_1 - Q) - (Q_1 - Q_2) = Q_2 - Q \tag{7.31}$$

The combined device absorbs energy as heat $(Q_2 - Q)$ from the reservoir at T_2, leaving the reservoir at T_1 unaffected, and does an equivalent amount work. Thus, it becomes a perpetual motion machine of the second kind, which violates the second law of thermodynamics. Hence, the assumption that the efficiency of the irreversible engine is greater than the efficiency of the reversible engine is incorrect. Therefore

$$\eta_R \geq \eta_I$$

7.9.2 Carnot theorem 2

All reversible heat engines operating between the two given thermal reservoirs have the same efficiency. The efficiency of a reversible heat engine does not depend on the working fluid; it depends only on the temperature of the reservoirs between which it operates.

To prove the Carnot theorem 2, let us assume that the efficiency of the reversible engine R_1 is greater than the efficiency of the reversible engine R_2. The engine R_1 absorbs energy as heat (Q_1) from the constant temperature thermal reservoir at T_1, does work W_{R1} and rejects energy as heat (Q) to the reservoir at T_2. The engine R_2 absorbs energy as heat (Q_1) from the reservoir at T_1, does work W_{R2} and rejects energy as heat (Q_2) to the reservoir at T_2 as shown in Fig. 7.15. Then

$$W_{R1} = Q_1 - Q$$

$$W_{R2} = Q_1 - Q_2$$

$$\eta_{R1} = \frac{W_{R1}}{Q_1} = 1 - \frac{Q}{Q_1} \tag{7.32}$$

and

$$\eta_{R2} = \frac{W_{R2}}{Q_1} = 1 - \frac{Q_2}{Q_1} \tag{7.33}$$

By assumption,

$$\eta_{R1} > \eta_{R2} \tag{7.34}$$

Then

$$1 - \frac{Q}{Q_1} > 1 - \frac{Q_2}{Q_1} \quad \text{or} \quad Q < Q_2 \tag{7.35}$$

Therefore

$$W_{R1} > W_{R2}$$

Since $W_{R1} > W_{R2}$, R_2 can be run as a heat pump utilizing part of the work done by R_1. The combination of the two devices is as shown in Fig. 7.15.

The net work done by the device is given by

$$W_{R1} - W_{R2} = (Q_1 - Q) - (Q_1 - Q_2) = Q_2 - Q$$

The resulting device absorbs energy as heat ($Q_2 - Q$) from the reservoir at temperature T_2, without requiring any interaction with the second reservoir, and delivers an equivalent amount of work. This is in violation of the Kelvin–Planck statement of the second law of thermodynamics. Hence the assumption that $\eta_{R1} > \eta_{R2}$ is incorrect. Therefore

$$\eta_{R2} \geq \eta_{R1} \tag{7.36}$$

Now let us assume that the reversible engine R_2 is more efficient than the reversible engine R_1. Then the reversible engine R_1 can be run as a heat pump, utilizing part of the work done by R_2. By following the similar argument we can arrive at the result, that:

$$\eta_{R1} \geq \eta_{R2} \tag{7.37}$$

From Eqs. (7.36) and (7.37) it can be concluded that

$$\eta_{R1} = \eta_{R2} \tag{7.38}$$

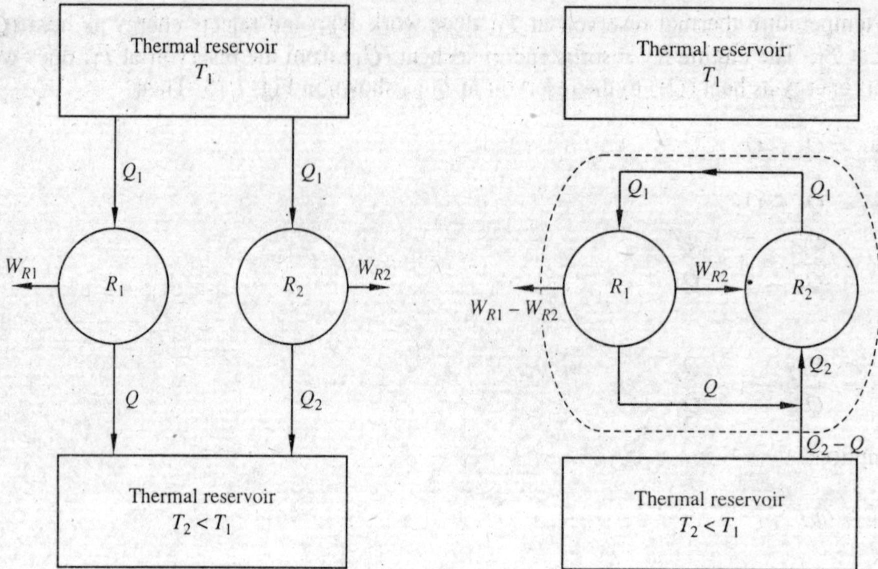

Fig. 7.15 *Illustration to prove the Carnot theorem 2*

Stated in words, *all reversible engines operating between the two given thermal reservoirs have the same efficiency.*

The Carnot theorem 2 further states that the efficiency of a reversible engine does not depend on the working fluid, but depends only on the temperature of the reservoirs between which it operates. That is, the efficiency of a reversible engine cannot be increased by changing the working fluid. That is, by using different working fluids in the reversible engines R_1 and R_2, η_{R1} cannot be made greater than η_{R2} or *vice-versa*. To prove this, let us assume that the working fluid in the engine R_1 is different from the working fluid in the engine R_2 such that $\eta_{R1} > \eta_{R2}$. By following the argument presented above, we arrive at the conclusion that $\eta_{R2} \geq \eta_{R1}$. The argument can be repeated again by assuming now that $\eta_{R2} > \eta_{R1}$ to arrive at the conclusion that $\eta_{R1} \geq \eta_{R2}$. Therefore, $\eta_{R1} = \eta_{R2}$, irrespective of the working fluids used in these engines. Therefore, the efficiency of a reversible engine does not depend on the nature of the working fluid, but depends on the temperature of the reservoirs only.

7.10 Thermodynamic Temperature Scale

The fact that the efficiency of a reversible engine does not depend on the nature of the working fluid, but depends only on the temperature of the two thermal reservoirs between which it operates can be made use of to define an absolute temperature scale or thermodynamic temperature scale. Since the absolute temperature scale is defined based on the Carnot theorems, the establishment of a thermodynamic temperature scale can also be considered as a consequence of the second law of thermodynamics.

To define the thermodynamic temperature scale, consider three isotherms the temperatures of which, as measured on this scale, are θ_1, θ_2 and θ_3, such that $\theta_1 > \theta_2 > \theta_3$. Let two reversible adiabatics intersect the three isotherms as shown in Fig. 7.16. In Fig. 7.16(b) AB, CD, EF are the three isotherms and ACE and BDF are the reversible adiabatics. Then ABDCA, CDFEC and ABDFECA represent three Carnot cycles. The three Carnot engines can be operated between the reservoirs with temperatures θ_1 and θ_2 ; θ_2 and θ_3 ; θ_1 and θ_3, respectively. Let the reversible engine 1 absorb energy as heat (Q_1) from the reservoir at θ_1, and reject energy as heat (Q_2) to the reservoir at θ_2. Let the reversible engine 2 absorb energy as heat (Q_2) from the reservoir at θ_2, and reject energy as heat (Q_3) to the reservoir at θ_3. Let W_1 and W_2 represent the work done by the engines 1 and 2, respectively. Since engine 2 absorbs the same amount of energy which is rejected by engine 1, the reservoir at θ_2 is unaffected. Let a third Carnot engine absorb energy as heat (Q_1) from the reservoir at θ_1, and reject energy as heat (Q_3) to the reservoir at θ_3.

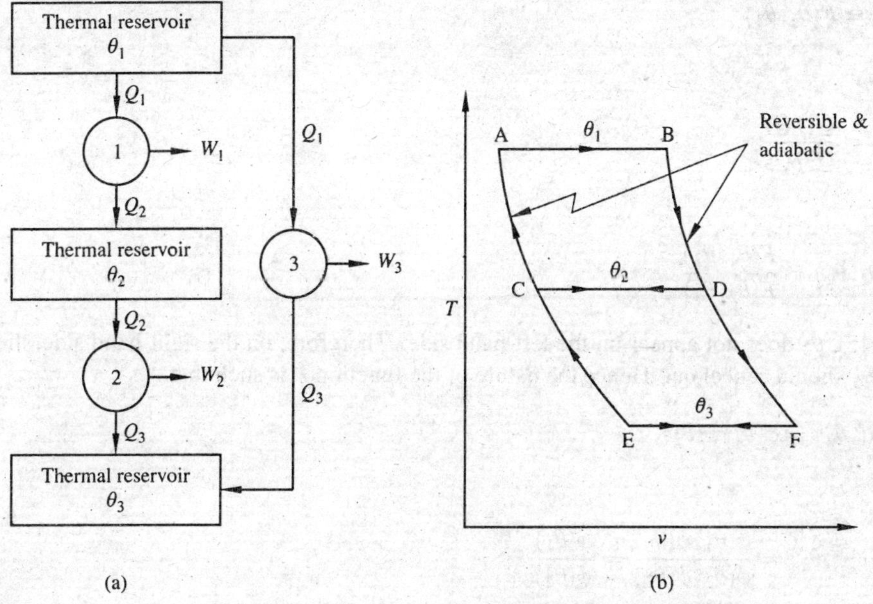

Fig. 7.16 *(a) Schematic arrangement of three Carnot engines (b) T–v diagram of three Carnot cycles*

The work done by the engine 3 is given by

$$W_3 = W_1 + W_2 \tag{7.39}$$

Then

$$\eta_1 = 1 - \frac{Q_2}{Q_1} = f(\theta_1, \theta_2) \tag{7.40}$$

or

$$\frac{Q_2}{Q_1} = 1 - f(\theta_1, \theta_2)$$

or

$$\frac{Q_1}{Q_2} = F(\theta_1, \theta_2) \tag{7.41}$$

where f and F represent two different functions.

Similarly

$$\eta_2 = 1 - \frac{Q_3}{Q_2} = f(\theta_2, \theta_3)$$

or

$$\frac{Q_2}{Q_3} = F(\theta_2, \theta_3) \tag{7.42}$$

$$\frac{Q_1}{Q_3} = F(\theta_1, \theta_3) \tag{7.43}$$

Then

$$\frac{Q_1}{Q_2} = \frac{Q_1/Q_3}{Q_2/Q_3} \tag{7.44}$$

or

$$F(\theta_1, \theta_2) = \frac{F(\theta_1, \theta_3)}{F(\theta_2, \theta_3)} \tag{7.45}$$

In Eq. (7.45), θ_3 does not appear on the left-hand side. Therefore, on the right-hand side, the dependence on θ_3 should cancel out. Hence the nature of the function F is such that

$$F(\theta_1, \theta_2) = \phi(\theta_1)\psi(\theta_2) \tag{7.46}$$

Then

$$\phi(\theta_1)\psi(\theta_2) = \frac{\phi(\theta_1)\psi(\theta_3)}{\phi(\theta_2)\psi(\theta_3)} = \frac{\phi(\theta_1)}{\phi(\theta_2)}$$

Therefore

$$\psi(\theta_2) = \frac{1}{\phi(\theta_2)} \tag{7.47}$$

Hence

$$\frac{Q_1}{Q_2} = F(\theta_1, \theta_2) = \frac{\phi(\theta_1)}{\phi(\theta_2)} \tag{7.48}$$

There is no further restriction on the function. The choice of the function ϕ now defines the temperature scale. Let us make a simple choice of the function ϕ, that is

$$\phi(\theta) = \theta \tag{7.49}$$

Then

$$\frac{Q_1}{Q_2} = \frac{\theta_1}{\theta_2} \tag{7.50}$$

The above equation states that the ratio of the energy absorbed to the energy rejected as heat by a Carnot engine operating between a source and a sink, is equal to the ratio of the absolute temperatures of the source and the sink. This equation can be used to determine the temperature of any reservoir by assigning a numerical value to one easily reproducible thermal reservoir (say, the triple point of water) and by measuring the heat interactions Q_1 and Q_2. The temperature measured by this method is independent of the working fluid since the efficiency of the Carnot engine does not depend on the nature of the working fluid. Thus, an absolute temperature scale can be established.

Recall that the efficiency of an engine operating between two thermal reservoirs is given by

$$\eta = 1 - \frac{Q_2}{Q_1}$$

Since

$$\frac{Q_2}{Q_1} = \frac{\theta_2}{\theta_1}$$

we get,

$$\eta = 1 - \frac{\theta_2}{\theta_1}$$

Since the efficiency of no heat engine can be greater than one, the lowest temperature in this scale is zero ($\theta_2 = 0$). Hence it is called the *absolute temperature scale*.

Kelvin originally proposed a temperature scale where the choice of the function was taken as

$$\phi(\theta) = \exp(\theta)$$

then

$$\frac{Q_1}{Q_2} = \frac{\phi(\theta_1)}{\phi(\theta_2)} = \frac{\exp(\theta_1)}{\exp(\theta_2)}$$

or

$$\ln \frac{Q_1}{Q_2} = \theta_1 - \theta_2$$

In 1954, at the tenth Conference on Weights and Measures held in Paris, the temperature of the triple point of water was assigned a value of 273.16 K. One can use this as the standard point and run a Carnot engine using water at its triple point as one of the thermal reservoirs and measure the heat interactions. Then

$$\frac{Q_1}{Q_2} = \frac{\theta_1}{\theta_2}$$

or

$$\theta_1 = \theta_2 \frac{Q_1}{Q_2}$$

Equation (7.50) which expresses the ratio of heat interactions to the ratio of absolute temperatures can be used to express the efficiency of a Carnot engine and the COP of a Carnot refrigerator or heat pump in terms of the absolute temperatures of the thermal reservoirs between which the devices operate. Then the thermal efficiency of a Carnot engine is given by

$$\eta = 1 - \frac{Q_2}{Q_1} = 1 - \frac{\theta_2}{\theta_1} \tag{7.51}$$

$$COP_R = \frac{Q_L}{Q_H - Q_L} = \frac{\theta_L}{\theta_H - \theta_L} \tag{7.52}$$

$$COP_{HP} = \frac{Q_H}{Q_H - Q_L} = \frac{\theta_H}{\theta_H - \theta_L} \tag{7.53}$$

7.10.1 Equivalence of absolute and ideal gas temperature scales

Consider a Carnot heat engine using an ideal gas as the working fluid. Let the high temperature reservoir be available at T_1 and the low temperature reservoir at T_2. It was shown earlier that the efficiency of a Carnot engine employing an ideal gas as the working fluid is given by

$$\eta = \frac{W}{Q_1} = \frac{Q_1 - Q_2}{Q_1} = 1 - \frac{Q_2}{Q_1} = 1 - \frac{T_2}{T_1}$$

or

$$\frac{Q_1}{Q_2} = \frac{T_1}{T_2} \tag{7.54}$$

But the absolute temperature scale is defined by the relation

$$\frac{Q_1}{Q_2} = \frac{\theta_1}{\theta_2}$$

Therefore

$$\frac{\theta_1}{\theta_2} = \frac{T_1}{T_2} \tag{7.55}$$

Since the numerical value (273.16) assigned to the triple point of water can be chosen to be the same for both the ideal gas scale and the absolute temperature scale, $\theta = T$. Therefore, the absolute temperature is numerically equal to the temperature measured by the ideal gas thermometer. Hereafter, the common symbol T will be used to represent the temperature.

***Example* 7.1** In recent years, geothermal power generation has received much attention. In such power plants, underground sources of hot water, or steam, are used as a high temperature reservoir and the ambient atmosphere as the low temperature reservoir. Suppose the hot water is available at 420 K and the ambient temperature is at 300 K, what is the maximum possible efficiency of a heat engine when it is operated between these reservoirs?

Solution It was shown that the efficiency is maximum for a reversible engine and all reversible engines have the same efficiency when operated between the same thermal reservoirs. The efficiency of a reversible engine operating between two constant temperature reservoirs does not depend on the working fluid, but depends only on the temperatures of the reservoirs between which it works. For a Carnot engine the efficiency is given by

$$\eta = 1 - \frac{T_2}{T_1} = 1 - \frac{300}{420} = 0.2857$$

Therefore, the maximum possible efficiency = 0.2857.

***Example* 7.2** An inventor claims that his heat engine delivers work at the rate of one kW, and absorbs energy as heat at the rate of 65 kJ/min from a source at 1127°C when the ambient temperature is 27°C. Would you agree with this claim?

Solution The maximum possible efficiency of a heat engine is the Carnot efficiency

$$\eta = 1 - \frac{T_2}{T_1} = 1 - \frac{300}{1400} = 0.7857$$

Efficiency claimed by the inventor is

$$\frac{W}{Q_1} = \frac{60}{65} = 0.923$$

Since the inventor's claim of efficiency is greater than the Carnot efficiency, the claim cannot be true.

***Example* 7.3** It is proposed to construct a refrigeration plant for a cold storage to be maintained at −3°C. The ambient temperature is 27°C. If 5×10^6 kJ/h of energy has to be continuously removed from the cold storage, estimate the minimum power required to run the refrigerator.

Solution The work required is minimum only when the refrigerator is a reversible device. The COP of a reversible refrigerator operating between the two thermal reservoirs at T_H and T_L is given by

$$COP_R = \frac{Q_L}{Q_H - Q_L} = \frac{T_L}{T_H - T_L} = \frac{270}{300 - 270} = 9$$

$$COP_R = \frac{Q_L}{W} \quad \text{or} \quad W = \frac{Q_L}{COP_R}$$

Hence

$$W = \frac{5 \times 10^6/3600}{9} = 154.32 \text{ kW}$$

***Example* 7.4** Liquid helium boils at 4.2 K at atmospheric pressure. Its latent heat of vaporization is 83.3 kJ/kmol. A Carnot refrigeration cycle is to be used to produce one kmol of liquid helium from saturated vapor at the same temperature. If the ambient temperature is 305 K, calculate the *COP* of the refrigerator and the work input.

Solution

$$COP_R = \frac{T_L}{T_H - T_L} = \frac{4.2}{305 - 4.2} = 0.013\,96$$

$$W = \frac{Q_L}{COP_R} = \frac{83.3}{0.013\,96} = 5967 \text{ kJ}$$

***Example* 7.5** To preserve certain drugs, it is essential to maintain them at −23°C in a refrigerator. If a refrigerator, which has a 200 W compressor, is used for this purpose, estimate the fraction of time the compressor runs in a cold country, where the ambient temperature is 18°C, for a cooling rate of 40,000 kJ/d. If the same refrigerator is used in a tropical country, where the ambient temperature is 37°C in summer, estimate the fraction of time the compressor runs for the same cooling rate.

Solution In cold country

$$COP_R = \frac{T_L}{T_H - T_L} = \frac{250}{291 - 250} = 6.098 = \frac{Q_L}{W}$$

or

$$W = \frac{Q_L}{COP_R} = \frac{4 \times 10^4}{6.098} = 6560 \text{ kJ/d}$$

Compressor work = 200 W = 200 J/s = 200 × 3600 × 24 = 17280 kJ/d

Fraction of time the compressor runs in cold country is

$$\frac{6560}{17280} = 0.3796$$

In tropical country

$$COP_R = \frac{250}{310 - 250} = 4.167 = \frac{Q_L}{W}$$

or

$$W = \frac{4 \times 10^4}{4.167} = 9599 \text{ kJ/d}$$

Hence the fraction of time the compressor runs in the tropical country

$$= \frac{9599}{17280} = 0.555$$

***Example* 7.6** In a particular ice making plant, a refrigerator absorbs energy as heat from water at 0°C producing ice and rejects energy as heat to the ambient atmosphere at 27°C. If a heat engine, which uses a source at 150°C and the ambient atmosphere as sink is used to run the refrigerator, calculate:

(a) the energy removed as heat from the cold body by the refrigerator for each kJ of energy taken in by the engine. Assume both the engine and the refrigerator to be reversible.

(b) Energy rejected to the ambient atmosphere for each kJ of energy absorbed by the engine from the source.

Solution A schematic arrangement of the heat engine and the refrigerator is shown in Fig. 7.17.

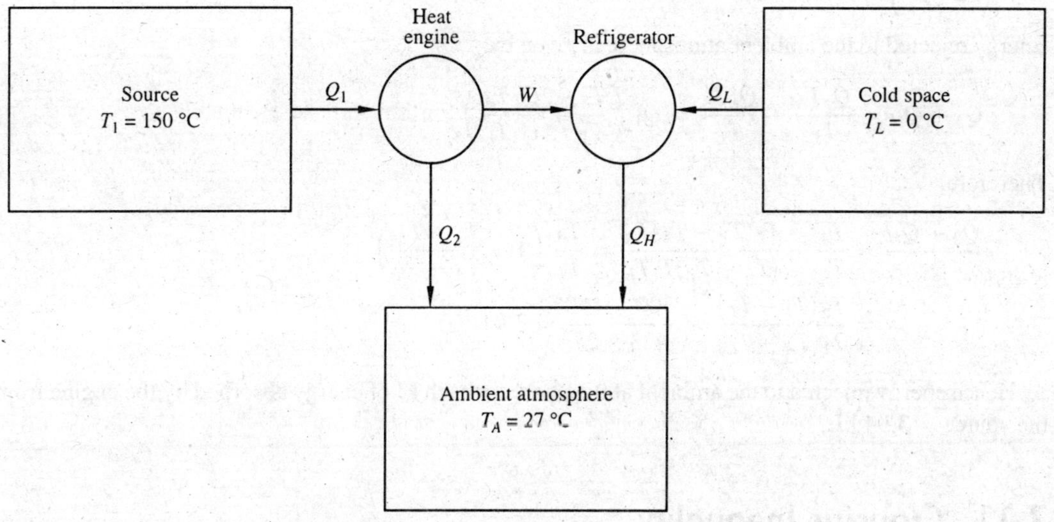

Fig. 7.17 *Sketch for Example 7.6*

(a)

$$COP_R = \frac{Q_L}{W} = \frac{T_L}{T_A - T_L}$$

or

$$Q_L = \frac{T_L W}{T_A - T_L}$$

$$\eta = \frac{W}{Q_1} = \frac{T_1 - T_A}{T_1} \quad or \quad W = \frac{(T_1 - T_A)Q_1}{T_1}$$

Therefore,

$$Q_L = \frac{T_L W}{T_A - T_L} = \frac{T_L(T_1 - T_A)Q_1}{(T_A - T_L)T_1}$$

or

$$\frac{Q_L}{Q_1} = \frac{T_L(T_1 - T_A)}{(T_A - T_L)T_1} = \frac{273 \times 123}{423 \times 27} = 2.94$$

Then energy removed from the cold body by the refrigerator for each kJ of energy absorbed by the heat engine $= Q_L/Q_1 = 2.94$.

(b)

$$\frac{Q_2}{Q_1} = \frac{T_A}{T_1} \quad \text{and} \quad \frac{Q_H}{Q_L} = \frac{T_A}{T_L}$$

$$Q_2 = \frac{Q_1 T_A}{T_1} \quad \text{and} \quad Q_H = \frac{Q_L T_A}{T_L}$$

Energy rejected to the ambient atmosphere is given by:

$$Q_2 + Q_H = \frac{Q_1 T_A}{T_1} + \frac{Q_L T_A}{T_L} = Q_1\left(\frac{T_A}{T_1} + \frac{Q_L T_A}{Q_1 T_L}\right)$$

Therefore,

$$\frac{Q_2 + Q_H}{Q_1} = \frac{T_A}{T_1} + \frac{T_L(T_1 - T_A)T_A}{(T_A - T_L)T_1 T_L} = \frac{T_A}{T_1}\left(1 + \frac{T_1 - T_A}{T_A - T_L}\right)$$

$$= \frac{T_A(T_1 - T_L)}{T_1(T_A - T_L)} = \frac{300 \times 150}{423 \times 27} = 3.94$$

Hence energy rejected to the ambient atmosphere for each kJ of energy absorbed by the engine from the source = 3.94 kJ.

7.11 Clausius Inequality

So far in the discussion of the second law of thermodynamics attention was focused mainly on heat engines and refrigeration systems. Although these devices are very important, in many cases one has to analyze the processes other than heat engines and refrigerators. In such instances the analysis cannot be limited to cycles only. For this purpose let us consider a heat engine cycle and make some important observations regarding the heat interactions, and temperature levels at which the heat interactions take place. This will help us in dealing with a general cyclic process.

7.11.1 Observation about cycles

It was shown that for a Carnot cycle, or for a reversible engine,

$$\frac{Q_1}{Q_2} = \frac{T_1}{T_2}$$

or for a reversible engine

$$\frac{Q_1}{T_1} - \frac{Q_2}{T_2} = 0 \tag{7.56}$$

If the usual sign convention is adopted, that is, energy flow as heat into the system taken as positive and energy flow as heat out of the system taken as negative, the above equation becomes

$$\frac{Q_1}{T_1} + \frac{Q_2}{T_2} = 0 \quad \text{or} \quad \sum \frac{Q_i}{T_i} = 0 \tag{7.57}$$

Now let us consider an irreversible engine, which absorbs energy as heat (Q_1) from a reservoir at T_1 and rejects energy as heat (Q_2') to a reservoir at T_2. Then according to Carnot theorem 1, the efficiency of the irreversible engine is less than the efficiency of a reversible engine. Therefore,

$$1 - \frac{Q_2'}{Q_1} < 1 - \frac{Q_2}{Q_1}$$

$$1 - \frac{Q_2'}{Q_1} < 1 - \frac{T_2}{T_1} \quad \text{or} \quad \frac{Q_2'}{Q_1} > \frac{T_2}{T_1}$$

or

$$\frac{Q_2'}{T_2} > \frac{Q_1}{T_1} \tag{7.58}$$

Making use of the sign convention, this can be written as

$$\frac{Q_1}{T_1} + \frac{Q_2'}{T_2} < 0$$

or for an irreversible engine

$$\sum \frac{Q}{T} < 0 \tag{7.59}$$

Replacement of a reversible process by an equivalent process

Now let us consider situations other than heat engines where we deal with a system which undergoes a cyclic change. If the cycle can be split into a large number of heat engine cycles, the above observation can be made use of in relating the heat interactions with the absolute temperature at which the heat interactions occur in the actual cycle. The cycle may consist of processes other than isothermal and reversible adiabatic processes. It can be shown that any reversible process can be approximated by a series of reversible isothermal and reversible adiabatic processes. Consider a system undergoing a reversible process 1–2 as shown in Fig. 7.18. The same change of state can be achieved by following a reversible adiabatic process 1–*a*, a reversible isothermal process *abc*, and a reversible adiabatic process *c*–2, if the areas 1*ab* and *bc*2 are equal. This change can be done if the heat and work interactions along the paths 1–2 and 1*abc*2 are equal.

Application of the first law of thermodynamics for the process 1*abc*2 gives

$$U_2 - U_1 = Q_{1abc2} - W_{1abc2} \tag{7.60}$$

Consider the cycle 1*abc*2*b*1. The net work of the cycle is zero, since area 1*ab* = area *bc*2. Then,

$$\oint dW = W_{1abc2} + W_{2b1} = 0$$

or

$$W_{1abc2} = -W_{2b1} = W_{1b2} \tag{7.61}$$

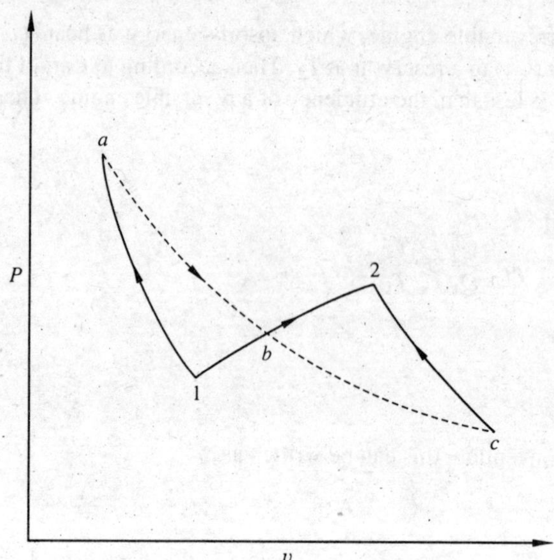

Fig. 7.18 *P–v diagram showing the replacement of process* 1 – 2 *by* 1*abc*2

That is, the work interaction along the path 1*abc*2 is equal to the work interaction along the path 1*b*2.

The heat interaction along the path 1*abc*2 is

$$Q_{1abc2} = Q_{1a} + Q_{abc} + Q_{c2} = Q_{abc} \tag{7.62}$$

since 1 – *a* and *c* – 2 are reversible adiabatic paths. Then Eq. (7.60) reduces to

$$U_2 - U_1 = Q_{abc} - W_{1b2} \tag{7.63}$$

Application of the first law of thermodynamics for the process 1*b*2 gives

$$U_2 - U_1 = Q_{1b2} - W_{1b2} \tag{7.64}$$

Comparing Eqs. (7.63) and (7.64), we obtain

$$Q_{abc} = Q_{1b2}$$

That is, the heat interaction along the reversible path 1*b*2 is equal to that along the isothermal path *abc*. Therefore, the given reversible process can be replaced by a series of reversible adiabatic and reversible isothermal processes. This is a significant observation which helps us to replace a reversible process, in which the temperature may change in an arbitrary manner, by a series of reversible isothermal and reversible adiabatic processes.

7.11.2 Clausius inequality

Consider a system undergoing a reversible cycle as shown in Fig. 7.19. The given cycle may be subdivided by drawing a family of reversible adiabatic lines. Every two adjacent adiabatic lines may be joined by two reversible isotherms, such that the heat interaction along the reversible path is equal to the heat interaction along the reversible isothermal path, and the work interaction along the reversible path is equal to the work interaction along the reversible adiabatic and the reversible isothermal path. That is,

$$Q_{ab} = Q_{a1b1} \quad \textit{and} \quad Q_{cd} = Q_{c1d1}$$

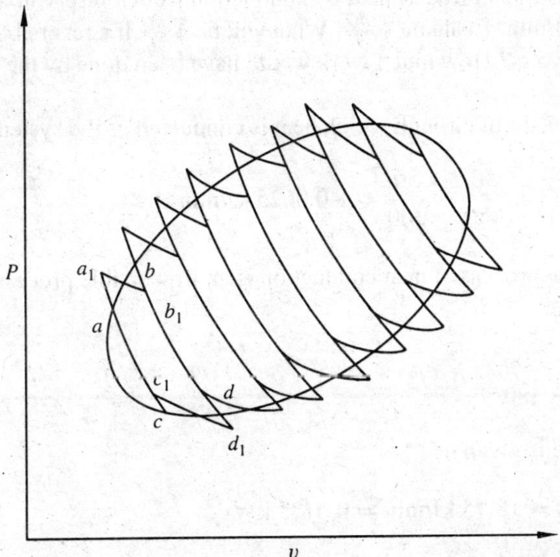

Fig. 7.19 *Reversible cycle approximated as a combination of several reversible adiabatic and reversible isothermal processes*

Then $a1b1d1c1$ is a Carnot cycle which has two reversible isothermal and two reversible adiabatic processes. Therefore, the original reversible cycle has been split into a family of Carnot cycles. By drawing a large number of Carnot cycles, the original reversible cycle can be closely approximated. For every Carnot cycle $\oint \frac{dQ}{T} = 0$. Therefore, for the given reversible cycle one can write

$$\oint \frac{dQ}{T} = 0 \tag{7.65}$$

If the original cycle is irreversible, one can follow the above argument and observe that some of the small cycles are reversible and some are irreversible. For the reversible cycles $\oint \frac{dQ}{T} = 0$, and for the irreversible cycles $\oint \frac{dQ}{T} < 0$. Then for the original irreversible cycle

$$\oint \frac{dQ}{T} < 0 \tag{7.66}$$

The above observation can be generalized and stated as

$$\oint \frac{dQ}{T} \leq 0 \tag{7.67}$$

This is the Clausius inequality. The Clausius-inequality states that *whenever a system undergoes a cyclic change, the algebraic sum of all the heat interactions divided by the absolute temperature at which the heat interactions are taking place, considered over the entire cycle, is less than or equal to zero*. The equality sign holds good if the cycle is completely reversible; and the inequality sign holds good if the cycle is irreversible. The Clausius inequality is a consequence of the second law of thermodynamics, and at times the inequality itself is used in place of the second law.

***Example* 7.7** Energy is transferred as heat by conduction from a reservoir at 480 K to a reservoir at 300 K at the rate of 50 kJ/min. Evaluate $\oint \frac{dQ}{T}$. What will be $\oint \frac{dQ}{T}$ if a reversible heat engine is operated between these two reservoirs? How much work would have been done by the engine?

Solution Choose the medium through which heat is conducted as the system.

$$\oint \frac{dQ}{T} = \frac{Q_1}{T_1} - \frac{Q_2}{T_2} = \frac{50}{480} - \frac{50}{300} = -0.0625 \text{ kJ/min K}$$

Here $\oint \frac{dQ}{T} < 0$, hence the process of heat conduction is an irreversible process.

For a reversible engine

$$\eta = 1 - \frac{T_2}{T_1} = 1 - \frac{300}{480} = 0.375$$

Therefore work delivered is given by

$$\eta Q_1 = 0.375 \times 50 = 18.75 \text{ kJ/min} = 0.3125 \text{ kW}$$

$$Q_2 = Q_1 - W = 50 - 18.75 = 31.25 \text{ kJ/min}$$

$$\oint \frac{dQ}{T} = \frac{50}{480} - \frac{31.25}{300} = 0$$

Note: $\oint \frac{dQ}{T} = 0$ for a reversible engine since $Q_1/Q_2 = T_1/T_2$. This result can be as well predicted without performing any calculations.

***Example* 7.8** A reversible heat engine shown in Fig. 7.20 operates between three constant temperature reservoirs at 600 K, 400 K and 300 K. It receives 2500 kJ of energy as heat from the reservoir at 600 K and does 1000 kJ of work. Determine the heat interactions with the other two reservoirs.

Solution The first law of thermodynamics gives

$$W = Q_1 - Q_2 - Q_3$$

$$1000 = 2500 - Q_2 - Q_3 \quad \text{or} \quad Q_2 + Q_3 = 1500 \text{ kJ} \tag{A}$$

The second law (in the form of Clausius inequality) gives for a reversible heat engine $\oint \frac{dQ}{T} = 0$.

Therefore,

$$\oint \frac{dQ}{T} = \frac{2500}{600} - \frac{Q_2}{400} - \frac{Q_3}{300} = 0$$

or

$$\frac{Q_2}{4} + \frac{Q_3}{3} = \frac{2500}{6} \tag{B}$$

Solving Eqs. (A) and (B) one can obtain $Q_2 = 1000$ kJ and $Q_3 = 500$ kJ.

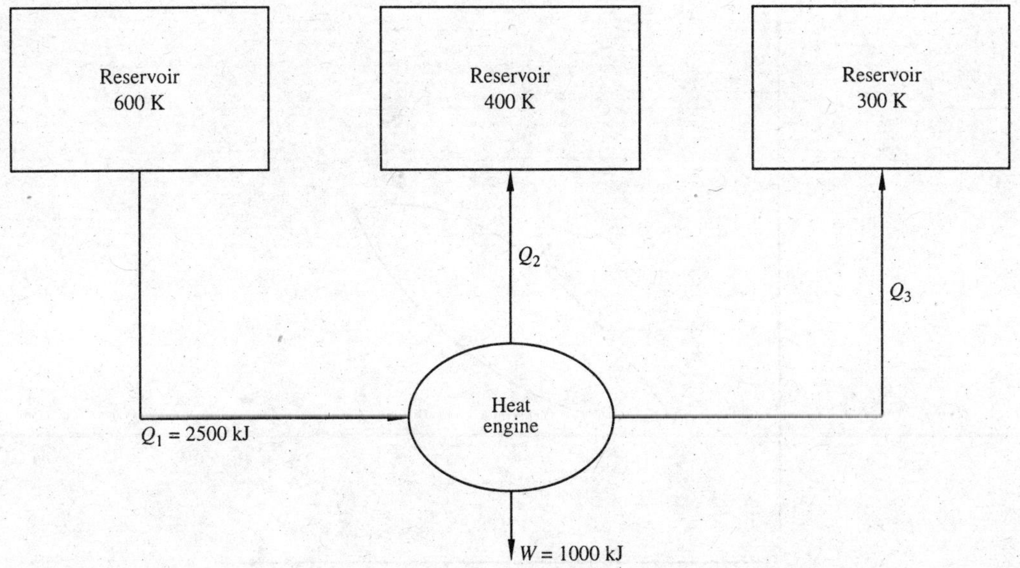

Fig. 7.20 *Sketch for Example 7.8*

7.12 Entropy

The second law of thermodynamics can be qualitatively applied to certain processes to judge whether the process is feasible or not while Clausius inequality can be used to analyze the cyclic processes in a quantitative manner. The second law became a law of wider applicability when Clausius introduced the property called entropy. Entropy is an important property in thermodynamics which stipulates the feasibility of a certain process. By evaluating the entropy change, one can explain why spontaneous processes occur in a particular direction only.

Consider a system in initial state 1. Let the system be taken from state 1 to state 2 along a reversible path $1A2$, and then be restored to its initial state by following another reversible path $2B1$ (see Fig. 7.21). Then the two paths put together form a reversible cycle $1A2B1$. Apply the Clausius inequality to this reversible cycle and obtain

$$\int_{1A2B1} \frac{dQ}{T} = 0$$

or

$$\int_{1A2} \frac{dQ}{T} + \int_{2B1} \frac{dQ}{T} = 0$$

or

$$\int_{1A2} \frac{dQ}{T} = -\int_{2B1} \frac{dQ}{T} = \int_{1B2} \frac{dQ}{T} \tag{7.68}$$

(Since the path $2B1$ is reversible, the limits of the integral can be reversed).

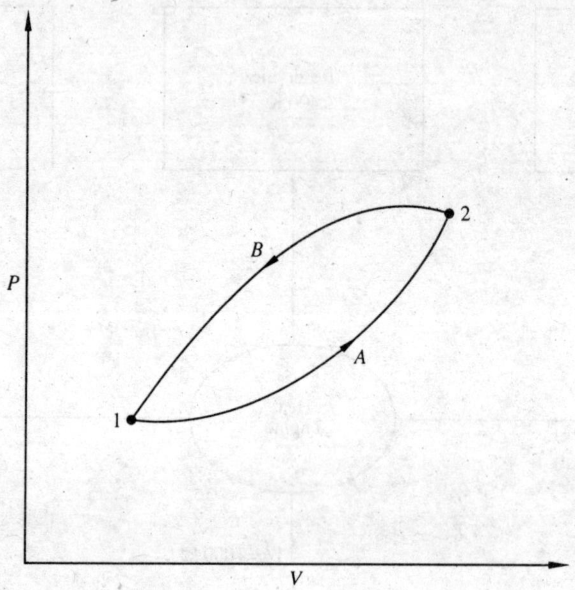

Fig. 7.21 *A reversible cycle on P–v diagram*

That is, $\int \frac{dQ}{T}$ has the same value whether the path followed is $1A2$ or $1B2$. One can connect the states 1 and 2 by several reversible paths, and see that $\int \frac{dQ}{T}$ has the same value irrespective of the path, as long as the paths are reversible. Therefore $(\frac{dQ}{T})_R$ is an exact differential of some function which we now identify as entropy. Hence one can say that there exists a function S, called entropy, the change in which is given by

$$S_2 - S_1 = \Delta S = \int_1^2 dS = \int_1^2 \left(\frac{dQ}{T}\right)_R \tag{7.69}$$

This follows from the Clausius inequality as a consequence of the second law of thermodynamics. Therefore, for a reversible process

$$dS = \left(\frac{dQ}{T}\right)_R \tag{7.70}$$

If the two terminal states—the initial and final states—are connected by a reversible process, the change in the entropy of the system can be evaluated by making use of the above equation. We also know that entropy is a property of the system and hence it is a state function. The change in the value of the property is independent of the path followed by the system. Therefore, as long as the two terminal states of the system are the same, the entropy change is the same regardless of how the system is taken from the initial state to the final state. That is, the change in the entropy of the system is independent of the path followed by the system. If a system undergoes a change of state by an irreversible process, then one can devise a reversible process connecting the given initial and final states and evaluate the entropy change for the reversible process. The entropy change of the system by the actual irreversible process is the same as that evaluated for the reversible process.

7.13 Calculation of Entropy Change

When calculating the entropy change of a system which undergoes a change in its state, the following facts should be kept in mind and adhered to:

1. $dS = dQ/T$ for reversible processes only.
2. Entropy is a state function. The entropy change of a system is determined by its initial and final states only, irrespective of how the system has changed its state.
3. In analyzing irreversible processes, it is not necessary to make a direct analysis of the actual process. One can substitute the actual process by an imaginary reversible process connecting the final state to the initial state, and the entropy change for the imaginary reversible process can be evaluated. The entropy change of the system by the actual irreversible process is the same as for the imaginary reversible process.

The calculation of entropy change is demonstrated below through several examples.

7.13.1 Absorption of energy by a constant temperature bath

Suppose a certain amount of energy is added as heat to a constant temperature bath. The energy addition may be carried out reversibly or irreversibly by bringing it into contact with a thermal reservoir or by performing work on the bath. The actual process can be replaced by a reversible path in which an equivalent amount of energy is added to the bath. Then the entropy change of the bath is given by

$$\Delta S = \int \left(\frac{dQ}{T}\right)_R$$

***Example* 7.9** The contents of a large constant temperature reservoir which is maintained at 500 K is continuously stirred by a paddle wheel driven by an electric motor. Estimate the entropy change of the reservoir if the paddle wheel is operated for 2 hours by a 250 W motor.

Solution The constant temperature reservoir is sufficiently large and is maintained at 500 K. Considering the contents of the reservoir as the system, work is done on the system. The paddle wheel work is converted into internal energy because of fluid friction. This is an irreversible process. One can replace this irreversible process by a process in which an equivalent amount of energy is added to the system at constant temperature. If dQ is the amount of energy added during this reversible process (T = constant), the entropy change of the system is:

$$\Delta S = \int \left(\frac{dQ}{T}\right)_R = \frac{Q}{T} = \frac{0.25 \times 2(3600)}{500} = 0.6 \text{ kJ/K}$$

7.13.2 Heating or cooling of matter

The heating of matter by bringing it into contact with a body at a higher temperature is an irreversible process. To evaluate the entropy change, one can replace the irreversible process by an imaginary reversible process connecting the given initial and final states of the system. For this assumed reversible process, the entropy change can be evaluated. The heating can be carried out either at constant pressure or at constant volume. From the first law of thermodynamics, we know that

- $Q = \Delta U$ for a constant volume heating or cooling process.
- $Q = \Delta H$ for constant pressure heating or cooling process.

In these cases it may be noted that the heat interaction is equal to the change in a property (U or H) of the system. These changes in the property of the system can be evaluated if the specific heats C_v and C_P of the substance are known. Therefore, the entropy change associated with the heating or cooling of matter is given by:

For a constant pressure process

$$\Delta S = \int \left(\frac{dQ}{T}\right)_R = m\int_{T_1}^{T_2} \frac{C_P dT}{T} \tag{7.71}$$

For a constant volume process

$$\Delta S = \int \left(\frac{dQ}{T}\right)_R = m\int_{T_1}^{T_2} \frac{C_v dT}{T} \tag{7.72}$$

If the specific heats C_P and C_v are assumed to be constant, this can be integrated to yield:

For a constant pressure process

$$\Delta S = mC_P \ln\left(\frac{T_2}{T_1}\right) \tag{7.73}$$

For a constant volume process

$$\Delta S = mC_v \ln\left(\frac{T_2}{T_1}\right) \tag{7.74}$$

If the specific heat of the substance varies with temperature, the expression relating the specific heat to temperature is substituted and integrated to obtain the entropy change. For example if

$$C_P = a + bT + cT^2 \tag{7.75}$$

Then

$$\Delta S = \int_{T_1}^{T2} \frac{m(a + bT + cT^2)dT}{T}$$

$$\Delta S = m\left(a\ln\left(\frac{T_2}{T_1}\right) + b(T_2 - T_1) + \frac{c(T_2^2 - T_1^2)}{2}\right) \tag{7.76}$$

***Example* 7.10** One mole of air (ideal gas) contained in a rigid vessel is heated from 300 K to 370 K by placing the container on a hot plate. Determine the entropy change of air.

Solution We know that

$$C_v = \frac{R}{\gamma - 1} = \frac{8.314}{1.4 - 1} = 20.785 \text{ J/mol K}$$

The entropy change for a constant volume heating process is given by

$$\Delta s = C_v \ln\left(\frac{T_2}{T_1}\right)$$

Therefore

$$\Delta s = 20.785 \ln\left(\frac{370}{300}\right) = 4.359 \text{ J/mol K}$$

7.13.3 Phase change

Pure substances undergo phase changes at constant temperature and pressure. During the phase change if energy is added to the system isothermally from a constant temperature reservoir, the process is reversible. Though the melting of ice and boiling of water is conducted with finite temperature differences, we can replace it by the above reversible process. The entropy changes associated with these processes are given by

$$S_{sf} = \int \frac{dq}{T} = \frac{h_{sf}}{T} \tag{7.77}$$

$$S_{fg} = \int \frac{dq}{T} = \frac{h_{fg}}{T} \tag{7.78}$$

***Example* 7.11** Ice melts at 0°C with a latent heat of fusion of 334.92 kJ/kg. At atmospheric pressure water boils at 100°C with a latent heat of vaporization of 2256.94 kJ/kg. Calculate the entropy change associated with the fusion and vaporization processes.

Solution

$$S_{sf} = \int \frac{dq}{T} = \frac{h_{sf}}{T} = \frac{334.92}{273} = 1.2268 \text{ kJ/kg K}$$

$$S_{fg} = \int \frac{dq}{T} = \frac{h_{fg}}{T} = \frac{2256.94}{373} = 6.0508 \text{ kJ/kg K}$$

7.13.4 Adiabatic mixing

When a hot substance is adiabatically mixed with a cold substance, the hot material cools to the final temperature and the cold material heats to the same final temperature. The resulting final temperature can be estimated by applying the principle of conservation of energy. The entropy change of each substance can be evaluated by applying the equation

$$\Delta S = \int_{T_1}^{T_2} \frac{dQ}{T}$$

The total entropy change is then obtained by the addition of the two individual entropy changes.

***Example* 7.12** A lump of steel weighing 30 kg at a temperature of 427°C is dropped in 150 kg of oil at 27°C. The specific heats of the steel and oil are 0.46 kJ/kg K and 2.5 kJ/kg K respectively. Estimate the entropy change of the steel, the oil and the system consisting of oil and the lump of steel.

Solution Let T be the final temperature after the steel comes to equilibrium with oil. Then energy lost by the lump of steel = Energy gained by oil, that is

$$(mC_P\Delta T)_{\text{steel}} = (mC_P\Delta T)_{\text{oil}}$$

$$30(0.46)(700 - T) = 150(2.5)(T - 300) \quad \text{or} \quad T = 314.2 \text{ K}$$

$$\Delta S_{\text{steel}} = \int \frac{mC_P dT}{T} = mC_P \ln \frac{T}{T_1}$$

$$= 30(0.46) \ln \frac{314.2}{700} = -11.0545 \text{ kJ/K}$$

$$\Delta S_{\text{oil}} = \int \frac{mC_P dT}{T} = mC_P \ln \frac{T}{T_2}$$

$$= 150(2.5) \ln \frac{314.2}{300} = 17.3427 \text{ kJ/K}$$

Hence

$$\Delta S_{\text{uni}} = \Delta S_{\text{steel}} + \Delta S_{\text{oil}} = -11.0545 + 17.3427 = 6.2882 \text{ kJ/K}$$

7.13.5 Change of state for an ideal gas

If an ideal gas undergoes a change of state from P_1, v_1, T_1 to P_2, v_2, T_2, the change in entropy can be calculated by devising a reversible path connecting the initial and final states. There are a large number of paths by which the gas can be taken from the initial state to the final state. Let us consider two such convenient paths shown in Fig. 7.22 and evaluate the entropy change for both the paths.

Path 1*a*2

The gas in state 1 is heated at constant pressure till the temperature T_2 is attained and then it is brought reversibly and isothermally to the final pressure P_2. 1–*a* is a reversible constant pressure process and *a*–2 is a reversible isothermal process.

$$\Delta s_{1a} = \int \frac{dq}{T} = \int_{T_1}^{T_2} \frac{C_P dT}{T} = C_P \ln \frac{T_2}{T_1} \tag{7.79}$$

$$\Delta s_{a2} = \int \frac{dq}{T} = \int \frac{(du + Pdv)}{T} = \int \frac{Pdv}{T} = R \ln \frac{v_2}{v_a} \tag{7.80}$$

(Since $du = 0$ for an isothermal process). For the isothermal process $a2$:

$$P_2 v_2 = P_a v_a = P_1 v_a \quad \text{or} \quad \frac{v_2}{v_a} = \frac{P_1}{P_2} \tag{7.81}$$

Then

$$\Delta s_{a2} = -R \ln \frac{P_2}{P_1} \tag{7.82}$$

Therefore

$$\Delta s = \Delta s_{1a} + \Delta s_{a2} = C_P \ln \frac{T_2}{T_1} - R \ln \frac{P_2}{P_1} \tag{7.83}$$

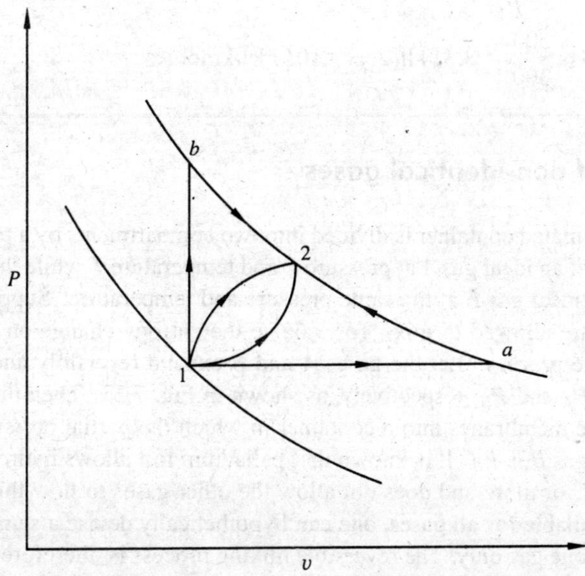

Fig. 7.22 *Two different reversible paths connecting the initial state 1 and the final state 2 on P–v diagram*

Path 1b2

The gas initially in state 1 is heated at constant volume to the final temperature T_2, and then it is reversibly and isothermally changed to the final pressure P_2. 1–b is a reversible constant volume process and b–2 is a reversible isothermal process.

$$\Delta s_{1b} = \int \frac{dq}{T} = \int_{T_1}^{T_2} \frac{C_v dT}{T} = C_v \ln \frac{T_2}{T_1} \tag{7.84}$$

$$\Delta s_{b2} = \int \frac{dq}{T} = \int \frac{(du + Pdv)}{T} = \int \frac{Pdv}{T} = R \ln \frac{v_2}{v_b} = R \ln \frac{v_2}{v_1} \tag{7.85}$$

Therefore

$$\Delta s = \Delta s_{1b} + \Delta s_{b2} = C_v \ln \frac{T_2}{T_1} + R \ln \frac{v_2}{v_1} \tag{7.86}$$

It can be easily verified that both the Eqs. (7.83) and (7.86) give the same result.

***Example* 7.13** Determine the entropy change when one kmol of an ideal gas ($C_P = 7R/2$) is taken from the initial state of 1 bar and 300 K to 2 bar and 500 K.

Solution Initial state: $P_1 = 1$ bar; $T_1 = 300$ K
Final state : $P_2 = 2$ bar; $T_2 = 500$ K
Hence

$$\Delta s = C_P \ln \frac{T_2}{T_1} - R \ln \frac{P_2}{P_1}$$

$$= \frac{7}{2} \times 8.314 \ln \frac{500}{300} - 8.314 \ln 2 = 9.1017 \text{ kJ/kmol K}$$

7.13.6 Mixing of non-identical gases

Suppose a rigid and insulated container is divided into two compartments by a partition. One compartment contains N_A mol of an ideal gas A at pressure P and temperature T while the second compartment contains N_B mol of an ideal gas B at the same pressure and temperature. Suppose the partition is removed and the gases are allowed to mix. To evaluate the entropy change on mixing , let us devise the following reversible process. Let the gases A and B expand reversibly and isothermally to their final partial pressures P_A and P_B, respectively, as shown in Fig. 7.23. Then the gases are introduced through semipermeable membranes into a container in which the partial pressure of gas A is P_A and the partial pressure of gas B is P_B. It is known that palladium foil allows hydrogen to pass through at a temperature of 200°C or more and does not allow the other gases to flow through it. Though such membranes are not available for all gases, one can hypothetically devise a semipermeable membrane which is permeable to one gas only. The reversible mixing process is, therefore, done in the following two steps.

(a) Reversible isothermal expansion of the gases to their respective partial pressures.
(b) Forcing the gases reversibly through semipermeable membranes.

For gas A:

$$(\Delta S_1)_A = \int \frac{dQ}{T} = \int \frac{dH - VdP}{T} = -\int \frac{VdP}{T} = -N_A R \ln \frac{P_A}{P} \tag{7.87}$$

Similarly, for gas B:

$$(\Delta S_1)_B = -N_B R \ln \frac{P_B}{P} \tag{7.88}$$

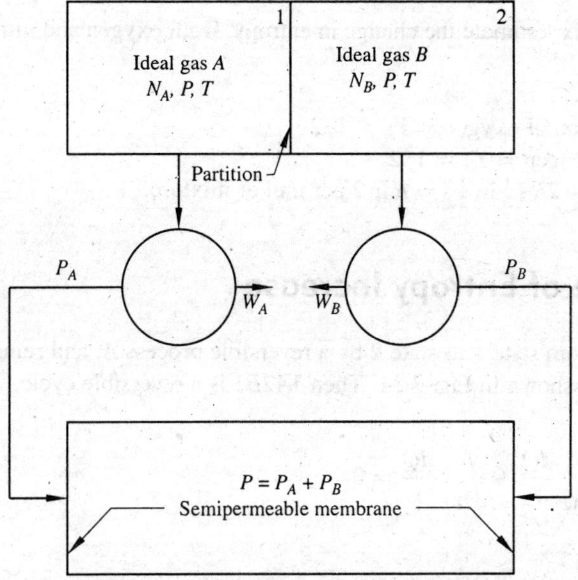

Fig. 7.23 *Adiabatic mixing of non-identical gases*

where,

$$P_A = \frac{PN_A}{N_A + N_B} = PY_A \tag{7.89}$$

and

$$P_B = PY_B \tag{7.90}$$

Y_A and Y_B represent the mole fractions of gases A and B, respectively, in the final mixture.

Hence

$$\Delta S_1 = -N_A R \ln Y_A - N_B R \ln Y_B \tag{7.91}$$

$$\Delta S_2 = 0 \tag{7.92}$$

Therefore, entropy change associated with the mixing of non-identical gases is:

$$\Delta S = -N_A R \ln Y_A - N_B R \ln Y_B$$

$$\Delta s = -R(Y_A \ln Y_A - Y_B \ln Y_B) = -R \sum Y_i \ln Y_i \tag{7.93}$$

where s is the molar entropy. Since $Y_i < 1$, Δs is always positive.

***Example* 7.14** A rigid and insulated container is divided into two compartments by a partition. One compartment contains one kmol of oxygen at 100 kPa pressure and 300 K while the second compartment contains one kmol of nitrogen at 100 kPa pressure and 300 K. If the partition is removed and the gases are allowed to mix, estimate the change in entropy. Both oxygen and nitrogen may be treated as ideal gases.

Solution Mole fraction of oxygen $= Y_A = 1/2$
Mole fraction of nitrogen $= Y_B = 1/2$
Therefore $\Delta s_{Mix} = -2R\left(\frac{1}{2}\ln\frac{1}{2}\right) = R\ln 2$ per mol of mixture.

7.14 Principle of Entropy Increase

Let a system change from state 1 to state 2 by a reversible process A, and return to state 1 by another reversible process B as shown in Fig. 7.24. Then $1A2B1$ is a reversible cycle. Therefore, the Clausius inequality gives

$$\int_{1A2B1} \frac{dQ}{T} = \int_{1A2} \frac{dQ}{T} + \int_{2B1} \frac{dQ}{T} = 0 \tag{7.94}$$

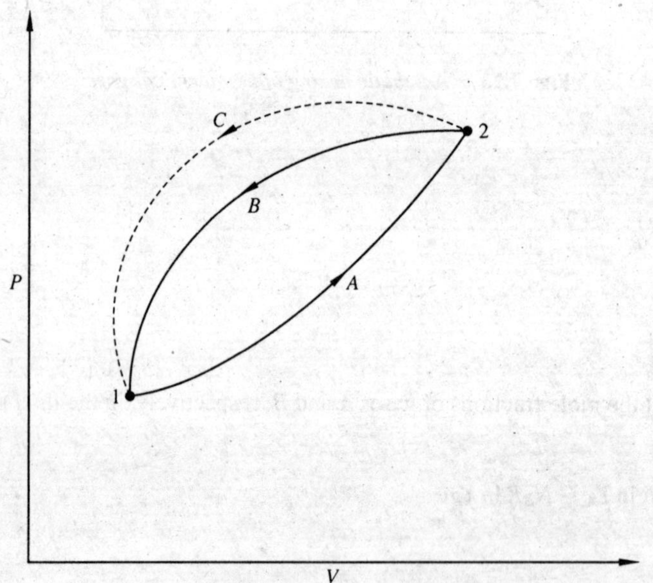

Fig. 7.24 *A reversible cycle 1A2B1 and an irreversible cycle 1A2C1 on P–v diagram*

If the system is restored to the initial state 1 from state 2 by an irreversible process C, then $1A2C1$ is an irreversible cycle. Then the Clausius inequality gives

$$\int_{1A2C1} \frac{dQ}{T} = \int_{1A2} \frac{dQ}{T} + \int_{2C1} \frac{dQ}{T} < 0 \tag{7.95}$$

Subtracting Eq. (7.95) from Eq. (7.94) and rearranging, we get

$$\int_{2B1} \frac{dQ}{T} > \int_{2C1} \frac{dQ}{T} \tag{7.96}$$

Since process $2B1$ is a reversible process,

$$\int_{2B1} \frac{dQ}{T} = \int_2^1 dS \tag{7.97}$$

Therefore

$$\int_2^1 dS > \int_{2C1} \frac{dQ}{T} \tag{7.98}$$

In general, one can write

$$dS \geq \frac{dQ}{T} \tag{7.99}$$

where the equality sign holds good for a reversible process and the inequality sign holds good for an irreversible process.

Now let us apply the above result to evaluate the entropy change of the universe when a system interacts with the surroundings and exchanges energy as heat with the surroundings. Let T_{sur} and T_{sys} be the temperatures of the surroundings and the system, with a finite temperature difference ($T_{sur} > T_{sys}$). The process is clearly an irreversible process. We are interested in evaluating the entropy change of the universe, that is, the combination of the system and the surroundings associated with the irreversible process. Let dQ represent the energy transfer as heat from the surroundings to the system during the given irreversible process(see Fig. 7.25). Then

$$dS_{sys} = \frac{dQ}{T_{sys}}$$

$$dS_{sur} = -\frac{dQ}{T_{sur}}$$

$$dS_{uni} = dS_{sys} + dS_{sur} = \frac{dQ}{T_{sys}} - \frac{dQ}{T_{sur}} > 0 \tag{7.100}$$

or

$$\Delta S_{uni} > 0 \tag{7.101}$$

If the system is isolated, that is, when there is no interaction between the system and its surroundings, there is no change in the entropy of the surroundings and hence for an isolated system

$$\Delta S \geq 0 \tag{7.102}$$

Therefore, *the entropy of an isolated system either increases, or in the limit remains constant.* The equality sign holds good when the process undergone by the system is reversible; the inequality sign holds if there is any irreversibility present in the process. This statement is usually called the principle of entropy increase.

The principle of entropy increase is very important because it dictates the direction of the spontaneous processes. *Irreversible or spontaneous processes can occur only in that direction for which the entropy of the universe, or that of an isolated system, increases.* These processes cannot occur in the direction of decreasing entropy. The above observations may be summarized as:

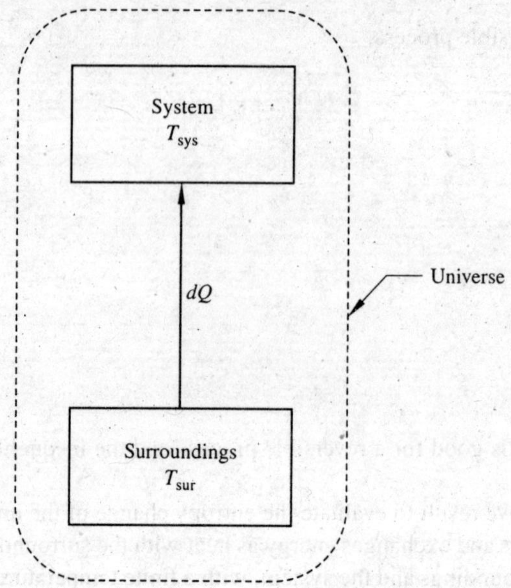

Fig. 7.25 *Entropy change of universe for an irreversible process*

For an isolated system,

- $\Delta S > 0$ for irreversible processes.
- $\Delta S = 0$ for reversible processes.
- $\Delta S < 0$ implies the process is impossible.

***Example* 7.15** One kg of superheated steam at 2 bar and 200°C contained in a piston–cylinder assembly is kept at ambient conditions of 300 K till the steam is condensed to saturated liquid at constant pressure. Calculate the change in the entropy of the universe associated with this process.

Solution Initial state of the steam: Superheated steam at 2 bar and 200°C.

$$h_1 = 2870.5 \text{ kJ/kg} \quad \text{and} \quad s_1 = 7.5072 \text{ kJ/kg K}$$

Final state : Saturated liquid at 2 bar.

$$h_2 = 504.70 \text{ kJ/kg} \quad \text{and} \quad s_2 = 1.5301 \text{ kJ/kg K}$$

Hence

$$\Delta s_{\text{steam}} = s_2 - s_1 = 1.5301 - 7.5072 = -5.9771 \text{ kJ/kg K}$$

For a constant pressure process: $q = \Delta h$

Therefore

$$q = h_2 - h_1 = 504.70 - 2870.5 = -2365.8 \text{ kJ}$$

Entropy change of the surroundings

$$= \Delta S_{\text{sur}} = \frac{Q}{T_{\text{sur}}} = \frac{2365.8}{300} = 7.886 \text{ kJ/K}$$

Hence

$$\Delta S_{\text{uni}} = \Delta S_{\text{sys}} + \Delta S_{\text{sur}} = -5.9771 + 7.886 = 1.9089 \text{ kJ/K}$$

$\Delta S_{\text{uni}} > 0$ and hence the process is irreversible and feasible.

7.15 Temperature–Entropy Diagram

It was already noticed that the entropy change of a system is given by $dS = \left(\frac{dQ}{T}\right)_R$. During a reversible process, the energy transfer as heat to the system from the surroundings is given by $dQ = TdS$. Hence, if T and S are chosen as independent variables, then the integral $\oint TdS$ is the area under the curve (see Fig. 7.26).

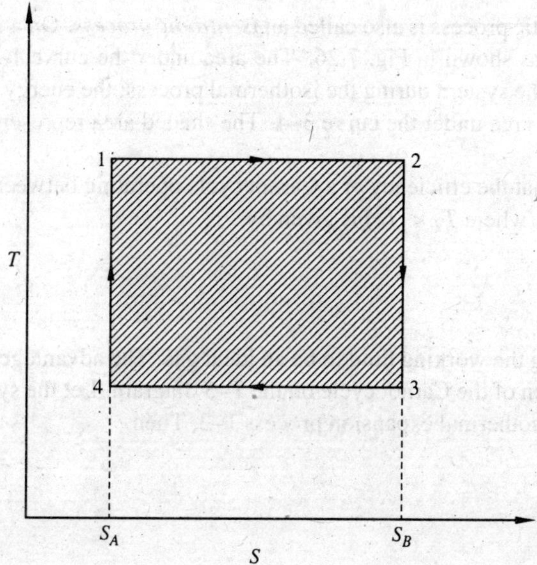

Fig. 7.26 *T–s diagram of a Carnot cycle*

The first law of thermodynamics gives

$$dU = dQ - dW$$

Also, for a reversible process, we can write

$$dQ = TdS \quad \text{and} \quad dW = PdV \tag{7.103}$$

Therefore,

$$dU = TdS - PdV \tag{7.104}$$

For a cyclic process, the above equation reduces to

$$\oint TdS = \oint PdV \tag{7.105}$$

For a cyclic process $\oint TdS$ represents the net heat interaction which is equal to the net work done by the system. Hence, the area enclosed by a cycle on a T–S diagram represents the net work done by a system.

For a reversible adiabatic process, we know that

$$dQ = 0 = TdS$$

Therefore

$$dS = 0 \quad \text{or} \quad S_2 = S_1 \tag{7.106}$$

Hence, a reversible adiabatic process is also called an *isentropic process*. On a T–S diagram, the Carnot cycle can be represented as shown in Fig. 7.26. The area under the curve 1–2 represents the energy absorbed as heat (Q_1) by the system during the isothermal process; the energy rejected as heat (Q_2) by the system is given by the area under the curve 3–4. The shaded area represents the net work done by the system.

It was already shown that the efficiency of a Carnot cycle operating between two thermal reservoirs at temperatures T_1 and T_2 (where $T_2 < T_1$) is given by

$$\eta = 1 - \frac{T_2}{T_1}$$

This was derived assuming the working fluid to be an ideal gas. The advantage of the T–S diagram can be realized by an inspection of the Carnot cycle on the T–S diagram. Let the system change its entropy from S_A to S_B during the isothermal expansion process 1–2. Then

$$q_1 = T_1(S_B - S_A)$$

and

$$Q_2 = T_2(S_B - S_A)$$

Then

$$\eta = \frac{Q_1 - Q_2}{Q_1} = \frac{(T_1 - T_2)(S_B - S_A)}{T_1(S_B - S_A)} = 1 - \frac{T_2}{T_1}$$

This demonstrates that in the analysis of many engineering problems, the temperature–entropy diagram is quite useful and convenient.

***Example* 7.16** One kg of superheated steam at 1.00 MPa and 300°C contained in a cylinder–piston assembly undergoes a reversible and adiabatic expansion to a pressure of 100 kPa. Calculate the work done by the steam.

Solution The initial and final states, and the process followed by the steam is shown in Fig. 7.27 on a Mollier diagram.

Initial state: Superheated steam at 1.00 MPa, 300°C

$$h_1 = 3052.1 \text{ kJ/kg}; \quad v_1 = 0.2580 \text{ m}^3/\text{kg}; \quad s_1 = 7.1251 \text{ kJ/kg K}$$

$$u_1 = 3052.1 - 1 \times 10^3(0.2580) = 2794.1 \text{ kJ/kg}$$

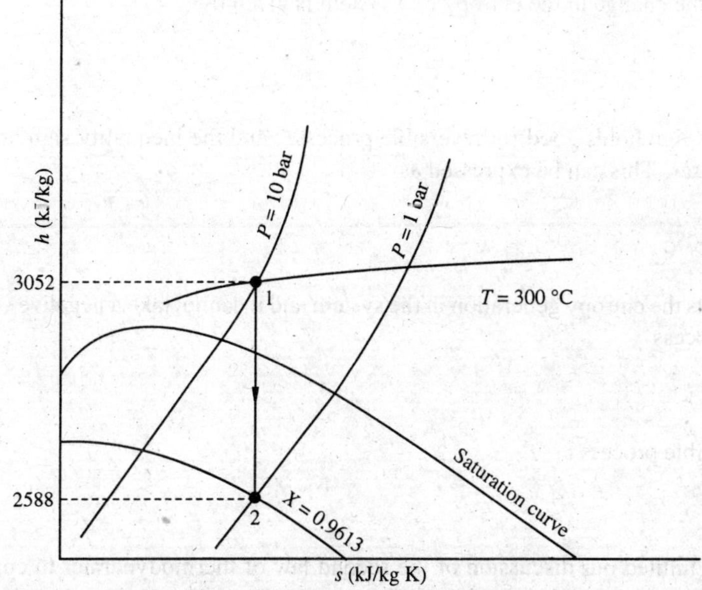

Fig. 7.27 *Sketch for example 7.16*

Final State: Steam at 100 kPa

Since the process is reversible and adiabatic, $s_2 = s_1$

At 100 kPa pressure,

$$s_f = 1.3027 \text{ kJ/kg K}; \quad s_g = 7.3598 \text{ kJ/kg K}$$

$$h_f = 417.54 \text{ kJ/kg}; \quad h_g = 2675.4 \text{ kJ/kg}$$

$$v_f = 0.001\,043\,4 \text{ m}^3/\text{kg}; \quad v_g = 1.6940 \text{ m}^3/\text{kg}$$

$$s_2 = s_1 = 7.1251 = 7.3598X_2 + (1 - X_2)1.3027$$

Therefore, $X_2 = 0.9613$

$$h_2 = 2675.4 \times 0.9613 + (1 - 0.9613)417.54 = 2588.02 \text{ kJ/kg}$$

$$v_2 = 1.6940 \times 0.9613 + (1 - 0.9613)0.001\,043\,4 = 1.6285 \text{ m}^3/\text{kg}$$

$$u_2 = h_2 - P_2 v_2 = 2588.02 - 100(1.6285) = 2425.17 \text{ kJ/kg}$$

For an adiabatic process, $\Delta u = -W$

Therefore, work done by the steam is given by

$$W = u_1 - u_2 = 2794.1 - 2425.17 = 368.93 \text{ kJ/kg}$$

7.16 Second Law Analysis of a Control Volume

It was shown that the change in the entropy of a system is given by

$$dS \geq \frac{dQ}{T}$$

where the equality sign holds good for reversible processes and the inequality sign holds good for all irreversible processes. This can be expressed as

$$\Delta S = \frac{dQ}{T} + S_G \tag{7.107}$$

where S_G represents the entropy generation in the system and it cannot take a negative value. Therefore, for a reversible process

$$S_G = 0 \tag{7.108}$$

and for an irreversible process

$$S_G > 0 \tag{7.109}$$

So far we have limited our discussion of the second law of thermodynamics to control mass only. Now the analysis can be extended to flow processes or to control volumes. The procedure used in writing the first law of thermodynamics to a control volume is adopted below. Consider a control volume shown in Fig. 7.28 through which material flows continuously in and out. In order to apply the second law of thermodynamics, let us identify a system such that its mass remains constant during a given process. At time t, the system constitutes both the mass inside the control volume and the mass about to enter the control volume during a small time interval dt (that is the mass occupying the region A). At time $(t+dt)$, the system is the mass inside the control volume as well as the mass in the region B. During the time interval dt, the system undergoes a change in configuration and receives energy as heat $(\dot{Q}dt)$ from the surroundings.

At time t, the entropy of the system is

$$m(t)s(t) + \dot{m}_i s_i dt = S(t) + \dot{m}_i s_i dt \tag{7.110}$$

At time $(t + dt)$, the entropy of the system is

$$m(t + dt)s(t + dt) + \dot{m}_e s_e dt = S(t + dt) + \dot{m}_e s_e dt \tag{7.111}$$

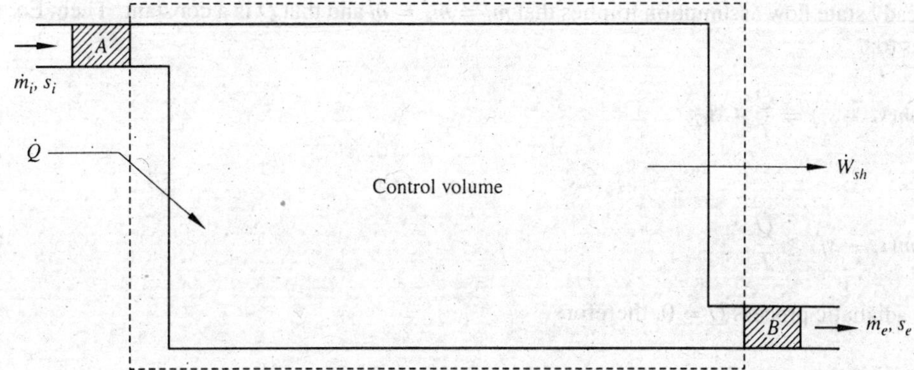

Fig. 7.28 *Control volume for second law analysis*

where, $m(t)$ = mass inside the control volume at time t
$s(t)$ = specific entropy of matter inside the control volume at time t
$m(t+dt)$ = mass inside the control volume at time $(t+dt)$
$s(t+dt)$ = specific entropy of mass inside the control volume at time $(t+dt)$
$S(t)$ = entropy of control volume at time t
$S(t+dt)$ = entropy of control volume at time $(t+dt)$

Then

$$S(t+dt) + \dot{m}_e s_e dt - \{S(t) + \dot{m}_i s_i dt\} = \frac{\dot{Q}dt}{T} + \dot{S}_G dt$$

or

$$\frac{S(t+dt) - S(t)}{dt} + \dot{m}_e s_e - \dot{m}_i s_i = \frac{\dot{Q}}{T} + \dot{S}_G$$

$$\frac{dS}{dt} + \dot{m}_e s_e - \dot{m}_i s_i = \frac{\dot{Q}}{T} + \dot{S}_G \tag{7.112}$$

This can be rearranged as

$$\frac{dS}{dt} = \left(\dot{m}_i s_i + \frac{\dot{Q}}{T}\right) - \dot{m}_e s_e + \dot{S}_G \tag{7.113}$$

or, stated in words,

The rate of accumulation of entropy = (Rate of inflow of entropy − Rate of outflow of entropy + Rate of generation of entropy).

7.16.1 Steady state steady flow processes

Equation (7.113) can be simplified if one assumes steady flow and steady state conditions. Under these conditions there is no change in the entropy of the control volume with respect to time, that is, $\frac{dS}{dt} = 0$.

The steady state flow assumption implies that $\dot{m}_i = \dot{m}_e = \dot{m}$ and that $\dot{Q}$ is a constant. Then, Eq. (7.113) reduces to:

$$\dot{m}(s_e - s_i) = \frac{\dot{Q}}{T} + \dot{S}_G \tag{7.114}$$

or

$$\dot{m}(s_e - s_i) \geq \frac{\dot{Q}}{T} \tag{7.115}$$

For an adiabatic process $\dot{Q} = 0$, therefore

$$s_e \geq s_i \tag{7.116}$$

If, however, the process is reversible and adiabatic then

$$s_e = s_i \tag{7.117}$$

***Example* 7.17** A steam turbine receives steam at 8 bar and 250°C. The steam expands in a reversible adiabatic process and leaves the turbine at 1 bar. The power output of the turbine is 20 MW. Determine the rate of steam flow into the turbine.

Solution The first law of thermodynamics for a control volume gives

$$\dot{m}_e\left(h_e + \frac{V_e^2}{2} + gZ_e\right) - \dot{m}_i\left(hi + \frac{V_i^2}{2} + gZ_i\right) = \dot{Q} - \dot{W}_s - \frac{dE}{dt}$$

Since the turbine is operating under steady state, steady flow and adiabatic conditions,

$$\dot{m}_i = \dot{m}_e = \dot{m};\ \ \dot{Q} = 0 \ \ and \ \ \frac{dE}{dt} = 0$$

Ignoring the potential energy and the kinetic energy changes, the first law reduces to:

$$h_e - h_i = -\frac{\dot{W}_s}{\dot{m}}$$

For a reversible adiabatic process, the second law of thermodynamics for a control volume gives

$$s_e = s_i$$

From steam tables: $s_i = 7.0248$ kJ/kg K and $h_i = 2947.95$ kJ/kg

$$s_e = s_i = s_g X_e + s_f(1 - X_e)$$

or

$$7.0248 = 7.3598 X_e + 1.3027(1 - X_e)$$

Therefore, $X_e = 0.9447$

$$h_e = h_g X_e + h_f(1 - X_e)$$
$$h_e = 2675.4 \times 0.9447 + 417.54(1 - 0.9447) = 2550.54 \text{ kJ/kg}$$

$$h_e - h_i = -\frac{\dot{W}_s}{m}$$

$$2550.54 - 2947.95 = -\frac{20 \times 10^3}{\dot{m}} \quad \text{or} \quad \dot{m} = 50.33 \text{ kg/s}$$

Therefore, the required flow rate of steam through the turbine is 50.33 kg/s. The problem can also be solved using a Mollier diagram as shown in Fig. 7.29.

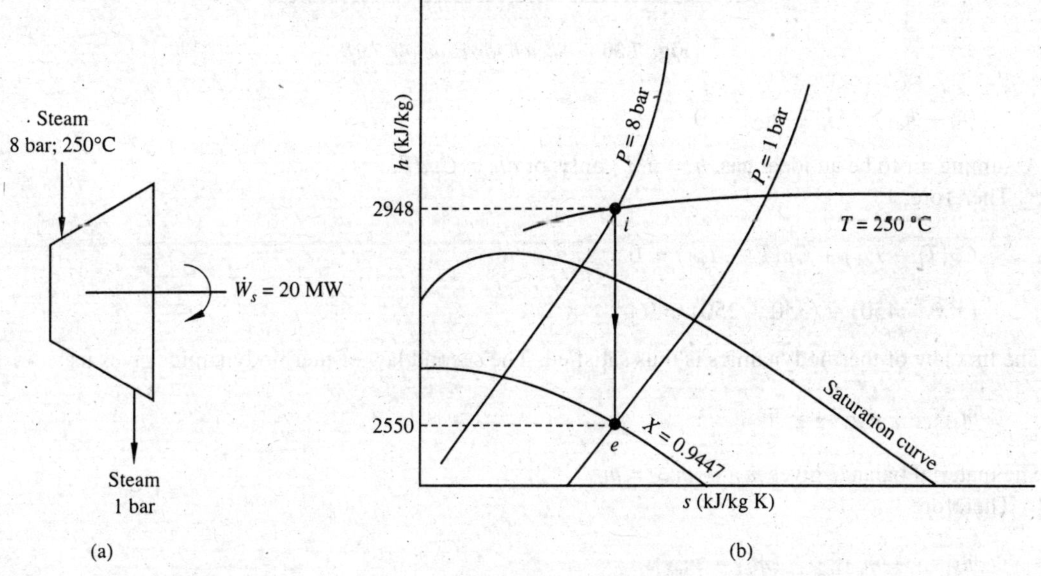

Fig. 7.29 *Sketch for Example 7.17 (a) control volume (b) Mollier diagram*

***Example* 7.18** An inventor claims to have designed a device shown in Fig. 7.30 which takes in 2 kmol/s of air at 250 kPa and 350 K and delivers equal amounts of hot stream and cold stream of air. The hot stream is at 100 kPa and 450 K while the cold stream of air is at 100 kPa and 250 K. He further claims that his device does not require any additional energy as heat or work to operate. Judge whether such a device is feasible or not. Consider air to be an ideal gas with $C_P = 7R/2$.

Solution The first law of thermodynamics for a control volume gives

$$\dot{m}_i h_i = \dot{m}_{e1} h_{e1} + \dot{m}_{e2} h_{e2}$$

or

$$2h_i = h_{e1} + h_{e2}$$

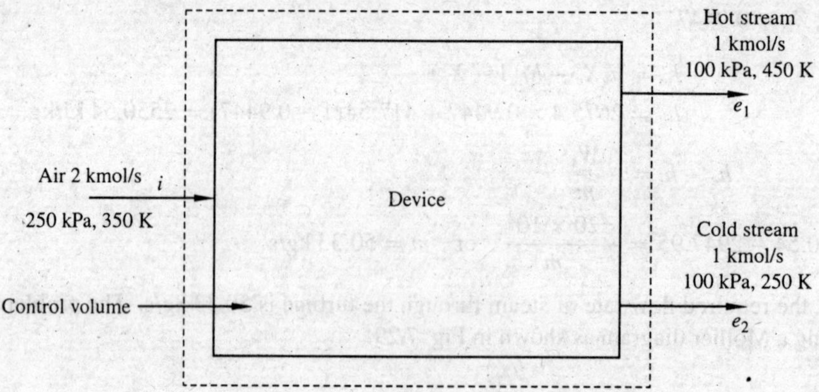

Fig. 7.30 *Sketch for Example 7.18*

$$(h_i - h_{e1}) + (h_i - h_{e2}) = 0$$

Assuming air to be an ideal gas, $h = h(T)$ only, or $dh = C_P dT$.

Therefore,

$$C_P(T_i - T_{e1}) + C_P(T_i - T_{e2}) = 0$$

$$(350 - 450) + (350 - 250) = 0$$

The first law of thermodynamics is thus satisfied. The second law of thermodynamics gives us

$$\dot{m}_{e1}s_{e1} + \dot{m}_{e2}s_{e2} \geq \dot{m}_i s_i$$

The material balance gives $\dot{m}_{e1} + \dot{m}_{e2} = \dot{m}_i$

Therefore

$$\dot{m}_{e1}s_{e1} + \dot{m}_{e2}s_{e2} \geq (\dot{m}_{e1} + \dot{m}_{e2})s_i$$

or

$$\dot{m}_{e1}(s_{e1} - s_i) + \dot{m}_{e2}(s_{e2} - s_i) \geq 0$$

The change in entropy for an ideal gas is given by

$$\Delta s = C_P \ln \frac{T_2}{T_1} - R \ln \frac{P_2}{P_1}$$

Hence

$$\dot{m}_{e1}(s_{e1} - s_i) = C_P \ln \frac{450}{350} - R \ln \frac{100}{250}$$

and

$$\dot{m}_{e2}(s_{e2} - s_i) = C_P \ln \frac{250}{350} - R \ln \frac{100}{250}$$

Then

$$\dot{m}_{e1}(s_{e1} - s_i) + \dot{m}_{e2}(s_{e2} - s_i) = C_P \ln \frac{450 \times 250}{350 \times 350} - R \ln \frac{100 \times 100}{250 \times 250}$$

$$= 12.7581 \text{ kJ/K} > 0$$

Hence the device satisfies the second law of thermodynamics also. Therefore the device is thermodynamically feasible.

7.17 Available Energy

When a system interacts with the surroundings and does some work, and at the same time receives energy in the form of heat, it is the first law of thermodynamics that relates the change in the energy of the system to the heat and work interactions across its boundaries. The work done by the system can be evaluated if the heat interaction and the change in the energy of the system are known. However, the first law does not specify the amount of work that can be done by the system when it receives a certain amount of energy as heat. In other words, the first law of thermodynamics does not specify what fraction of the energy received by the system in the form of heat can be converted into work. For that matter, the first law does not forbid even the complete conversion of the energy absorbed as heat into work. It is the second law of thermodynamics that forbids the complete conversion of energy absorbed as heat into work. It stipulates that during a cyclic process part of the energy received by the system in the form of heat has to be rejected to low temperature surroundings, thus setting a limit on the extent of the conversion of energy absorbed as heat into work.

To determine that part of the energy which can be converted into work by an engine, let us consider a thermal reservoir at constant temperature T from which a quantity of energy Q is being absorbed as heat. A Carnot engine may be operated between the reservoir at T and the surroundings at temperature T_0 to convert part of the energy taken in as heat into work. The amount of work that can be obtained is given by

$$W = Q\eta = Q\left(1 - \frac{T_0}{T}\right) = (Q - T_0 \mid \Delta S \mid) \tag{7.118}$$

where, $\mid \Delta S \mid$ is the change in the entropy of the reservoir that supplies the energy. The Carnot cycle is shown on a $T - S$ diagram in Fig. 7.31(a).

The area 1–2–3–4 represents the portion of the energy delivered as work by the reversible engine and is called the *available energy*. The area 43BA represents that portion of the energy which is discarded to the ambient atmosphere. This portion of energy is not available for conversion into work and is called *unavailable energy*. Therefore, the available energy represents the portion of the energy supplied as heat which can be converted into work by means of a reversible engine.

The result that $W = Q - T_0 \mid \Delta S \mid$ is true also for the case of a finite body which acts as a source, the temperature of which varies during the interaction as shown below.

Now let us consider a finite body from which energy in the form of heat can be absorbed by a heat engine. As the energy is withdrawn from the finite body, the temperature of the body keeps on decreasing. This process can be carried out using a large number of differential Carnot engines. For a differential Carnot cycle, let dQ represent the energy absorbed at temperature T. The differential

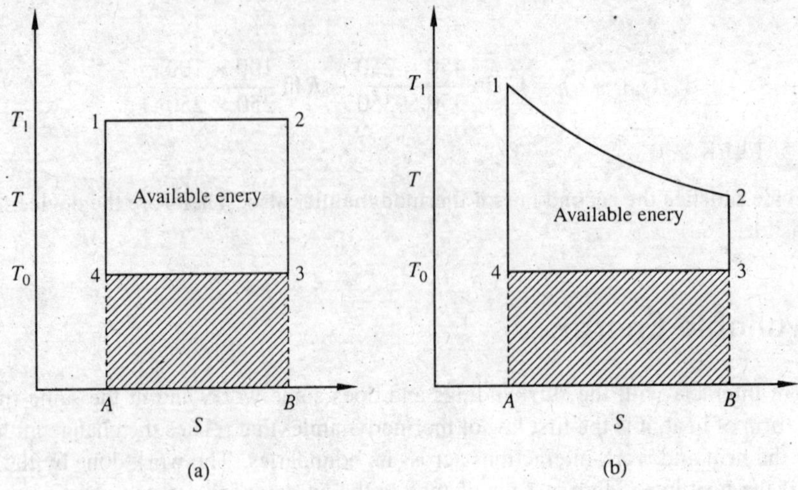

Fig. 7.31 *Available and unavailable energy (a) when a constant temperature thermal reservoir is used as a source (b) when a finite body is used as a source*

amount of work done by the engine is given by

$$dW = dQ\eta = dQ\left(1 - \frac{T_0}{T}\right)$$

If the Carnot engine works till the temperature of the body attains a value T_2, the total work done by the reversible engine is given by

$$W = \int dQ\eta = \int_{T1}^{T_2} dQ\left(1 - \frac{T_0}{T}\right) = \int dQ - T_0 \mid \Delta S \mid$$

$$W = Q - T_0 \mid \Delta S \mid = \text{Available energy} \tag{7.119}$$

Therefore, if a certain amount of energy (Q) is absorbed as heat from a finite body, then $Q - T_0 \mid \Delta S \mid$ represents the available portion of the energy which can be converted into work.

7.18 Loss in Available Energy

Suppose a certain amount of energy Q is transferred as heat from a body at constant temperature T_1 to a body at constant temperature T_2 (where $T_1 > T_2$). Because of the finite temperature difference between the two interacting bodies, the energy transfer is an irreversible process. This irreversibility leads to a loss in the available energy. This loss in the available energy would not have occurred, had a reversible engine been employed to transfer the energy from the body at T_1 to the body at T_2. Suppose the same quantity of energy Q is absorbed by a reversible engine. Before the energy transfer takes place the energy is available at temperature T_1. After the energy transfer the energy is available at a lower temperature T_2. Let the ambient be at a temperature T_0.

Then the initial available energy is given by

$$Q\left(1-\frac{T_0}{T_1}\right) \tag{7.120}$$

and the final available energy is given by

$$Q\left(1-\frac{T_0}{T_2}\right) \tag{7.121}$$

Therefore, the loss in the available energy is:

$$Q\left(1-\frac{T_0}{T_1}\right)-Q\left(1-\frac{T_0}{T_2}\right)=T_0\left(\frac{Q}{T_2}-\frac{Q}{T_1}\right)=T_0\Delta S_{\text{uni}} \tag{7.122}$$

where, ΔS_{uni} is the change in the entropy of the universe. It implies that whenever $\Delta S_{\text{uni}} > 0$, there is a loss in the available energy.

The second law of thermodynamics states that $\Delta S_{\text{uni}} \geq 0$, where the equality is true for reversible processes and the inequality is true for all spontaneous or irreversible, processes. Therefore, all spontaneous or irreversible processes are associated with a loss in available energy. Note that, though the same quantity of energy as heat is available before and after the irreversible energy transfer, the available energy is less at T_2 compared to the available energy at T_1. That is, there is a degradation in the quality of the energy and internal energy at lower temperature is considered as degraded energy. We have seen that all the energy absorbed as heat cannot be converted into work, That is, the internal energy of a reservoir or a body is a low quality energy. On the other hand, work is considered as the highest quality of energy, because it is interchangeable to the other forms with cent percent energy conversion efficiency. It is possible to convert work completely into internal energy but not *vice versa*.

***Example* 7.19** In a certain process a vapor is condensed at 500°C by transferring energy to water, which in turn is vaporized at 250°C. The resulting water vapor is used in a Carnot engine with the ambient atmosphere at 25°C as its sink. Determine the fraction of the available energy lost due to the irreversible energy transfer.

Solution Let Q represent the quantity of energy transferred as heat from the condensing vapor to water.

Available energy from condensing vapor is given by

$$Q\left(1-\frac{298}{773}\right)$$

Loss in available energy due to irreversible process is given by

$$T_0\Delta S_{\text{uni}}=298\left(\frac{Q}{523}-\frac{Q}{773}\right)$$

Fraction of available energy lost due to irreversible process

$$=\frac{298\left(\frac{Q}{523}-\frac{Q}{773}\right)}{Q\left(1-\frac{298}{773}\right)}=0.3$$

7.19 Availability

An engineer is interested in knowing the maximum useful work that can be obtained from a system when it is in a given state. Obviously, the work obtained will be maximum only when the final state of the system is in equilibrium with the surroundings. That is, the system attains a state of thermal equilibrium and mechanical equilibrium with the surroundings. In other words, there is no possibility of obtaining further work from the system due to interaction with the surroundings. That is, the system attains a *dead state* where its pressure is equal to the ambient pressure P_0, and its temperature is equal to the ambient temperature T_0. A part of the total work done by the system is spent in pushing back the surrounding atmosphere, which is not available to raise a mass in the gravitational field. Therefore, the maximum useful work that can be obtained from a system is less than the maximum work done by the system during a given process.

The availability of a given system is defined as *the maximum useful work that can be obtained in a process in which the system comes to equilibrium with the surroundings or attains the dead state.* Clearly, the availability of a system depends on the conditions of the system as well as those of the surroundings.

7.19.1 Availability for non-flow process

Part of the total work delivered by the system, is used in "pushing back" the ambient atmosphere when the system comes into equilibrium with the surroundings. Since the system comes to equilibrium with the ambient conditions, its final pressure is P_0. Let V_1 and V_0 represent the initial and final volumes of the system, respectively. Then the change in the volume of the system is $(V_0 - V_1)$. This change in the volume of the system is resisted by the ambient pressure P_0. Then the work done in pushing back the ambient atmosphere is $P_0(V_0 - V_1)$. Hence, the useful work that can be obtained from the system is the maximum work less the work done in pushing back the ambient atmosphere. Then,

$$\text{Availability} = W_{\text{useful}} = W_{\text{max}} - P_0(V_0 - V_1) \tag{7.123}$$

To determine the availability of a system, let us consider a system shown in Fig. 7.32, which interacts with the ambient at T_0.

Application of the first and the second laws of thermodynamics to the system gives

$$Q - W = U_0 - U_1 \tag{7.124}$$

and

$$\Delta S_{\text{uni}} = (S_0 - S_1) - \frac{Q}{T_0} \geq 0$$

or

$$Q \leq T_0(S_0 - S_1) \tag{7.125}$$

Maximum work can be obtained from a reversible process for which $\Delta S_{\text{uni}} = 0$. Hence, Eq. (7.125) becomes

$$Q = T_0(S_0 - S_1) \tag{7.126}$$

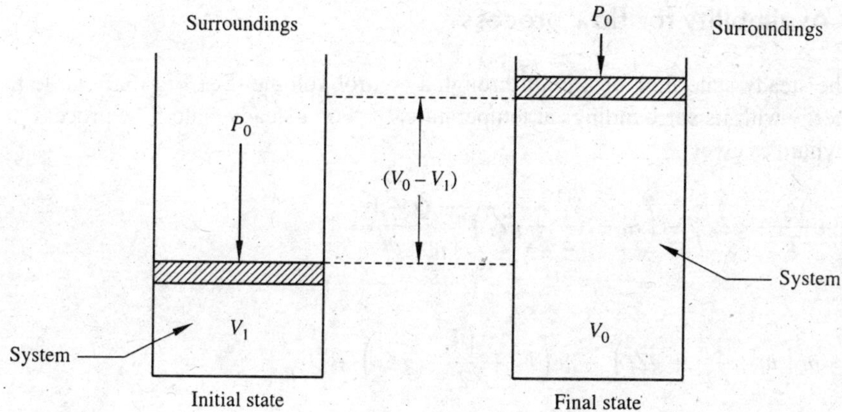

Fig. 7.32 *A system interacting with the surroundings at T_0 and P_0*

Combining Eqs. (7.124) and (7.126), we get

$$W_{max} = (U_1 - U_0) - T_0(S_1 - S_0) \tag{7.127}$$

$$\text{Availability} = \phi = W_{useful} = W_{max} - P_0(V_0 - V_1)$$

$$= (U_1 - T_0S_1) - (U_0 - T_0S_0) - P_0(V_0 - V_1)$$

$$\phi = (U_1 + P_0V_1 - T_0S_1) - (U_0 + P_0V_0 - T_0S_0) \tag{7.128}$$

If a system undergoes a change of state from the initial state 1 (where the availability is ϕ_1 to the final state 2 (where the availability is ϕ_2), the change in the availability or the change in the maximum useful work associated with the process is $\phi_2 - \phi_1$.

***Example* 7.20** A steam generator in an industry produces superheated steam at 20 bar and 300°C. Sometimes steam at 1 bar and 200°C is required to carry out a specified process. It is planned to expand the superheated steam and obtain some useful work before it is supplied to carry out the process. Determine the maximum useful work per kg of steam that can be obtained during the expansion process. The ambient conditions are 1 bar and 25°C.

Solution The maximum useful work that can be obtained from the steam

$$= (\phi_1 - \phi_2) = (u_1 + P_0v_1 - T_0s_1) - (u_2 + P_0v_2 - T_0s_2)$$

From steam tables:

$h_1 = 3025.0$ kJ/kg; $v_1 = 0.1255$ m^3/kg; $s_1 = 6.7696$ kJ/kg K

$h_2 = 2875.4$ kJ/kg; $v_2 = 2.1720$ m^3/kg; $s_2 = 7.8349$ kJ/kg K

$u_1 = 3025.0 - 20 \times 100(0.1255) = 2774$ kJ/kg

$u_2 = 2875.4 - 100(2.1720) = 2658.2$ kJ/kg

Therefore, $\phi_1 - \phi_2 =$

$$\left[2774 + 10^5(0.1255) \times 10^{-3} - 298(6.7696)\right] - \left[2658.2 + 10^5(2.1720) \times 10^{-3} - 298(7.8349)\right]$$

$$= 341.56 \text{ kJ/kg}$$

7.19.2 Availability for flow process

Consider the steady state flow of a fluid through a control volume (See Fig. 6.3) while the fluid exchanges energy with its surroundings at temperature T_0. For a steady state flow process, the first law of thermodynamics gives

$$\left(h_e + \frac{V_e^2}{2} + gZ_e\right) - \left(h_i + \frac{V_i^2}{2} + gZ_i\right) = \frac{\dot{Q} - \dot{W}_s}{\dot{m}}$$

or

$$\dot{W}_s = \dot{m}\left(h_i + \frac{V_i^2}{2} + gZ_i\right) - \dot{m}\left(h_e + \frac{V_e^2}{2} + gZ_e\right) + \dot{Q} \tag{7.129}$$

The second law of thermodynamics applied to the control volume for reversible process gives

$$\Delta S_{\text{uni}} = \dot{m}(s_e - s_i) - \frac{\dot{Q}}{T_0} = 0$$

or

$$\dot{Q} = \dot{m}T_0(s_e - s_i) \tag{7.130}$$

Combining Eqs. (7.129) and (7.130), we get the maximum shaftwork as

$$\dot{W}_s = \dot{m}\left(h_i + \frac{V_i^2}{2} + gZ_i\right) - \dot{m}\left(h_e + \frac{V_e^2}{2} + gZ_e\right) - \dot{m}T_0(s_i - s_e) \tag{7.131}$$

or

$$\dot{W}_s = \dot{m}\left[(h_i + KE_i + PE_i - T_0 s_i) - (h_e + KE_e + PE_e - T_0 s_e)\right]$$

Since no part of the shaft work is spent in pushing back the atmosphere, there is no difference between the maximum useful work, in a steady flow process, and the maximum work.

The dead state for a nonflow process has been defined in terms of thermal and mechanical equilibrium with the surroundings. A dead state for steady flow through a control volume implies not only thermal and mechanical equilibrium with the surroundings but also the kinetic energy is zero relative to the surroundings and its elevation is the same as that of the surroundings. The availability of a flowing fluid in a steady state flow is defined as the maximum work output which can be obtained as the fluid is changed reversibly from a given state to the dead state while exchanging heat with the surroundings. Therefore, the stream availability ψ is given by

$$\psi = (h + KE + PE - T_0 s) - (h_0 + PE_0 - T_0 s_0) \tag{7.132}$$

where the subscript 0 refers to the values in the dead state.

Quite often an engineer encounters a situation in which the material undergoes a change of state from the initial state 1 to the final state 2 during a flow process. In such cases the useful power that can be obtained from the material during a specified change of state is given by $(\psi_2 - \psi_1)$. So far we have considered the heat interaction with the surroundings at T_0. Many industrial processes involve heat interaction with other bodies instead of the surroundings. In analyzing such cases we can consider the combination of both the interacting bodies as our system and evaluate the change in the availability of the system as shown in the following example.

***Example* 7.21** In a desalination plant, drinking water is produced by condensing steam in a heat-exchanger. Steam at 0.1 MPa and 150°C enters the heat-exchanger shown in Fig. 7.33 at a rate of 10,000 kg/h and emerges as a liquid at 40°C. Cooling water enters the heat-exchanger at 25°C and leaves at 35°C. Estimate the loss in availability in the heat exchange process. The ambient temperature is 300 K.

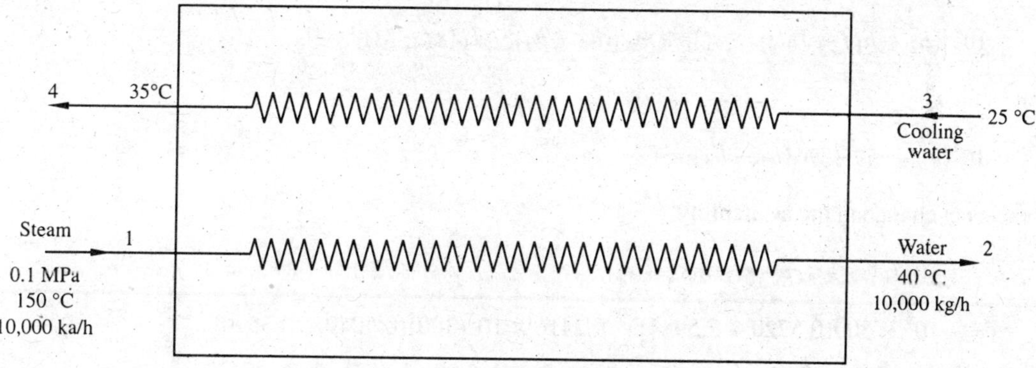

Fig. 7.33 *Sketch for Example 7.21*

Solution The first law of thermodynamics for the control volume gives us

$$\dot{m}_1 h_1 + \dot{m}_3 h_3 = \dot{m}_2 h_2 + \dot{m}_4 h_4$$

The conservation of mass gives us $\dot{m}_1 = \dot{m}_2$ and $\dot{m}_3 = \dot{m}_4$. The heat-exchanger is considered to be adiabatic, and the changes in the kinetic energy and potential energy terms are ignored.
From steam tables:

$$h_1 = 2775.8 \text{ kJ/kg} \quad \text{and} \quad s_1 = 7.5984 \text{ kJ/kg K}$$
$$h_2 = 167.45 \text{ kJ/kg} \quad \text{and} \quad s_2 = 0.5721 \text{ kJ/kg K}$$
$$h_3 = 104.77 \text{ kJ/kg} \quad \text{and} \quad s_3 = 0.3670 \text{ kJ/kg K}$$
$$h_4 = 146.56 \text{ kJ/kg} \quad \text{and} \quad s_4 = 0.5049 \text{ kJ/kg K}$$

Therefore,

$$10^4(2775.8) + \dot{m}_3(104.77) = 10^4(167.45) + \dot{m}_4(146.56)$$

or

$$\dot{m}_3 = 6.2416 \times 10^5 \text{ kg/h}$$

Change in the availability of steam

$$= 10^4(\psi_2 - \psi_1) = 10^4[(h_2 - T_0 s_2) - (h_1 - T_0 s_1)]$$
$$= 10^4[(h_2 - h_1) - T_0(s_2 - s_1)]$$

Change in the availability of cooling water

$$= \dot{m}_3(\psi_4 - \psi 3) = \dot{m}_3[(h_4 - h_3) - T_0(s_4 - s_3)]$$

Net change in the availability

$$10^4[(h_2 - h_1) - T_0(s_2 - s_1)] + \dot{m}_3[(h_4 - h_3) - T_0(s_4 - s_3)]$$

But

$$10^4(h_2 - h_1) + \dot{m}_3(h_4 - h_3) = 0$$

Hence, net change in the availability

$$= -10^4 T_0(s_2 - s_1) - \dot{m}_3 T_0(s_4 - s_3)$$
$$= -10^4 \times 300(0.5720 - 7.5984) - 6.2416 \times 10^5(300)(0.5049 - 0.3670)$$
$$= -4.7423 \times 10^6 \text{ kJ}$$

Therefore, the loss in availability = 4.7423×10^6 kJ.

7.20 Irreversibility

All spontaneous or naturally occurring processes are irreversible. They cannot take place in both the directions. The second law of thermodynamics states that spontaneous processes occur in that direction for which the entropy change of the universe associated with the process is positive. An engineer is naturally interested in knowing the extent to which the processes are irreversible so that efforts can be made to reduce the irreversibility in the process.

The useful work that can be obtained from a system during a given change of state is maximum only when the process is carried out reversibly. The work done (W) by a system in an actual process is less than the useful work. The irreversibility (I) of a process is defined as the difference between the useful work and the actual work. That is

$$I = W_{\text{useful}} - W \tag{7.133}$$

or

$$I = \phi - W$$

or

$$I = (U_1 - T_0 S_1 + P_0 V_1) - (U_0 - T_0 S_0 + P_0 V_0) - W$$

or

$$I = -\Delta U + T_0 \Delta S - P_0 \Delta V - W \qquad (7.134)$$

Considering the system and its surroundings, as a composite system, the work done by the composite system under adiabatic conditions, is given by

$$W = -\Delta U_{\text{composite system}} = -\Delta U - \Delta U_{\text{sur}}$$

or

$$\Delta U_{\text{sur}} = -\Delta U - W \qquad (7.135)$$

(since there is no energy exchange as heat between the composite system and the universe.) Application of the two laws of thermodynamics to the surroundings gives

$$\Delta U_{\text{sur}} = T_0 \Delta S_{\text{sur}} - P_0 \Delta V_{\text{sur}}$$

or

$$\Delta U_{\text{sur}} = -\Delta U - W = T_0 \Delta S_{\text{sur}} - P_0 \Delta V_{\text{sur}} \qquad (7.136)$$

Substituting Eq. (7.136) in Eq. (7.134), we get

$$I = T_0 \Delta S_{\text{sur}} - P_0 \Delta V_{\text{sur}} + T_0 \Delta S - P_0 \Delta V$$

or

$$I = T_0(\Delta S + \Delta S_{\text{sur}}) - P_0(\Delta V + \Delta V_{\text{sur}}) \qquad (7.137)$$

As the ambient atmosphere with which the system is interacting is sufficiently large, the net change in the volume of the composite system is zero. That is,

$$\Delta V + \Delta V_{\text{sur}} = 0$$

Then Eq. (7.137) reduces to

$$I = T_0(\Delta S + \Delta S_{\text{sur}}) = T_0 \Delta S_{\text{uni}} \qquad (7.138)$$

Therefore, the irreversibility of a given process is equal to the product of the temperature of the surroundings, and the net change in the entropy of the interacting systems. If there is any irreversibility in the process, ΔS_{uni} is positive and hence the irreversibility (I) is positive. If the system undergoes a reversible process, then ΔS_{uni} is zero and, the irreversibility of the process is zero. The irreversibility represents the loss in the work.

7.21 Summary

All naturally occurring or spontaneous processes proceed in one direction only, and never in the opposite direction. However, the first law of thermodynamics does not deny the feasibility of a spontaneous process reversing itself. The direction of a spontaneous process is governed by the second law of thermodynamics. Any statement of the second law in essence denies the feasibility of a spontaneous process reversing itself. The two important statements of the second law of thermodynamics are discussed.

The Kelvin–Planck statement of the second law of thermodynamics reads: *It is impossible to construct a cyclically operating device such that it produces no other effect than the absorption of energy as heat from a single thermal reservoir and performs an equivalent amount of work.* A device which could perform work solely by absorbing energy as heat from a body is called a Perpetual Motion Machine of the Second Kind (PMMSK). Kelvin–Planck statement of the second law tells us that it is impossible to construct a PMMSK.

The Clausius statement of the second law of thermodynamics reads: *It is impossible to construct a cyclically operating device which will produce no other effect than the transfer of energy as heat from a low temperature body to a high temperature body.* The equivalence of the Kelvin–Planck and Clausius statements of the second law of thermodynamics is discussed.

An engineer is interested in reversible processes since they yield maximum work in engines and require minimum work in devices such as refrigerator. Hence, it is necessary to recognize reversible processes. A criterion based on the second law of thermodynamics can be used to distinguish between reversible and irreversible processes. The criterion that can be used to recognize processes as reversible or irreversible is to see if the assumption of reversibility of a particular process leads to violation of the second law of thermodynamics. If it does, the process is said to be irreversible. This criterion is applied to a few processes to judge whether the processes are irreversible or not. The processes which can be idealized as reversible are: motion without friction; expansion or compression with no pressure difference between the system and its surroundings; and energy transfer as heat with no temperature difference across the system boundary.

The idea of a reversible cycle was first introduced by Carnot. The Carnot cycle consists of two reversible isothermal processes and two reversible adiabatic processes. Carnot proposed two theorems regarding the efficiency of reversible engines and they are the consequences of the second law of thermodynamics. The Carnot theorems state:

1. No heat engine operating between the two given thermal reservoirs can be more efficient than a reversible engine operating between the same two thermal reservoirs.
2. All reversible engines operating between the two given thermal reservoirs have the same efficiency. The efficiency of a reversible engine does not depend on the working fluid; it depends only on the temperatures of the reservoirs between which it operates.

The absolute temperature scale was established as a consequence of the Carnot theorems. The equivalence of the thermodynamic or absolute temperature scale and ideal gas temperature scale is discussed.

An important consequence of the second law of thermodynamics is the Clausius inequality:

$$\oint \frac{dQ}{T} \leq 0$$

which leads to the concept of entropy. Entropy is an important property in thermodynamics which stipulates the feasibility of a given process. By evaluating the entropy change one can determine the direction in which a process would proceed. The change in the entropy of a system is given by:

$$dS = \left(\frac{dQ}{T}\right)_R \quad \text{for reversible processes}$$

Since entropy is a state function the entropy change of a system is determined by its initial and final states only, irrespective of how the system has changed its state. Therefore, in analyzing irreversible processes, one can substitute the actual process by an imaginary reversible process connecting the final state to the initial state, and the entropy change for the imaginary reversible process can be evaluated. The calculation of entropy change is illustrated through several examples.

Another important consequence of the second law of thermodynamics is the principle of entropy increase which reads: *the entropy of an isolated system either increases or, in the limit, remains constant.*

Application of the second law of thermodynamics to a control volume gives:

$$\frac{dS}{dt} + \dot{m}_e s_e - \dot{m}_i s_i = \frac{\dot{Q}}{T} + \dot{S}_G$$

For steady state process, the second law for a control volume reduces to:

$$\dot{m}(s_e - s_i) \geq \frac{\dot{Q}}{T}$$

The available energy is that portion of energy supplied as heat which can be converted into work by means of a reversible engine. The available energy is given by $Q - T_0 \mid \Delta S \mid$ irrespective of whether the source of energy is infinite or finite. When a certain quantity of energy Q is transferred as heat from one system to another, the loss in available energy is equal to $T_0 \Delta S_{\text{uni}}$, where T_0 is the ambient temperature. For all spontaneous or irreversible processes $\Delta S_{\text{uni}} > 0$. Therefore, every irreversible process is associated with a loss in available energy.

Availability is defined as the maximum useful work that can be obtained in a process in which the system comes to equilibrium with the surroundings. The availability function for a non-flow process is given by

$$\phi = (u + P_0 v - T_0 s) - (u_0 + P_0 v_0 - T_0 s_0)$$

The availability of a flowing fluid, for a steady flow process is given by:

$$\psi = (h + KE + PE - T_0 s) - (h_0 + PE_0 - T_0 s_0)$$

The irreversibility I of a process is defined as the difference between the useful work and the actual work. The irreversibility of a process is given by:

$$I = T_0 \Delta S_{\text{uni}}$$

7.22 Review Questions

1. What are the limitations of the first law of thermodynamics?
2. Does the first law of thermodynamics deny the possibility of a spontaneous process proceeding in the reverse direction on its own? Explain with the help of an example.
3. What kind of questions cannot be answered by the application of the first law of thermodynamics?

4. What is a heat engine? Give some examples of heat engines.
5. What is a source and what is a sink?
6. Define the thermal efficiency of a heat engine.
7. Distinguish between a heat pump and a refrigerator.
8. Define the Coefficient Of Performance (COP) of a refrigerator.
9. How is the COP of a heat pump related to the COP of a refrigerator?
10. Name the characteristics which are common to all the heat engines.
11. State the Kelvin–Planck statement of the second law of thermodynamics.
12. State the Clausius statement of the second law of thermodynamics.
13. Can the thermal efficiency of a heat engine be unity? If not specify why.
14. Is it possible to devise a device which absorbs energy as heat from a single constant temperature reservoir and convert all of it into work? Explain.
15. What is a Perpetual Motion Machine of the Second Kind?
16. Does the second law of thermodynamics specify the direction of a process?
17. Is it possible to devise a refrigerator or heat pump, the COP of which is infinity? If not specify the reasons.
18. Prove that violation of the Kelvin–Planck statement leads to violation of the Clausius statement of the second law of thermodynamics.
19. Prove that violation of the Clausius statement leads to violation of the Kelvin–Planck statement of the second law of thermodynamics.
20. Name a few reversible processes and a few irreversible processes.
21. What is the criterion to identify an irreversible process?
22. Sketch the Carnot heat engine cycle on a P–v diagram.
23. Sketch the Carnot heat pump or refrigerator cycle on a P–v diagram.
24. Is it possible to operate a heat engine on a Carnot cycle? What purpose does a Carnot engine serve?
25. Derive a relation to estimate the thermal efficiency of a Carnot heat engine which employs an ideal gas as its working medium.
26. State the Carnot theorems.
27. Prove that no heat engine can be more efficient than a reversible heat engine when both of them are operating between the same thermal reservoirs.
28. Prove that all reversible heat engines operating between the same set of thermal reservoirs will have the same efficiency.
29. Is it possible to increase the efficiency of a reversible heat engine by changing the working fluid while the temperatures of the source and sink are held constant? If not specify the reason.
30. What is the basis for the establishment of thermodynamic temperature scale? Why is it called the absolute temperature scale?
31. Show that a given reversible process can be replaced by a combination of reversible isothermal and reversible adiabatic processes.
32. State the Clausius inequality.
33. How is entropy defined?
34. If a system undergoes an irreversible process, how does one estimate its entropy change?
35. What is the change in entropy if a mass m of a liquid is heated from T_1 to T_2 at constant pressure?

36. It is known that C_P of a substance varies with temperature. Is it possible to use the mean heat capacity, which is defined as

$$C_{Pm} = \frac{\int_{T_1}^{T_2} C_P dT}{T_2 - T_1}$$

for calculating the entropy change of a substance when it undergoes a constant pressure process?
37. What is the change in entropy if one mole of an ideal gas changes its state from (T_1, P_1) to (T_2, P_2)?
38. What is the change in entropy when N_A moles of an ideal gas A at T and P are mixed with N_B moles of an ideal gas B at T and P?
39. State the principle of entropy increase.
40. How does one judge whether a process is reversible or irreversible based on the entropy change?
41. Can the entropy of a system ever decrease? If yes, give an example.
42. Show that a reversible adiabatic process is an isentropic process.
43. Sketch a Carnot heat engine cycle on a T–s diagram and making use of this diagram, derive a relation to determine the thermal efficiency of a Carnot heat engine.
44. Apply the second law of thermodynamics to a control volume and derive a relation, which can be used for second law analysis of a flow process.
45. If an adiabatic turbine is operating under steady state conditions, what is the relation between the entropies of the inflowing and outflowing fluid?
46. What is entropy generation? Can the entropy generation be negative? Explain.
47. What is available energy and what is unavailable energy?
48. What is meant by degradation of energy?
49. Define the dead state for a system.
50. Define the dead state for a flowing fluid.
51. What is useful work?
52. Define the availability for a system?
53. Define the stream availability or the availability of a flowing fluid in a steady state flow.
54. What is irreversibility?
55. What is the relation between irreversibility and entropy generation?

7.23 Problems

7.1 (a) Does the Clausius statement prohibit the transfer of energy as heat from a body at a low temperature to a body at a high temperature?
(b) A domestic refrigerator transfers energy as heat from the cold space to the ambient atmosphere at a higher temperature while work is done (usually in the form of electrical energy) on it. Does the refrigerator violate the Clausius statement of the second law of thermodynamics?

7.2 Discuss whether the following processes are reversible or not. If they are not reversible, specify the source of irreversibility.

(a) Saturated liquid water at 100°C is evaporated to saturated steam at the same temperature in a boiler.

(b) Liquid water at 30°C is continuously stirred by a paddle wheel till the water attains a temperature of 40°C.
(c) An ideal gas is compressed from the initial pressure P_1 to the final pressure P_2 in an insulated cylinder-piston assembly.
(d) An ideal gas is heated at a constant pressure from the initial temperature T_1 to the final temperature T_2.
(e) An ideal gas A at temperature T and pressure P is mixed with another ideal gas B at the same temperature and pressure in an insulated tank.
(f) One kg of ice at 0°C is mixed with 5 kg of liquid water at 30°C in an insulated container.
(g) Melting of ice at 0°C in a container placed in a constant-temperature bath which is maintained at 0°C.

7.3 In a particular process a system receives 50 kJ of energy as heat from a thermal reservoir and performs 10 kJ of work. Is it possible to restore the system to its initial state by following an adiabatic process?

7.4 One kg of air (ideal gas) expands isothermally at 350 K from 10 bar to 1 bar. During the expansion process, energy is transferred to the air from a single thermal reservoir. Calculate the work done by the air and the energy transferred to the air. Does this process violate the laws of thermodynamics?

7.5 Make use of the Kelvin–Planck statement of the second law of thermodynamics to prove that the COP of an irreversible refrigerator is less than that of a reversible refrigerator when both the devices operate between the same set of thermal reservoirs.

7.6 Make use of the second law of thermodynamics to prove that all reversible heat pumps operating between the same set of two thermal reservoirs have the same COP.

7.7 Prove that two reversible adiabatic lines cannot intersect.

7.8 A Carnot engine delivers 10 kW of power while absorbing energy as heat from a source at 1000°C. It rejects energy as heat to a sink at 27°C. Determine the energy absorbed and the energy rejected per second by the engine.

7.9 Determine the temperature in °C at which energy is added to a Carnot heat engine which rejects 100 kJ/s of energy to a sink at 290 K while delivering 50 kW power. Also calculate the rate of energy absorption.

7.10 A Carnot engine uses air (an ideal gas) as the working medium. The volume of the air at the start of the isothermal expansion process is $V_1 = 0.1\ m^3$. The volume of the air at the end of the expansion process is $V_2 = 0.25\ m^3$. The volume of the air at the start of the isothermal compression process is $V_3 = 8\ m^3$. Determine the efficiency of the Carnot engine.

7.11 Two Carnot engines A and B operate in series between a high-temperature reservoir at 1500 K and a low-temperature reservoir at 300 K. Engine A absorbs energy from the high-temperature reservoir and rejects energy as heat to engine B. Engine B rejects energy as heat to the low-temperature reservoir at 300K. Both the engines have the same efficiency. Determine the temperature level at which engine A rejects energy as heat. If engine A absorbs 1000 kJ of energy as heat from the high-temperature reservoir, calculate the work done by engine A and engine B.

7.12 Calculate the coefficient of performance of a Carnot refrigerator operating between −20°C and 30°C.

7.13 If the thermal efficiency of a reversible heat engine operating between two thermal reservoirs is η, determine the COP of a reversible refrigerator, operating between the same reservoirs, in terms of η.

7.14 A Carnot engine with 50 percent thermal efficiency is used to drive a Carnot refrigerator having a coefficient of performance of 5. Determine the energy absorbed by the engine for each kJ of energy absorbed from the cold body by the refrigerator.

7.15 The efficiency of a reversible heat engine which rejects energy at a rate of 500 kW to a reservoir at 20°C is 0.6. Calculate the power output of the engine and the temperature of the source which supplies energy as heat to the engine.

7.16 The efficiency of a Carnot engine can be increased either by decreasing the sink temperature while keeping the source temperature constant, or by increasing the source temperature while keeping the sink temperature constant. Which of the above two possibilities is more effective?

7.17 A refrigerator maintains a freezer compartment at −10°C by absorbing energy at a rate of 5 kJ/s from the freezer. If the ambient temperature is at 37°C, determine the minimum power required to operate the refrigerator.

7.18 A reversible heat engine receives energy Q_1 from a thermal reservoir at T_1 and rejects energy Q_2 to a reservoir at T_2. The work output of the engine is used to drive a reversible heat pump which removes energy Q_L from a reservoir at T_L and rejects energy Q_H to the reservoir at T_2. Determine the value of Q_H/Q_1 in terms of T_1, T_2 and T_L.

7.19 Two reversible engines are working between three thermal reservoirs at temperature T_1, T_2 and T_3, respectively. The first engine absorbs energy as heat from the reservoir at T_1 and rejects energy as heat to the reservoir at T_2. The second engine uses the reservoirs at temperature T_2 and T_3 as source and sink, respectively. The second engine absorbs the same quantity of heat which is rejected by the first engine. Find a relation among T_1, T_2 and T_3 if:

(a) the thermal efficiencies of both the engines are the same, and
(b) the work delivered by both the engines are the same.

7.20 A reversible heat engine operates with three thermal reservoirs and delivers 50 kW power. The engine absorbs 1000 kJ/s from a reservoir at 600 K. The temperatures of the other reservoirs are 400 K and 300 K. Determine the rates of energy exchange with the reservoirs at 400 K and 300 K and the thermal efficiency of the engine.

7.21 A Carnot refrigerator operates between the thermal reservoirs at −25°C and 35°C. If the capacity of the unit is 5 tons of refrigeration, determine:

(a) the power consumption, and
(b) the coefficient of performance of the refrigerator.

A ton of refrigeration is defined as the heat absorption rate of 211 kJ/min.

7.22 A Carnot heat pump operates between the temperature levels −5°C and 25°C. If the power consumption of the heat pump is 5 kW, determine:

(a) the rate at which energy is rejected at 25°C, and
(b) the Coefficient Of Performance.

7.23 Two Carnot heat pumps A and B are arranged in series. The heat pump A receives heat from a low temperature reservoir at −25°C and rejects to an intermediate reservoir at temperature T. The heat pump B receives the heat from the reservoir at T and rejects energy at the rate of 40 kJ/s to reservoir at 127°C. The amounts of energy rejected by the heat pump A and the energy absorbed by heat pump B are the same so that the reservoir at T is unaffected. If the two heat pumps have the same COP, determine:

(a) the temperature T of the intermediate reservoir

(b) the COP of the heat pumps, and
(c) the rate of energy absorption from the reservoir at −25°C.

7.24 A Carnot heat engine operating between the temperature levels 1200 K and 400 K absorbs energy from the source at a rate of 400 kJ/s. Sixty percent of the power delivered by the engine is used to operate a Carnot heat pump which rejects energy to the reservoir at 400 K. If the heat pump absorbs energy at a rate of 600 kJ/s from a low temperature at T, determine:

(a) the temperature T of the reservoir, and
(b) the total rate of energy rejection by both the engine and the heat pump to the reservoir at 400 K.

7.25 An inventor claims to have developed a heat engine which absorbs 500 kJ of energy as heat from a reservoir at 227°C, delivers 350 kJ of work and rejects 150 kJ of energy as heat to reservoir at 27°C. Judge whether the claim is theoretically feasible or not.

7.26 An inventor claims to have developed a heat pump which absorbs 100 kJ of energy as heat from a reservoir at 7°C and rejects 150 kJ of heat to a reservoir at 27°C while it consumes 50 kJ of work. Judge whether the device is feasible or not.

7.27 An inventor claims to have devised an engine which absorbs energy as heat from a high-temperature source at 1200°C and rejects energy as heat to the ambient atmosphere at 27°C. He further claims that the energy conversion efficiency of the engine is 0.95. Would you support his claim?

7.28 A Carnot refrigerator operates between a cold storage space at −20°C and the ambient atmosphere at 30°C. It is required to remove 100 kJ of energy per second from the cold storage. Determine the power required to operate the refrigerator.

7.29 It is required to heat an auditorium in winter using a heat pump. The ambient temperature is 3°C and the auditorium is to be maintained at 20°C. If the energy losses as heat from the auditorium are estimated at 100 kJ/s, determine the minimum power required to operate the heat pump.

7.30 An engine working between a thermal reservoir at 350°C and the ambient atmosphere at 25°C, drives a refrigerator which operates between the ambient atmosphere and a cold space at −20°C. Determine the ratio of the energy absorbed as heat by the engine to the energy absorbed as heat from the cold space by the refrigerator.

7.31 An inventor claims to have developed a refrigerator which maintains the cold space at −20°C when the ambient atmosphere is at 37°C. He further claims that the coefficient of performance of the refrigerator is 6. As Chief Patent Officer, would you issue a patent for the refrigerator? Suppose he changes his claim and quotes the coefficient of performance as 4, would you agree with the revised claim?

7.32 Two Carnot refrigerators A and B operate in series. The refrigerator A absorbs energy at the rate of 1 kJ/s from a body at 300 K and rejects energy as heat to a body at T. The refrigerator B absorbs the same quantity of energy which is rejected by the refrigerator A from the body at T, and rejects energy as heat to a body at 1000 K. If both the refrigerators have the same COP, calculate:

(a) the temperature T of the body
(b) the *COP* of the refrigerators, and
(c) the rate at which energy is rejected as heat to the body at 1000 K.

7.33 A reversible refrigerating unit consumes 3 kW power to absorb 7.5 kJ/s from a cold body. If the device is used as a heat engine between the same temperature levels, how much power does it produce when it absorbs 30 kJ/s from the hot source?

7.34 A heat pump is used in a house for heating in winter and for cooling in summer. It is desired to maintain the house at 25°C throughout the year. The energy losses as heat from the roof and walls is estimated at 50 kW per degree temperature difference between the inside and outside. The outside temperature is 3°C in winter and 37°C in summer. Determine the power required to operate the device in winter and in summer.

7.35 It is desired to use a reversible refrigerator to maintain certain drugs at a low temperature in a cold box. The energy flow as heat through the walls from the surroundings to the cold box is estimated at 140 kJ per hour per degree temperature difference between the surroundings and the cold box. If the power input to the refrigerator is 0.5 kW when the surrounding temperature is 37°C, determine the temperature of the cold box.

7.36 The latent heat of vaporization of helium is 83.3 kJ/kmol at 4.2 K. Suppose helium is available as saturated vapor at a temperature of 4.2 K and it is desired to produce liquid helium at the same temperature. Determine the minimum work required to produce 1 kmol of liquid helium if the ambient temperature is 37°C. Also, estimate the coefficient of performance of a Carnot refrigerator operating between the ambient temperature and the liquid helium temperature.

7.37 The lowest temperature which has been achieved till date is 0.0014 K. Suppose a sample is to be maintained at that temperature. The energy losses as heat from the sample is estimated at 50 J/s and the ambient temperature is 35°C. Suppose a heat engine which uses a source at 400°C and the ambient atmosphere as the sink to drive a refrigerator in order to maintain the sample at the required temperature. Determine the power required to operate the refrigerator, and the ratio of the energy absorbed by the heat engine from the source to the energy absorbed by the refrigerator from the sample.

7.38 A Carnot heat engine absorbs 1000 kJ of energy as heat from a source at 500°C, delivers 500 kJ of work, and rejects the remaining energy as heat to a sink. Determine the sink temperature, the change in the entropy of the source, the change in the entropy of the sink, and the change in the entropy of the universe.

7.39 A heat engine operates between three thermal reservoirs. It absorbs 260 kJ of energy as heat from a reservoir at 52°C and rejects 100 kJ and 88 kJ of energy as heat to reservoirs at 27°C and 2°C, respectively. It delivers 72 kJ of work. Determine whether the heat engine violates the laws of thermodynamics or not.

7.40 A reversible heat engine uses three reservoirs at temperatures 1000 K, 800 K and 600 K. The engine absorbs 2000 kJ/s and 1200 kJ/s as heat from the reservoirs at 1000 K and 800 K, respectively. Calculate the power output of the engine and the rate at which energy is rejected as heat to the third reservoir.

7.41 One kmol of air undergoes a cycle composed of the following three reversible processes: a constant pressure expansion from 0.1 MPa, and 300 K to 400 K; a constant volume cooling to 300 K; an isothermal compression to restore the air to the initial state at 0.1 MPa. Sketch the P–V diagram for the cycle and determine $\oint dQ$, $\oint dW$ and $\oint dS$. Also estimate the change in entropy of the air for each of the processes.

7.42 An experiment has been designed to determine the thermal conductivity of steel. In this, one end of the steel rod is placed in contact with a steam chamber 100°C while the other end is kept in contact with cooling water at 25°C. Energy is transferred as heat by conduction from the steam chamber to the cooling water. If 1000 kJ of energy is transferred per hour as heat, determine the total entropy change of the steam chamber and the cooling water. Suppose a reversible engine is used to transfer energy as heat from the steam chamber to the cooling water how much power would the engine have delivered?

7.43 The latent heat of fusion of ice at 0°C is 334.92 kJ/kg. Determine the change in entropy of one kg ice as it melts to liquid at 0°C, when :

(a) energy is added reversibly as heat to a mixture of ice and water at 0°C; and
(b) the mixture of ice and water is stirred by means of a paddle wheel.

7.44 The latent heat of vaporization of water is 2256.94 kJ/kg, and 1938.53 kJ/kg at 100°C and 200°C, respectively. Determine the entropy of vaporization of water at 100°C and 200°C. Compare the results with the data given in the steam tables.

7.45 Determine the change in specific entropy if air undergoes a process from 1 bar and 300 K to 5 bar and 600 K.

7.46 When air at 4 bar and 500 K undergoes a process where the final temperature is 420 K, the change in specific entropy is −0.096 kJ/kg K. Determine the final pressure of air.

7.47 Derive the following relation for the entropy change of an ideal gas

$$\Delta s = C_P \ln\left(\frac{v_2}{v_1}\right) + C_v \ln\left(\frac{P_2}{P_1}\right)$$

7.48 An insulated cylinder–piston assembly contains one kg of saturated Freon-12 of quality 0.6 at a temperature of 10°C. A paddle wheel is inserted into the cylinder and rotated until the Freon-12 reaches a state of dry saturated vapor at the same pressure. Determine the work done on Freon-12 and the change in the entropy of the Freon-12.

7.49 A rigid and insulated cylinder of volume 1 m^3 contains steam of quality 0.6 at a temperature of 120°C. A paddle wheel is inserted into the cylinder and rotated till the steam reaches a temperature of 200°C. Determine the work done on the steam and the change in the entropy of the steam.

7.50 Given that the entropy of saturated liquid water at 100 kPa is equal to 1.3027 kJ/kg K, and the specific heat at constant pressure for steam is 1.992 kJ/kg K, estimate the entropy of superheated steam at 100 kPa and 200°C and compare the result with the value given in the steam tables.

7.51 A rigid tank of volume 0.1 m^3 containing nitrogen at 7 bar and 127°C is exposed to the surroundings at 27°C. If the nitrogen cools to 37°C, assuming that nitrogen is an ideal gas, calculate:

(a) the entropy change of nitrogen
(b) the entropy change of the surroundings, and
(c) the entropy change of the universe.

7.52 One mole of oxygen at 1 bar and 47°C is compressed in a piston–cylinder assembly to 4 bar and 127°C by doing 5 kJ of work when the surrounding temperature is 27°C. Determine:

(a) the entropy change of oxygen
(b) the entropy change of the surroundings, and
(c) the entropy change of the universe.

7.53 Nitrogen is throttled from 3 bar and 600 K to 1.5 bar pressure. Determine the entropy change of the gas. Is the process reversible?

7.54 An insulated tank initially contains 5 kg liquid water at 35°C. If 1 kg ice at 0°C is dropped into the tank while the tank is held at 1 bar pressure, determine the final equilibrium temperature and the entropy change associated with the process. The latent heat of fusion of water at 0°C is 334.92 kJ/kg. Assume that C_P of liquid water is 4.188 kJ/kg K.

7.55 A 20 kg iron casting ($C_P = 0.452$ kJ/kg K) at 727°C is dropped in a tank containing 10 kg water at 30°C. The tank is insulated and open to the atmosphere. Determine:

(a) the entropy change of the iron casting, and

(b) the entropy change of water.

7.56 A lead casting of mass 20 kg at 200°C is quenched in 30 kg liquid water at 30°C. The specific heats of lead and water are 0.13 kJ/kg K and 4.2 kJ/kg K, respectively. Determine the entropy change of lead casting and water. Assume that there are no heat losses during quenching.

7.57 An aluminium block of 10 kg at 400°C is dropped in a vessel containing 10 kg liquid water at 30°C. The vessel is open to the ambient atmosphere at 1 bar. If the specific heats of aluminium and water are 0.896 kJ/kg K and 4.2 kJ/kg K, respectively, calculate the entropy change of aluminium block and the universe.

7.58 If a 10 kg block of ice at −5°C is dumped in an open and insulated tank containing 50 kg water at 30°C, determine the final equilibrium temperature and the total entropy change. The specific heat of ice is 0.5 kJ/kg K.

7.59 Air enters a reversible, adiabatic nozzle at 2.5 bar and 127°C with a velocity of 5 m/s and exits at 1 bar. Assuming air to behave as an ideal gas, determine the exit velocity of air.

7.60 When air is throttled, its entropy increases by 0.87 kJ/kmol K. Determine the ratio of the initial pressure to the final pressure of the air.

7.61 A mass m of liquid at temperature T_1 is mixed with an equal amount of the same liquid at temperature T_2. If the specific heat of the liquid is C, show that the total entropy change because of the mixing process is given by:

$$\Delta S = 2mC \ln \left(\frac{(T_1 + T_2)/2}{\sqrt{T_1 T_2}} \right)$$

7.62 One kg of saturated liquid water at 100 kPa is mixed with 5 kg of superheated steam at 100 kPa and 300°C in an insulated container. If the pressure remains constant during mixing, calculate the entropy generation due to the mixing process.

7.63 In a counter-current flow heat exchanger the hot fluid and the cold fluid enter the heat exchanger at opposite ends. It is desired to heat 10,000 kg/h of water from 25°C to 75°C by the following methods. Determine the change in the entropy of the hot fluid, the cold fluid and the universe in each case.

(a) By condensing saturated steam at 100 kPa

(b) By condensing saturated steam at 3 MPa

(c) By using superheated steam at 3 MPa and 300°C which leaves the heat exchanger as saturated steam at 3 MPa.

7.64 It is desired to design a heat exchanger to cool 1 kmol of air from 240°C to 140°C by another stream of cold air which is originally at 30°C. Equal amounts of cold air and hot air will be used. Consider the heat exchanger to be adiabatic.

(a) Calculate the entropy change of the cold air, hot air and ΔS(total) for a co-current heat exchanger.

(b) Calculate the entropy change of the cold air, hot air and ΔS(total) for a counter-current flow heat exchanger.

(c) Calculate the entropy change of cold air, hot air and ΔS(total) for a counter-current flow heat exchanger if the cold air is originally at 140°C.

7.65 Two kmol of oxygen at 500 kPa and 300 K is in an insulated container which is connected through a valve to a second container (also insulated) filled with 3 kmol of nitrogen at 250 kPa and 400 K. The valve is opened, and the gases are allowed to mix. Calculate the entropy change for the mixing process. Consider both nitrogen and oxygen as ideal gases.

7.66 Two identical bodies of mass m are initially at temperature T_1 and T_2. They are used as source and sink, respectively, for a heat engine. If the specific heat of the bodies is C, determine the maximum work that can be obtained.

7.67 Steam at 40 bar and 500°C enters an adiabatic turbine at the rate of 5 kg/s and exits at 1 bar. Determine the maximum power that can be developed by the turbine.

7.68 It is required to compress 3 kg/s of air from 1 bar and 27°C to 5 bar in a compressor. Assuming that the compressor is well insulated, determine the minimum power required to operate the compressor and the temperature of the exit air.

7.69 A rigid and insulated vessel of 1 m^3 contains air (ideal gas) at 200 kPa and 30°C. A valve on the vessel is opened allowing the air pressure inside the vessel to fall quickly to 100 kPa. Assuming that the air left inside the vessel has undergone a reversible and adiabatic expansion, determine the mass of air that has escaped from the vessel. Compare the result with the value obtained by applying the first law of thermodynamics to the transient flow process.

7.70 The following experiment suggested by Desormes and Clement is used to measure the specific heat ratio of a gas. Initially the gas is contained in a rigid vessel at a pressure P_1, which is slightly higher than the atmospheric pressure P_0, and at the ambient temperature. A valve on the vessel is then opened for a short time allowing the gas pressure to fall to the atmospheric pressure and then closed. After allowing some time for the gas remaining inside the vessel to come to thermal equilibrium with the atmosphere, the pressure inside the vessel is measured as P_2. Assuming that the gas inside the vessel is ideal, develop an expression to determine γ in terms of P_0, P_1 and P_2.

7.71 Air flows through a nozzle at the rate of 1 kg/s. The inlet conditions are 3 bar and 250°C. The velocity of the gas at the inlet is 10 m/s. The gas leaves the nozzle at 1.5 bar. Assuming the process to be reversible and adiabatic, find the exit velocity and the exit area of the nozzle. Air may be treated as an ideal gas.

7.72 A steam turbine receives steam at 3 MPa and 300°C. The steam expands in a reversible and adiabatic process and leaves the turbine at 10 kPa. The power output of the turbine is 10 MW. Determine the flow rate of steam through the turbine.

7.73 Steam enters an insulated nozzle at 3 bar and 200°C with negligible velocity and leaves at 1 bar. Determine the exit velocity and the state of the steam assuming that the steam undergoes reversible adiabatic expansion.

7.74 Feed water heaters are used in thermal power plants operating on regenerative Rankine cycle. In this the water is preheated before pumping it into the boiler for steam generation. In a particular feed water heater, saturated water vapor at 500 kPa is mixed with liquid water at 500 kPa and 30°C to obtain saturated liquid at the same pressure. Determine the flow rates of the entering liquid water and saturated steam at 500 kPa and the entropy generation. Assume the process to be steady-flow and adiabatic.

7.75 A capillary tube is used in place of a throttle valve in a domestic refrigerator. Saturated liquid Freon-12 at 40°C enters the capillary and leaves at a pressure of 180 kPa. Determine the entropy change per kg of Freon-12 associated with the flow through the capillary tube.

7.76 One kg ice block at $-10°C$ is brought into contact with 5 kg copper block at 30°C in an insulated container. Determine the change in entropy of: (a) ice block (b) copper block and (c) the universe. Specific heat of ice and copper are 2.057 kJ/kg K and 0.385 kJ/kg K, respectively. The latent heat of fusion of water at 0°C is 334.92 kJ/kg.

7.77 One kg of supercooled water at $-10°C$ freezes into ice at $-10°C$ and 100 kPa pressure. The specific heats of ice and water are 2.057 kJ/kg K and 4.2 kJ/kg K, respectively. The latent heat of fusion of water at 0°C is 334.92 kJ/kg. Determine the change in the entropy of water.

7.78 While tabulating the thermodynamic properties of water (steam tables) the triple point of water is taken as a reference state where the entropy of liquid water is assigned a value of zero. As an approximation, the specific heats of steam and liquid water may be taken as 1.8646 kJ/kg K and 4.2 kJ/kg K, respectively. The latent heat of vaporization of water at 100°C is 2256.94 kJ/kg. Estimate the entropy of superheated steam at 3.0 MPa and 300°C and compare the result with the value given in the steam tables.

7.79 Air is flowing steadily through a thermally insulated duct. The pressure and temperature of the air measured at two different stations are as follows:

	Station A	Station B
Pressure	130 kPa	100 kPa
Temperature	50°C	13°C

Considering air as an ideal gas, determine the direction of flow of the air in the duct.

7.80 In a Hilsch Vortex tube, high pressure air at 600 kPa and 294 K is fed tangentially at the centre of a round tube. The tube does not exchange energy either as heat or work with the ambient atmosphere. The air splits into two streams. The hot air stream leaves the tube at 100 kPa and 355 K while the cold air stream emerges at 100 kPa and 233 K. Determine the flow rates of the hot and the cold air streams if the flow rate of the entering air is one kmol/s. Also, determine the total entropy change of air.

7.81 An inventor claims to have designed a flow device to increase the superheat of steam. He does not provide the details but claims that he could feed steam at 200 kPa and 200°C and obtain steam at 100 kPa and 300°C. He also claims that the device yields saturated steam at 200 kPa and 44.6 per cent of the feed steam will be converted into superheated steam. It receives no energy either as work or heat from the surroundings. Would you agree with the claim? Justify your answer based on thermodynamic grounds.

7.82 Air (ideal gas) at 500 kPa and 27°C enters a vortex tube shown in Fig. 7.34 at the centre. Stream S_1 leaves the right end of the tube at 100 kPa and 350°C and stream S_2 leaves the left end of the tube at 100 kPa and $-150°C$. These two streams are used as source and sink, respectively to run a reversible engine till their temperature is reduced to T_f. Determine:

(a) The flow rates of streams S_1 and S_2 if the flow rate of the entering air is 1 kmol/s.
(b) The change in the entropy of air associated with the flow through the vortex tube.
(c) The final temperature T_f of the air leaving the engine.
(d) The work done (in one second) by the engine.

7.83 One kmol/s of air (ideal gas) at 500 kPa and 27°C enters a reversible and adiabatic turbine and leaves at 100 kPa. One kmol/s of air at 500 kPa and 27°C enters a vortex tube and stream S_1 leaves the tube at 100 kPa and 177°C while stream S_2 leaves the tube at 100 kPa and $-123°C$. It is desired to cool stream S_2, while rejecting energy as heat to stream S_1 using a reversible heat

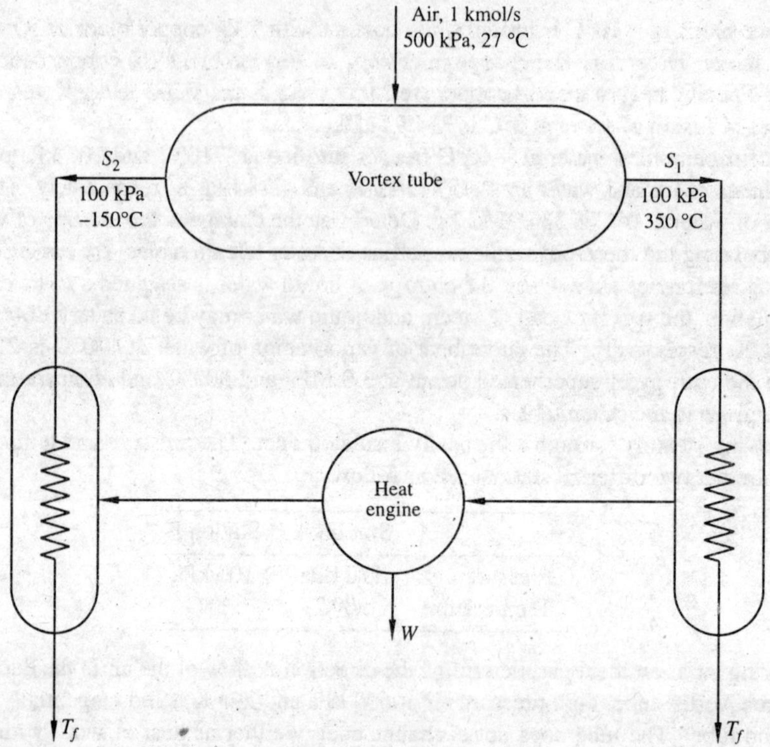

Fig. 7.34 *Sketch for Problem 7.82*

pump as shown in Fig. 7.35. The work delivered by the turbine is used to operate the heat pump. Determine:

(a) The work done (in one second) by the turbine
(b) The flow rates of streams S_1 and S_2.
(c) The change in the entropy of air associated with the flow through the vortex tube, and
(d) The final temperatures of the cold stream (T_1) and the hot stream (T_H).

7.84 Air at 5 bar and 27°C enters a reversible and adiabatic turbine shown in Fig. 7.36 at the rate of one kmol/s and emerges at 1 bar. The air leaving the turbine is split into two equal streams. These two streams then act as source and sink for a Carnot heat pump driven by the turbine, and leaves at temperature T_1 and T_2. Determine:

(a) The temperature of the air leaving the turbine,
(b) The work done by the turbine, and
(c) The temperature T_1 and T_2.

7.85 At present steam is condensed in a heat exchanger to produce pure water. The inlet and outlet conditions of the steam and cooling water are shown in Fig. 7.37. It is proposed to replace the existing heat exchanger by a reversible device which not only does the same job requiring less quantity of cooling water but also delivers some work. Determine: (a) The saving in cooling water requirement by the proposed device, and (b) The power output of the reversible device.

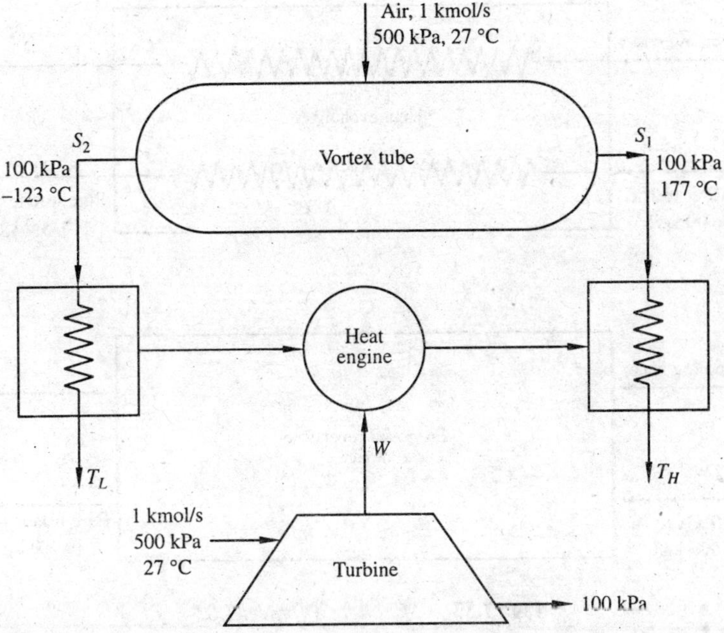

Fig. 7.35 *Sketch for Problem 7.83*

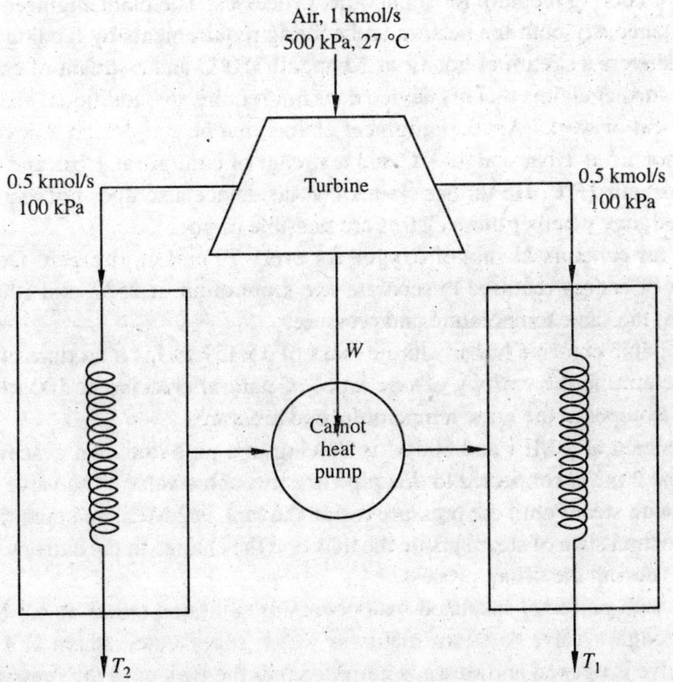

Fig. 7.36 *Sketch for Problem 7.84*

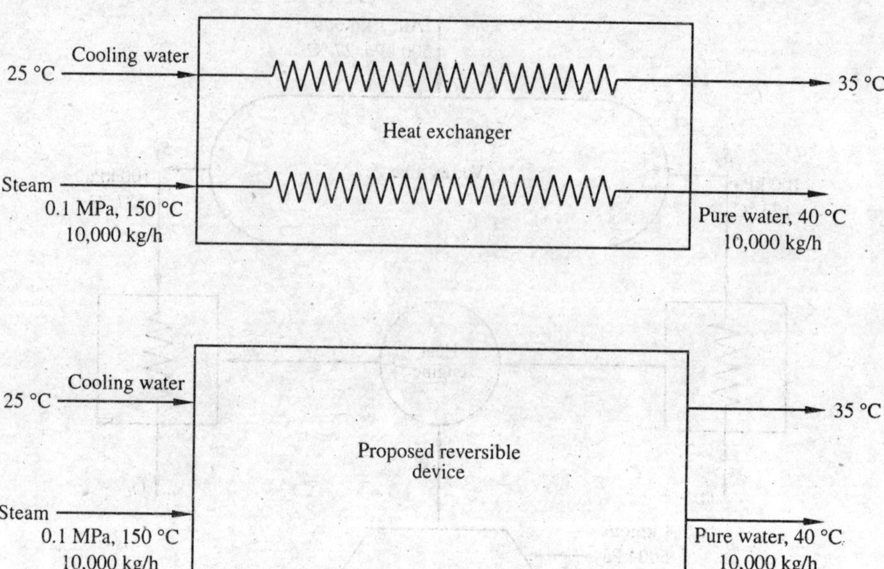

Fig. 7.37 *Sketch for Problem 7.85*

7.86 In a particular chemical process industry air at 5 bar and 27°C is available. The plant uses air as a heating or cooling medium for some other processes. The plant engineer claims that he can satisfy simultaneously both the heating and cooling requirements by feeding the air to a vortex tube which delivers a stream of hot air at 1 bar and 350°C and a stream of cold air at 1 bar and −150°C. He further claims that his device does not require any additional energy input either in the form of heat or work. Another engineer claims that he can design a device which delivers a stream of hot air at 1 bar and 400°C and a stream of cold air at 1 bar and −200°C by using the air at 5 bar and 27°C. He further claims that his device also does not require any additional energy input. Judge whether these claims are possible or not.

7.87 Atmospheric air contains 21 mol of oxygen for every 79 mol of nitrogen. Determine the minimum amount of energy required to separate one kmol of air at 25°C and 100 kPa into its pure components at the same temperature and pressure.

7.88 Naturally available chlorine has an atomic mass of 35.457 and is a mixture of the isotopes Cl_2^{35} and Cl_2^{37}. Determine the entropy of one kmol of natural chlorine at 100 kPa and 25°C with respect to its isotopes at the same temperature and pressure.

7.89 Superheated steam at 2 MPa and 500°C is flowing in a pipe line. An evacuated and insulated tank of volume 2 m^3 is connected to this pipe line through a valve. The valve is opened and the tank is filled with steam until the pressure inside the tank is 2 MPa, and then the valve is closed. Determine the final state of steam inside the tank and the change in the entropy within the control volume (tank) during the filling process.

7.90 A 2 m^3 tank with perfectly insulated walls contains saturated steam at 1.0 MPa. The tank is connected through a valve to steam mains in which superheated steam at 4 MPa and 400°C flows. The valve is opened and steam is admitted into the tank until the pressure inside the tank is 4.0 MPa. Determine the mass of steam that enters the tank and the change in the entropy of the tank during the process.

7.91 In a certain Carnot cycle, energy as heat is supplied at 250°C and rejected at 30°C. The working fluid is water which, while receiving energy as heat, evaporates from liquid at 250°C to vapor at the same temperature. The entropy change for this process is 3.2773 kJ/kg K. Assuming the cycle to operate on a unit mass of water, determine:

(a) the energy supplied as heat
(b) the available energy
(c) the unavailable energy, and
(d) the change in available energy when the sink temperature is lowered to 20°C.

7.92 Ten kg of liquid water in a saturated state at a pressure of 0.4 MPa is converted into steam at the same pressure in a heat-exchanger. For this purpose air at a pressure of 200 kPa and at a temperature of 500°C enters the heat-exchanger and leaves at 300°C. Determine the change in the available energy due to the heat transfer process. The ambient temperature is 25°C. C_P of air is 1.0047 kJ/kg K.

7.93 Consider two identical blocks of copper, each having a mass of 100 kg and a specific heat of 0.3831 kJ/kg K. One block is available at 1200 K and the second block is available at 600 K. The two blocks are brought into thermal contact with each other and allowed to reach a state of thermal equilibrium. Determine the loss in available energy if the ambient temperature is 300 K.

7.94 Determine the maximum amount of useful work that can be obtained from one kmol of hot gases at 700°C and 0.1 MPa if the ambient temperature is 27°C and the ambient pressure is 100 kPa. Assume that the molar heat capacity of the gases at constant pressure is 29.1783 kJ/kmol K.

7.95 A heat-exchanger is designed to cool 100 kmol/h of CO_2 from 400°C to 240°C by another stream of CO_2 which is initially at 30°C. Equal amounts of cold CO_2 and hot CO_2 enter at the same end of the heat-exchanger (co-current flow). Assume that the heat-exchanger is adiabatic and the ambient temperature is 25°C. Determine:

(a) The entropy change of the hot stream and cold stream.
(b) The change in availability of the hot stream.
(c) The change in the availability of the cold stream, and
(d) The loss in the available energy because of the energy transfer from the hot stream to the cold stream.

7.96 A heat-exchanger is designed to cool CO_2 from 400°C to 50°C by another stream of CO_2 which is initially at 50°C. The two streams enter the heat-exchanger at opposite ends (counter-current flow). Equal amounts of the hot stream and the cold stream flow through the heat exchanger. If the flow rates of the streams are 10 kmol/m and the ambient temperature is 300 K, determine:

(a) the entropy change of the hot stream and the cold stream,
(b) the change in the availability of the hot stream,
(c) the change in the availability of the cold stream, and
(d) the loss in the available energy due to the energy exchange between the two streams.

7.97 A rigid tank of volume 1 m^3 contains saturated steam at 1 bar. Energy is added as heat to the tank until the pressure rises to 2 bar while the surroundings are at 1 bar and 25°C. Calculate (a) the quantity of energy added, and (b) the availability in the final state.

7.98 A rigid tank of volume 5 m^3 contains air at 1 bar and 300 K. Energy is transferred as heat to the air from a reservoir at 500 K. The surroundings are at 1 bar and 300 K. Determine (a) the quantity of energy that can be transferred to the air, and (b) the fraction of energy transferred is available energy.

7.99 Steam at 5 bar and 500°C contained in a piston-cylinder assembly is exposed to the ambient atmosphere and cooled at constant pressure until the temperature reaches 200°C while the ambient atmosphere is at 27°C and 1 bar. Calculate the loss in available energy associated with the process.

8 Thermodynamic Relations

Learning Objectives

This chapter will enable the readers to:

- Understand the importance of Maxwell relations.
- Obtain Maxwell relations from the thermodynamic properties u, h, a and g.
- Know the thermodynamic mnemonic diagram.
- Know the thermodynamic potentials.
- Use the Jacobian method to derive thermodynamic relations.
- Understand the Clapeyron equation and Kirchoff equation.
- Estimate the latent heat of phase transition at any temperature.
- Derive the Gibbs phase rule.

In the earlier chapters we discussed the first and the second laws of thermodynamics in detail and solved some problems of technical importance by applying these laws. In the thermodynamic analysis of the problems, the properties of the substances which are available in the form of tables, charts, thermodynamic diagrams, or equations of state, were used. In Chapter 3 the thermodynamic properties were classified into two groups, namely, (i) measurable properties such as pressure, volume and temperature (P, V, T), and (ii) properties which cannot be measured directly such as internal energy, entropy, enthalpy, etc. The relationships among the measurable properties were discussed in Chapter 3. It was also stated that an important task in thermodynamics is to find the relationships which express the non-measurable properties in terms of the measurable properties. In this chapter we shall study these relationships.

8.1 Maxwell's Relations

The calculation of entropy change of a system was extensively discussed in Chapter 7. It was shown that the entropy change of an ideal gas can be estimated by either of the following equations.

$$\Delta s = C_P \ln\left(\frac{T_2}{T_1}\right) - R\ln\left(\frac{P_2}{P_1}\right)$$

$$\Delta s = C_v \ln\left(\frac{T_2}{T_1}\right) + R\ln\left(\frac{v_2}{v_1}\right)$$

By devising a different reversible path, connecting the final state with the initial state, the entropy change for an ideal gas can be estimated by the following equation also.

$$\Delta s = C_P \ln\left(\frac{v_2}{v_1}\right) + C_v \ln\left(\frac{P_2}{P_1}\right) \tag{8.1}$$

When an ideal gas undergoes a change of state from P_1, v_1, T_1 to P_2, v_2, T_2, the entropy change of the gas can be estimated by any of the above three equations. Depending on the independent variables, we choose the appropriate equation for estimating Δs. However, these relations can be used only for an ideal gas and not for any other substance.

The above equations express the entropy in terms of two independent variables which can be easily measured. Since entropy plays an important role in thermodynamic analysis of processes, a knowledge of the estimation of entropy change of a substance is essential. A differential change in the entropy associated with the differential change in the independent variables can be expressed as:

$$s = s(T, P)$$

$$ds = \left(\frac{\partial s}{\partial T}\right)_P dT + \left(\frac{\partial s}{\partial P}\right)_T dP \tag{8.2}$$

$$s = s(T, v)$$

$$ds = \left(\frac{\partial s}{\partial T}\right)_v dT + \left(\frac{\partial s}{\partial v}\right)_T dv \tag{8.3}$$

$$s = s(P, v)$$

$$ds = \left(\frac{\partial s}{\partial P}\right)_v dP + \left(\frac{\partial s}{\partial v}\right)_P dv \tag{8.4}$$

These relations can be integrated to obtain Δs. To integrate these relations, a knowledge of the six partial derivatives of entropy, namely, $(\partial s/\partial T)_P$; $(\partial s/\partial T)_v$; $(\partial s/\partial P)_T$; $(\partial s/\partial P)_v$; $(\partial s/\partial v)_T$ and $(\partial s/\partial v)_P$, is essential. Of these, the partial derivatives of entropy with respect to temperature are related to the specific heats C_P and C_v by the equations:

$$C_P = T\left(\frac{\partial s}{\partial T}\right)_P \tag{8.5}$$

$$C_v = T\left(\frac{\partial s}{\partial T}\right)_v \tag{8.6}$$

The remaining four partial derivatives of entropy are expressed in terms of the measurable quantities by equations known as the Maxwell relations and are discussed below.

Combining the first and second laws of thermodynamics, we get

$$du = Tds - Pdv$$

Since u is a property, its differential must be exact. Applying the criterion for exactness, we get

$$\left(\frac{\partial T}{\partial v}\right)_s = -\left(\frac{\partial P}{\partial s}\right)_v \tag{8.7}$$

The term $(\partial T/\partial v)_s$ represents the slope of an isentropic curve on a T–v diagram and this slope can be easily measured or estimated. This is related to the partial derivative of entropy with respect to pressure at constant volume. Therefore, Eq. (8.7) is a Maxwell relation.

Since $h = u + Pv$, or $H = U + PV$, we get

$$dh = du + Pdv + vdP \quad \text{or} \quad dH = dU + Pdv + VdP \tag{8.8}$$

Substituting Eq. (7.104) in Eq. (8.8), we get

$$dh = Tds + vdP \quad \text{or} \quad dH = TdS + VdP \tag{8.9}$$

We shall define two other thermodynamic properties of matter. The Helmholtz function a or A is defined as

$$a = u - Ts \quad \text{or} \quad A = U - TS \tag{8.10}$$

Then

$$dA = dU - TdS - SdT \quad \text{or} \quad da = du - Tds - sdT \tag{8.11}$$

Substituting Eq. (7.104) in Eq. (8.11), we get

$$dA = -PdV - SdT \quad \text{or} \quad da = -Pdv - sdT \tag{8.12}$$

The Gibbs free energy G or g is defined as

$$G = H - TS \quad \text{or} \quad g = h - Ts \tag{8.13}$$

Then

$$dG = dH - TdS - SdT \quad \text{or} \quad dg = dh - Tds - sdT \tag{8.14}$$

Substituting for dH from Eq. (8.9) in Eq. (8.14), we get

$$dG = -SdT + VdP \quad \text{or} \quad dg = -sdT + vdP \tag{8.15}$$

Note that the independent variables of U, H, A and G include P, v, T and s only. The equations (8.9), (8.12) and (8.15) are sometimes referred to as the *Gibbsian Equations*. Applying the criterion for exactness to Eqs. (8.9), (8.12) and (8.15), we get

$$\left(\frac{\partial T}{\partial P}\right)_s = \left(\frac{\partial v}{\partial s}\right)_P \tag{8.16}$$

$$\left(\frac{\partial P}{\partial T}\right)_v = \left(\frac{\partial s}{\partial v}\right)_T \tag{8.17}$$

$$\left(\frac{\partial s}{\partial P}\right)_T = -\left(\frac{\partial v}{\partial T}\right)_P \tag{8.18}$$

The Eqs. (8.7) and (8.16)–(8.18) are the well known Maxwell relations.

8.2 Mnemonic Diagram

The differential expressions for the thermodynamic properties U, H, A and G and the above Maxwell relations can be remembered conveniently in terms of a thermodynamic mnemonic diagram. The diagram consists of a square with two diagonal arrows pointing upwards. The sides of the square are labeled with the thermodynamic properties A, G, H and U. Starting from the top the properties are placed in alphabetical order clockwise as shown in Fig. 8.1. The independent variables of the thermodynamic properties are placed in the corners such that each property is flanked by its own independent variables.

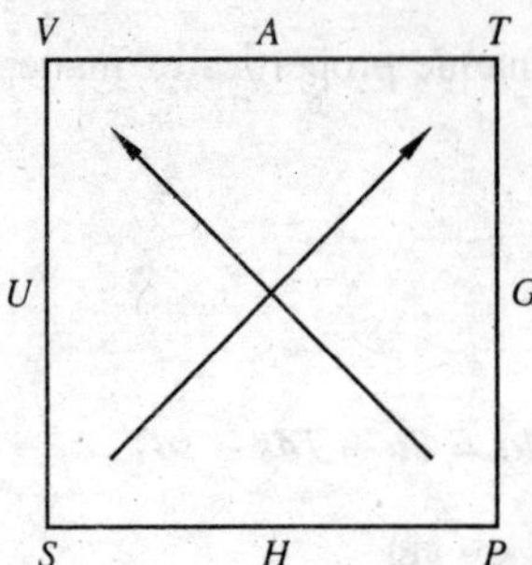

Fig. 8.1 *A thermodynamic mnemonic diagram*

The differential expression for the property can be expressed in terms of the differentials of its independent variables, where the coefficients associated with the independent variables are indicated by the diagonal arrow. The sign associated with the coefficient is positive if the arrow is pointing away from the independent variable, whereas the sign associated with the coefficient is negative if the arrow is pointing towards the independent variable. Thus, we can write as

$$du = (\text{sign})(\text{Coefft})ds + (\text{sign})(\text{Coefft})dv$$

$$du = (\text{sign})Tds + (\text{sign})Pdv$$

$$du = +Tds - Pdv \quad \text{or} \quad du = Tds - Pdv \tag{8.19}$$

Similarly, for the property g we can write:

$$dg = (\text{sign})(\text{Coefft})dT + (\text{sign})(\text{Coefft})dP$$

$$dg = (\text{sign})sdT + (\text{sign})vdP$$

$$dg = -sdT + vdP \tag{8.20}$$

While writing the Maxwell relations we need to concentrate on the direction of the arrows and the independent variables only. By observing Fig. 8.1, we can write the first Maxwell relation as

$$\left(\frac{\partial v}{\partial s}\right)_P = \left(\frac{\partial T}{\partial P}\right)_s \tag{8.21}$$

Since both the arrows are pointing in the same direction (upwards), there is no need to change the sign. However, if one arrow is pointing upwards and the other arrow is pointing downwards, the equation should carry a negative sign. To write the second Maxwell relation, rotate the diagram clockwise by 90° as shown in Fig. 8.2.

$$\left(\frac{\partial s}{\partial P}\right)_T = -\left(\frac{\partial v}{\partial T}\right)_P \tag{8.22}$$

Similarly, the other two Maxwell relations also can be written.

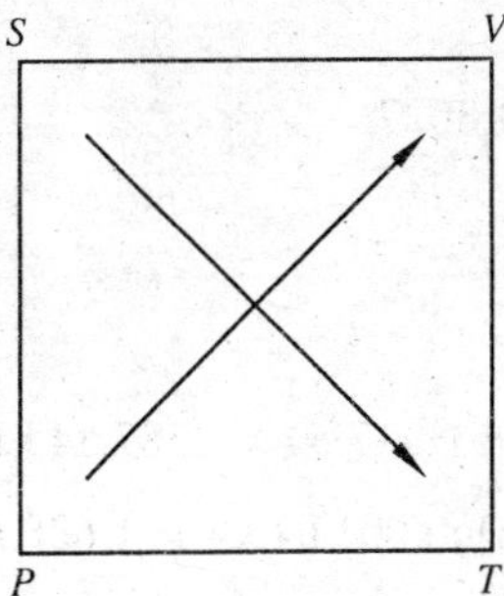

Fig. 8.2 *Rotated thermodynamic mnemonic diagram*

8.3 Thermodynamic Potentials

8.3.1 Internal energy

Suppose a system undergoes a change of state during a given adiabatic process. Application of the first law of thermodynamics to the system gives

$$dU = -dW \tag{8.23}$$

Therefore, the work done by a system during an adiabatic process is equal to the decrease in the internal energy of the system. Since the decrease in internal energy represents the potential to do work by the system under adiabatic conditions, it is called a thermodynamic potential. Now if the work is done on the system during an adiabatic process, it is exactly equal to the increase in the internal energy of the system.

***Example* 8.1** One kg of superheated steam at 10 bar and 300°C contained in a cylinder piston assembly is allowed to undergo an adiabatic expansion to a pressure of 1 bar. Calculate the work that can be obtained during the expansion process.

Solution The work that can be obtained from a system, if it undergoes an adiabatic process, is equal to the decrease in the thermodynamic potential u.

$$W = -\Delta u$$

Initial values of u and s (from the steam tables) are:

$$u_1 = 3052.1 \times 10^3 - 10 \times 10^5 \times 0.2580 = 2794.1 \text{ kJ/kg}$$

$$s_1 = 7.1251 \text{ kJ/kg K}$$

For a reversible and adiabatic expansion, $s_2 = s_1$ or,

$$s_2 = s_1 = X_2 s_g + (1 - X_2) s_f$$

$$7.1251 = 7.3598 X_2 + (1 - X_2) 1.3027$$

or,

$$X_2 = 0.961$$

$$h_2 = 2675.4 \times 0.961 + (1 - 0.961) 417.54 = 2587.34 \text{ kJ/kg}$$

$$v_2 = 1.694 \times 0.961 + (1 - 0.961) 0.001\,043\,4 = 1.628 \text{ m}^3/\text{kg}$$

$$u_2 = 2587.34 \times 10^3 - 1 \times 10^5 \times 1.628 = 2424.54 \text{ kJ/kg}$$

$$u_1 - u_2 = 2794.1 - 2424.54 = 369.56 \text{ kJ/kg}$$

or

$$W = 369.56 \text{ kJ}$$

8.3.2 Helmholtz potential

Now let us consider a system which interacts with the surroundings. The work done by the system during a given process is given by:

$$dW = -dU + dQ \tag{8.24}$$

Let the surroundings be maintained at a constant temperature T_0. Let dS and dS_0 represent the changes in the entropy of the system and the surroundings during the interaction, respectively. From the second law of thermodynamics

$$dS + dS_0 \geq 0$$

The change in the entropy of the surroundings is given by

$$dS_0 = -\frac{dQ}{T_0}$$

Therefore,

$$dS - \frac{dQ}{T_0} \geq 0$$

$$dQ \leq T_0 dS \tag{8.25}$$

Then the work done by the system is given by

$$dW = -dU + dQ \leq T_0 dS - dU$$

or,

$$W \leq (U_1 - U_2) - T_0(S_1 - S_2) \tag{8.26}$$

If the initial and final temperatures of the system are equal to the surrounding temperature ($T_1 = T_2 = T_0$), then

$$W \leq (U_1 - U_2) - (T_1 S_1 - T_2 S_2)$$

$$W \leq (U_1 - T_1 S_1) - (U_2 - T_2 S_2)$$

or,

$$W \leq A_1 - A_2 \tag{8.27}$$

Therefore, *the maximum work that can be obtained from a system during a given process, in which the initial and final temperatures of the system are equal to the surrounding temperature, is equal to the decrease in the Helmholtz function*. The maximum work is delivered only when the process is reversible. The work delivered is less than the decrease in the Helmholtz function, if there is any irreversibility present in the process. Since the decrease in the Helmholtz function represents the potential to do work by the system, it is also a thermodynamic potential.

Now if work is performed on the system, the minimum work required during a reversible process is equal to the increase in the Helmholtz function. On the other hand, if the process is irreversible, the work required is greater than the increase in the Helmholtz function. Quite often the Helmholtz function is called Helmholtz free energy or Helmholtz potential.

***Example* 8.2** It is desired to compress one kmol of an ideal gas from an initial state of 100 kPa and 300 K to a final state of 1 MPa and 300 K in a piston–cylinder assembly. Determine the minimum amount of work required for the compression process if the surroundings are at 300 K.

Solution The initial and final states of the system are at 300 K which is also the temperature of the surroundings. Therefore, the minimum work required is equal to the increase in the Helmholtz potential or Helmholtz function.

$$W_{\text{min}} = \Delta a = a_2 - a_1 = (u_2 - T_0 s_2) - (u_1 - T_0 s_1)$$

For an ideal gas $u = u(T)$ only. Since $T_1 = T_2$, we get $u_1 = u_2$.

Therefore,

$$W_{\text{min}} = -T_0(s_2 - s_1)$$

$$s_2 - s_1 = C_P \ln\left(\frac{T_2}{T_1}\right) - R \ln\left(\frac{P_2}{P_1}\right) = -8.314 \ln\left(\frac{1 \times 10^6}{100 \times 10^3}\right) = -19.1437 \text{ kJ/kmol K}$$

$$W_{min} = -T_0\Delta s = 300 \times 19.1437 = 5743.11 \text{ kJ/kmol K}$$

We have seen that the thermodynamic potentials U and A represent the work that can be obtained from a system under specified conditions. Some of the other important thermodynamic potentials are discussed below.

8.3.3 Enthalpy

Recall that while discussing the first law of thermodynamics, enthalpy was defined as $H = U + PV$. It was also stated that enthalpy is a form of energy, since both U and PV represent energy. To see how enthalpy is related to the work obtained, let us consider the flow of a material through a control volume. The first law of thermodynamics for a control volume is given by

$$\dot{m}_e\left(h_e + \frac{V_e^2}{2} + gZ_e\right) - \dot{m}_i\left(h_i + \frac{V_i^2}{2} + gZ_i\right) = \dot{Q} - \dot{W}_s - \frac{dE}{dt}$$

Consider an adiabatic, steady state, steady flow process and let us ignore the changes in the kinetic energy and the potential energy terms. Under these conditions the first law of thermodynamics reduces to

$$h_e - h_i = -\frac{\dot{W}_s}{\dot{m}}$$

or,

$$\dot{W}_s = -\dot{m}(h_e - h_i) = -(\dot{H}_e - \dot{H}_i) = -\Delta\dot{H} \tag{8.28}$$

Therefore, the rate of work that can be obtained from an adiabatic flow process is exactly equal to the rate of decrease in the enthalpy of the flowing fluid.

***Example* 8.3** Calculate the power output from an adiabatic and reversible steam turbine which is fed with superheated steam at 8 bar and 250°C at the rate of 1 kg/s. The steam leaves the turbine at a pressure of 1 bar.

Solution Since the steam flows continuously through the turbine (control volume) under adiabatic conditions, the rate of work that can be obtained is equal to the decrease in the enthalpy of the steam.

$$\dot{W} = -\Delta\dot{H}$$

$$h_i = 2947.95 \text{ kJ/kg}; \quad \text{and} \quad s_i = 7.0248 \text{ kJ/kg K}$$

For a reversible and adiabatic process: $s_e = s_i$

Hence

$$s_e = 7.0248 = 7.3598X_e + (1 - X_e)1.3027$$

or,

$$X_e = 0.945$$

$$h_e = 2675.4 \times 0.945 + (1 - 0.945)417.54 = 2551.22 \text{ kJ/kg}$$

Then,

$$\dot{W} = \dot{H}_i - \dot{H}_e = 2947.95 - 2551.22 = 396.73 \text{ kJ/s}$$

Therefore, the power output = 396.73 kW.

8.3.4 Gibbs free energy

To examine how the decrease in the Gibbs free energy is related to the work obtained under the specified conditions, consider a steady state, steady flow of a material through a control volume. Let us ignore the changes in the potential energy and kinetic energy terms. Under these conditions, the first law of thermodynamics for a control volume reduces to

$$\dot{m}(h_e - h_i) = \dot{H}_e - \dot{H}_i = \dot{Q} - \dot{W}_s$$

or,

$$\dot{W}_s = \dot{Q} - (\dot{H}_e - \dot{H}_i)$$

If the control volume receives energy in the form of heat from the surroundings at temperature T_0, the change in the entropy of the surroundings is given by

$$\Delta\dot{S}_0 = -\frac{\dot{Q}}{T_0}$$

The second law of thermodynamics gives

$$\Delta\dot{S} + \Delta\dot{S}_0 \geq 0$$

$$\Delta\dot{S} - \frac{\dot{Q}}{T_0} \geq 0$$

or,

$$\dot{Q} \leq T_0 \Delta\dot{S}$$

Therefore,

$$\dot{W}_s \leq T_0(\dot{S}_e - \dot{S}_i) - (\dot{H}_e - \dot{H}_i) \tag{8.29}$$

If the material flows through the control volume such that its temperature is equal to that of the surroundings ($T_i = T_e = T_0$), then the above equation reduces to

$$\dot{W}_s \leq (\dot{H}_i - T_i\dot{S}_i) - (\dot{H}_e - T_e\dot{S}_e)$$

or,

$$\dot{W}_s \leq \dot{G}_i - \dot{G}_e \tag{8.30}$$

The maximum power that can be obtained from a flow process, in which the initial and final temperatures of the flowing fluid are the same as that of the surroundings, is equal to the rate of decrease in the Gibbs free energy of the flowing fluid. If there is any irreversibility in the process, the power that can be obtained is less than just the decrease in the Gibbs free energy. Since the Gibbs free energy represents the potential to do work, it is also a thermodynamic potential.

***Example* 8.4** The separation of a mixture into pure components is an extremely important operation in chemical industry. A thermodynamic analysis of the separation processes provides valuable data in the design of separation equipment. Determine the minimum power required to separate one kmol/h of air at 100 kPa and 300 K into its pure components, nitrogen and oxygen, at 100 kPa and 300 K. The air may be treated as an ideal gas. The surroundings are at 300 K.

Solution The air flows continuously through the separation equipment. The inlet and outlet temperature of the air is equal to the surrounding temperature ($T_i = T_e = T_0$). Therefore, the minimum power required is equal to the increase in the Gibbs free energy.

$$\dot{W}_{\text{min}} = \Delta\dot{g} = \dot{g}_e - \dot{g}_i = (\dot{h}_e - T_0\dot{s}_e) - (\dot{h}_i - T_0\dot{s}_i)$$

For an ideal gas $h = h(T)$ only. Since $T_i = T_e$ we have $\Delta\dot{h} = 0$

$$\dot{g}_e - \dot{g}_i = T_0(\dot{s}_i - \dot{s}_e)$$

The initial state is a mixture of N_2 and O_2 at 100 kPa and 300 K and the final state consists of pure components at the same temperature and pressure. Therefore,

$$(\dot{s}_i - \dot{s}_e) = -\Delta\dot{s}_{\text{mix}} = R\sum X_i \ln X_i$$

or,

$$\dot{g}_e - \dot{g}_i = RT_0\sum X_i \ln X_i$$

$$= 8.314(300)\,(0.79\ln 0.79 + 0.21\ln 0.21) = -1281.91\text{ J/mol.}$$

Therefore,

$$\dot{W} = \frac{1281.91\times 10^3}{3600} = 356.09\text{ W}$$

The above example illustrates that a very small amount of power is required to separate one kmol/h of air into its pure components if the process is carried out reversibly. However the large amount of power consumed in industry for the above separation is indicative of poor thermodynamic efficiency. Hence, there is a significant scope for improvement of the separation processes used in the chemical industry.

8.4 Mathematical Preliminaries

Consider a single component system. The properties which are widely used for such a system are P, v, T, s, u, h, a and g. It is possible to express any property as a function of two independent properties. For example, one can write

$$s = s(T, P) \tag{8.31}$$

Then

$$ds = \left(\frac{\partial s}{\partial T}\right)_P dT + \left(\frac{\partial s}{\partial P}\right)_T dP \tag{8.32}$$

Dividing Eq. (8.32) by dT with v held constant, we obtain

$$\left(\frac{\partial s}{\partial T}\right)_v = \left(\frac{\partial s}{\partial T}\right)_P + \left(\frac{\partial s}{\partial P}\right)_T \left(\frac{\partial P}{\partial T}\right)_v \tag{8.33}$$

That is, $(\partial s/\partial T)_v$ is expressed in terms of three partial derivatives $(\partial s/\partial T)_P$, $(\partial s/\partial P)_T$ and $(\partial P/\partial T)_v$. Thus, it is always possible to express one derivative in terms of three other derivatives. For a single component system, the number of partial derivatives of the type $(\partial s/\partial P)_v$ that exist among the eight properties of interest (P, v, T, s, u, h, a and g) are $3! \times {}^8C_3 = 336$. Among the 336 partial derivatives that we come across in thermodynamic analysis, the number of possible relations is equal to the number of ways of choosing 4 out of 336 derivatives, ${}^{336}C_4 = 5.216 \times 10^8$. Perhaps it is needless to say that it is humanly impossible to remember these many relations. Therefore, we shall develop a systematic method by which one can manipulate these derivatives without much confusion. One such method is based on the use of Jacobians. The method of partial differentials also can be used. In this section a few important mathematical preliminaries are recapitulated.

Suppose there exists a relationship among the three variables x, y and z as

$$f(x, y, z) = 0 \tag{8.34}$$

Then one can write

$$z = z(x, y) \tag{8.35}$$

and

$$dz = \left(\frac{\partial z}{\partial x}\right)_y dx + \left(\frac{\partial z}{\partial y}\right)_x dy \tag{8.36}$$

If x and y are in turn functions of the new variables a and b, and are given by

$$x = x(a, b) \tag{8.37}$$

and

$$y = y(a, b) \tag{8.38}$$

then

$$dx = \left(\frac{\partial x}{\partial a}\right)_b da + \left(\frac{\partial x}{\partial b}\right)_a db \tag{8.39}$$

$$dy = \left(\frac{\partial y}{\partial a}\right)_b da + \left(\frac{\partial y}{\partial b}\right)_a db \tag{8.40}$$

Substituting Eqs. (8.39) and (8.40) in Eq. (8.36), we get

$$dz = \left(\frac{\partial z}{\partial x}\right)_y \left[\left(\frac{\partial x}{\partial a}\right)_b da + \left(\frac{\partial x}{\partial b}\right)_a db\right] + \left(\frac{\partial z}{\partial y}\right)_x \left[\left(\frac{\partial y}{\partial a}\right)_b da + \left(\frac{\partial y}{\partial b}\right)_a db\right] \tag{8.41}$$

or

$$dz = \left[\left(\frac{\partial z}{\partial x}\right)_y \left(\frac{\partial x}{\partial a}\right)_b + \left(\frac{\partial z}{\partial y}\right)_x \left(\frac{\partial y}{\partial a}\right)_b\right] da + \left[\left(\frac{\partial z}{\partial x}\right)_y \left(\frac{\partial x}{\partial b}\right)_a + \left(\frac{\partial z}{\partial y}\right)_x \left(\frac{\partial y}{\partial b}\right)_a\right] db \tag{8.42}$$

Substituting Eqs. (8.37) and (8.38) in Eq. (8.35), we get

$$z = z(a, b) \tag{8.43}$$

Then

$$dz = \left(\frac{\partial z}{\partial a}\right)_b da + \left(\frac{\partial z}{\partial b}\right)_a db \tag{8.44}$$

Comparing Eqs. (8.41) and (8.44), we get

$$\left(\frac{\partial z}{\partial a}\right)_b = \left(\frac{\partial z}{\partial x}\right)_y \left(\frac{\partial x}{\partial a}\right)_b + \left(\frac{\partial z}{\partial y}\right)_x \left(\frac{\partial y}{\partial a}\right)_b \tag{8.45}$$

$$\left(\frac{\partial z}{\partial b}\right)_a = \left(\frac{\partial z}{\partial x}\right)_y \left(\frac{\partial x}{\partial b}\right)_a + \left(\frac{\partial z}{\partial y}\right)_x \left(\frac{\partial y}{\partial b}\right)_a \tag{8.46}$$

Now let us consider the special case, where $x = x(a, y)$ and $y = y$. Then

$$\left(\frac{\partial y}{\partial a}\right)_y = 0 \quad \text{and} \quad \left(\frac{\partial y}{\partial y}\right)_y = 1 \tag{8.47}$$

Substituting Eqs. (8.47) and $b = y$ in Eq. (8.45), we get

$$\left(\frac{\partial z}{\partial a}\right)_y = \left(\frac{\partial z}{\partial x}\right)_y \left(\frac{\partial x}{\partial a}\right)_y \tag{8.48}$$

Equation (8.48) is known as the *chain rule of partial differentiation* which is true for any number of changes or transformations as long as the same variable is held constant. Substituting Eqs. (8.47) and $b = y$ in Eq. (8.46), we get

$$\left(\frac{\partial z}{\partial y}\right)_a = \left(\frac{\partial z}{\partial x}\right)_y \left(\frac{\partial x}{\partial y}\right)_a + \left(\frac{\partial z}{\partial y}\right)_x \tag{8.49}$$

Equation (8.49) is equivalent to dividing Eq. (8.36) by dy holding a constant. From Eq. (8.34) we can get

$$x = x(y, z) \tag{8.50}$$

$$\text{and } y = y(x, z) \tag{8.51}$$

Then from Eqs. (8.50) and (8.51), we can write

$$dx = \left(\frac{\partial x}{\partial y}\right)_z dy + \left(\frac{\partial x}{\partial z}\right)_y dz \tag{8.52}$$

and

$$dy = \left(\frac{\partial y}{\partial x}\right)_z dx + \left(\frac{\partial y}{\partial z}\right)_x dz \tag{8.53}$$

Substituting Eq. (8.53) in Eq. (8.57), we get

$$dx = \left(\frac{\partial x}{\partial y}\right)_z \left[\left(\frac{\partial y}{\partial x}\right)_z dx + \left(\frac{\partial y}{\partial z}\right)_x dz\right] + \left(\frac{\partial x}{\partial z}\right)_y dz \tag{8.54}$$

or

$$dx = \left(\frac{\partial x}{\partial y}\right)_z \left(\frac{\partial y}{\partial x}\right)_z dx + \left[\left(\frac{\partial x}{\partial y}\right)_z \left(\frac{\partial x}{\partial z}\right)_y\right] dz \tag{8.55}$$

In Eq. (8.55) the independent variables are x and z and hence, we get

$$\left(\frac{\partial x}{\partial y}\right)_z \left(\frac{\partial y}{\partial x}\right)_z = 1 \quad \text{or} \quad \left(\frac{\partial x}{\partial y}\right)_z = \frac{1}{\left(\frac{\partial y}{\partial x}\right)_z} \tag{8.56}$$

and

$$\left(\frac{\partial x}{\partial y}\right)_z \left(\frac{\partial y}{\partial z}\right)_x + \left(\frac{\partial x}{\partial z}\right)_y = 0 \tag{8.57}$$

or

$$\left(\frac{\partial x}{\partial y}\right)_z \left(\frac{\partial y}{\partial z}\right)_x \left(\frac{\partial z}{\partial x}\right)_y = -1 \tag{8.58}$$

Equation (8.56) is commonly known as the *reciprocal relation* and Eq. (8.58) is called the *cyclic relation*. The chain rule of partial differentiation, the reciprocal relation and the cyclic relation are extensively used in deriving the thermodynamic property relations.

8.5 Jacobian Method

The thermodynamic property relations can also be derived through Jacobian methods. If

$$X = X(p, r) \tag{8.59}$$

$$\text{and } Y = Y(p, r) \tag{8.60}$$

then the Jacobian of X, Y with respect to p, r is defined as:

$$J\left[\frac{X,\ Y}{p,\ r}\right] = \begin{vmatrix} \frac{\partial X}{\partial p} & \frac{\partial X}{\partial r} \\ \frac{\partial Y}{\partial p} & \frac{\partial Y}{\partial r} \end{vmatrix} = \left(\frac{\partial X}{\partial p}\frac{\partial Y}{\partial r} - \frac{\partial X}{\partial r}\frac{\partial Y}{\partial p}\right) \tag{8.61}$$

In general $J\left[\frac{X,\ Y}{p,\ r}\right]$ is written as $\frac{[X,\ Y]}{[p,\ r]}$

The Jacobians obey the following properties.

$$\frac{[X,\ Y]}{[p,\ r]}\frac{[p,\ r]}{[s,\ t]} = \frac{[X,\ Y]}{[s,\ t]} \tag{8.62}$$

$$\left(\frac{\partial X}{\partial Y}\right)_Z = \frac{[X,\ Z]}{[Y,\ Z]} \tag{8.63}$$

$$[X,\ Y] = -[Y,\ X] \tag{8.64}$$

$$[X,\ X] = 0 \tag{8.65}$$

If $Z = Z(X, Y)$, then

$$dZ = \left(\frac{\partial Z}{\partial X}\right)_Y dX + \left(\frac{\partial Z}{\partial Y}\right)_X dY = MdX + NdY \tag{8.66}$$

or

$$dZ = \frac{[Z, Y]}{[X, Y]}dX + \frac{[Z, X]}{[Y, X]}dY$$

or

$$[X, Y]\,dZ = [Z, Y]\,dX - [Z, X]\,dY$$

$$[X, Y]\,dZ + [Y, Z]\,dX + [Z, X]\,dY = 0 \tag{8.67}$$

Suppose we divide the Eq. (8.66) by dA, holding B constant, we get

$$\left(\frac{\partial Z}{\partial A}\right)_B = \left(\frac{\partial Z}{\partial X}\right)_Y \left(\frac{\partial X}{\partial A}\right)_B + \left(\frac{\partial Z}{\partial Y}\right)_X \left(\frac{\partial Y}{\partial A}\right)_B$$

This can be expressed in the Jacobian notation as:

$$\left[\frac{Z, B}{A, B}\right] = \left[\frac{Z, Y}{X, Y}\right]\left[\frac{X, B}{A, B}\right] + \left[\frac{Z, X}{Y, X}\right]\left[\frac{Y, B}{A, B}\right] \tag{8.68}$$

$$[Z, B]\,[X, Y] = [Z, Y]\,[X, B] - [Z, X]\,[Y, B]$$

or

$$[X, Y]\,[Z, B] + [Y, Z]\,[X, B] + [Z, X]\,[Y, B] = 0 \tag{8.69}$$

Then Eq. (8.68) can be rewritten as:

$$\left[\frac{Z, B}{A, B}\right] = M\left[\frac{X, B}{A, B}\right] + N\left[\frac{Y, B}{A, B}\right]$$

or

$$[Z, B] = M\,[X, B] + N\,[Y, B] \tag{8.70}$$

We can apply the above property of the Jacobians to the thermodynamic potentials and write them as:

$$du = Tds - Pdv; \quad [u,X] = T[s,X] - P[v,X] \tag{8.71}$$

$$dh = Tds + vdP; \quad [h,X] = T[s,X] + v[P,X] \tag{8.72}$$

$$da = -sdT - Pdv; \quad [a,X] = -s[T,X] - P[v,X] \tag{8.73}$$

$$dg = -sdT + vdP; \quad [g,X] = -s[T,X] + v[P,X] \tag{8.74}$$

Now let us return to the main task of expressing the non-measurable quantities in terms of the measurable quantities. It was already stated that the Maxwell relations provide us a means of expressing the partial derivatives of entropy in terms of measurable quantities. It can be easily verified that in the Jacobian notation, all the Maxwell relations are identical to the relation

$$[T,s] = [P,v] \tag{8.75}$$

The variation in any thermodynamic property associated with the simultaneous change in the independent variables can be expressed in terms of the measurable properties. In addition to P, v and T, the specific heats C_P, C_v and the coefficients of thermal expansion, β, and isothermal compressibility, κ, can also be easily measured. The coefficient of thermal expansion and the isothermal compressibility are defined as

$$\beta = \frac{1}{v}\left(\frac{\partial v}{\partial T}\right)_P$$

or

$$\beta = \frac{1}{v}\left[\frac{v,P}{T,P}\right] \tag{8.76}$$

and

$$\kappa = -\frac{1}{v}\left(\frac{\partial v}{\partial P}\right)_T = \frac{1}{v}\left[\frac{v,T}{T,P}\right] \tag{8.77}$$

Also

$$\frac{\beta}{\kappa} = -\left(\frac{\partial v}{\partial T}\right)_P\left(\frac{\partial P}{\partial v}\right)_T = \left(\frac{\partial P}{\partial T}\right)_v = \left[\frac{P,v}{T,v}\right] \tag{8.78}$$

The specific heats are defined by the relations:

$$C_P = T\left(\frac{\partial s}{\partial T}\right)_P = T\left[\frac{s,P}{T,P}\right] \tag{8.79}$$

and

$$C_v = T\left(\frac{\partial s}{\partial T}\right)_v = T\left[\frac{s,v}{T,v}\right] \tag{8.80}$$

There are no definite procedures and rules for handling the partial derivatives. One can apply the above properties of the Jacobians in expressing the partial derivatives in terms of the measurable quantities. The following guidelines may be adopted in deriving the thermodynamic relations.

1. Express the partial derivative in the Jacobian notation.
2. If any Jacobian contains only one of the thermodynamic potentials u, h, a or g, they can be eliminated through the relations (8.71) to (8.74).
3. If any Jacobian contains two thermodynamic potentials, it can be reduced to Jacobians having only one thermodynamic potential by substituting a cyclic relation similar to Eq. (8.69) in which two independent measurable variables of the involved potentials are included.
4. If the entropy s appears in any of the Jacobians, it can be eliminated either in terms of the Maxwell relations (8.75) or in terms of the specific heats (8.79) and (8.80).
5. If any Jacobian contains P, v and T, they can be eliminated by substituting the definitions of β and κ.

8.6 Entropy Relations

The entropy of a substance can be expressed in terms of the independent variables (T, P) or (T, v) and these relations are quite useful in estimating the entropy change when the system changes its state from (P_1, v_1, T_1) to (P_2, v_2, T_2). Such relations have been developed already for the case of an ideal gas. The relations presented below are applicable to all substances.

8.6.1 Entropy as a function of T and P

Consider entropy as a function of the independent variables T and P, that is,

$$s = s(T, P) \tag{8.81}$$

Then the differential change in the entropy associated with the simultaneous change in the independent variables can be written as:

$$ds = \left(\frac{\partial s}{\partial T}\right)_P dT + \left(\frac{\partial s}{\partial P}\right)_T dP \tag{8.82}$$

Expressing the partial differentials in the Jacobian notation, we get:

$$ds = \left[\frac{s, P}{T, P}\right] dT + \left[\frac{s, T}{P, T}\right] dP \tag{8.83}$$

Eliminating the Jacobians containing entropy in terms of the Maxwell relations

$$[T, s] = [P, v]$$

and the specific heat

$$C_P = \frac{T\,[s, P]}{[T, P]}$$

we obtain

$$ds = \frac{C_P dT}{T} - \left[\frac{P, v}{P, T}\right] dP \tag{8.84}$$

Substituting Eq. (8.76) in Eq. (8.84) gives:

$$ds = \frac{C_P dT}{T} - \beta v dP$$

or

$$Tds = C_P dT - \beta v T dP \qquad (8.85)$$

The above relation can also be obtained by the method of partial differentials as given below:

Substituting the definitions of C_P and the Maxwell relation

$$\left(\frac{\partial s}{\partial P}\right)_T = -\left(\frac{\partial v}{\partial T}\right)_P$$

in Eq. (8.82), we get

$$ds = \frac{C_P dT}{T} - \left(\frac{\partial v}{\partial T}\right)_P dP$$

Substituting the definition of β in the above relation and rearranging, we get

$$Tds = C_P dT - \beta v T dP$$

8.6.2 Entropy as a function of T and v

Now let us consider entropy as a function of the independent variables T and v, that is,

$$s = s(T, v) \qquad (8.86)$$

Then,

$$ds = \left(\frac{\partial s}{\partial T}\right)_v dT + \left(\frac{\partial s}{\partial v}\right)_T dv \qquad (8.87)$$

or

$$ds = \left[\frac{s, v}{T, v}\right] dT + \left[\frac{s, T}{v, T}\right] dv \qquad (8.88)$$

Eliminating the Jacobians containing entropy (s) using the definition for C_v and the Maxwell relation (8.75), we obtain

$$ds = \frac{C_v dT}{T} + \left[\frac{v, P}{v, T}\right] dv \qquad (8.89)$$

Next, eliminating the Jacobian containing P, v and T using the relation (8.78), we obtain

$$ds = \frac{C_v dT}{T} + \frac{\beta}{\kappa} dv$$

or

$$Tds = C_v dT + \frac{T\beta}{\kappa} dv \tag{8.90}$$

To obtain the above relation by the method of partial differentials, substitute Eqs. (8.17) and (8.6) in Eq. (8.87). Then, we get

$$ds = \frac{C_v dT}{T} + \left(\frac{\partial P}{\partial T}\right)_v dv \tag{8.91}$$

Now substituting Eq. (8.78) in the above equation gives:

$$ds = \frac{C_v dT}{T} + \frac{\beta}{\kappa} dv \tag{8.92}$$

or

$$Tds = C_v dT + \frac{T\beta}{\kappa} dv$$

The Eqs. (8.85) and (8.90) are sometimes referred as *Tds* equations.

8.7 Enthalpy and Internal Energy Relations

The enthalpy of a substance is usually measured in terms of the independent variables T and P. A relationship expressing the enthalpy as a function of these variables is quite useful in estimating the change in enthalpy. Let us assume the following relationship between enthalpy, pressure and temperature:

$$h = h(T, P)$$

Then

$$dh = \left(\frac{\partial h}{\partial T}\right)_P dT + \left(\frac{\partial h}{\partial P}\right)_T dP \tag{8.93}$$

or

$$dh = \left[\frac{h, P}{T, P}\right] dT + \left[\frac{h, T}{P, T}\right] dP \tag{8.94}$$

Eliminating the Jacobians containing enthalpy by using the relations

$$[h, P] = T\,[s, P] + v\,[P, P] = T\,[s, P] \tag{8.95}$$

$$[h, T] = T\,[s, T] + v\,[P, T] \tag{8.96}$$

We obtain:

$$dh = T\left[\frac{s, P}{T, P}\right] dT + \left[T\left[\frac{s, T}{P, T}\right] + v\right] dP \tag{8.97}$$

Next, eliminating the Jacobians containing s by making use of the definition of C_P and the Maxwell relation

$$[T,s] = [P,v]$$

we get

$$dh = C_P dT + \left[v - T\left[\frac{P,v}{P,T}\right]\right] dP \tag{8.98}$$

Finally we can eliminate the Jacobian containing P, v, and T by substituting Eq. (8.76) to get:

$$dh = C_P dT + (v - T\beta v)dP \tag{8.99}$$

The above relationship can also be obtained by the method of partial differentials as presented below. To obtain the partial derivative of h, let us write the differential of h in terms of the differentials of its independent variables as

$$dh = Tds + vdP$$

Dividing the above relation by dP, holding T constant, we obtain

$$\left(\frac{\partial h}{\partial P}\right)_T = T\left(\frac{\partial s}{\partial P}\right)_T + v \tag{8.100}$$

Now we can substitute the Maxwell relation

$$\left(\frac{\partial s}{\partial P}\right)_T = -\left(\frac{\partial v}{\partial T}\right)_P$$

in Eq. (8.100) to get

$$\left(\frac{\partial h}{\partial P}\right)_T = v - T\left(\frac{\partial v}{\partial T}\right)_P \tag{8.101}$$

Therefore,

$$dh = C_P dT + \left[v - T\left(\frac{\partial v}{\partial T}\right)_P\right] dP \tag{8.102}$$

On substituting the definition of β in the above equation it reduces to:

$$dh = C_P dT + (v - T\beta v)dP$$

Similarly the internal energy of a substance is usually expressed as a function of the independent variables T and v. Such a relation is highly useful in estimating the change in the internal energy of the system.

***Example* 8.5** Derive a relation for the change of internal energy when the independent variables T and v of the substance undergo a simultaneous change. Express the result in terms of the measurable quantities $P, v, T, \beta, \kappa, C_P$ and C_v only. Simplify the relation for an ideal gas.

Solution Let us consider a relation of the form

$$u = u(T, v)$$

Then

$$du = \left(\frac{\partial u}{\partial T}\right)_v dT + \left(\frac{\partial u}{\partial v}\right)_T dv \tag{8.103}$$

or

$$du = \left[\frac{u, v}{T, v}\right] dT + \left[\frac{u, T}{v, T}\right] dv \tag{8.104}$$

Eliminating the Jacobians containing the potential u using the following relations

$$[u, v] = T\,[s, v] - P\,[v, v] = T\,[s, v] \tag{8.105}$$

$$[u, T] = T\,[s, T] - P\,[v, T] \tag{8.106}$$

we get

$$du = T\left[\frac{s, v}{T, v}\right] dT + \left[T\left[\frac{s, T}{v, T}\right] - P\right] dv \tag{8.107}$$

The Jacobians containing s can be eliminated by substituting the definition of C_v and the Maxwell relation to obtain

$$du = C_v dT + \left[T\left[\frac{v, P}{v, T}\right] - P\right] dv \tag{8.108}$$

Next, the Jacobian containing P, v and T can be eliminated by substituting Eq. (8.78) to obtain

$$du = C_v dT + \left(\frac{T\beta}{\kappa} - P\right) dv \tag{8.109}$$

The above relation can also be derived by the method of partial differentials as shown below. To evaluate $(\partial u/\partial v)_T$, let us consider the differential of u in terms of the differentials of its independent variables

$$du = Tds - Pdv$$

Dividing the above relation with dv, holding T constant, we obtain

$$\left(\frac{\partial u}{\partial v}\right)_T = T\left(\frac{\partial s}{\partial v}\right)_T - P$$

Then, substitute the Maxwell relation

$$\left(\frac{\partial s}{\partial v}\right)_T = \left(\frac{\partial P}{\partial T}\right)_v$$

and the definition of specific heat at constant volume in the above equation to obtain:

$$du = C_v dT + \left[T\left(\frac{\partial P}{\partial T}\right)_v - P\right] dv$$

Next, on substituting the relation

$$\left(\frac{\partial P}{\partial T}\right)_v = \frac{\beta}{\kappa}$$

the above equation reduces to

$$du = C_v dT + \left[\frac{T\beta}{\kappa} - P\right] dv$$

For an ideal gas

$$\beta = \frac{1}{T} \tag{8.110}$$

$$\kappa = \frac{1}{P} \tag{8.111}$$

Hence,

$$du = C_v dT + \left[T\left(\frac{1}{T}\right)P - P\right] dv = C_v dT \tag{8.112}$$

That is, for an ideal gas:

$$\left(\frac{\partial u}{\partial v}\right)_T = 0 \tag{8.113}$$

$$\left(\frac{\partial u}{\partial P}\right)_T = \left(\frac{\partial u}{\partial v}\right)_T \left(\frac{\partial v}{\partial P}\right)_T = 0 \tag{8.114}$$

Therefore, the internal energy of an ideal gas is independent of v and P and is a function of T only.

***Example* 8.6** Show that

$$\left(\frac{\partial u}{\partial T}\right)_P = C_P - Pv\beta$$

Solution

$$\left(\frac{\partial u}{\partial T}\right)_P = \left[\frac{u, P}{T, P}\right] = T\left[\frac{s, P}{T, P}\right] - P\left[\frac{v, P}{T, P}\right]$$

$$= C_P - Pv\beta$$

Alternatively, one can start with the relation

$$du = Tds - Pdv$$

Then

$$\left(\frac{\partial u}{\partial T}\right)_P = T\left(\frac{\partial s}{\partial T}\right)_P - P\left(\frac{\partial v}{\partial T}\right)_P$$

Substituting the definitions of C_P and β, we obtain:

$$\left(\frac{\partial u}{\partial T}\right)_P = C_P - Pv\beta$$

***Example* 8.7** Show that

$$\left(\frac{\partial u}{\partial v}\right)_P = \frac{C_P}{v\beta} - P$$

Solution

$$\left(\frac{\partial u}{\partial v}\right)_P = \left[\frac{u, P}{v, P}\right] = T\left[\frac{s, P}{v, P}\right] - P\left[\frac{v, P}{v, P}\right]$$

$$= T\left[\frac{s, P}{v, P}\right] - P = T\left[\frac{s, P}{T, P}\right]\left[\frac{T, P}{v, P}\right] - P$$

$$= \frac{C_P}{\beta v} - P$$

8.8 Specific Heat Relations

Consider the ratio of specific heats C_P and C_v:

$$\gamma = \frac{C_P}{C_v} = T\left[\frac{s, P}{T, P}\right]\frac{1}{T}\left[\frac{T, v}{s, v}\right]$$

$$= \left[\frac{s, P}{s, v}\right]\left[\frac{T, v}{T, P}\right] = \left(\frac{\partial P}{\partial v}\right)_s\left(\frac{\partial v}{\partial P}\right)_T$$

or

$$\left(\frac{\partial P}{\partial v}\right)_s = \gamma\left(\frac{\partial P}{\partial v}\right)_T \tag{8.115}$$

That is, on a P–v diagram the slope of a reversible adiabatic process is equal to γ times the slope of a reversible isothermal process.

***Example* 8.8** Derive a relation for the difference between the two specific heats C_P and C_v and show that it is always positive.

Solution The specific heats C_P and C_v are defined as:

$$C_P = T\left(\frac{\partial s}{\partial T}\right)_P \quad \text{and} \quad C_v = T\left(\frac{\partial s}{\partial T}\right)_v$$

From the above relations it is obvious that we should consider the relation among the four variable s, P, T and v.

$$[s,P]\,[T,v] + [P,T]\,[s,v] + [T,s]\,[P,v] = 0 \tag{8.116}$$

Dividing the above relation throughout by $[T,P]\,[T,v]$, we get

$$\left[\frac{s,P}{T,P}\right] - \left[\frac{s,v}{T,v}\right] + \left[\frac{T,s}{T,P}\right]\left[\frac{P,v}{T,v}\right] = 0$$

or

$$\frac{C_P - C_v}{T} = -\left[\frac{T,s}{T,P}\right]\left[\frac{P,v}{T,v}\right] \tag{8.117}$$

Substitute the Maxwell relation $[T,s] = [P,v]$ in the above equation to get:

$$\frac{C_P - C_v}{T} = \left[\frac{v,P}{T,P}\right]\left[\frac{P,v}{T,v}\right] \tag{8.118}$$

Then, substitute Eqs. (8.76) and (8.78) in the above relation to obtain:

$$C_P - C_v = \frac{Tv\beta^2}{\kappa} \tag{8.119}$$

The above relation can also be derived by the method of partial differentials as presented below. From the definitions of specific heats we can write:

$$C_P - C_v = T\left[\left(\frac{\partial s}{\partial T}\right)_P - \left(\frac{\partial s}{\partial T}\right)_v\right]$$

Consider $s = s(T,v)$. Then

$$ds = \left(\frac{\partial s}{\partial T}\right)_v dT + \left(\frac{\partial s}{\partial v}\right)_T dv$$

Dividing the above relation by dT, holding P constant, we obtain

$$\left(\frac{\partial s}{\partial T}\right)_P = \left(\frac{\partial s}{\partial T}\right)_v + \left(\frac{\partial s}{\partial v}\right)_T \left(\frac{\partial v}{\partial T}\right)_P$$

or

$$\left(\frac{\partial s}{\partial T}\right)_P - \left(\frac{\partial s}{\partial T}\right)_v = \left(\frac{\partial s}{\partial v}\right)_T \left(\frac{\partial v}{\partial T}\right)_P$$

Therefore,

$$C_P - C_v = T\left(\frac{\partial s}{\partial v}\right)_T \left(\frac{\partial v}{\partial T}\right)_P$$

Now, substitute the Maxwell relation

$$\left(\frac{\partial s}{\partial v}\right)_T = \left(\frac{\partial P}{\partial T}\right)_v$$

in the above relation to get

$$C_P - C_v = T\left(\frac{\partial P}{\partial T}\right)_v \left(\frac{\partial v}{\partial T}\right)_P$$

The cyclic relation in P, v and T is given by

$$\left(\frac{\partial P}{\partial T}\right)_v \left(\frac{\partial T}{\partial v}\right)_P \left(\frac{\partial v}{\partial P}\right)_T = -1$$

or

$$\left(\frac{\partial P}{\partial T}\right)_v = -\left(\frac{\partial P}{\partial v}\right)_T \left(\frac{\partial v}{\partial T}\right)_P$$

Therefore,

$$C_P - C_v = -T\left(\frac{\partial P}{\partial v}\right)_T \left(\frac{\partial v}{\partial T}\right)_P \left(\frac{\partial v}{\partial T}\right)_P$$

or

$$C_P - C_v = \frac{Tv\beta^2}{\kappa}$$

The isothermal compressibility is always positive and hence $C_P - C_v > 0$. For liquids and solids β is usually small and κ is negligible. Therefore, $(C_P - C_v)$ for these substances is close to zero. Hence, the specific heats of solids and liquids are usually tabulated without specifying C_P or C_v.

***Example* 8.9** Estimate the rise in temperature if liquid water at 25°C is compressed isentropically from 0.1 MPa to 10 MPa. For liquid water $C_P = 4.2$ kJ/kg K and $\beta = 2.07 \times 10^{-4}$ K^{-1}.

Solution The rise in temperature with a rise in pressure for an adiabatic compression is given by [left as an exercise to show this relation (see Problem 8.20)]

$$\left(\frac{\partial T}{\partial P}\right)_s = \frac{Tv\beta}{C_P}$$

At 25°C, the specific volume of liquid water is $v_f = 001\ 002\ 9$ m^3/kg. Therefore,

$$\left(\frac{\partial T}{\partial P}\right)_s = \frac{298 \times 0.001\ 002\ 9 \times 2.07 \times 10^{-4}}{4.2 \times 10^3}$$

$$= 1.473 \times 10^{-8} = \left(\frac{\Delta T}{\Delta P}\right)_s$$

Hence, the temperature rise ΔT is given by:

$$\Delta T = 1.473 \times 10^{-8} \times 9.9 \times 10^6 = 0.15°\text{C}$$

8.9 Joule–Thompson Coefficient

The Joule–Thompson experiment was discussed in an earlier Chapter and it was shown that throttling is an isenthalpic process. A knowledge of the Joule–Thompson coefficient is important in the design of air- conditioning and refrigeration equipment. The Joule–Thompson coefficient can be estimated if the equation of state of the substance is known. Let us derive a relationship to estimate the Joule–Thompson coefficient in terms of the measurable quantities.

The Joule–Thompson coefficient is defined as:

$$\mu_{JT} = \left(\frac{\partial T}{\partial P}\right)_h$$

or

$$\mu_{JT} = \left[\frac{T, h}{P, h}\right] = \left[\frac{h, T}{h, P}\right] \tag{8.120}$$

Eliminating the Jacobians containing the potential h by making use of the relations

$$[h, P] = T\,[s, P] + v\,[P, P] = T\,[s, P]$$

$$[h, T] = T\,[s, T] + v\,[P, T]$$

we get

$$\mu_{JT} = \left[\frac{s, T}{s, P}\right] + \frac{v}{T}\left[\frac{P, T}{s, P}\right] \tag{8.121}$$

Now we can eliminate the Jacobians containing entropy by substituting equations (8.75) and (8.79) to obtain:

$$\mu_{JT} = \frac{T}{C_P}\left[\frac{v, P}{T, P}\right] + \frac{v}{C_P}\left[\frac{P, T}{T, P}\right] \tag{8.122}$$

The Jacobian containing P, v and T can be eliminated by substituting equation (8.76) to obtain:

$$\mu_{JT} = \frac{v}{C_P}(T\beta - 1) \tag{8.123}$$

It is also possible to derive the above relation by the method of partial differentials as shown below. Consider the functional relationship:

$$f(h, T, P) = 0$$

Then the cyclic relation in the variables h, T and P is given by:

$$\left(\frac{\partial h}{\partial T}\right)_P \left(\frac{\partial T}{\partial P}\right)_h \left(\frac{\partial P}{\partial h}\right)_T = -1$$

This relation can be written as:

$$\left(\frac{\partial T}{\partial P}\right)_h = -\left(\frac{\partial h}{\partial P}\right)_T \left(\frac{\partial T}{\partial h}\right)_P = -\frac{(\partial h/\partial P)_T}{(\partial h/\partial T)_P}$$

Next, substitute the relation

$$\left(\frac{\partial h}{\partial P}\right)_T = v - T\left(\frac{\partial v}{\partial T}\right)_P$$

in the above relation to get:

$$\mu_{JT} = \frac{1}{C_P}\left[T\left(\frac{\partial v}{\partial T}\right)_P - v\right]$$

We know that

$$\beta = \frac{1}{v}\left(\frac{\partial v}{\partial T}\right)_P$$

Therefore,

$$\mu_{JT} = \frac{1}{C_P}(\beta v T - v) = \frac{v}{C_P}(T\beta - 1)$$

For an ideal gas, we know that $\beta = T$ and hence $\mu_{JT} = 0$.

8.10 Clapeyron Equation

The thermodynamic properties of fluids were discussed at length in Chapter 3. In the pressure versus temperature or phase diagram, the solid, the liquid and the vapor regions were shown. Now let us investigate the nature of the lines separating the various phases on a phase diagram (P versus T diagram). That is, we would like to find out the dependence of pressure on equilibrium temperature when two phases coexist. Along a phase transition line, the pressure and temperature are not independent of each other because the system is univariant (for a single component simple system). That is, only one intensive property can be varied independently. Consider a single component simple system containing two phases (say liquid and vapor) in a state of equilibrium. When the system is in a state of equilibrium it has to satisfy the criteria of thermal, mechanical and chemical equilibria. That is, the temperature of the two phases has to be identical, the pressure of the two phases has to be equal and the chemical potential also should be same in both the phases. Since the phase transition occurs at constant temperature and pressure, the Gibbs free energy can be conveniently used in analyzing the problem. Then the change in Gibbs free energy is given by

$$dg = -s dT + v dP$$

Consider a system, consisting of a liquid phase at state 1 and a vapor phase at state 1′, in a state of equilibrium. If the temperature of the system is changed from T_1 to T_2 along the vaporization curve as shown in Fig. 8.3, the system attains the state 2, and 2′, where 2 represents the saturated liquid in equilibrium with the vapor in state 2′.

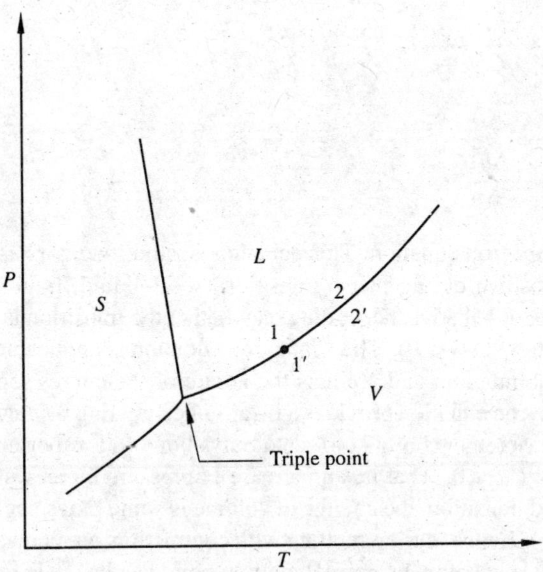

Fig. 8.3 *Phase diagram*

For the phase transition from 1 to 1′

$$dg = -sdT + vdP = 0$$

or,

$$g_f = g_g \quad \text{or} \quad g_1 = g_{1'}$$

Similarly , the Gibbs free energy of the liquid phase at state 2 is equal to the Gibbs free energy of the vapor phase at state 2′. In reaching state 2 from state 1, the change in the Gibbs free energy of the liquid phase is given by

$$g_2 - g_1 = -s_f dT + v_f dP \tag{8.124}$$

Similarly, the change in the Gibbs free energy of the vapor phase in reaching the state 2′ from state 1′ is given by

$$g_{2'} - g_{1'} = -s_g dT + v_g dP \tag{8.125}$$

Therefore,

$$-s_f dT + v_f dP = -s_g dT + v_g dP$$

$$\left(\frac{\partial P}{\partial T}\right)_{sat} = \frac{s_g - s_f}{v_g - v_f} \tag{8.126}$$

where the subscript *sat* implies that the derivative is along the saturation curve. It was shown earlier that the entropy change associated with phase transition can be calculated by the equation

$$s_g - s_f = \frac{h_g - h_f}{T} = \frac{h_{fg}}{T}$$

Hence,

$$\left(\frac{\partial P}{\partial T}\right)_{sat} = \frac{h_{fg}}{T v_{fg}} = \frac{\Delta h}{T \Delta v} \tag{8.127}$$

which is known as the Clapeyron equation. This equation is valid for any phase change.

Since Δh is always positive during phase transition (solid–liquid, liquid–vapor or solid–vapor), $(\partial P/\partial T)_{sat}$ will be positive or negative, depending on whether the transition is accompanied by expansion ($\Delta v > 0$) or contraction ($\Delta v < 0$). The Clapeyron equation is applicable to any change of state: fusion, vaporization or sublimation, and predicts the nature of the curves separating the phases on a phase diagram. The Clapeyron equation provides a means of estimating the latent heats associated with phase transition, from measurements of P–v–T data only. For solid–vapor or liquid–vapor transition $\Delta v > 0$ and hence $(\partial P/\partial T) > 0$. That is, an increase in pressure increases the boiling point of the liquid. For the solid–liquid transition, the change in volume is some times negative (especially for water, bismuth and antimony). Hence, during melting with contraction, an increase in pressure lowers the melting point. Therefore, ice at below its normal melting point can be made to melt by the application of pressure and this explains the phenomenon of regelation.

The Clapeyron equation can be put in a simple form if we make certain approximations. Let us consider the liquid–vapor phase transition at low pressures. Then the vapor phase may be approximated as an ideal gas. Moreover, the volume of the liquid phase is negligible compared to the volume of the vapor phase ($v_g \gg v_f$) and hence $v_{fg} = \Delta v = v_g = RT/P$. Under these conditions the Clapeyron equation becomes

$$\left(\frac{\partial P}{\partial T}\right)_{sat} = \frac{h_{fg} P}{RT^2}$$

$$\text{or} \quad \frac{d \ln P}{dT} = \frac{h_{fg}}{RT^2} \tag{8.128}$$

which is known as the Clausius-Clapeyron equation.

Further, if we assume that h_{fg} is constant over a small temperature range, the above equation can be integrated to get:

$$\ln\left(\frac{P_2}{P_1}\right) = \frac{h_{fg}}{R}\left(\frac{1}{T_1} - \frac{1}{T_2}\right) \tag{8.129}$$

or

$$\ln P = -\frac{h_{fg}}{RT} + \text{Const.} \tag{8.130}$$

Therefore, a plot of $\ln P$ versus $1/T$ yields a straight line the slope of which is equal to $-(h_{fg}/R)$. The vapor pressure of most of the substances agrees fairly well with the above relation.

***Example* 8.10** The weight placed on the steam exhaust port of a domestic pressure cooker never allows the pressure to build up above 200 kPa inside the cooker. Determine the temperature at which water boils in the pressure cooker. The latent heat of vaporization of water is 2256.94 kJ/kg at 100°C and $P = 101.325$ kPa. Assume that no other data is available.

Solution In absence of any other information regarding v_g and v_f the Clausius-Clapeyron equation may be used to determine the saturation temperature corresponding to the given pressure.

$$\ln\left(\frac{P_2}{P_1}\right) = \frac{h_{fg}}{R}\left(\frac{1}{T_1} - \frac{1}{T_2}\right)$$

$$\ln\frac{200}{101.325} = \frac{2256.94 \times 18}{8.314}\left(\frac{1}{373} - \frac{1}{T_2}\right)$$

$$0.68 = 4886.33\left(\frac{1}{373} - \frac{1}{T_2}\right)$$

$$T_2 = 393.42 \text{ K} \quad \text{or} \quad 120.42°\text{C}$$

(From steam tables, the saturation temperature at 200 kPa is 120.23°C).

8.11 Kirchoff Equation

The latent heat of vaporization and the latent heat of fusion vary with temperature. From the steam tables it can be observed that the latent heat of vaporization at 100°C is 2256.94 kJ/kg whereas at the critical point it is equal to zero. The latent heat of vaporization decreases as the saturation temperature is increased and at the critical point where the two phases are indistinguishable, it reduces to zero. This is true for all substances. Hence, while dealing with large variation of temperature and pressure, the latent heat of phase transition cannot be assumed to be constant. Kirchoff relation which predicts the effect of temperature on the latent heat of phase transition is derived below.

Consider the vaporization of a liquid at constant temperature and pressure as shown in Fig. 8.4. The latent heat of vaporization associated with the phase change 1 to 1′ is $(h_{g1} - h_{f1})$ at temperature T. When the saturation temperature is raised to $(T + dT)$, the latent heat of vaporization is $(h_{g2} - h_{f2})$. The change in the latent heat associated with a change in the saturation temperature is given by:

$$dh_{fg} = dh_g - dh_f \tag{8.131}$$

The variation in the enthalpy associated with the variation in the independent variables T and P is given by:

$$dh = C_P dT + v(1 - \beta T)dP \tag{8.132}$$

$$\text{or} \quad dh_{fg} = (C_{Pg} - C_{Pf}) + \left[(v_g - v_f) - (v_g\beta_g - v_f\beta_f)T\right] dP$$

$$\text{or} \quad \frac{\partial h_{fg}}{\partial T} = (C_{Pg} - C_{Pf}) + \left[(v_g - v_f) - (v_g\beta_g - v_f\beta_f)T\right]\left(\frac{\partial P}{\partial T}\right)_{\text{sat}} \tag{8.133}$$

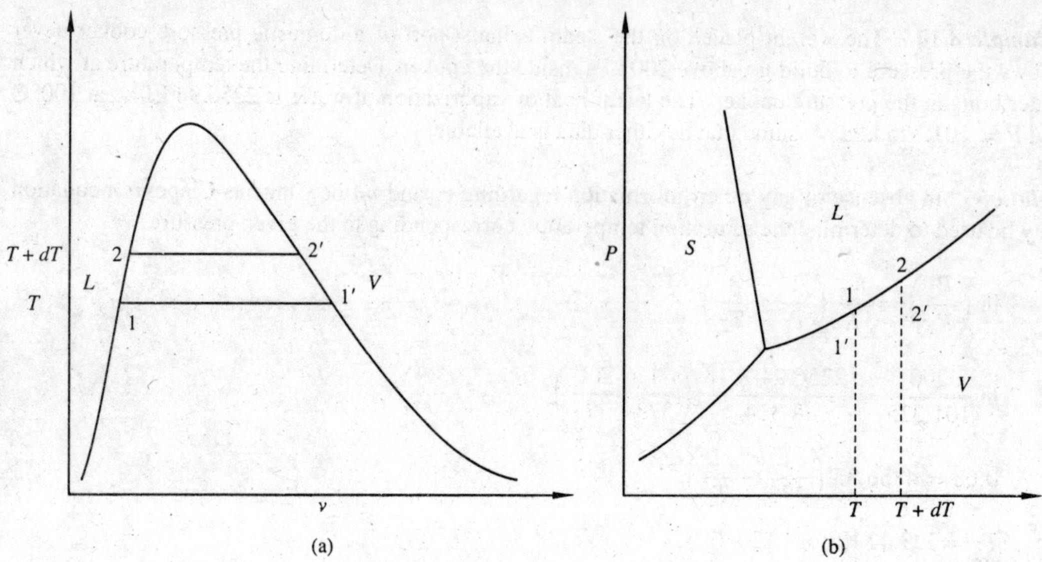

Fig. 8.4 *(a) T–v diagram indicating the vaporization process. (b) Phase diagram showing the change of state*

Substituting for $(\partial P/\partial T)_{\text{sat}}$ from the Clapeyron equation (8.127), we get:

$$\frac{\partial h_{fg}}{\partial T} = (C_{Pg} - C_{Pf}) - \frac{(v_g\beta_g - v_f\beta_f)h_{fg}}{v_g - v_f} + \frac{h_{fg}}{T} \tag{8.134}$$

This is known as the Kirchoff relation. This can be rewritten as:

$$\frac{\partial(h_{fg}/T)}{\partial T} + \frac{(v_g\beta_g - v_f\beta_f)}{v_g - v_f}\frac{h_{fg}}{T} = \frac{C_{Pg} - C_{Pf}}{T} \tag{8.135}$$

The Kirchoff relation can be transformed into a more convenient form by making some approximations. For solid to liquid transition, it is a reasonably good approximation to assume that the molar heat capacity and the molar volume are constant in each phase and the coefficient of volume expansion β is negligible for each phase. With these approximations, the Kirchoff equation (8.135) reduces to:

$$\frac{\partial(h_{sf}/T)}{\partial T} = \frac{C_{Pf} - C_{Ps}}{T} \tag{8.136}$$

where h_{sf} is the latent heat of fusion.

For the transition from liquid phase to vapor phase, the molar volume of the liquid phase can be neglected compared to the molar volume of the vapor phase ($v_g \gg v_f$) and $\beta_g \gg \beta_f$. Further, the vapor phase may be approximated as an ideal gas. Then $\beta_g = 1/T$. Since $\beta_g \gg \beta_f$ and $v_g \gg v_f$, it is clear that $v_g\beta_g \gg v_f\beta_f$ and hence, $v_f\beta_f$ can be neglected compared to $v_g\beta_g$. With these approximations the Kirchoff equation (8.134) reduces to:

$$\left(\frac{\partial h_{fg}}{\partial T}\right) = C_{Pg} - C_{Pf} \tag{8.137}$$

***Example* 8.11** The latent heat of vaporization of water at 100°C is 2256.94 kJ/kg. Estimate the latent heat of vaporization of water at 120°C and compare with the value given in the steam tables. The specific heats of liquid water and vapor are $C_{Pf} = 4.23$ kJ/kg K and $C_{Pg} = 1.55$ kJ/kg K, respectively.

Solution The Kirchoff equation for the liquid to vapor transition is given by:

$$\left(\frac{\partial h_{fg}}{\partial T}\right) = C_{Pg} - C_{Pf}$$

On integrating the above equation, we get:

$$\begin{aligned} h_{fg2} - h_{fg1} &= \left(C_{Pg} - C_{Pf}\right)(T_2 - T_1) \\ &= (1.55 - 4.23)20 = -53.6 \end{aligned}$$

Therefore,

$$h_{fg} \text{ at } 120^\circ\text{C} = 2256.94 - 53.6 = 2203.34 \text{ kJ/kg}$$

The steam tables gives the value as 2202.28 kJ/kg.

***Example* 8.12** One kg liquid water at −10°C and is converted into ice at the same temperature. Calculate the entropy change accompanying the process. The latent heat of fusion of water at 0°C is 334.92 kJ/kg. The specific heats of liquid water and ice are 4.186 kJ/kg K and 2.093 kJ/kg K, respectively.

Solution The Kirchoff relation can be used to determine the latent heat of fusion at −10°C, which in turn can be used to determine the entropy change associated with the process. Kirchoff relation for the solid to liquid transition is given by:

$$\frac{\partial (h_{sf}/T)}{\partial T} = \frac{C_{Pf} - C_{Ps}}{T}$$

This can be integrated to obtain:

$$\begin{aligned} \left(\frac{h_{sf}}{T}\right)_2 &= \left(\frac{h_{sf}}{T}\right)_1 + \left(C_{Pf} - C_{Ps}\right) \ln \frac{T_2}{T_1} \\ &= \frac{334.92}{273.15} + (4.186 - 2.093) \ln \frac{263.15}{273.15} = 1.1481 \end{aligned}$$

Then,

$$\Delta s = -\frac{h_{sf}}{T} = -1.1481 \text{ kJ/kg K}$$

8.12 Gibbs Phase Rule

The state postulate tells us that the number of thermodynamic properties that can be varied independently is equal to the number of relevant reversible work modes plus one. Hence, for a single component simple system the number of variables that can be varied independently is equal to two. That is, two independent variables are to be specified for the complete description of the state of such a system. However, the situation is not that simple when we deal with a complex system consisting of several phases and several components.

Let us consider a system consisting of C components which are distributed among all the P phases. Since each phase contains C components, each phase is completely specified by the variables temperature, pressure and $(C-1)$ mole fractions of the components (since the sum of all the mole fractions must add to unity). That is, there are $(C+1)$ variables. Then, for the entire system, $P(C+1)$ variables are needed for complete description. If the system is existing in a state of thermodynamic equilibrium, the criteria for equilibrium tells that the following equalities between the variables must be satisfied.

$$T^1 = T^2 = T^3 = \ldots = T^P \tag{8.138}$$

That is, there are $(P-1)$ equations in temperature.

$$P^1 = P^2 = P^3 = \ldots = P^P \tag{8.139}$$

That is, there are $(P-1)$ equations in pressure.

$$\mu_1^1 = \mu_1^2 = \mu_1^3 = \ldots = \mu_1^P \tag{8.140}$$

$$\mu_2^1 = \mu_2^2 = \mu_2^3 = \ldots = \mu_2^P \tag{8.141}$$

$$\mu_c^1 = \mu_c^2 = \mu_c^3 = \ldots = \mu_c^P \tag{8.142}$$

That is, there are $(P-1)$ equations among each of the chemical potentials μ_i. The superscripts $1, 2, \ldots, P$ denote the phases and the subscripts $1, 2, \ldots, C$ denote the components. Thus the total number of relations existing between the variables is $(P-1)(C+2)$. The additional variables over and above the number of relationships existing among them would represent the degrees of freedom of the system. If F denotes the degrees of freedom of the system (that is F variables can be varied independently), then

$$F = P(C+1) - (P-1)(C+2) \tag{8.143}$$

or

$$P + F = C + 2 \tag{8.144}$$

which is the celebrated Gibbs phase rule.

The phase rule provides a relationship among the degrees of freedom of the system, the number of phases and the number of components. The relationship is very simple and its derivation is based on the criterion of equilibrium which in turn is based on the laws of thermodynamics.

8.12.1 Single component system

Let us apply the phase rule for a single component system. Then $F = C + 2 - P = 3 - P$. Therefore, there are 3 possibilities.

When $P = 3$; $F = 0$: That is, the system is invariant. Therefore, for a single component system the maximum number of phases that can coexist are equal to three. Such a system has no degrees of freedom; that is, no property of the system can be varied independently. The system is present at the triple point and the temperature and pressure of the triple point are fixed.

When $P = 2$, $F = 1$: The system is univariant; that is, it has only one degree of freedom when the system contains two phases in a state of equilibrium. Consider the saturation curve shown on the phase diagram shown in Fig. 8.5. When the saturated liquid is in equilibrium with the saturated vapor, specification of either saturation pressure or saturation temperature fixes the other variable. Along the sublimation, fusion and vaporization curves the system is univariant.

If $P = 1$; $F = 2$. When the system is present in only one phase, the degrees of freedom of such a system is equal to two, or the system is bivariant. That is, two intensive properties can be varied independently, and the system lies in any of the regions indicated as solid, liquid or vapor in the phase diagram. That is, two independent variables P and T are to be specified for a complete description of the system.

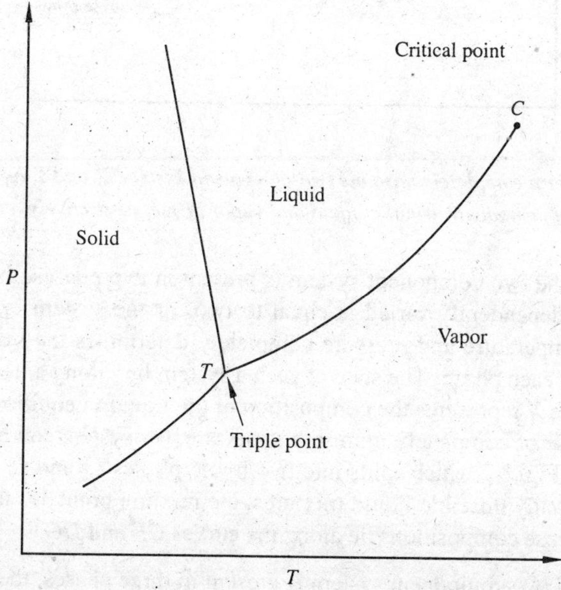

Fig. 8.5 *Phase diagram for a single component system*

8.12.2 Two component system

If $C = 2$; $F = 4 - P$, there are four possibilities. When $P = 1$; $F = 3$. If a two component system is present in a single phase, it has 3 degrees of freedom, or the system is said to be trivariant. For a complete description of such a system, three independent variables, like pressure, temperature and composition (mole fraction of one of the components), are to be specified. In Figs. 8.6 to 8.8 , the system may be present either as a liquid or vapor, whose specification requires a knowledge of T, P and Y.

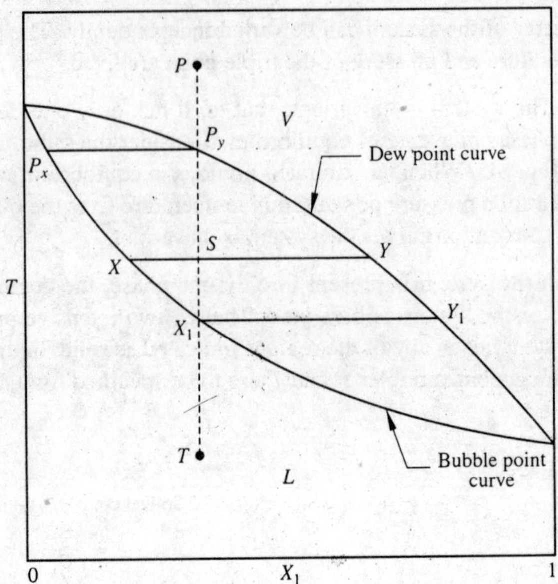

Fig. 8.6 *Phase diagram for a completely miscible two-component system. X and Y represent the mole fraction of component 1 in the liquid and vapor phase, respectively*

When $P = 2$; $F = 2$. If the two component system is present in two phases, the number of intensive properties that can be independently varied is equal to two, or the system is bivariant. Therefore, specifying the system temperature and pressure completely determines the state of the system, and hence, the composition of each phase. The state of such a system lies along a tie-line ($X - Y$) shown in the phase diagrams, where X represents the composition of the liquid in equilibrium with the vapor of composition Y. In the case of completely immiscible mixture, the system can be present in the liquid region, say at point M in Fig. 8.7 which splits into two liquid phases L_1 and L_2 of pure components 1 and 2. In the case of partially miscible liquid mixtures, the mixture point M, shown in Fig. 8.8 splits into two liquid phases whose compositions lie along the curves CF and DG.

When $P = 3$; $F = 1$. If a two component system is present in three phases, the number of degrees of freedom of such a system is equal to one, or the system is univariant. Once the pressure is specified, the temperature at which the three phases can coexist is fixed and the compositions of all the three phases are also fixed. In Fig. 8.7 and 8.8 the state of such a system is indicated by point E where the compositions of the two liquid phases are L_1 and L_2, which are in a state of equilibrium with the vapor of composition E.

Consider the miscible system in the vapor phase at point P as shown in Fig. 8.6. If the closed system is cooled at constant pressure, it follows the path P–T indicated by the dotted line till it reaches the state P_y, at which point the vapor becomes saturated, and the first drop of the liquid formed has the composition given by P_x. The liquid of composition P_x is in equilibrium with the vapor of composition P_y. If the mixture is allowed to cool, the vapor phase composition moves along the dew point curve (P_y—Y) and the liquid composition moves along the bubble point curve (P_x—X). Further cooling brings the system to state S, consisting of two phases - a liquid phase of composition X in equilibrium with a vapor phase of composition Y. Further cooling brings the system to state T which is in the liquid

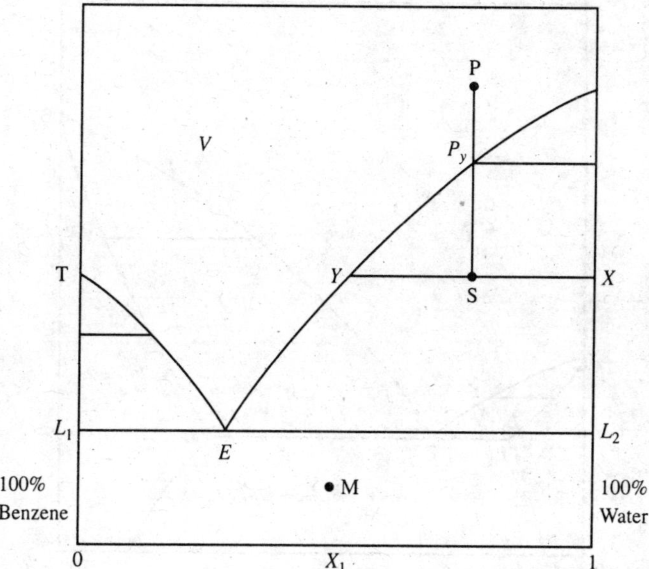

Fig. 8.7 *Phase diagram for a completely immiscible two-component system at a constant pressure*

phase. If the liquid mixture of composition T is heated in a closed container, the path followed by the system is T–P. The first bubble formed corresponds to the composition Y_1 which is in equilibrium with the liquid of composition X_1. On continued heating the vapor phase composition moves along the dew point curve Y_1–P_y while the liquid phase composition moves along the bubble point curve $X_1 - P_x$.

Now consider the completely immiscible system (benzene–water) shown in Fig. 8.7. If the vapor in state P is cooled at constant pressure in a closed container, the mixture becomes saturated when it reaches the state P_y and the first drop of liquid formed is pure water. Further cooling separates the liquid water, and the vapor composition moves along the curve $P_y - Y$. On the other hand, if the initial state of the vapor is to the left of E, cooling results in the condensation of liquid benzene.

8.13 Summary

In thermodynamic analysis of processes we deal with a large number of partial derivatives of the type $(\partial s/\partial T)_v)$. A systematic method of deriving the required partial derivatives is essential to apply the principles of thermodynamics in analyzing the processes. For this purpose, one can use either the partial derivatives method or the Jacobian method. The Jacobian method of deriving the required partial derivatives is discussed in detail.

The Maxwell relations express the partial derivatives of entropy with respect to P and v in terms of measurable quantities. The Maxwell relations are discussed and a thermodynamic mnemonic diagram which helps in writing the Maxwell relations with ease is presented.

The change in pressure associated with a change in equilibrium temperature for the coexistence of two phases is given by Clapeyron equation. The Clapeyron equation is applicable to any phase transition. It predicts the nature of the curves separating the phases on a phase diagram. The Clapeyron

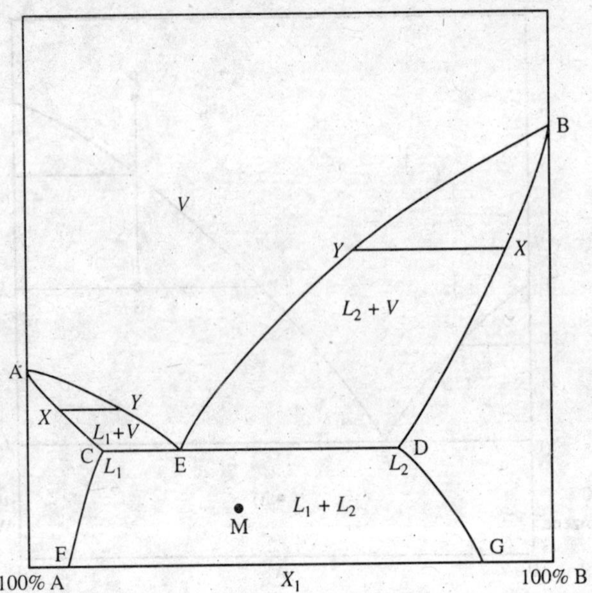

Fig. 8.8 *Phase diagram for a partially miscible two-component system at a constant pressure*

equation provides a means of estimating the latent heat of vaporization from a knowledge of P–v–T data of the substance.

The latent heats of fusion and vaporization change with temperature. The dependence of the latent heat on temperature is given by Kirchoff equation. Making use of the thermodynamic criteria for equilibrium, the Gibbs phase rule is derived.

8.14 Review Questions

1. What are the Maxwell relations and what is their necessity?
2. Sketch the thermodynamic mnemonic diagram and explain how it can be used to obtain Maxwell relations.
3. State the combined (mathematical) statement of the two laws of thermodynamics.
4. What are the independent variables for the properties U, H, A and G?
5. Define the Helmholtz function and Gibbs free energy.
6. Which thermodynamic properties represent the maximum work that can be obtained from a system which interacts with the surroundings such that the initial and final temperatures of the system are the same as that of the surroundings?
7. Show that the work done by a system during an adiabatic process is equal to the decrease in the internal energy of the system.
8. Which thermodynamic properties are called thermodynamic potentials and why are they named so?
9. Why the Gibbs free energy is called a thermodynamic potential?
10. Define the Jacobian $\left[\frac{X,Y}{A,B}\right]$.

11. State some properties of Jacobians.
12. State the Maxwell relations in Jacobian notation.
13. What is Clapeyron equation?
14. What assumptions are made in obtaining the Clausius-Clapeyron equation from Clapeyron equation?
15. Is the Clapeyron equation applicable for solid–liquid and solid–vapor phase transitions also?
16. Is the Clausius-Clapeyron equation applicable for solid–liquid and solid–vapor phase transitions also? Explain.
17. Is it possible to estimate the latent heat of vaporization of a liquid without performing calorimetric measurements? If so, explain the method.
18. What is regelation?
19. Does the melting point of ice increase with increasing pressure? Explain.
20. What is Kirchoff equation and what does it predict?

8.15 Problems

8.1 For a pure substance the entropy can be expressed as a function of the two independent variables P and v. Derive the following relation using the method of partial derivatives.

$$Tds = \left(\frac{\kappa C_v}{\beta}\right) dP + \left(\frac{C_P}{v\beta}\right) dv$$

8.2 Rework Problem 8.1 using the method of Jacobians.

8.3 Following the method of partial derivatives, derive the following partial derivatives of the internal energy.

$$\left(\frac{\partial u}{\partial T}\right)_v = C_P - \frac{Tv\beta^2}{\kappa}$$

$$\left(\frac{\partial u}{\partial P}\right)_v = \left(\frac{C_P\kappa}{\beta}\right) - Tv\beta$$

8.4 Following the method of partial derivatives, derive the relations to estimate the values of

$$\left(\frac{\partial u}{\partial P}\right)_T \quad \left(\frac{\partial u}{\partial v}\right)_T$$

in terms of measurable quantities.

8.5 Rework Problem 8.3 using the method of Jacobians.

8.6 Rework Problem 8.4 using the method of Jacobians.

8.7 The enthalpy of a pure substance can be expressed as a function of any two of the independent variables P, v and T. Using the method of partial derivatives, derive the following partial derivatives of enthalpy in terms of the measurable properties, whose knowledge enables us to estimate the change in the enthalpy associated with the change in the independent variables:

$$\left(\frac{\partial h}{\partial v}\right)_T \quad \left(\frac{\partial h}{\partial P}\right)_T \quad \left(\frac{\partial h}{\partial P}\right)_v \quad \left(\frac{\partial h}{\partial T}\right)_v$$

8.8 Using the method of partial derivatives, derive the relations to estimate the values of

$$\left(\frac{\partial h}{\partial v}\right)_P \quad \left(\frac{\partial h}{\partial T}\right)_P$$

8.9 Rework Problem 8.7 using the method of Jacobians.

8.10 Rework Problem 8.8 using the method of Jacobians.

8.11 Derive the following relations by following the method of partial derivatives.

$$\left(\frac{\partial h}{\partial s}\right)_P = T \quad \left(\frac{\partial h}{\partial s}\right)_v = T\left(1 + \frac{v\beta}{C_v\kappa}\right)$$

$$\left(\frac{\partial h}{\partial v}\right)_s = -\frac{\gamma}{\kappa} \quad \left(\frac{\partial h}{\partial s}\right)_T = T - \frac{1}{\beta}$$

8.12 Using the method of partial derivatives, show that

$$a - T\left(\frac{\partial a}{\partial T}\right)_v = u$$

$$g - T\left(\frac{\partial g}{\partial T}\right)_P = h$$

8.13 Rework Problem 8.11 using the method of Jacobians.

8.14 Rework Problem 8.12 using the method of Jacobians.

8.15 Using the method of partial derivatives, prove that

$$\left(\frac{\partial u}{\partial P}\right)_s = \frac{Pv\kappa}{\gamma} \quad \left(\frac{\partial u}{\partial T}\right)_s = \frac{PC_v\kappa}{T\beta}$$

8.16 Rework Problem 8.15 using the method of Jacobians.

8.17 Using the method of partial derivatives, show that

$$\left(\frac{\partial T}{\partial P}\right)_h = \frac{v}{C_P}(T\beta - 1) \quad \left(\frac{\partial T}{\partial s}\right)_u = \frac{T}{C_v}\left(1 - \frac{T\beta}{P\kappa}\right)$$

8.18 Rework Problem 8.17 using the method of Jacobians.

8.19 Show that

$$\left(\frac{\partial a}{\partial s}\right)_T = -\frac{P\kappa}{\beta} \quad \left(\frac{\partial g}{\partial v}\right)_T = -\frac{1}{\kappa}$$

8.20 Show that the slope of a reversible adiabatic process on a temperature versus pressure diagram is given by

$$\left(\frac{\partial T}{\partial P}\right)_s = \frac{Tv\beta}{C_P}$$

8.21 Suppose an ideal gas is compressed reversibly and adiabatically, how does the temperature ratio depend on the pressure ratio?

8.22 Given $\beta = 2.07 \times 10^{-4}\ \text{K}^{-1}$ and $\kappa = 4.85 \times 10^{-4}\ \text{MPa}^{-1}$ for liquid water. Estimate the rise in pressure if liquid water is heated at constant volume from 25°C to 75°C.

8.23 Suppose liquid water at 25°C is isentropically compressed such that the volume is decreased by 10%. Estimate the rise in the temperature of the water. It is known that $\beta = 2.07 \times 10^{-4}\ \text{K}^{-1}$ and $\kappa = 4.85 \times 10^{-4}\ \text{MPa}^{-1}$ for liquid water.

8.24 Prove that:

$$\left(\frac{\partial C_v}{\partial v}\right)_T = T\left(\frac{\partial^2 P}{\partial T^2}\right)_v$$

8.25 Prove that:

$$\left(\frac{\partial C_P}{\partial P}\right)_T = -T\left(\frac{\partial^2 v}{\partial T^2}\right)_P$$

8.26 Determine the isothermal compressibility of an ideal gas, a Clausius gas, and a van der Waals' gas, respectively.

8.27 Derive an expression to calculate the change in the internal energy for one mole of a nonideal gas when it is changed from the initial state P_1, T_1 to the final state P_2, T_2. List the measurable data required to carry out a numerical calculation.

8.28 Derive an expression to estimate the change in the enthalpy of Clausius gas when it is changed from the initial state P_1, T_1 to the final state P_2, T_2.

8.29 Show that the Joule–Thompson coefficient of a van der Waals gas is:

$$\mu_{JT} = \frac{v}{C_P}\left[\frac{2a(v-b)^2 - TRbv^2}{RTv^3 - 2a(v-b)^2}\right]$$

8.30 Obtain a relationship between C_P and C_v for a gas that obeys the equation of state $P(v-b) = RT$.

8.31 For a gas that obeys the equation of state $P(v-b) = RT$ determine the relations between T and v and between P and v for an isentropic process.

8.32 For a van der Waals gas

(a) Prove that C_v is a function of temperature only.

(b) Derive an expression for $C_P - C_v$.

8.33 Show that for a van der Waals gas:

$$du = C_v dT + \left(\frac{a}{v^2}\right) dv$$

$$ds = \frac{C_v dT}{T} + \frac{R dv}{v-b}$$

8.34 Prove that:

$$du = (C_P - Pv\beta)dT + v(P\kappa - \beta T)dP$$

8.35 Prove that a gas which has the equation of state $P(v-b) = RT$ does not exhibit a change in C_P as the pressure is changed.

8.36 The P–v–T behavior of a particular gas is represented by $Z = 1 + \frac{BP}{RT}$ where B is a function of temperature only. Derive expressions to evaluate $\left(\frac{\partial h}{\partial P}\right)_T$ and $\left(\frac{\partial s}{\partial P}\right)_T$.

8.37 If the P–v–T behavior of a substance is expressed as $Z = 1 + \frac{B}{v} + \frac{C}{v^2}$ where B and C are functions of temperature, derive an expression to determine $\left(\frac{\partial s}{\partial v}\right)_T$.

8.38 Prove that $\left(\frac{\partial \beta}{\partial P}\right)_T = -\left(\frac{\partial \kappa}{\partial T}\right)_P$.

8.39 Derive expressions to evaluate the slopes of constant pressure and constant volume lines on a T–S diagram for an ideal gas, and show that the constant volume line is steeper.

8.40 Evaluate the slopes of a constant pressure process and a constant volume process on a enthalpy versus pressure diagram and on a enthalpy versus entropy diagram respectively for an ideal gas. Redraw the cycle shown in Fig. 8.9 on h–P and h–s diagrams.

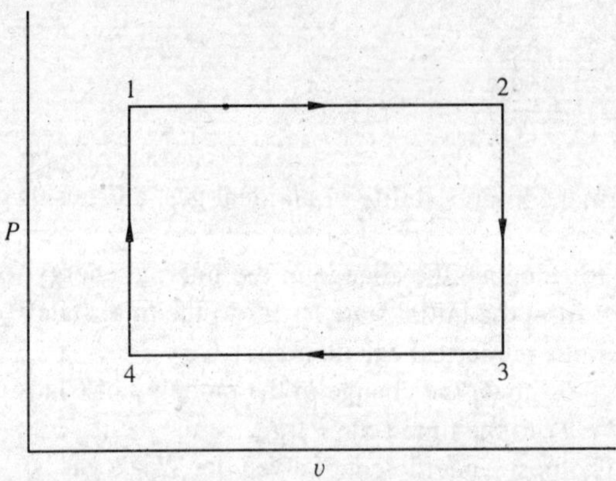

Fig. 8.9 *Sketch for Problem 8.40*

8.41 An air standard Diesel cycle is shown in Fig. 8.10. Assume that an ideal gas with constant specific heats is employed as the working medium for the cycle. Evaluate the slopes of the various processes of the cycle on enthalpy versus pressure diagram and sketch the air standard Diesel cycle on h–P diagram.

8.42 The velocity of sound or sonic velocity is given by

$$c = \sqrt{-v^2\left(\frac{\partial P}{\partial v}\right)_s}$$

Given that for liquid water at 30°C, $\rho = 995.65$ kg/m^3, $\beta = 303.1 \times 10^{-6}$ K^{-1} and $\kappa = 44.77 \times 10^{-6}$ bar^{-1}, estimate the velocity of sound in liquid water at 30°C.

8.43 For copper at 500 K, $\rho = 8930$ kg/m^3; $\beta = 54.2 \times 10^{-6}$ K^{-1}; $C_P = 0.3831$ kJ/kg K and $\kappa = 0.837 \times 10^{-11}$ m^2/N. Determine the percent error in C_v that would result if it were assumed that $C_P = C_v$.

8.44 For liquid water at 30°C, $\rho = 995.65$ kg/m^3; $\beta = 303.1 \times 10^{-6}$ K^{-1}; $\kappa = 44.77 \times 10^{-6}$ bar^{-1}, and $C_P = 4.18$ kJ/kg K. Estimate the per cent error in C_v of liquid water at 30°C that would result if it were assumed that $C_P = C_v$.

8.45 The specific volumes of ice and water at 0°C are 1.091×10^{-3} m^3/kg and 1.0002×10^{-3} m^3/kg, respectively. The latent heat of fusion of water at 0°C is 333.4 kJ/kg. Determine the pressure change required to lower the melting point of ice by 0.5°C.

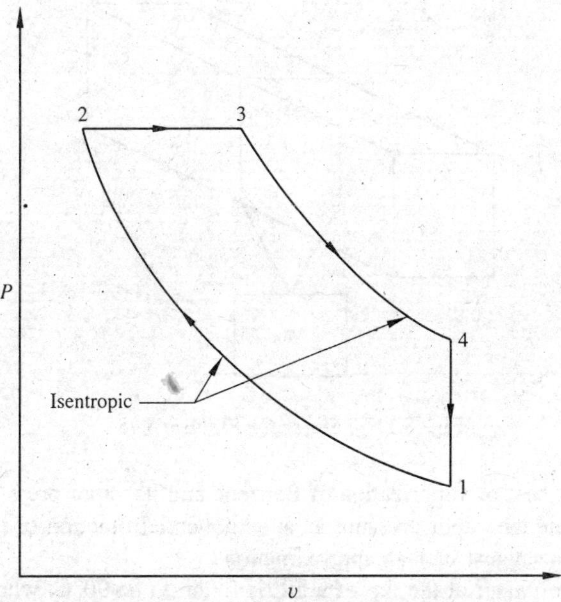

Fig. 8.10 *An air-standard Diesel cycle*

8.46 At 0°C the specific volumes of ice and water are 1.091×10^{-3} m³/kg and 1.0002×10^{-3} m³/kg, respectively. The latent heat of fusion of water at 0°C is 333.4 kJ/kg. Estimate the melting temperature of ice at 200 bar and at 500 bar.

8.47 The decrease in the melting point of ice with increase in pressure is largely responsible for our ability to skate on ice. At the line of contact between the skate and the ice a high pressure is exerted by the person who is skating because of the small area of contact. Therefore, ice melts and provides a thin layer of liquid water which acts as a lubricant and helps in skating. Suppose a man of mass 75 kg desires to skate on ice. The area of contact between the skates and ice is 20 mm². Will he be able to skate on ice which is at -2°C ?

8.48 If a thin wire loaded with a mass m is passed over a block of ice, the wire cuts through the ice block and does not leave a trace of its path. This phenomenon is termed as regelation. It is desired to demonstrate the phenomenon of regelation in the laboratory. A thin wire whose area of contact with the surface of the ice is 2 mm² is used to suspend the mass as shown in Fig. 8.11. Determine the mass to be suspended if the ice is at -3°C.

8.49 If the saturation pressure of a particular substance is given by $\ln P = A - \frac{B}{T}$, where A and B are constants, obtain expressions for h_{fg} and s_{fg} in terms of P–v–T data and the constant B.

8.50 Carbon tetrachloride boils at 76.75°C at 101 kPa. The latent heat of vaporization of carbon tetrachloride is 194.8 kJ/kg. Determine the boiling point of carbon tetrachloride at 200 kPa.

8.51 The measured vapor pressure of benzene is as follows:

T°C	7.6	15.4	26.1	42.2	60.6	80.1
P (kPa)	5.333	7.999	13.332	26.664	53.329	101.325

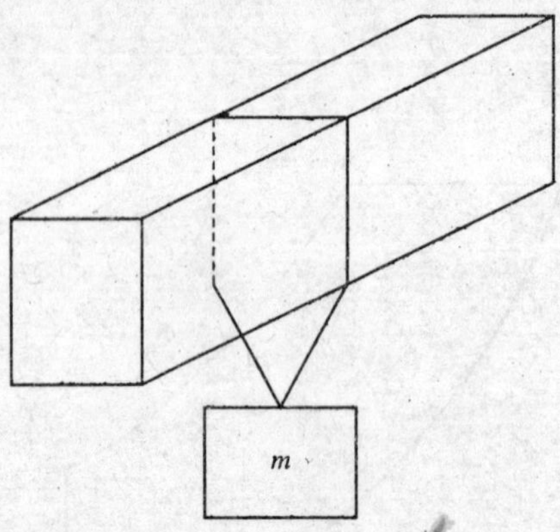

Fig. 8.11 *Sketch for Problem 8.48*

Estimate the latent heat of vaporization of Benzene and its vapor pressure at 120°C. Is it possible to approximate the vapor pressure as an exponential function of temperature? Does the Clapeyron equation suggest such an approximation?

8.52 The boiling point of water at the top of a hill is found to be 90°C, whereas it boils at 99.6°C with the latent heat of vaporization of 2256.94 kJ/kg at the foot of the hill, where the pressure is 101.325 kPa. Assuming that the atmosphere is locally isothermal at 300 K (that is the pressure variation with the specific volume of the air is given by the relation $Pv = P_0 v_0$, where the subscript zero denotes the conditions at the foot of the hill) estimate the height of the hill.

8.53 Rework Problem 8.52 if the relation between pressure and specific volume is given by $Pv^\gamma =$ constant.

8.54 The latent heat of vaporization of water at 100°C is 2256.94 kJ/kg. Estimate the latent heat of vaporization of water at 130°C and compare with the value given in the steam tables. The specific heats of the liquid and vapor are given as $C_{Pf} = 4.23$ kJ/kg K and $C_{Pg} = 1.55$ kJ/kg K.

8.55 In deriving the Gibbs phase rule it was assumed that all the C components are present in all the P phases. Derive the phase rule if a particular component is absent from a particular phase. Determine the number of degrees of freedom when an aqueous solution of sodium chloride is in equilibrium with its vapor. (Sodium chloride is present only in the liquid phase and not in the vapor phase).

8.56 One kmol of an ideal gas ($\gamma = 1.4$) at 6 MPa and 1000 K is contained in a cylinder. Determine the maximum work that can be obtained if the gas is allowed to undergo a reversible and an adiabatic expansion to 600 kPa.

8.57 One kg of steam with quality 0.75 at 100 kPa is contained in a cylinder piston assembly. It is desired to compress the steam adiabatically to 1.0 MPa. Determine the minimum amount of work required to compress the steam, and the final temperature of the steam.

8.58 In a steam power plant, superheated steam at 3 MPa and 350°C enters an adiabatic turbine at a rate of 2 kg/s. The steam leaves the turbine at 10 kPa. Calculate the maximum possible output power of the turbine.

8.59 In a steam power plant the saturated liquid water at 10 kPa enters an adiabatic pump at a rate of 2 kg/s. The pump delivers the water to a boiler at 3 MPa. Calculate the minimum power input to the pump.

8.60 Natural chlorine is a mixture of 75% Cl_2^{35} and 25% Cl_2^{37}. It is desired to separate natural chlorine at 100 kPa and 300 K into its pure isotopes at the same temperature and pressure. The surroundings are at 300 K. Calculate the minimum amount of work required to separate one kmol/s of chlorine.

8.61 One kmol of an ideal gas ($\gamma = 1.4$) is available at 500 kPa and 300 K. The air can be allowed to expand to a pressure of 100 kPa while in contact with the surroundings at 300 K. The final temperature of the air is 300 K. Determine the maximum amount of work that can be obtained.

8.62 A tank of 2 m^3 volume is divided into two equal compartments by a partition. An ideal gas is present in both the compartments at the conditions shown in Fig. 8.12. The tank is in contact with the surroundings at 300 K. If the partition is allowed to move till the pressures on both the sides of the partition are equal, determine the amount of work done by the system.

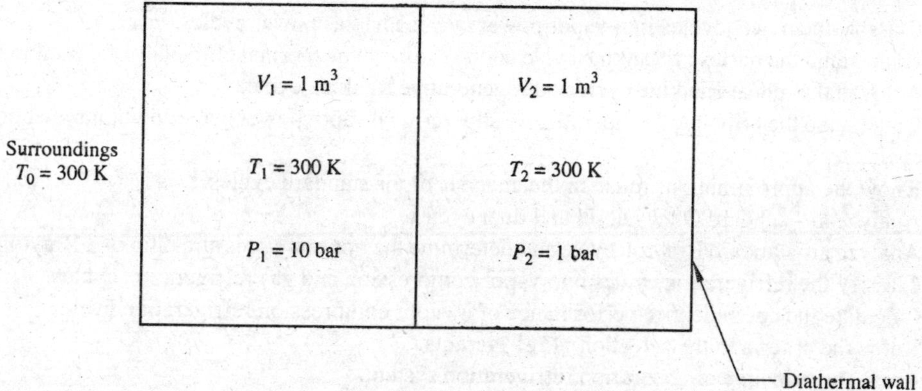

Fig. 8.12 *Sketch for Problem 8.62*

9 Power and Refrigeration Cycles

Learning Objectives

This chapter will enable the readers to

- Classify the power cycles into vapor power cycles and gas power cycles.
- Understand the basics of Rankine cycle and to estimate the thermal efficiency of a Rankine cycle.
- Analyze the reheat Rankine cycle and regenerative Rankine cycle.
- Understand the principles of operation of supercritical vapor power cycle and binary vapor power cycle.
- Know the approximations made in the analysis of air-standard cycles.
- Analyze air-standard Otto, Diesel and dual cycles.
- Analyze air-standard Brayton cycle and determine the optimum pressure ratio of a Brayton cycle.
- Classify the refrigeration cycles into vapor compression and gas refrigeration cycles.
- Calculate the coefficient of performance of a vapor compression refrigeration cycle.
- Know the criteria in the selection of refrigerants.
- Know the principle of absorption refrigeration system.

9.1 Classification of Power Cycles

Historically, man's endeavour for continuous conversion of internal energy of a substance into work led to the development of thermodynamics. Therefore, an important application of thermodynamics is the analysis of power cycles through which the energy absorbed as heat can be continuously converted into work. A thermodynamic analysis of the heat engine cycles provides valuable information regarding the design of new cycles or for improving the existing cycles.

Refrigeration and production of low temperature, where energy in the form of heat is removed from a low temperature body and rejected to a high temperature sink, is another important field of interest. In this case, the devices employed are usually called refrigerators. The second law of thermodynamics sets a limit on the maximum possible coefficient of performance of refrigerators. A thermodynamic analysis of refrigeration cycles is as important as the analysis of heat engine cycles. In this chapter, the power cycles and refrigeration cycles are discussed with emphasis only on the thermodynamic analysis of the cycles. No attempt is made to describe the equipment.

In a heat engine, or a power cycle, the main objective is to convert the energy available in the fuels into mechanical energy or electrical energy. In a simple steam power plant, the fuel is burned and the hot flue gases are used to convert water into steam in a steam generator. The steam is subsequently expanded in a turbine to obtain work.

In some heat engines, like internal combustion engines, the fuel is burned in the engine cylinder itself and the hot combustion products are expanded to obtain work. In all the heat engines, most of the processes employed are irreversible, but for the purpose of thermodynamic analysis these processes are idealized as reversible.

Some of the energy conversion systems, such as steam power plant operate in a cycle in which the working fluid undergoes a series of processes and returns to the initial state periodically. In others, such as internal combustion engines, the working fluid does not undergo a cyclic change. At the end of the expansion process the working fluid is rejected into the atmosphere and these cycles are usually referred to as open cycles.

Different types of working fluids are employed in the power plants. The nature of the working fluids can be classified into two groups: vapors and gases. The power cycles are accordingly classified into two groups as:

- Vapor power cycles, in which the working fluid undergoes a phase change during the cyclic process, and
- Gas power cycles, in which the working fluid does not undergo any phase change.

In the thermodynamic analysis of power cycles, our chief interest lies in estimating the energy conversion efficiency, or the thermal efficiency. There are several types of power cycles in common use and it is not intended to discuss all of them in this book. A few common power cycles are analyzed to illustrate the principles. Once the principles and the method of analysis are known, the thermodynamic analysis of any power cycle can be done without any difficulty.

9.2 Vapor Power Cycles

A major fraction of the electric power is produced in the country by thermal power stations. The working fluid used in these power plants is water, which undergoes a series of processes involving a phase change during the cycle while periodically returning to the original state. Though water is available in abundance and is inexpensive, the substance is made to undergo a cyclic change. The turbines in the power plant require large quantity of steam which is produced in steam generators at high pressure. Since corrosion and scaling in steam generators and turbines are very important, it is necessary to employ pure water. One cannot afford to reject large quantities of such pure water and hence it is usually condensed and recycled (closed cycle).

9.2.1 Carnot cycle as a vapor power cycle

The Carnot cycle which consists of only reversible processes, has the highest possible efficiency of all the real cycles operating between two thermal reservoirs. Therefore, a study of the Carnot cycle employing vapor as a working medium provides a good basis for the study of vapor power cycles. Consider the Carnot cycle shown in Fig. 9.1. Wet steam at state 1 is isentropically compressed to the saturated liquid (state 2). The saturated liquid undergoes a phase change at constant temperature and pressure and leaves the steam generator as saturated vapor (state 3). Then the saturated vapor is allowed to expand isentropically in a turbine and leaves the turbine at state 4. Finally, the turbine effluent is condensed to state 1 thus completing the cycle.

Energy is added to the working fluid at constant temperature T_1, while energy is rejected at constant temperature T_2 in the condenser. We know that the efficiency of a Carnot engine is given by

$$\eta = \frac{\text{Net work done}}{\text{Energy supplied}} = 1 - \frac{T_2}{T_1}$$

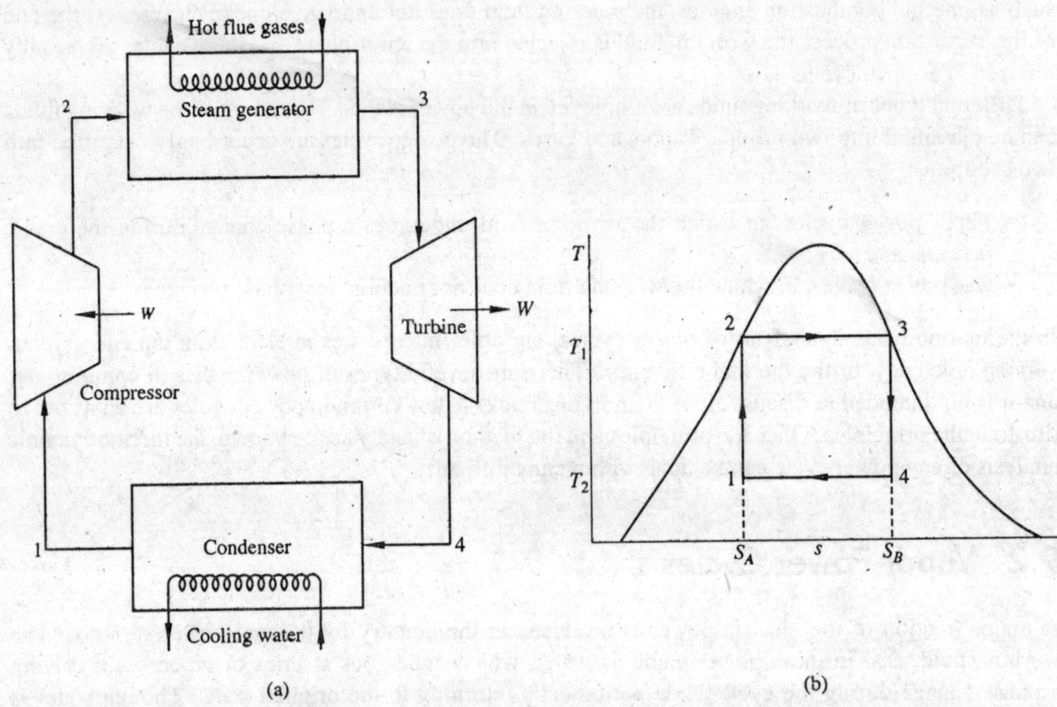

Fig. 9.1 *(a) Schematic diagram of Carnot cycle (b) T–s diagram of Carnot cycle*

It is not practical to operate a power plant on a Carnot cycle because of the difficulties associated with isentropic compression of wet steam in state 1. Moreover, the rate of delivery of work by such a plant will be so low that it is not worthwhile to pursue it. Though the isentropic compression of the wet steam in state 1 is not desirable, the energy addition and rejection processes at constant temperature and pressure in the steam generator and the condenser are highly desirable. These desirable processes are retained and the undesirable process is replaced by a technically feasible process in devising the Rankine cycle discussed below.

9.3 Ideal Rankine Cycle

Professor W.J. Rankine (1820–1872), a contemporary of Joule (1818–1889) and Clausius (1822–1888), who made significant contributions to the engineering and scientific development of that period, modified the Carnot cycle and presented a technically feasible cycle. Since the major difficulty in the Carnot cycle is associated with the isentropic compression of wet steam in state 1 (which is a two-phase mixture), the process is replaced by the isentropic compression of the saturated liquid (only one phase) which can be achieved by employing a pump. A schematic diagram of the Rankine cycle is shown in Fig. 9.2. The saturated liquid water at low pressure coming out of the condenser in state 1, is isentropically compressed by a pump and fed to the steam generator. The constant pressure energy addition as heat takes place in the steam generator and saturated steam leaves at state 4. The saturated steam enters the turbine and undergoes an isentropic expansion to state 5 at low pressure. The low pressure steam goes to a condenser where the constant pressure energy rejection takes place and the condensate leaves as saturated liquid in state 1, thus completing the cycle. In Fig. 9.3(a) the Rankine cycle is shown on a T–s diagram. It can be observed that the isentropic expansion of the steam continuously decreases its quality in going from state 4 to state 5. Presence of excessive moisture content causes serious erosion of the turbine blades which is highly undesirable. To overcome this, modern steam power plants produce superheated steam which is fed to the turbine for subsequent expansion. The Rankine cycle with superheat is shown in Fig. 9.3(b).

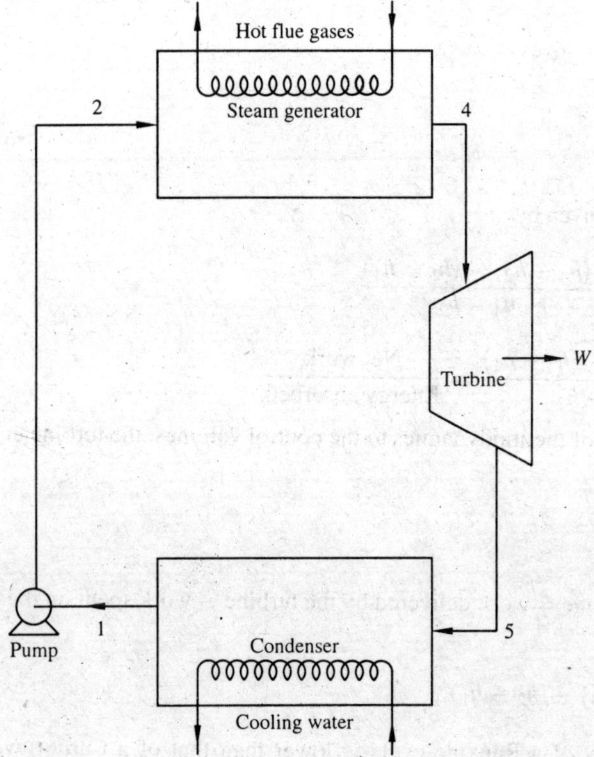

Fig. 9.2 *Schematic diagram of a simple Rankine cycle*

Let q_1 represent the energy transferred per unit mass of the working fluid during the energy addition process. Let q_2 denote the energy transferred per unit mass of the working fluid during the energy rejection process. Application of the first law of thermodynamics to the control volumes, steam generator

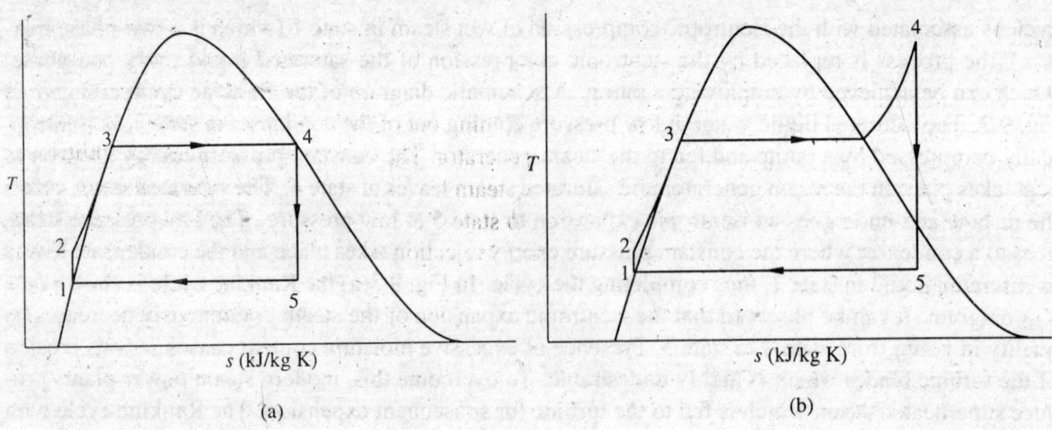

Fig. 9.3 *(a) T–s diagram of simple ideal Rankine cycle (b) T–s diagram of ideal Rankine cycle with superheat*

or boiler and condenser, gives

$$q_1 = h_4 - h_2 \tag{9.1}$$

$$q_2 = h_5 - h_1 \tag{9.2}$$

Thermal efficiency is given by

$$\eta = \frac{q_1 - q_2}{q_1} = \frac{(h_4 - h_2) - (h_5 - h_1)}{h_4 - h_2} \tag{9.3}$$

$$\eta = \frac{(h_4 - h_5) - (h_2 - h_1)}{h_4 - h_2} = \frac{\text{Net work}}{\text{Energy absorbed}} \tag{9.4}$$

Applying the first law of thermodynamics to the control volumes, the turbine and the pump, yields:

$$W_T = h_4 - h_5 \tag{9.5}$$

$$W_P = h_2 - h_1 \tag{9.6}$$

Therefore, net work done = work delivered by the turbine − work spent on the pump

or

$$W_{\text{net}} = (h_4 - h_5) - (h_2 - h_1) \tag{9.7}$$

The thermal efficiency of a Rankine cycle is lower than that of a Carnot cycle operating between the same temperature levels. This is primarily because of the fact that the energy transfer as heat in the steam generator does not take place at constant temperature in the Rankine cycle. The average temperature between the states 2 and 3 is less than the temperature at which vaporization takes place. Particularly, in the case of Rankine cycle with superheat, the maximum temperature corresponds to state 4 which is much above the temperature of vaporization at which a major fraction of the energy addition takes place.

9.3.1 Practical Rankine cycle

In actual practice, the pump and the turbine cannot be operated under isentropic conditions because of irreversibilities. Therefore, processes 1–2 and 4–5 are non-isentropic. Applying the second law of thermodynamics to the control volumes—the turbine and the pump—we find that the entropy of the exit fluid is greater than the entropy of the entering fluid. That is $s_e > s_i$.

Therefore,

$$s'_2 > s_1 \tag{9.8}$$

$$s'_5 > s_4 \tag{9.9}$$

The practical Rankine cycle is shown in Fig. 9.4.

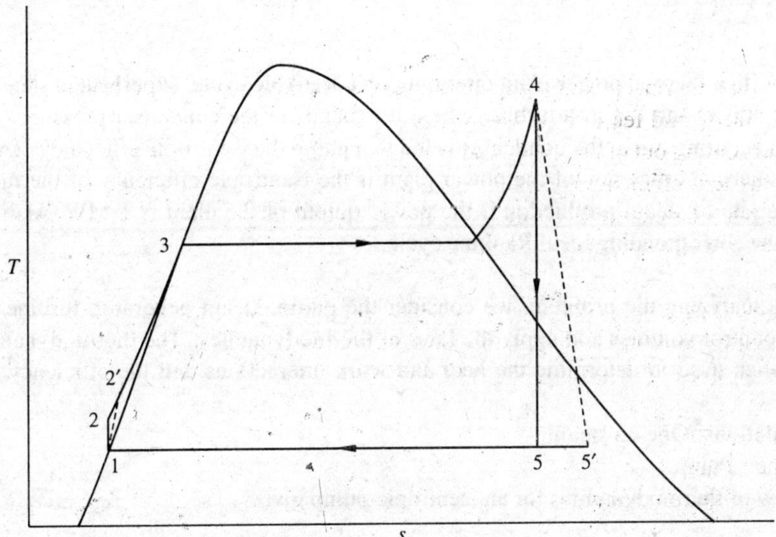

Fig. 9.4 *T–s diagram of practical Rankine cycle*

1–2′: Irreversible path followed by the working fluid in the actual pump
1–2: Path followed by the working fluid in the case of an isentropic pump
4–5′: Irreversible path followed by steam in the actual turbine
4–5: Path followed by steam in the case of an isentropic turbine

It can be observed from Fig. 9.4 that

$$h'_5 > h_5 \quad \text{and} \quad h'_2 > h_2$$

Therefore,

$$h_4 - h'_5 < h_4 - h_5 \quad \text{and} \quad h'_2 - h_1 > h_2 - h_1$$

That is, the actual work delivered by the turbine is less than the work delivered by an isentropic turbine. The work spent on the actual pump is greater than the work spent on an isentropic pump. Because of the irreversibilities in the turbine and the pump, the actual efficiency of a practical Rankine cycle is less than the efficiency of an ideal Rankine cycle.

The performance of an actual turbine or pump is usually expressed in terms of the isentropic efficiency. The isentropic efficiency of a turbine is defined as the ratio of the work delivered by the practical turbine to the work that could be obtained if the turbine were isentropic. That is,

$$\eta_T = \frac{h_4 - h_5'}{h_4 - h_5} \tag{9.10}$$

Similarly, the isentropic efficiency of a pump is defined as the ratio of the work required by the isentropic pump to the work required by the practical pump. That is,

$$\eta_P = \frac{h_2 - h_1}{h_2' - h_1} \tag{9.11}$$

***Example* 9.1** In a thermal power plant operating on a Rankine cycle, superheated steam is produced at 3 MPa and 300°C and fed to a turbine where it expands to the condenser pressure of 5 kPa. The saturated liquid coming out of the condenser is fed to a pump the isentropic efficiency of which is 0.80. Calculate the thermal efficiency of the power plant if the isentropic efficiency of the turbine is 0.85. Determine the rate of steam production if the power output of the plant is 1 MW. Also calculate the efficiency of the corresponding ideal Rankine cycle.

Solution In analyzing the problem, we consider the pump, steam generator, turbine and the condenser as the control volumes and apply the laws of thermodynamics. The thermodynamic properties of steam are then used to determine the heat and work interactions and the efficiency of the power plant.

Basis of calculations: One kg steam.

Control volume : Pump

The first law of thermodynamics for an isentropic pump gives

$$W_P = h_2 - h_1 = \Delta h$$

The second law of thermodynamics for an isentropic pump gives

$$s_2 = s_1$$

Since $dh = Tds + vdP$, we get $dh = vdP$ for an isentropic process.

Considering saturated liquid water as incompressible,

$$W_P = \int vdP = v(P_2 - P_1)$$

From saturated steam tables at 5 kPa:

$$h_1 = 137.77 \text{ kJ/kg}, \quad \text{and} \quad v_1 = 0.001\,005\,2 \text{ m}^3/\text{kg}$$

Therefore

$$W_P = 0.001\,005\,2(3 - 0.005) \times 10^6 = 3.01 \text{ kJ/kg}$$

$$\eta_P = \frac{h_2 - h_1}{h_2' - h_1} = 0.8$$

$$h_2' - h_1 = \frac{h_2 - h_1}{0.8} = \frac{3.01}{0.8} = 3.76 \text{ kJ/kg}$$

$$h_2' = h_1 + (h_2' - h_1) = 137.77 + 3.76 = 141.53 \text{ kJ/kg}$$

Control volume: Steam generator

$$q_1 = h_4 - h_2$$

From superheated steam tables at 3 MPa and 300°C:

$$h_4 = 2995.1 \text{ kJ/kg} \quad \text{and} \quad s_4 = 6.5422 \text{ kJ/kg K}$$

Hence

$$q_1 = 2995.1 - 141.53 = 2853.57 \text{ kJ/kg.}$$

Control volume: Turbine.

The first law of thermodynamics for an isentropic turbine gives

$$W_T = h_4 - h_5$$

The second law of thermodynamics for an isentropic turbine gives

$$s_5 = s_4$$

At $P = 5$ kPa:

$$s_f = 0.4763 \text{ kJ/kg K}; \quad h_f = 137.77 \text{ kJ/kg}$$

$$s_g = 8.3960 \text{ kJ/kg K}; \quad h_g = 2561.6 \text{ kJ/kg}$$

$$s_5 = s_4 = X_5 s_g + (1 - X_5) s_f$$

$$6.5422 = 8.3960 X_5 + (1 - X_5) 0.4763 \quad \text{or} \quad X_5 = 0.766$$

Therefore,

$$h_5 = h_g X_5 + (1 - X_5) h_f = 2561.6 \times 0.766 + (1 - 0.766) \times 137.77$$

or $h_5 = 1994.42$ kJ/kg

$$\eta_T = \frac{h_4 - h_5'}{h_4 - h_5} = 0.85$$

$$h_4 - h_5' = 0.85(2995.1 - 1994.42) = 850.58 \text{ kJ/kg}$$

Thermal efficiency $= \eta =$ Net work output/energy absorbed

$$\eta = \frac{(h_4 - h_5') - (h_2' - h_1)}{h_4 - h_2'} = \frac{850.58 - 3.76}{2853.57} = 0.297$$

Power output per kg of steam $= 850.58 - 3.76 = 846.82$ kJ

$$\text{Steam production rate for 1 MW power output} = \frac{1 \times 10^6}{846.82 \times 10^3} = 1.18 \text{ kg/s}$$

Thermal efficiency of an ideal Rankine cycle

$$= \frac{(h_4 - h_5) - (h_2 - h_1)}{h_4 - h_2} = \frac{(2995.1 - 1994.42) - 3.01}{2995.1 - (137.77 + 3.01)} = 0.35$$

9.4 Reheat Cycle

The thermal efficiency of any power cycle can be increased by increasing the temperature at which energy is added while maintaining the sink temperature at the same value. The efficiency can also be increased by decreasing the sink temperature while the temperature at which energy addition takes place is maintained at a constant value. The sink temperature is determined by the environment, or the availability of cooling water. Any effort to reduce the sink temperature requires the performance of additional work and hence it is not an economical alternative to follow. Therefore, to increase the efficiency, the temperature at which energy addition takes place can be raised. In a Rankine cycle, the maximum temperature corresponds to the temperature of the superheated steam (state 4) leaving the steam generator (see Fig. 9.5). However, the major fraction of energy addition takes place during vaporization. Hence, the temperature at which vaporization takes place can be raised. Since during phase transition (vaporization) a rise in temperature is associated with a rise in pressure, the operating pressure of the steam generator has to be raised.

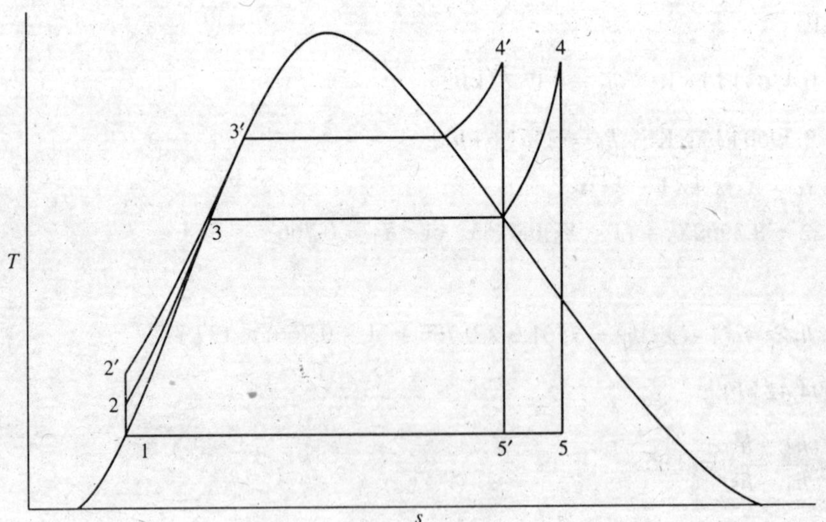

Fig. 9.5 *Rankine cycle with two different steam generator pressures*

Consider the Rankine cycle 1–2–3–4–5 shown in Fig. 9.5. The steam enters the turbine at state 4 and leaves at state 5 where the quality is sufficiently high. Suppose the steam generator pressure is increased to raise the average temperature at which energy addition takes place. The corresponding Rankine cycle is shown as 1–2′ – 3′ – 4′ – 5′. In this case the steam is raised to the same maximum temperature as in the previous case, but the average temperature at which energy addition takes place is raised. The maximum temperature in a steam generator is dictated by the material of construction of the steam generator and safe operating conditions. The maximum temperature normally allowed is 600°C. The superheated steam in state 4′ enters the turbine and exits at state 5′. In the low pressure region of the turbine the quality of the steam is low. If the moisture content in the steam is more than 10%, it causes serious erosion of the turbine blades. Therefore, this is not desirable. The reheat cycle is developed to obtain the advantage of increased efficiency by increasing the temperature at which energy is added, and at the same time to avoid the high moisture content in the low pressure region of the turbine.

The Rankine cycle with reheat is shown in Fig. 9.6. The steam leaving the steam generator at state 4 enters the high pressure section of the turbine where it expands to some intermediate pressure. Then the steam leaves the turbine at state 5, and enters the steam generator where it is reheated. The reheating of the steam is usually done to the original temperature of the steam. The reheated steam at state 6 enters the low pressure section of the turbine and expands to the condenser pressure. The steam at state 7 enters the condenser where it rejects energy to the cooling water, and leaves as condensate at state 1.

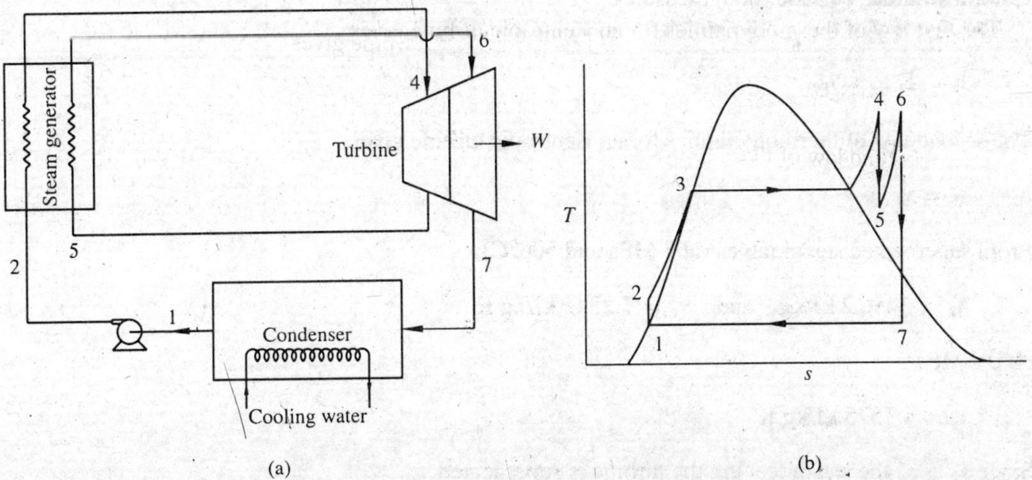

Fig. 9.6 *Rankine cycle with reheat modification (a) Schematic diagram (b) T–s diagram*

In large capacity thermal power plants the reheat modification is usually employed. In modern thermal power plants, several stages of reheating are employed. This modification is not economical in small capacity thermal power plants because of the high initial cost. Let us now evaluate the thermal efficiency of a Rankine cycle with reheat modification.

Total energy added in the steam generator per unit mass of the working fluid

$$= (h_4 - h_2) + (h_6 - h_5)$$

Energy rejected in the condenser $= (h_7 - h_1)$

Thermal efficiency is given by

$$\eta = \frac{(h_4 - h_2) + (h_6 - h_5) - (h_7 - h_1)}{(h_4 - h_2) + (h_6 - h_5)}$$

$$\eta = \frac{(h_4 - h_5) + (h_6 - h_7) - (h_2 - h_1)}{(h_4 - h_2) + (h_6 - h_5)} \tag{9.12}$$

***Example* 9.2** In a power plant, employing a Rankine cycle with reheat modification, the steam enters the turbine at 3 MPa and 500°C. After expansion to 0.6 MPa, the steam is reheated to 500°C and expanded in a second turbine to a condenser pressure of 5 kPa. The steam leaves the condenser as saturated liquid. Calculate the thermal efficiency of the plant if the isentropic efficiency of the pump is 0.6 and the isentropic efficiency of the turbine is 0.8.

Solution The Rankine cycle with the reheat modification is shown in Fig. 9.6. By applying the first and the second laws of thermodynamics to several units (control volumes) the problem can be solved in a manner similar to that employed in the Example 9.1.
Control volume: Turbine (high pressure)

The first law of thermodynamics for an isentropic turbine gives

$$W_{T1} = h_4 - h_5$$

The second law of thermodynamics for an isentropic turbine gives

$$s_5 = s_4$$

From superheated steam tables (at 3 MPa and 500°C):

$$h_4 = 3456.2 \text{ kJ/kg} \quad \text{and} \quad s_4 = 7.2345 \text{ kJ/kg K}$$

At 0.6 MPa ;

$$s_g = 6.7575 \text{ kJ/kg K.}$$

Since $s_4 > s_g$ the steam leaving the turbine is superheated.

By interpolation, we get $T_5 = 265.8$°C. Hence $h_5 = 2989.59$ kJ/kg (obtained by interpolation).

$$\eta_T = \frac{h_4 - h_5'}{h_4 - h_5} = 0.8$$

$$W_{T1} \text{ actual} = h_4 - h_5' = 0.8(3456.2 - 2989.59) = 373.29 \text{ kJ/kg}$$

and

$$h_5' = 3082.91 \text{ kJ/kg}$$

Control volume: Second turbine

The first and the second laws of thermodynamics for an isentropic turbine give

$$W_{T2} = h_6 - h_7 \quad \text{and} \quad s_6 = s_7$$

From steam tables at 0.6 MPa and 500°C:

$$h_6 = 3482.7 \text{ kJ/kg} \quad \text{and} \quad s_6 = 8.0027 \text{ kJ/kg K}$$

$$s_6 = s_7 = X_7 s_g + (1 - X_7) s_f$$

$$8.0027 = 8.3960 X_7 + (1 - X_7) 0.4763 \quad \text{or} \quad X_7 = 0.95$$

Read the following data from steam tables at 5 kPa:

$$h_f = 137.77 \text{ kJ/kg} \quad h_g = 2561.6 \text{ kJ/kg}$$

$$s_f = 0.4763 \text{ kJ/kg K} \quad s_g = 8.3960 \text{ kJ/kg K}$$

$$h_7 = h_g X_7 + (1 - X_7) h_f = 2561.6 \times 0.95 + (1 - 0.95) 137.77 = 2440.41 \text{ kJ/kg}$$

$$\eta_T = \frac{h_6 - h_7'}{h_6 - h_7} = 0.8$$

$$W_{T2} \text{ actual} = h_6 - h_7' = 0.8(3482.7 - 2440.41) = 837.03 \text{ kJ/kg}$$

Control volume: Pump

The first and the second laws of thermodynamics for an isentropic pump give

$$W_P = h_2 - h_1 \quad \text{and} \quad s_2 = s_1 \quad \text{or} \quad ds = 0$$

We know that $dh = Tds + vdp$. Therefore, $W_P = \Delta h = \int vdP$

Considering the liquid to be incompressible, we get

$$W_P = v(P_2 - P_1) = 0.001\,005\,2(3 - 0.005) \times 10^6 = 3.01 \text{ kJ/kg}$$

$$\eta_P = \frac{h_2 - h_1}{h_2' - h_1} = 0.6$$

Therefore

$$h_2' - h_1 = W_P \text{ actual} = \frac{3.01}{0.6} = 5.02 \text{ kJ/kg}$$

$$h_2' = h_1 + (h_2' - h_1) = 137.77 + 5.02 = 142.79 \text{ kJ/kg}$$

$$\eta = \frac{(h_4 - h_5') + (h_6 - h_7') - (h_2' - h_1)}{(h_4 - h_2') + (h_6 - h_5')}$$

$$\eta = \frac{373.29 + 837.03 - 5.02}{(3456.2 - 142.79) + (3482.7 - 3082.91)} = 0.325$$

9.5 Regenerative Cycle

In the reheat modification of the Rankine cycle, the effective temperature at which energy is added is increased by supplying a large portion of energy in the superheat region. The average temperature at which energy addition takes place can also be increased by preheating the feed water before it enters

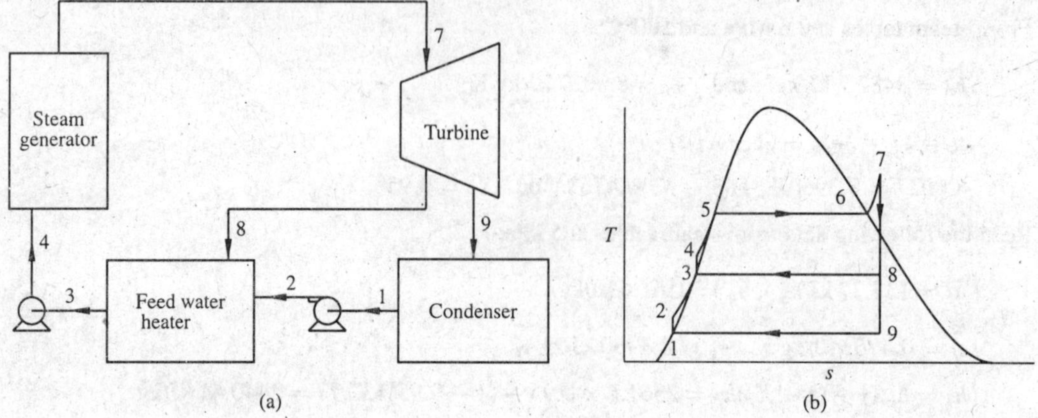

Fig. 9.7 *Regenerative Rankine cycle (a) Schematic diagram and (b) T–s diagram*

the steam generator. For this purpose part of the steam is withdrawn at some intermediate stage in the turbine and used to preheat the feed water. Because of this modification, the energy which would be lost in the condenser is being utilized in saving the energy requirements. A Rankine cycle using this type of feed water heating is called a regenerative cycle. A schematic regenerative cycle is shown in Fig. 9.7.

Superheated steam at state 7 enters the turbine. A part of the steam leaves the turbine at an intermediate stage (state 8). This steam enters the feed water heater. In a direct contact feed water heater, the steam is mixed with the condensate coming out of the condenser. The rest of the steam expands in the turbine to the condenser pressure. Then the low pressure steam leaves the turbine at state 9. The low pressure steam enters a condenser where it rejects energy to the cooling water and leaves at state 1. The condensate at state 1 enters a pump and leaves it at the feed water heater pressure (state 2), which is equal to the pressure of the steam that is withdrawn from the turbine at an intermediate stage. The condensate (2) mixes with the steam leaving the turbine at state 8. Under optimal conditions, the resultant mixture is saturated liquid at state 3. This preheated water enters the steam generator at state 4 through a pump.

Consider the feed water heater shown in Fig. 9.8 as the control volume and apply the first law of thermodynamics to obtain

$$\dot{m}_2 + \dot{m}_8 = \dot{m}_3 \tag{9.13}$$

and

$$\dot{m}_2 h_2 + \dot{m}_8 h_8 = \dot{m}_3 h_3$$

or

$$\dot{m}_3 h_3 = \dot{m}_8 h_8 + (\dot{m}_3 - \dot{m}_8) h_2$$

or

$$h_3 = \frac{\dot{m}_8}{\dot{m}_3} h_8 + \left(1 - \frac{\dot{m}_8}{\dot{m}_3}\right) h_2 \tag{9.14}$$

Let Y' denote the fraction of steam extracted from the turbine for preheating. Then

$$Y' = \frac{\dot{m}_8}{\dot{m}_3} \tag{9.15}$$

$$h_3 = Y'h_8 + (1 - Y')h_2 \tag{9.16}$$

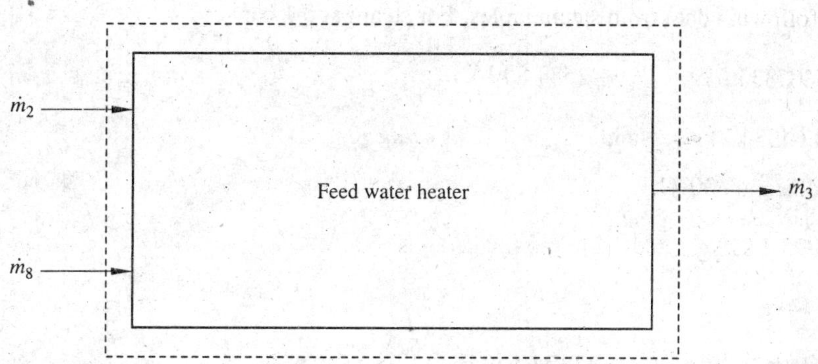

Fig. 9.8 *Direct contact feed water heater*

Equation (9.16) can be used to determine the fraction of steam to be withdrawn from the turbine at an intermediate state so that the resultant stream is saturated liquid at state 3, after mixing.

Energy added in the steam generator per unit mass of the working fluid

$$= q_1 = h_7 - h_4$$

Energy rejected in the condenser

$$= q_2 = (1 - Y')(h_9 - h_1)$$

The thermal efficiency η of the cycle is given by

$$\eta = \frac{(h_7 - h_4) - (1 - Y')(h_9 - h_1)}{h_7 - h_4} \tag{9.17}$$

The work output of the turbine is given by

$$(h_7 - h_8) + (1 - Y')(h_8 - h_9) \tag{9.18}$$

Work spent on the pumps is given by

$$(h_4 - h_3) + (1 - Y')(h_2 - h_1) \tag{9.19}$$

In modern power plants several stages of reheat and regeneration are employed to improve thermal efficiency.

Example **9.3** Consider a regenerative Rankine cycle in which steam is supplied to the turbine at 3 MPa and 300°C. A suitable fraction of steam is withdrawn from the turbine at 0.4 MPa and the remaining steam in the turbine is allowed to expand to the condenser pressure of 0.1 bar. This steam is then mixed with the condensate after it has been raised to 0.4 MPa and the mixture is fed to the feed pump. Calculate the thermal efficiency of the steam power plant.

Solution In the previous examples it was observed that the work spent on the pumps is very small. Hence, the work done in compressing the liquid water can be neglected.

Read the following data from steam tables. For steam at 0.1 bar

$$h_f = 191.83 \text{ kJ/kg}; \quad h_g = 2584.8 \text{ kJ/kg}$$

$$s_f = 0.6493 \text{ kJ/kg K} \quad \text{and} \quad s_g = 8.1511 \text{ kJ/kg K}$$

Steam at 3 MPa and 300°C:

$$h = 2995.1 \text{ kJ/kg} \quad \text{and} \quad s = 6.5422 \text{ kJ/kg K}$$

Steam at 0.4 MPa:

$$h_f = 604.67 \text{ kJ/kg} \quad h_g = 2737.6 \text{ kJ/kg}$$

$$s_f = 1.7764 \text{ kJ/kg K} \quad \text{and} \quad s_g = 6.8943 \text{ kJ/kg K}$$

$$h_2 = h_1 = 191.83 \text{ kJ/kg}$$

$$h_4 = h_3 = 604.67 \text{ kJ/kg}$$

$$h_7 = 2995.7 \text{ kJ/kg} \quad \text{and} \quad s_7 = 6.5422 \text{ kJ/kg K}$$

$$s_8 = s_7 = s_g X_8 + (1 - X_8) s_f$$

$$6.5422 = 6.8943 X_8 + (1 - X_8) 1.7764$$

Therefore, $X_8 = 0.931$

$$h_8 = 2737.6 \times 0.931 + (1 - 0.931) 604.67 = 2590.43 \text{ kJ/kg}$$

$$s_7 = s_9 = 6.5422 = 8.1511 X_9 + (1 - X_9) 0.6493$$

or $X_9 = 0.7855$

$$h_9 = 2584.8 \times 0.7855 + (1 - 0.7855) 191.83 = 2071.51 \text{ kJ/kg}$$

$$h_3 = Y' h_8 + (1 - Y') h_2$$

$$Y' = \frac{h_3 - h_2}{h_8 - h_2} = \frac{604.67 - 191.83}{2590.43 - 191.83} = 0.1721$$

$$\eta = \frac{(h_7 - h_4) - (1 - Y')(h_9 - h_1)}{h_7 - h_4}$$

$$\eta = \frac{(2995.1 - 604.67) - (1.0 - 0.1721)(2071.51 - 191.83)}{2995.1 - 604.67} = 0.349$$

9.6 Supercritical and Binary Vapor Power Cycles

9.6.1 Supercritical vapor power cycle

Some other types of steam power cycles have been developed with the view of improving the energy conversion efficiency. The efficiency of the cycle as discussed earlier can be increased by increasing the temperature at which energy addition takes place. Under normal operating conditions, there is a large temperature difference between the flue gases (1300°C) and the water (285.79°C) to be heated in a steam generator when it is operated at a pressure of 7.0 MPa. This large temperature difference during energy addition leads to large irreversibility. This irreversibility can be reduced by decreasing the temperature difference between the flue gases and water. Therefore, raising the temperature at which energy addition takes place is advantageous. However, the maximum temperature is governed by the metallurgical considerations, the maximum being 600°C. This temperature is much above the critical point of water (374.15°C). If the energy addition has to take place at 600°C, the steam generator pressure has to be higher than the critical pressure of water. The power cycle with this modification is called the supercritical vapor power cycle. In such a power cycle, the work spent in compressing the water to the steam generator pressure cannot be ignored while estimating the cycle efficiency. The supercritical power cycle is shown in Fig. 9.9.

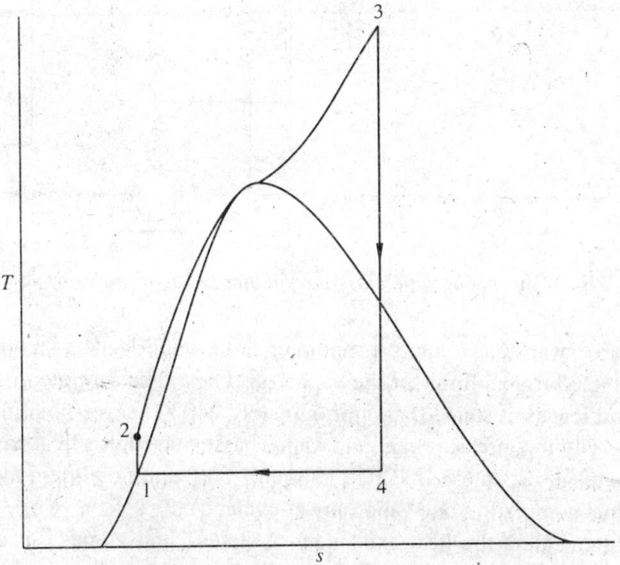

Fig. 9.9 *T–s diagram of supercritical Rankine cycle*

9.6.2 Binary vapor power cycle

Suppose the steam leaving the steam generator which is operated at high pressure is fed to a turbine for expansion. Then the moisture content of the steam increases in the low pressure region of the turbine, which is not desirable. Another disadvantage with water is that if the energy rejection has to take place at the ambient temperature (25 – 30°C), the condenser has to be maintained under vacuum. This requires elaborate and expensive arrangements. To overcome these disadvantages a binary vapor power cycle is developed in which two different fluids are used. A binary cycle employing mercury and water as the working fluids operates in the following manner (see Figs. 9.10 and 9.11).

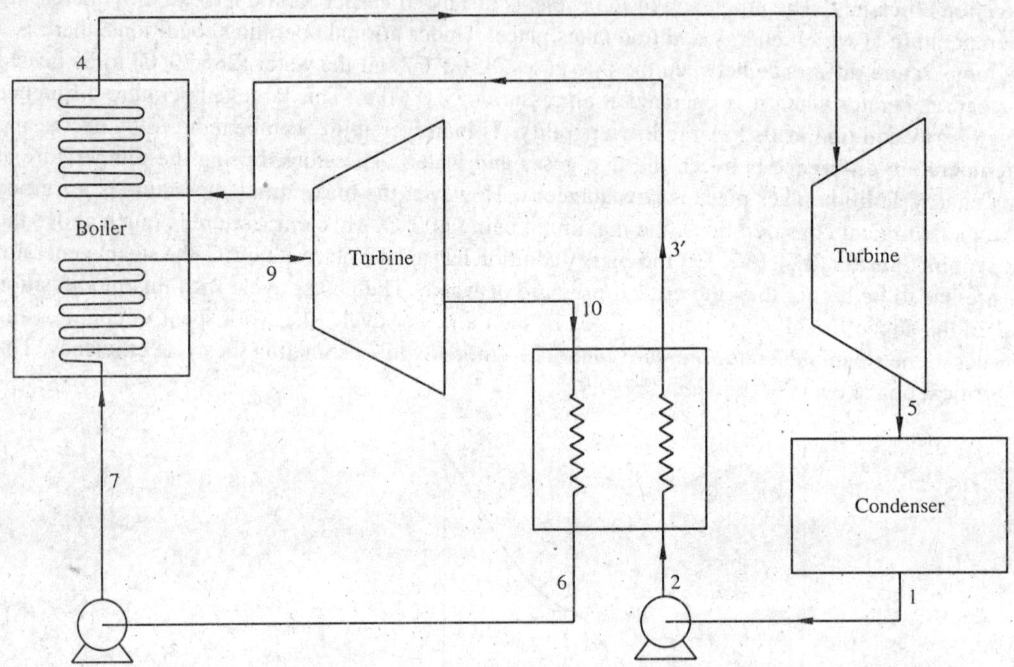

Fig. 9.10 *Schematic diagram of a binary vapor power cycle*

Saturated liquid mercury at state 7 enters a vaporizer and is vaporized at a temperature of $\simeq$ 550°C, which is close to the metallurgical limit for the vaporizer tubes. The saturated mercury vapor at state 9 enters the turbine and leaves at state 10 at approximately 240°C (corresponding saturation pressure for mercury is 7 kPa). The low pressure mercury vapor passes through a heat exchanger, transferring its energy to water and leaves at state 6 as saturated liquid. The saturated liquid mercury enters a pump and leaves at state 7 thus completing one vapor power cycle.

Saturated liquid water enters the heat exchanger at state 2 and comes out as saturated vapor at state 3′, which enters the vaporizer for superheating. The superheated steam leaves the vaporizer at state 4. This steam enters a second turbine, expands to the condenser pressure, and leaves at state 5. The low pressure steam condenses to saturated liquid water at state 1. The saturated liquid water at state 1 is pumped through a heat exchanger. The vaporization of water in the heat exchanger (from 2 to 3′) is achieved by condensing the low pressure mercury leaving the turbine.

Since the latent heat of vaporization of mercury is low compared to the latent heat of vaporization of water, a large quantity of mercury is needed to operate this power cycle. Moreover, mercury is quite expensive and toxic. Because of the cost and associated health hazards, the usage of binary vapor power cycles with mercury was discontinued after 1960's.

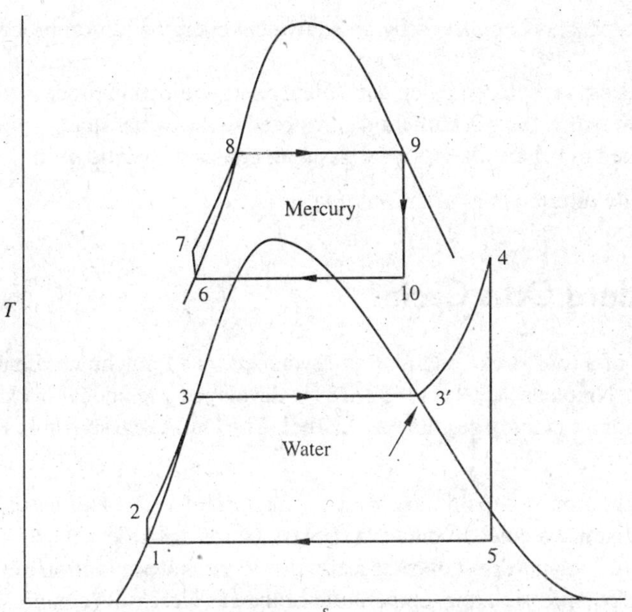

Fig. 9.11 *T–s diagram of a binary vapor power cycle*

9.7 Gas Power Cycles

So far we have discussed the Rankine cycle and its modifications. These cycles employ a fluid as the working medium, which undergoes a phase change during the cycle. These are called vapor power cycles. A large class of engines, however, employ a gas as the working medium and the cycles on which they operate are termed gas power cycles. The familiar examples of such gas power cycles are the Otto cycle and the Diesel cycle. These engines are some times called *internal combustion engines*, because energy is added in these engines through the process of combustion of the fuel in the engine cylinder, rather than from an outside source. The chemical energy of the fuel is converted into thermal energy which in turn is converted into work in these engines. The fuels used in these engines are gasoline, diesel oil, alcohol and gasohol (generally a 90–10 mixture of gasoline and ethanol). In this class of engines, the gas is predominantly air with some products of combustion like CO_2, H_2O, etc. The composition of the working fluid does not remain constant throughout the cycle. In the beginning of the cycle, the working fluid constitutes either air or a mixture of fuel vapor and air. But in the subsequent processes, the working fluid constitutes the combustion products. Though the engines operate in a mechanical cycle, the working fluid does not undergo a cyclic change. The working fluid is exhausted after the expansion process (power stroke) and a fresh mixture of fuel vapor and air is sucked in. Therefore, effectively the engines operate on open cycles.

For the purpose of thermodynamic analysis of such internal combustion engines, the following approximations are made:

1. The engine is assumed to operate on a closed cycle with a fixed mass of air which does not undergo any chemical change.

2. The combustion process is replaced by an equivalent energy addition process from an external source.
3. The exhaust process is replaced by an equivalent energy rejection process to external surroundings by means of which the working fluid is restored to the initial state.
4. The air is assumed to behave like an ideal gas with constant specific heat.

These cycles are usually referred to as *air-standard cycles*.

9.8 Air-Standard Otto Cycle

In 1862, the principle of a four stroke engine was developed by a Frenchman, Beau de Rochas. However, only in 1867 did Nikolaus A. Otto (1832–1891) developed a practical cycle which was named after him. The Otto engines employ gasoline as the fuel. The Otto cycle shown in Fig. 9.12 consists of the following processes:

- 0–1: Constant pressure suction during which a mixture of fuel vapor and air is drawn into the cylinder as the piston executes an outward stroke (suction stroke).
- 1–2: The mixture is compressed isentropically due to the inward motion of the piston. Because of the isentropic compression, the temperature of the gas increases (compression stroke).
- 2–3: The hot fuel vapor–air mixture is ignited by means of an electric spark. Since the combustion is instantaneous, there is not enough time for the piston to move outward. This process is approximated as a constant-volume energy addition process.
- 3–4: The hot combustion products undergo isentropic expansion and the piston executes an outward motion (power stroke).
- 4–1: The exhaust port opens and the combustion products are exhausted into the atmosphere. This process is conveniently approximated as a constant volume energy rejection process.
- 1–0: The remaining combustion products are exhausted by an inward motion of the piston at constant pressure (exhaust stroke).

Effectively there are four strokes in the cycle. They are suction, compression, expansion and exhaust strokes, respectively. From the P–v diagram it can be observed that the work done during the process 0–1 is exactly balanced by the work done during 1–0. Hence, for the purpose of thermodynamic analysis we need to consider only the cycle 1–2–3–4, which is the air- standard Otto cycle.

In the analysis of the power cycles, our interest lies in evaluating the thermal efficiency of the cycle. The thermal efficiency is defined as the ratio of the net work delivered to the energy absorbed. Thus

$$\eta = \frac{W_{\text{net}}}{Q_1} = \frac{Q_1 - Q_2}{Q_1}$$

where, Q_1 and Q_2 denote the energy absorbed and the energy rejected in the form of heat, respectively. Application of the first law of thermodynamics to processes 2–3 and 4–1 gives:

$$Q_1 = U_3 - U_2 = m(u_3 - u_2) = mC_v(T_3 - T_2)$$

$$Q_2 = U_4 - U_1 = m(u_4 - u_1) = mC_v(T_4 - T_1)$$

Therefore,

$$\eta = 1 - \frac{T_4 - T_1}{T_3 - T_2} \tag{9.20}$$

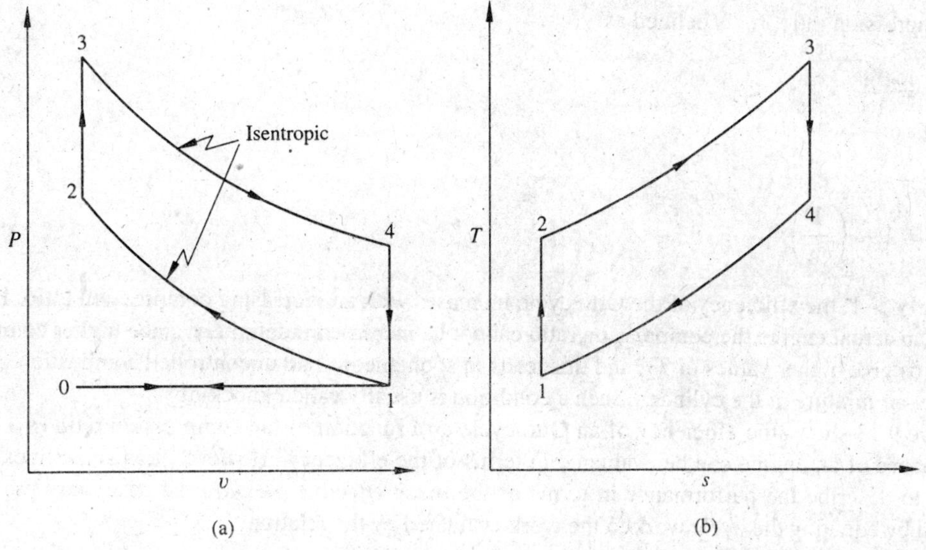

Fig. 9.12 *Air-standard Otto cycle (a) P–v diagram and (b) T–s diagram*

1–2 and 3–4 are isentropic processes for which $TV^{\gamma-1}$ = constant. Therefore

$$\frac{T_1}{T_2} = \left(\frac{V_2}{V_1}\right)^{\gamma-1} \tag{9.21}$$

$$\frac{T_3}{T_4} = \left(\frac{V_4}{V_3}\right)^{\gamma-1} \tag{9.22}$$

But

$$V_1 = V_4 \quad \text{and} \quad V_2 = V_3 \tag{9.23}$$

Hence

$$\frac{T_1}{T_2} = \frac{T_4}{T_3} \quad \text{or} \quad \frac{T_1}{T_4} = \frac{T_2}{T_3}$$

$$1 - \frac{T_1}{T_4} = 1 - \frac{T_2}{T_3}$$

or

$$\frac{T_4 - T_1}{T_4} = \frac{T_3 - T_2}{T_3}$$

$$\frac{T_4}{T_3} = \frac{T_1}{T_2} = \frac{T_4 - T_1}{T_3 - T_2} \tag{9.24}$$

Therefore,

$$\eta = 1 - \frac{T_1}{T_2} \tag{9.25}$$

The compression ratio, r_0, is defined as

$$r_0 = \frac{V_1}{V_2} \tag{9.26}$$

Hence,

$$\eta = 1 - \left(\frac{1}{r_0}\right)^{\gamma-1} \tag{9.27}$$

Since $\gamma > 1$, the efficiency of the Otto cycle increases with an increasing compression ratio. However, in an actual engine the compression ratio cannot be increased indefinitely, since higher compression ratios give higher values of T_2, and this leads to spontaneous and uncontrolled combustion of the gasoline–air mixture in the cylinder. Such a condition is usually called knocking.

Figure 9.13 shows the efficiency of an Otto cycle as a function of the compression ratio (r_0). The performance of an engine can be evaluated in terms of the efficiency. However, it is sometimes convenient to describe the performance in terms of the *mean effective pressure*, an imaginary pressure obtained by equating the cycle work to the work evaluated by the relation

$$W_{net} = P_m \int dV = P_m(V_1 - V_2) \tag{9.28}$$

That is, the mean effective pressure is defined as the net work divided by the displacement volume.

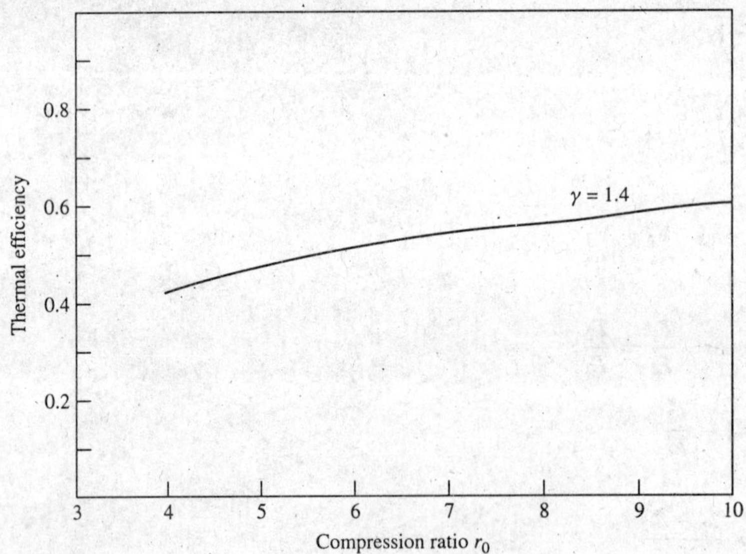

Fig. 9.13 *The dependence of thermal efficiency of an air-standard Otto cycle on the compression ratio*

That is,

$$P_m = \frac{W_{net}}{V_1 - V_2} \tag{9.29}$$

***Example* 9.4** Consider an air-standard Otto cycle with a compression ratio of 8. At the beginning of the compression stroke, the temperature is 300 K and pressure is 100 kPa. Thermal energy is supplied at 1840 kJ/kg. Given that C_v for air is 0.7176 kJ/kg K, determine:

(a) the temperature and pressure at the terminal points of all the processes
(b) the thermal efficiency
(c) the work done per kg air, and
(d) the mean effective pressure.

Solution It is given that $T_1 = 300$ K; $P_1 = 100$ kPa and $r_0 = \dfrac{V_1}{V_2} = 8$

(a) Process 1–2 is isentropic. Therefore,

$$Pv^\gamma = \text{const.}$$

$$T_2 = T_1 \left(\frac{V_1}{V_2}\right)^{\gamma-1} = 300(8)^{0.4} = 689.2 \text{ K}$$

$$P_2 = P_1 \left(\frac{V_1}{V_2}\right)^{\gamma} = 100(8)^{1.4} = 1837.9 \text{ kPa}$$

2–3 is a constant volume process. Hence,

$$q = u_3 - u_2 = C_v(T_3 - T_2)$$

Therefore,

$$1840 = 0.7176(T_3 - 689.2) \quad \text{or} \quad T_3 = 3253.5 \text{ K}$$

For a constant volume process,

$$\frac{P_3}{T_3} = \frac{P_2}{T_2}$$

$$P_3 = \frac{T_3 P_2}{T_2} = \frac{3253.5 \times 1837.9}{689.2} = 8676.1 \text{ kPa}$$

3–4 is an isentropic process. Hence

$$T_4 = T_3 \left(\frac{V_3}{V_4}\right)^{\gamma-1} = 3253.5 \left(\frac{1}{8}\right)^{0.4} = 1416.2 \text{ K}$$

$$P_4 = P_3 \left(\frac{V_3}{V_4}\right)^{\gamma} = 8676.1 \left(\frac{1}{8}\right)^{1.4} = 472.1 \text{ kPa}$$

(b) Thermal efficiency η is given by

$$\eta = 1 - \left(\frac{1}{r_0}\right)^{\gamma-1} = 1 - \left(\frac{1}{8}\right)^{0.4} = 0.565$$

(c) Work done = $q_1\eta = 1840(0.565) = 1039.6$ kJ/kg.

(d) $V_1 = \dfrac{NRT_1}{P_1} = \dfrac{8.314 \times 10^3 \times 300}{28.97 \times 10^5} = 0.861\ \text{m}^3/\text{kg}$

$V_2 = V_1/r_0 = 0.861/8 = 0.1076\ \text{m}^3/\text{kg}$

$P_m = 1039.6/(0.861 - 0.1076) = 1379.88$ kPa.

***Example* 9.5** In an air-standard Otto cycle, the conditions at the beginning of the compression stroke are 100 kPa and 37°C. The maximum pressure and temperature in the cycle are 7 MPa and 2700°C, respectively. Determine the compression ratio, the thermal efficiency and the net work/kg of air.

Solution Compression ratio,

$$r_0 = \frac{V_1}{V_2} = \frac{P_2T_1}{T_2P_1}$$

For a constant volume process,

$$\frac{P_2}{T_2} = \frac{P_3}{T_3}$$

Therefore,

$$r_0 = \frac{P_3T_1}{T_3P_1} = \frac{7(310.15)}{2973.15(0.1)} = 7.3022$$

Thermal efficiency, η is given by

$$\eta = 1 - \left(\frac{1}{r_0}\right)^{\gamma-1} = 1 - \left(\frac{1}{7.3022}\right)^{0.4} = 0.5485$$

$$T_2 = T_1\left(\frac{V_1}{V_2}\right)^{\gamma-1} = 310.15(7.3022)^{0.4} = 687\ \text{K}$$

Energy added as heat, q_1 is given by

$$q_1 = C_v(T_3 - T_2) = 0.7176(2973.15 - 687) = 1640.54\ \text{kJ/kg}$$

Net work, $W = q_1\eta = 1640.54(0.5485) = 899.84$ kJ/kg

9.9 Air-Standard Diesel Cycle

The Diesel cycle was developed by Rudolf Diesel (1858–1913) in 1893. This cycle is extensively used by large and small stationary engines. Most of the present day transport trucks, buses, a few locomotives and cars are operating on the Diesel cycle because of the low cost of the Diesel oil. In this cycle, high compression ratios can be employed using only air at the start. After the compression, the fuel is injected into the hot air. The combustion of the fuel occurs spontaneously. This type of engine is some times referred to as *compression ignition engine*. The Diesel cycle shown in Fig. 9.14 consists of the following processes.

- 0–1: Constant pressure suction during which fresh air is drawn into the cylinder as the piston executes the outward stroke.
- 1–2: The air is compressed isentropically. Usually the compression ratio in the Diesel cycle is much higher than in an Otto cycle. Because of the high compression ratio, the temperature of the gas at the end of the isentropic compression is so high that when fuel is injected, it gets ignited immediately.
- 2–3: The fuel (diesel oil) is injected into the hot compressed air at state 2 and the fuel undergoes a chemical reaction. The combustion of diesel oil in air is not as spontaneous as the combustion of gasoline, and the combustion is relatively slow. Hence, the piston starts moving outward as combustion takes place. The combustion process is conveniently approximated as a constant pressure energy addition process.
- 3–4: The combustion products undergo isentropic expansion and the piston executes an outward motion.
- 4–1: The combustion products are exhausted at constant volume when the discharge port opens. This is replaced by a constant volume energy rejection process.
- 1–0: The remaining combustion products are exhausted at constant pressure by the inward motion of the piston.

It can be seen from Fig. 9.14, that the net work of the processes 1–0 and 0–1 is zero. Hence they can be ignored for the purpose of thermodynamic analysis.

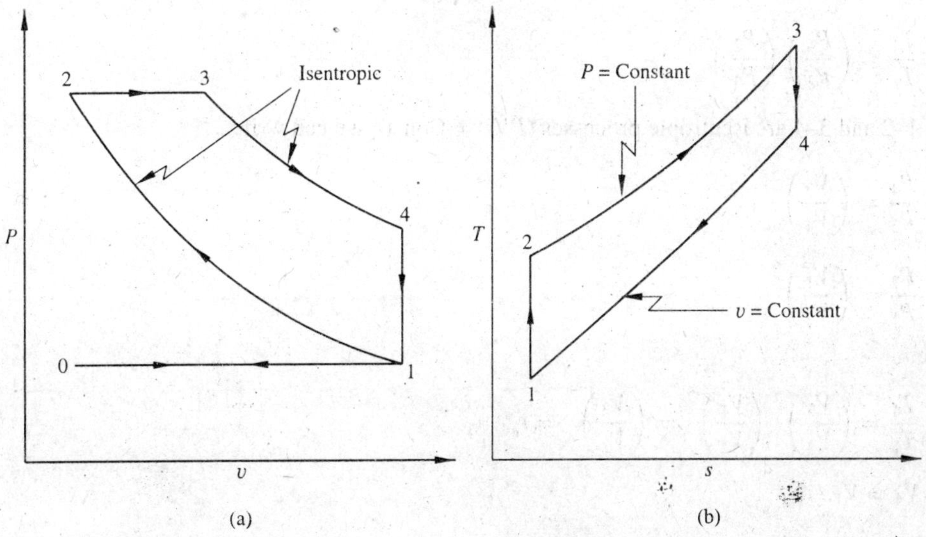

Fig. 9.14 *Air-standard Diesel cycle (a) P–v diagram and (b) T–s diagram*

Let us find an expression for estimating the thermal efficiency of the Diesel cycle. For this purpose, we have to evaluate the energy added and rejected as heat. In addition to the compression ratio, $r_0 = V_1/V_2$, another ratio called the cut-off ratio is used in the analysis of Diesel cycle. The *cut-off ratio* r_c is defined as the ratio of the volume at the end of the constant pressure energy addition process to the volume at the beginning of the energy addition process.

$$r_c = \frac{V_3}{V_2} \tag{9.30}$$

Energy added $= Q_1 = mC_P(T_3 - T_2)$
Energy rejected $= Q_2 = mC_v(T_4 - T_1)$
Therefore,

$$\eta = \frac{Q_1 - Q_2}{Q_1} = 1 - \frac{T_4 - T_1}{\gamma(T_3 - T_2)} = 1 - \frac{T_1(T_4/T_1 - 1)}{\gamma T_2(T_3/T_2 - 1)} \tag{9.31}$$

The process 1–2 is isentropic. Therefore,

$$\frac{T_1}{T_2} = \left(\frac{V_2}{V_1}\right)^{\gamma-1} = \left(\frac{1}{r_0}\right)^{\gamma-1} \tag{9.32}$$

The process 4–1 is a constant volume process. Therefore,

$$\frac{T_4}{T_1} = \frac{P_4}{P_1} = \left(\frac{P_4}{P_3}\right)\left(\frac{P_3}{P_1}\right) \tag{9.33}$$

But $P_2 = P_3$. Hence,

$$\frac{T_4}{T_1} = \left(\frac{P_4}{P_3}\right)\left(\frac{P_2}{P_1}\right) \tag{9.34}$$

Since 1–2 and 3–4 are isentropic processes ($PV^\gamma = \text{Const}$), we can write

$$\frac{P_4}{P_3} = \left(\frac{V_3}{V_4}\right)^{\gamma} \tag{9.35}$$

$$\frac{P_2}{P_1} = \left(\frac{V_1}{V_2}\right)^{\gamma} \tag{9.36}$$

Hence

$$\frac{T_4}{T_1} = \left(\frac{V_3}{V_4}\right)^{\gamma}\left(\frac{V_1}{V_2}\right)^{\gamma} = \left(\frac{V_3}{V_2}\right)^{\gamma} = r_c^{\gamma} \tag{9.37}$$

Since $V_4 = V_1$.

$$\eta = 1 - \frac{1}{\gamma r_0^{\gamma-1}}\left(\frac{r_c^{\gamma} - 1}{r_c - 1}\right) \tag{9.38}$$

***Example* 9.6** An engine operates on the air-standard Diesel cycle. The conditions at the start of the compression stroke are 323 K and 100 kPa while at the end of the stroke the pressure is 4.0 MPa. The energy absorbed is 600 kJ/kg of air. Given that C_P of air is 1.0047 kJ/kg K, calculate

(a) the compression ratio

(b) the cut-off ratio
(c) the thermal efficiency and
(d) the work done per kg air.

Solution (a) 1–2 is an isentropic process. Therefore,

$$\frac{V_1}{V_2} = r_0 = \left(\frac{P_2}{P_1}\right)^{1/\gamma} = \left(\frac{4}{0.1}\right)^{1/1.4} = 13.95$$

Compression ratio $= r_0 = 13.95$

(b)

$$T_2 = \frac{P_2 V_2 T_1}{P_1 V_1} = \frac{4(323)}{0.1(13.95)} = 926 \text{ K}$$

Energy absorbed $= q_1 = C_P(T_3 - T_2)$

$$600 = 1.0047(T_3 - 926) \quad \text{or} \quad T_3 = 1523.2 \text{ K}$$

2–3 is a constant pressure process. Therefore,

$$P_2 = P_3 = 4.0 \text{ MPa}$$

$$\frac{V_3}{V_2} = r_c = \frac{T_3}{T_2} = \frac{1523.2}{926} = 1.6449$$

The cut-off ratio $= r_c = 1.6449$

(c) The efficiency of the cycle is given by

$$\eta = 1 - \frac{1}{\gamma r_0^{\gamma-1}}\left(\frac{r_c^{\gamma} - 1}{r_c - 1}\right)$$

$$\eta = 1 - \frac{(1.6449)^{1.4} - 1}{1.4(13.95)^{0.4}(1.6449 - 1)} = 0.6112$$

(d) Work done $= \eta q_1 = 0.6112(600) = 366.72$ kJ/kg

***Example* 9.7** Consider an air-standard Diesel cycle with a compression ratio of 16. The conditions at the start of the compression stroke are 100 kPa and 37°C. The entropy change of the air during the energy addition process is 1.2 kJ/kg K. Determine the maximum temperature of the cycle, the cut-off ratio, the energy added as heat, and the thermal efficiency of the cycle.

Solution

$$T_2 = T_1 r_0^{\gamma-1} = 310.15(16)^{0.4} = 940.2 \text{ K}$$

$$s_3 - s_2 = C_P \ln\left(\frac{T_3}{T_2}\right)$$

$$1.2 = 1.0047 \ln \left(\frac{T_3}{940.2} \right) \quad \text{or,} \quad T_3 = 3104.1 \text{ K}$$

Cut-off ratio, r_c is given by

$$r_c = \frac{T_3}{T_2} = \frac{3104.1}{940.2} = 3.3015$$

Energy added as heat is given by:

$$q_1 = C_P(T_3 - T_2) = 1.0047(3104.1 - 940.2) = 2174.07 \text{ kJ/kg}$$

$$\eta = 1 - \frac{(3.3015)^{1.4} - 1}{1.4(16)^{0.4}(3.3015 - 1)} = 0.5574$$

Equation (9.38) shows that the thermal efficiency of a Diesel cycle can be increased either by increasing the compression ratio or by decreasing the cut-off ratio. Figure 9.15 shows the efficiency of a Diesel cycle as a function of the compression ratio and cut-off ratio for a gas with $\gamma = 1.4$. The compression ratios normally employed in Diesel engines vary from 14 to 17.

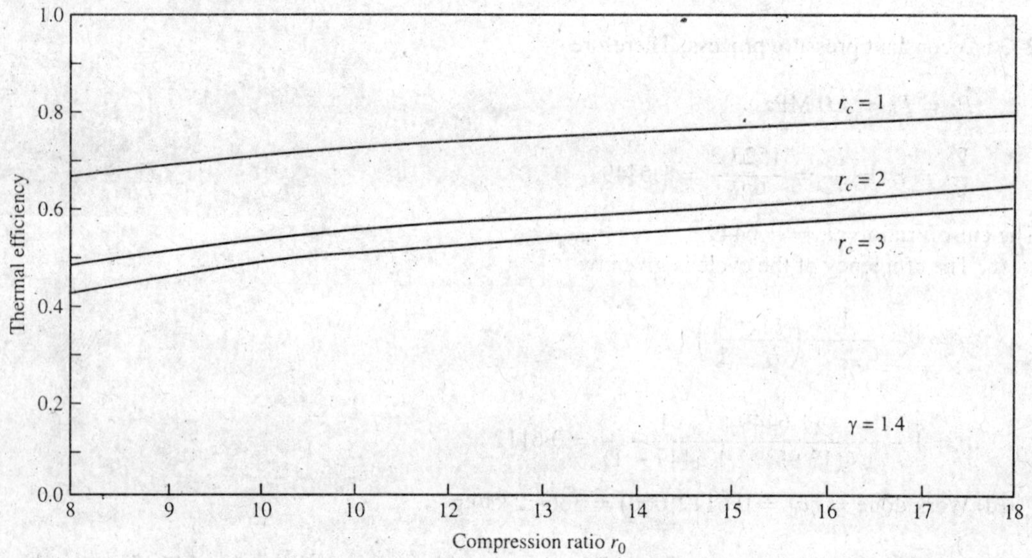

Fig. 9.15 *The dependence of the efficiency of an air-standard Diesel cycle on the compression ratio and cut-off ratio*

9.10 Comparison of Otto and Diesel Cycles

The Otto and Diesel cycles can be compared on the basis of either the compression ratio or the maximum temperature and pressure of the cycle. Figure 9.16 shows the Otto and Diesel cycles having the same compression ratio; both of them reject the same quantity of energy Q_2 to the surroundings. We know that

$$\int dQ = \int Tds$$

That is, the area under the curve on a T–s diagram represents the heat interaction. In Fig. 9.16, the Otto cycle is represented by 1234 and the Diesel cycle is represented by 123′4. Both the cycles have the same compression ratio v_2/v_1 and reject the same quantity of energy Q_2 during the constant volume process 4–1 (shown by the hatched area under the curve 4–1 on the T–s diagram). The efficiency of the cycle is given by

$$\eta = 1 - \frac{Q_2}{Q_1}$$

For the same value of Q_2, the efficiency will be higher for higher values of Q_1. On the T–s diagram, the area under the curve 2–3 represents Q_1 for the Otto cycle and the area under the curve 2–3′ represents Q_1 for the Diesel cycle. Since Q_1 for the Otto cycle is greater than that for the Diesel cycle, we find that

$$\eta_{\text{Otto}} > \eta_{\text{Diesel}}$$

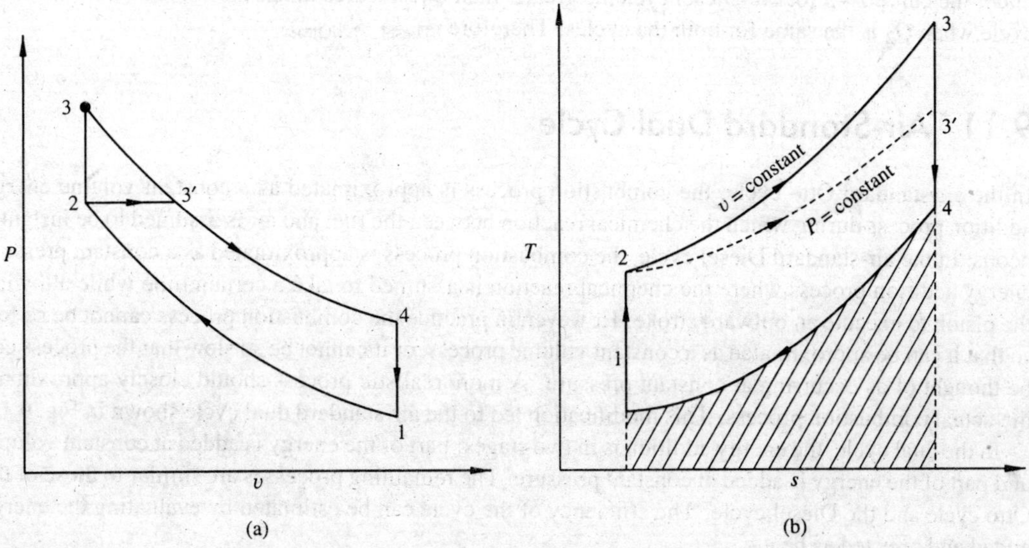

Fig. 9.16 *(a) P–v diagram for Otto cycle (1–2–3–4) and Diesel cycle (1–2–3′–4) (b) T–s diagram for Otto cycle (1–2–3–4) and Diesel cycle (1–2–3′–4)*

The Otto and Diesel cycles for the same maximum temperature and pressure (state 3) and for the same quantity of energy rejection Q_2 are shown in Fig. 9.17. The Otto cycle is represented by 1234 and the Diesel cycle is represented as 12′34. From the T–s diagram it can be observed that Q_1, the area

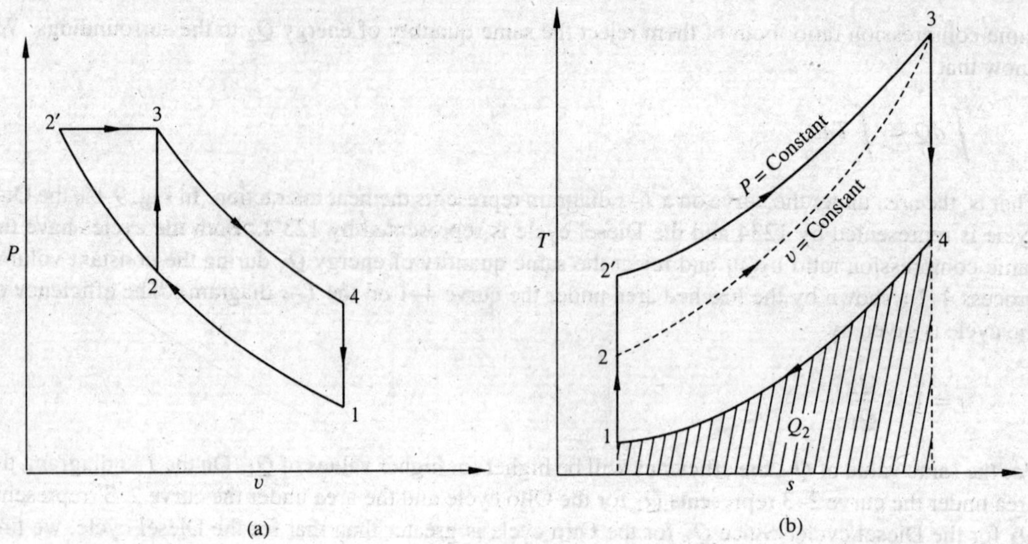

Fig. 9.17 *(a) P–v diagram for Otto cycle (1–2–3–4) and Diesel cycle (1–2′–3–4) (b) T–s diagram for Otto cycle (1–2–3–4) and diesel cycle (1–2′–3–4)*

under the curve 2′–3 for the Diesel cycle, is greater than Q_1, the area under the curve 2–3 for the Otto cycle when Q_2 is the same for both the cycles. Therefore $\eta_{\text{Diesel}} > \eta_{\text{Otto}}$.

9.11 Air-Standard Dual Cycle

In the air-standard Otto cycle, the combustion process is approximated as a constant volume energy addition process during which the chemical reaction between the fuel and air is assumed to be instantaneous. In the air-standard Diesel cycle, the combustion process is approximated as a constant pressure energy addition process where the chemical reaction is assumed to take a certain time while allowing the piston to execute an outward stroke. However, in practice the combustion process cannot be so fast so that it can be approximated as a constant volume process, or it cannot be so slow that the process can be thought of as occurring at constant pressure. A more realistic process should closely approximate the actual combustion process. This modification led to the air-standard dual cycle shown in Fig. 9.18.

In the dual cycle, the energy addition is in two stages: part of the energy is added at constant volume and part of the energy is added at constant pressure. The remaining processes are similar to those of the Otto cycle and the Diesel cycle. The efficiency of the cycle can be estimated by evaluating the energy added and rejected as heat.

Energy added q_1 is given by:

$$q_1 = C_v(T_3 - T_2) + C_P(T_4 - T_3) \tag{9.39}$$

Energy rejected q_2 is given by:

$$q_2 = C_v(T_5 - T_1) \tag{9.40}$$

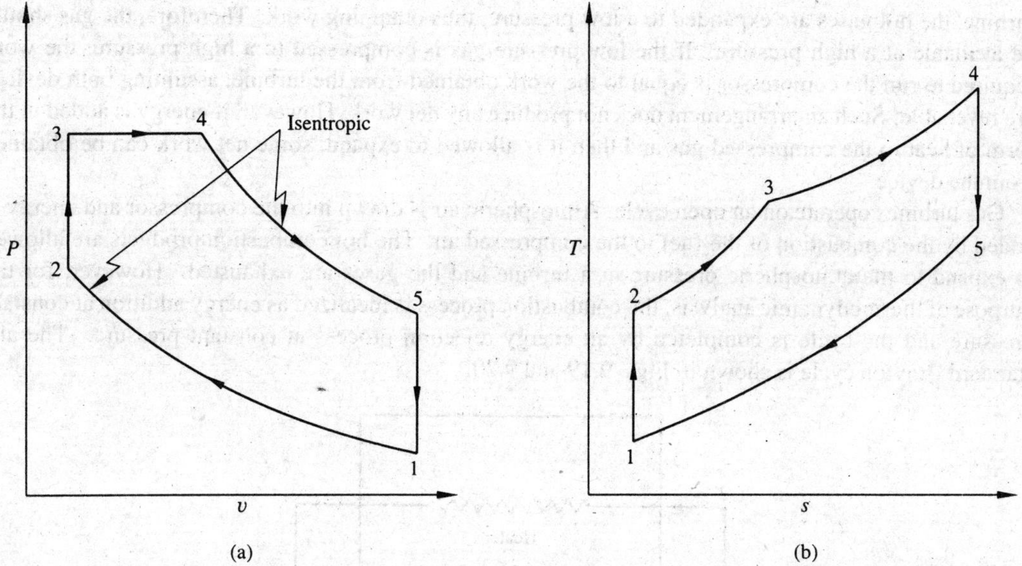

Fig. 9.18 *Air-standard dual cycle (a) P–v diagram and (b) T–s diagram*

Therefore, Thermal efficiency $= \eta = 1 - \dfrac{q_2}{q_1}$

$$\eta = 1 - \frac{C_v(T_5 - T_1)}{C_v(T_3 - T_2) + C_P(T_4 - T_3)}$$

$$\eta = 1 - \frac{(T_5 - T_1)}{(T_3 - T_2) + \gamma(T_4 - T_3)} \tag{9.41}$$

The efficiency can be expressed in terms of the following ratios also, and it is left as an exercise for the reader.

$$\text{Compression ratio} = r_0 = \frac{V_1}{V_2}$$

$$\text{Cut-off ratio} = r_c = \frac{V_4}{V_3} \tag{9.42}$$

The constant volume pressure ratio r_{vp} is defined as

$$r_{vp} = \frac{P_3}{P_2} \tag{9.43}$$

9.12 Air-Standard Brayton Cycle

The Brayton cycle was developed by George B. Brayton (1813–1892), a Boston engineer and was first used with reciprocating engines. This cycle is ideally suited and widely used for gas turbines. In a gas turbine, the hot gases are expanded to a low pressure, thus obtaining work. Therefore, the gas should be available at a high pressure. If the low pressure gas is compressed to a high pressure, the work required to run the compressor is equal to the work obtained from the turbine, assuming both devices are reversible. Such an arrangement does not produce any net work. However, if energy is added in the form of heat to the compressed gas and then it is allowed to expand, some net work can be obtained from the device.

Gas turbines operate on an open cycle. Atmospheric air is drawn into the compressor and energy is added by the combustion of the fuel to the compressed air. The hot combustion products are allowed to expand to the atmospheric pressure in a turbine and the gases are exhausted. However, for the purpose of thermodynamic analysis, the combustion process is idealized as energy addition at constant pressure and the cycle is completed by an energy rejection process at constant pressure. The air-standard Brayton cycle is shown in Figs. 9.19 and 9.20.

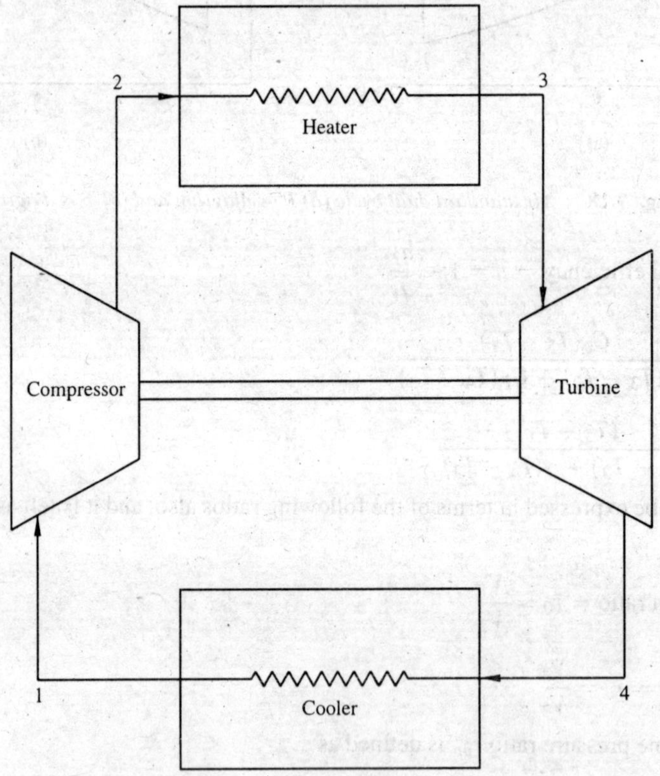

Fig. 9.19 *Schematic diagram of an air-standard Brayton cycle*

The air-standard Brayton cycle consists of:

- 1–2: Isentropic compression of the air.
- 2–3: Constant-pressure energy addition by the combustion of fuel.
- 3–4: Isentropic expansion of combustion products.
- 4–1: Constant-pressure energy rejection.

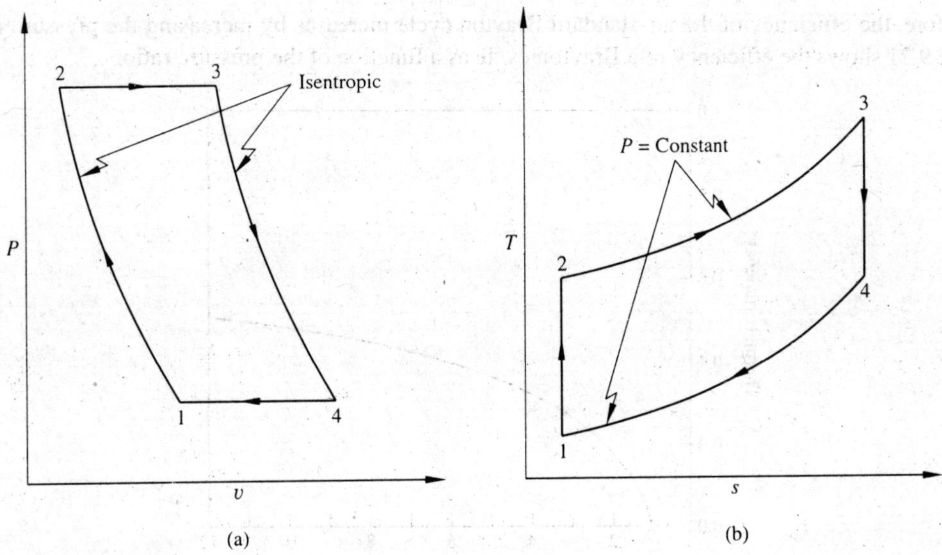

Fig. 9.20 *(a) P–v diagram and (b) T–s diagram of an air-standard Brayton cycle*

The efficiency of a Brayton cycle can be estimated in the usual way by evaluating the energy added and rejected as heat.

Energy added $= q_1 = C_P(T_3 - T_2)$

Energy rejected $= q_2 = C_P(T_4 - T_1)$

Thermal efficiency, η is given by:

$$\eta = 1 - \frac{T_4 - T_1}{T_3 - T_2}$$

$$\eta = 1 - \frac{T_1(T_4/T_1 - 1)}{T_2(T_3/T_2 - 1)} \tag{9.44}$$

For the Brayton cycle, the pressure ratio is defined as

$$r_p = \frac{P_2}{P_1} \tag{9.45}$$

Since $P_3/P_4 = P_2/P_1$ we obtain $T_3/T_2 = T_4/T_1$ Therefore

$$\eta = 1 - \frac{T_1}{T_2} \tag{9.46}$$

Since the process 1–2 is isentropic, we can write

$$\frac{P_2}{P_1} = \left(\frac{T_2}{T_1}\right)^{\gamma/\gamma-1} \tag{9.47}$$

The thermal efficiency can be expressed in terms of the pressure ratio as

$$\eta = 1 - \frac{1}{r_p^{\frac{\gamma-1}{\gamma}}} \tag{9.48}$$

Therefore, the efficiency of the air-standard Brayton cycle increases by increasing the pressure ratio. Figure 9.21 shows the efficiency of a Brayton cycle as a function of the pressure ratio.

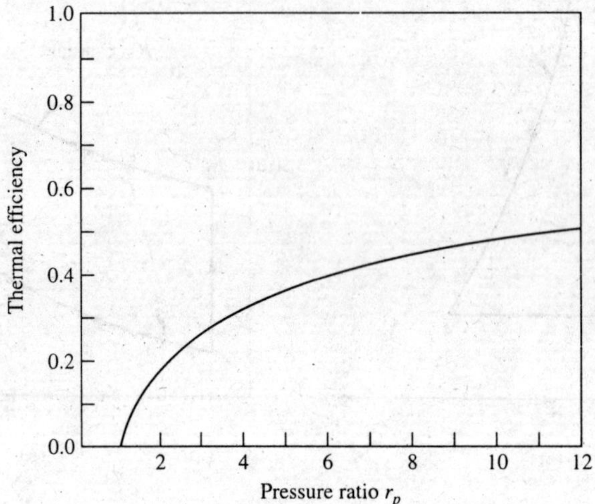

Fig. 9.21 *The dependence of the efficiency of a Brayton cycle on pressure ratio*

To increase the power output of the turbine, more energy can be added by the combustion of fuel at constant pressure. However, there is a metallurgical limit for the turbine blades and the maximum temperature in the cycle cannot exceed 1000°C. The metallurgical considerations and the ambient conditions limit the maximum (T_3) and minimum temperatures (T_1) of the cycle, respectively. Hence for fixed values of T_1 and T_3, there is an optimum pressure ratio which gives the maximum amount of work. Let us now evaluate the conditions required for optimum work output as follows.

$$\text{Net work done} = C_P(T_3 - T_4) - C_P(T_2 - T_1) \tag{9.49}$$

1–2 and 3–4 are isentropic processes, therefore

$$\frac{T_2}{T_1} = \left(\frac{P_2}{P_1}\right)^{\frac{\gamma-1}{\gamma}}$$

$$\frac{T_3}{T_4} = \left(\frac{P_3}{P_4}\right)^{\frac{\gamma-1}{\gamma}}$$

But

$$\frac{P_2}{P_1} = \frac{P_3}{P_4} = r_p$$

Hence

$$\frac{T_2}{T_1} = \frac{T_3}{T_4}$$

Then

$$W = C_P T_3(1 - T_4/T_3) - C_P T_1(T_2/T_1 - 1)$$

$$W = C_P T_3 \left(1 - \frac{1}{r_p^{\frac{\gamma-1}{\gamma}}}\right) - C_P T_1 \left(r_p^{\frac{\gamma-1}{\gamma}} - 1\right) \tag{9.50}$$

The optimum pressure ratio is obtained by differentiating W with respect to r_p, holding T_1 and T_3 constant, and setting it equal to zero. $\frac{dW}{dr_p} = 0$ gives

$$\text{optimum } r_p = \left(\frac{T_3}{T_1}\right)^{\frac{\gamma}{2\gamma-1}} \tag{9.51}$$

Example 9.8 In an air-standard Brayton cycle (see Fig. 9.20), the air enters the compressor at 300 K and 100 kPa. The pressure ratio is 8 and the maximum allowable temperature is 1300 K. Determine the temperature and pressure at each state in the cycle, the compressor work, the turbine work per kg of air and the cycle efficiency.

Solution

$$P_1 = 100 \text{ kPa}, \quad T_1 = 300 \text{ K}, \quad \frac{P_2}{P_1} = r_p = 8$$

Therefore $P_2 = 800$ kPa, and $T_2 = T_1 (r_p)^{\frac{\gamma-1}{\gamma}}$

$$T_2 = 300(8)^{0.4/1.4} = 543.4 \text{ K}$$

$$P_3 = P_2 = 800 \text{ kPa} \quad \text{and} \quad T_3 = T_{max} = 1300 \text{ K}$$

$$T_4 = T_3 \left(\frac{P_4}{P_3}\right)^{\frac{\gamma-1}{\gamma}} = 1300 \left(\frac{1}{8}\right)^{0.4/1.4} = 716.7 \text{ K}$$

$$P_4 = P_1 = 100 \text{ kPa}$$

Work done on the compressor is estimated by applying the first law to the compressor. That is

$$W_c = -(h_2 - h_1) = -C_P(T_2 - T_1)$$

$$= -1.0047(543.4 - 300) = -244.54 \text{ kJ/kg}$$

The work done by the turbine (W_T) is given by

$$W_T = h_3 - h_4 = C_P(T_3 - T_4)$$

$$= 1.0047(1300 - 716.7) = 586.04 \text{ kJ/kg}$$

Net work done $= 586.04 - 244.54 = 341.5$ kJ/kg

$$\text{Energy added} = q_1 = h_3 - h_2 = C_P(T_3 - T_2)$$

$$= 1.0047(1300 - 543.4) = 760.16 \text{ kJ/kg}$$

Thermal efficiency η is given by:

$$\eta = \frac{W_{net}}{q_1} = \frac{341.5}{760.16} = 0.4492$$

9.13 Air-Standard Stirling and Ericsson Cycles

9.13.1 Air-standard Stirling cycle

Even though the Stirling cycle and Ericsson cycle are not commonly used, a brief discussion of these cycles is presented below. These two cycles have certain common features. In both the cycles, the compression and expansion processes are carried out reversibly and isothermally and both the cycles use regenerators. The Stirling cycle is presented on the P–v and T–s diagrams in Fig. 9.22.

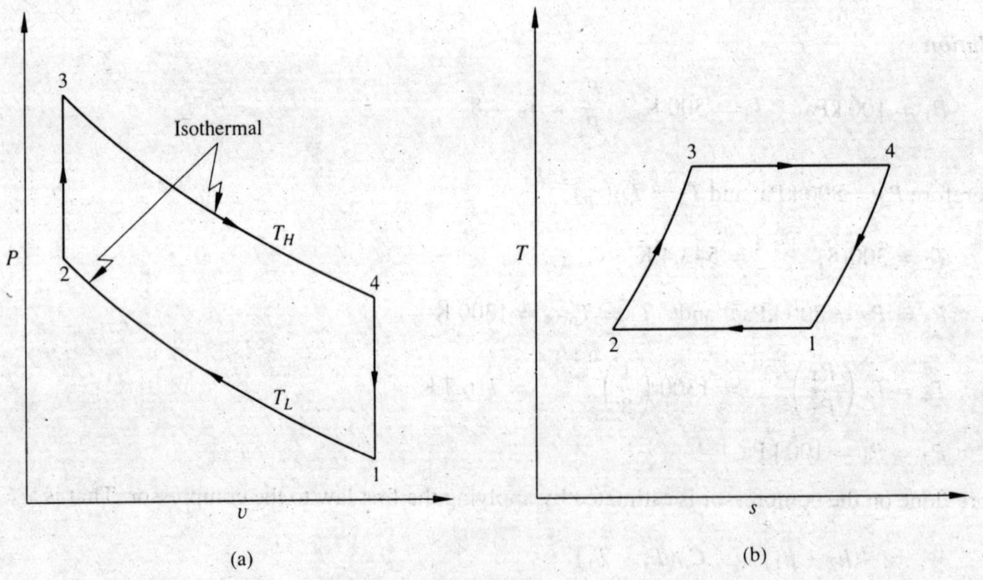

Fig. 9.22 *(a) P–v diagram and (b) T–s diagram of air-standard Stirling cycle*

In the Stirling cycle, energy addition takes place at a constant temperature from state 3 to state 4 when the gas undergoes expansion. Then the air enters a regenerator and is cooled at constant volume to reach state 1. The air in state 1 undergoes isothermal compression while rejecting energy and reaches state 2. The air then enters the regenerator and reaches state 3. The energy received from the regenerator during this process is $(u_3 - u_2)$ or $C_v(T_3 - T_2)$ taking air as an ideal gas. The regenerator receives energy $(u_4 - u_1)$ or $C_v(T_4 - T_1)$ which is equal to $C_v(T_3 - T_2)$ during the process 4–1 (since

$T_3 = T_4 = T_H$ and $T_2 = T_1 = T_L$). The main difficulty associated with the cycle is in the design and operation of the regenerator.

Let us estimate the thermal efficiency of the Stirling cycle. During the processes 2–3 and 4–1, there is no heat interaction with the surroundings. The heat interaction is present during the processes 1–2 and 3–4 only. By the application of the first law of thermodynamics, we get

$$q_1 = W_{34} = RT_H \ln(V_4/V_3)$$

$$q_2 = W_{21} = RT_L \ln(V_1/V_2)$$

Therefore thermal efficiency η is given by

$$\eta = 1 - \frac{T_L}{T_H} \tag{9.52}$$

It is interesting to note that the thermal efficiency of the Stirling cycle is the same as the efficiency of a Carnot cycle.

9.13.2 Air-standard Ericsson cycle

To overcome the difficulty associated with the constant volume regeneration, Ericsson developed a constant pressure regeneration cycle which is named after him. The air-standard Ericsson cycle is shown in Figs. 9.23 and 9.24. The thermal efficiency of the Ericsson cycle can be shown to be equal to $1 - T_L/T_H$ and is left to the reader to verify.

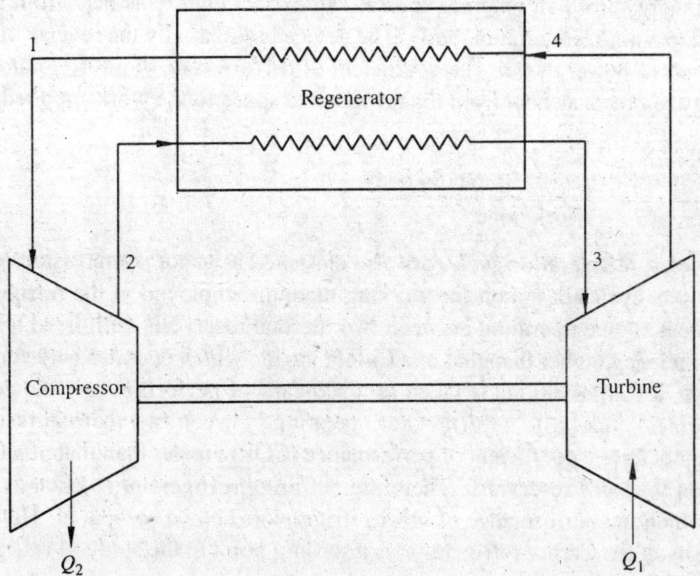

Fig. 9.23 *Schematic diagram of air-standard Ericsson cycle*

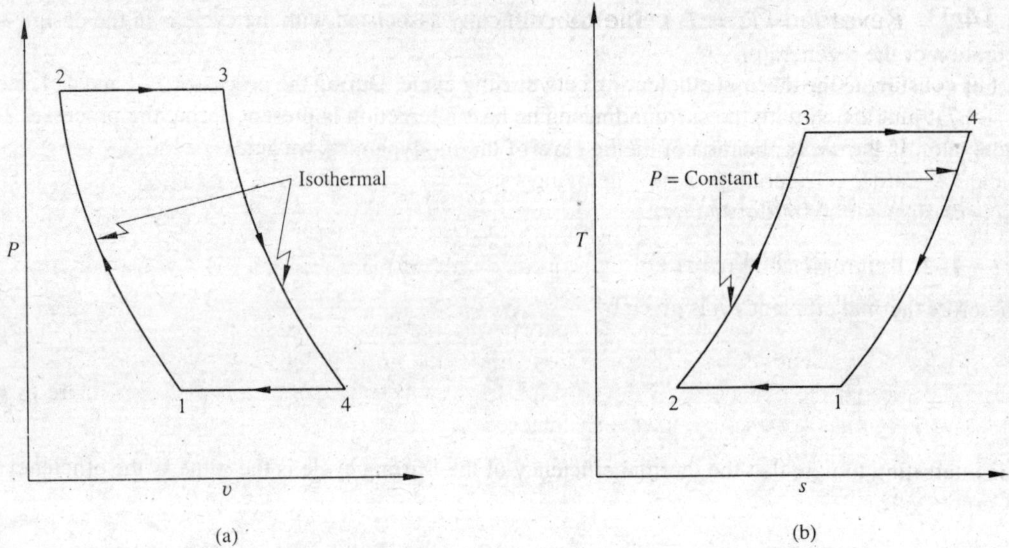

Fig. 9.24 *(a) P–v diagram and (b) T–s diagram of air-standard Ericsson cycle*

9.14 Refrigeration Cycles

Refrigeration is the production and maintenance of a temperature lower than the surrounding temperature. Such a process does not take place spontaneously. Hence, a device must be developed, which while absorbing energy from an external source, can extract energy as heat from a low temperature body and reject it to a high temperature body. The device is essentially the reverse of a heat engine and operates on a reversed power cycle. The coefficient of performance of a refrigerator is defined as the ratio of the energy removed as heat from the refrigerated space to the work supplied.

$$COP_R = \frac{\text{Energy extracted from cold body}}{\text{Work done}} \tag{9.53}$$

Like power cycles, refrigeration cycles are also classified as vapor compression refrigeration cycles and gas refrigeration cycles based on the working medium employed in the refrigerator. It has been observed that a heat engine operating between two thermal reservoirs with fixed temperatures cannot have a thermal efficiency higher than that of a Carnot engine which operates between the same thermal reservoirs. Hence, a Carnot engine is taken as a standard of perfection to evaluate the performance of other heat engines. Similarly, a refrigerator operating between two thermal reservoirs, with fixed temperatures, cannot have a coefficient of performance (COP) greater than that of a Carnot refrigerator operating between the same reservoirs. Therefore, a Carnot refrigerator is taken as a standard of perfection against which the performance of other refrigerators can be compared. Hence, it is natural to focus our attention on the Carnot refrigerator as a starting point in the study of refrigeration cycles.

9.14.1 Reversed Carnot cycle as a refrigeration cycle

Let us consider a Carnot engine operating between two reservoirs with temperatures T_H and T_L, where $T_H > T_L$. Let the working medium in the engine be a vapor which undergoes phase transition during the cycle. If the cycle is operated in the reverse order, the resulting cycle is a Carnot refrigeration cycle. A Carnot refrigeration cycle employing vapor as a working fluid is shown in Fig. 9.25. The cycle consists of the following processes.

- 1–2: Isentropic compression of the working fluid. At the end of the compression, the working fluid is saturated vapor at temperature T_H.
- 2–3: Isothermal energy rejection as heat (Q_H) to the surroundings at temperature T_H in a condenser. At the end of the energy rejection process the working fluid is saturated liquid.
- 3–4: Isentropic expansion of the working fluid to the low pressure. The temperature of the working fluid at the end of the expansion process is T_L.
- 4–1: Reversible isothermal energy transfer as heat (Q_L) from the low temperature reservoir at T_L to the working fluid.

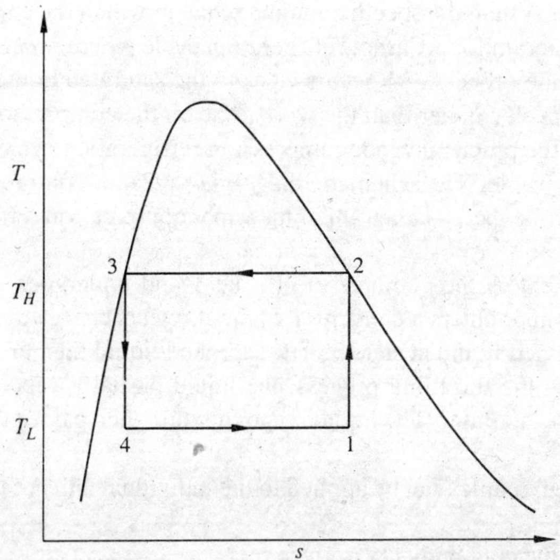

Fig. 9.25 *T–s diagram of a Carnot refrigeration cycle employing a vapor as a working fluid*

$$COP_R = \frac{Q_L}{W} = \frac{Q_L}{Q_H - Q_L} = \frac{T_L}{T_H - T_L}$$

The COP of the Carnot refrigerator indicates that the best performance of a refrigerator is obtained when the temperature difference between the reservoirs is as small as possible. Therefore, a refrigeration system designed for a relatively small temperature difference has higher COP compared to the refrigeration systems designed to operate with higher temperature differences.

9.15 Vapor Compression Refrigeration Cycle

The Carnot refrigeration cycle is not a practical cycle though the constant temperature energy rejection and absorption processes are desirable. The practical difficulty associated with the isentropic expansion of the saturated liquid is that the expander has to handle a two phase mixture of relatively low quality. Considering the compressor and the expander as the control volumes, the first and the second laws of thermodynamics can be written as

$$-\frac{\dot{W}}{\dot{m}_c} = h_2 - h_1 = -\int v dP \quad \text{and} \quad s_2 = s_1 \tag{9.54}$$

$$-(\dot{W}/\dot{m})_e = h_4 - h_3 = -\int v dP \quad \text{and} \quad s_4 = s_3 \tag{9.55}$$

where, $\dot{m}$ represents the mass flow rate of the working fluid (refrigerant). The above relations are written assuming that both the compressor and the expander are reversible. Even though the compressor and the expander work between the same pressure levels, the specific volume range in which the compressor works is much larger than the specific volume range in which the expander works. This can be observed from the T–s diagram for a Carnot refrigeration cycle where process 1–2 occurs close to the saturated vapor region while process 3–4 occurs close to the saturated liquid region. Hence, the work done by the expander is much smaller than the work done on the compressor. Generally the expander costs are high. Hence, in the practical vapor compression refrigeration cycle, the expander is replaced by an inexpensive throttle valve. The schematic diagram of a vapor compression refrigeration system is shown in Fig. 9.26(a) while the T–s diagram of the vapor compression refrigeration system is shown in Fig. 9.26(b).

Wet refrigerant vapor enters the compressor at state 1 and undergoes isentropic compression to state 2. The compressed vapor enters a condenser where it rejects energy as heat to the ambient atmosphere and leaves as saturated liquid at state 3. The saturated liquid then undergoes throttling through a valve to state 4. During the throttling process, the liquid partially vaporizes and attains the same temperature as that of the evaporator. The liquid–vapor mixture then passes through the evaporator and comes out at state 1.

The first law of thermodynamics can be applied to the individual units of the refrigerator to evaluate the COP.

Energy removed from the cold body $\dot{Q}_L$ is given by

$$\dot{Q}_L = \dot{m}(h_1 - h_4) \tag{9.56}$$

$$\text{Work done} = -\dot{W} = \dot{m}(h_2 - h_1) \tag{9.57}$$

$$COP_R = \frac{\dot{Q}_L}{\dot{W}} = \frac{h_1 - h_4}{h_2 - h_1} \tag{9.58}$$

Since the throttling process is isenthalpic, $h_3 = h_4$. But the process is irreversible. Therefore, $s_4 > s_3$. If the expansion of the liquid in state 3 is conducted reversibly, the final state would be 4′. Then the amount of energy removed as heat from the cold body is $T_L(s_1 - s_4')$. Because of the irreversibility, the energy removed from the cold body is $T_L(s_1 - s_4)$. Hence the shaded area $= T_L(s_4 - s_4')$ shown in Fig. 9.26(b) represents the loss in the refrigeration effect.

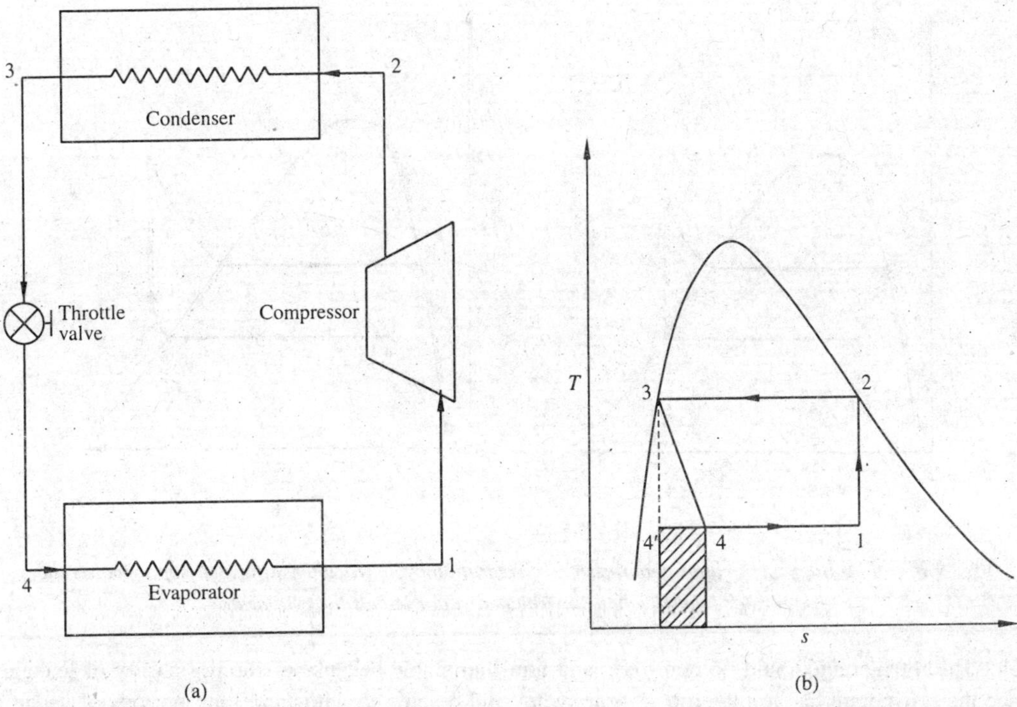

Fig. 9.26 *Vapor compression refrigeration cycle: (a) Schematic diagram and (b) T–s diagram*

In the ideal refrigeration cycle, the refrigerant leaving at state 1 is a mixture of liquid and vapor. However, in some cases it is possible that the refrigerant leaves the evaporator as a saturated vapor or a superheated vapor. In such cases, the refrigeration cycle appears as shown in Fig. 9.27.

9.16 Refrigerants

There are several refrigerants from which one has to choose depending on the design of a refrigerator. Of these fluorocarbon refrigerants are widely used. The important parameters that govern the selection of a refrigerant are the temperatures of the refrigerated space and that of the environment. In the vapor compression refrigeration system, the working fluid undergoes a phase change during the cycle. Is it possible to use any fluid as the working medium, or refrigerant, or are there any criteria in selecting a refrigerant? The first requirement is that the pressure at which the refrigerant enters the compressor should be slightly higher than the atmospheric pressure to make sure that no air leaks into the refrigerant. Then, it is also necessary that the boiling point of the refrigerant should be lower than the temperature at which the evaporator is to be maintained. Moreover, the refrigerant is compressed to a pressure where its saturation temperature is greater than the surrounding temperature, so that energy can be rejected to the surroundings. Moreover, to have reasonable heat transfer rates, a temperature difference of approximately 10°C is necessary between the refrigerant and the medium to which energy has to be rejected or absorbed from. Hence, the refrigerant critical pressure should be furthermore

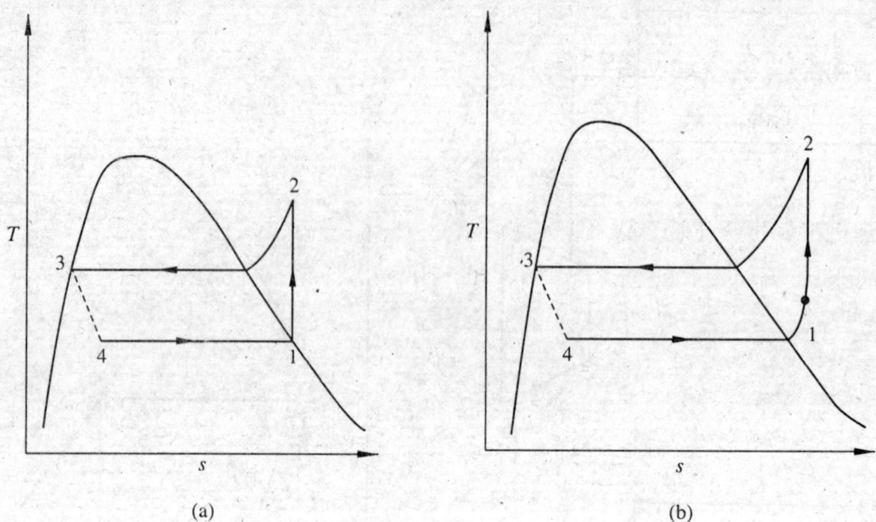

Fig. 9.27 *T–s diagram of vapor compression refrigeration cycle (a) when saturated vapor leaves the evaporator and (b) when superheated vapor leaves the evaporator*

high. The refrigerant should be non-toxic and non-flammable because of the possibility of leakage. Since the refrigerant has to undergo several cycles and be able to withstand long periods of usage it must be stable. Of course, it should also not be too expensive. In view of this, the commonly used refrigerants are ammonia, sulfur dioxide and Freons (halogenated hydrocarbons). The thermodynamic properties of Freon-12 (dichlorodifluoro methane CCl_2F_2) are presented in Appendices E and F.

Quite often the capacity of a refrigerating device is expressed in terms of tons of refrigeration. If a refrigerating device can convert 1 ton (2000 pounds mass = 907.18 kg) of water at 0°C into ice at the same temperature in 24 hours, its capacity is called one ton of refrigeration. Therefore,

$$\text{Ton of refrigeration} = \frac{907.18 \times 334.92}{24 \times 60} = 211 \text{ kJ/min} \tag{9.59}$$

(The latent heat of fusion of water at 0°C = 334.92 kJ/kg.)

***Example* 9.9** An ideal vapor compression refrigeration cycle shown in Fig. 9.26(b) with Freon-12 as the refrigerant is operated between the temperature levels −20°C and 40°C. If the flow rate of the refrigerant is 0.05 kg/s, calculate the COP of the refrigerator, the capacity of the refrigerator and the power required to run the refrigerator. Also calculate the COP of Carnot refrigerator working between the same temperature levels.

Solution The refrigerant is in the saturated vapor state at 40°C; hence,

$$h_2 = 203.1063 \text{ kJ/kg}, \quad s_2 = 0.6825 \text{ kJ/kg K}$$

At the evaporator temperature of −20°C:

$$s_f = 0.0736 \text{ kJ/kg K}; \quad \text{and} \quad s_g = 0.7094 \text{ kJ/kg K}$$

$$h_f = 17.9517 \text{ kJ/kg}, \quad \text{and} \quad h_g = 178.9017 \text{ kJ/kg}$$

Since the process 1–2 is isentropic:

$$s_2 = s_1 = s_g X_1 + (1 - X_1)s_f$$

$$0.6825 = 0.7094X_1 + (1 - X_1)0.0736 \quad \text{or} \quad X_1 = 0.958$$

$$h_1 = 0.958 \times 178.9017 + (1 - 0.958)17.9517 = 172.1418 \text{ kJ/kg}$$

The refrigerant leaves the condenser as saturated liquid at 40°C. Hence

$$h_3 = 75.1134 \text{ kJ/kg}.$$

3–4 is an isenthalpic process:

$$h_4 = h_3 = 75.1134 \text{ kJ/kg}$$

$$COP_R = \frac{h_1 - h_4}{h_2 - h_1} = \frac{172.1418 - 75.1134}{203.1063 - 172.1418} = 3.134$$

Capacity of the refrigerator (the rate of energy removal as heat from the evaporator)

$$= \dot{m}(h_1 - h_4) = 0.05(172.1418 - 75.1134) = 4.85 \text{ kW}.$$

Power required $= \dot{m}(h_2 - h_1) = 0.05(203.1063 - 172.1418) = 1.548$ kW

COP of Carnot refrigerator is given by:

$$\text{COP} = \frac{T_L}{T_H - T_L} = \frac{253}{60} = 4.217$$

***Example* 9.10** A vapor compression refrigeration system employing Freon-12 works between pressure limits of 150 kPa and 960 kPa. The energy transfer as heat from the condenser is found to be 300 kJ/min while the capacity of the refrigerator is 1 ton. The refrigerant vapor leaves the evaporator in the saturated state. Calculate:

(a) the refrigerant flow rate through the system in kg/min
(b) the energy input to the compressor, and
(c) the COP of the system.

Solution

(a) $h_1 = h_g$ at 150 kPa = 178.8372 kJ/kg
$h_4 = h_3$ at 960 kPa = 75.1013 kJ/kg
Energy absorbed from the evaporator

$$= \dot{m}(178.8372 - 75.1013) = 211 \text{ kJ/min} \quad \text{or} \quad \dot{m} = 2.034 \text{ kg/min}$$

(b) Energy transferred as heat from the condenser = Energy absorbed from the evaporator + work done on the compressor

or, $300 = 211 + \dot{W}$ or, $\dot{W} = 89$ kJ/min $= 1.483$ kW

(c)

$$COP_R = \frac{\text{Energy absorbed from evaporator}}{\text{work done on compressor}} = \frac{211}{89} = 2.371$$

9.17 Absorption Refrigeration System

In the vapor compression refrigeration system, the energy required for compressing the refrigerant from a low pressure to a high pressure is supplied in the form of work (electrical or mechanical). The work done is converted into thermal energy and is rejected to the surroundings along with the energy extracted from the evaporator.

Energy extracted from evaporator $= \dot{m}(h_1 - h_4)$
Work done on the compressor $= \dot{m}(h_2 - h_1)$
Energy rejected at the condenser $= \dot{m}(h_2 - h_3)$

$$= \dot{m}(h_1 - h_4) + \dot{m}(h_2 - h_1) = \dot{m}(h_2 - h_4)$$

and $h_4 = h_3$, since 3–4 is an isenthalpic process.

We have learned that all the energy absorbed in the form of heat by a cyclic engine cannot be converted into work, and a part of it is to be rejected to a low temperature sink. That is, complete conversion of thermal energy into work is not possible; but work can be completely converted into thermal energy. Therefore, work is hundred percent available energy whereas thermal energy is not. Hence, it is possible to effect some saving in the available energy if the energy required to operate the refrigerator is supplied in the form of heat rather than in the form of work.

In the absorption refrigeration system, refrigeration is produced mainly by the use of energy as heat. In such a system, the refrigerant is usually dissolved in a liquid. A common refrigerant is ammonia dissolved in water. A schematic diagram of an ammonia absorption refrigeration system is shown in Fig. 9.28.

A concentrated solution of ammonia is boiled in a vapor generator producing ammonia vapor at high pressure. The high pressure ammonia vapor is fed to a condenser where it is condensed to liquid ammonia by rejecting energy as heat to the surroundings. Then, the liquid ammonia is throttled through a valve to a low pressure. During throttling, ammonia is partially vaporized and its temperature decreases. This low temperature ammonia is fed to an evaporator where it is vaporized removing energy from the evaporator. Then, this low pressure ammonia vapor is absorbed in the weak solution of ammonia which leaves the generator. The resulting strong ammonia solution is pumped back to the vapor generator and the cycle is complete. A detailed analysis of the ammonia absorption refrigeration system requires a knowledge of the thermodynamic properties of a binary mixture as a function of its composition. Such a detailed analysis is beyond the scope of this book. However, the COP of the absorption refrigeration system can be evaluated by considering it as a combination of a heat pump and a heat engine as shown in Fig. 9.29.

Let Q_1 represent the quantity of energy supplied as heat in the generator at temperature T_1. This energy can be given to a reversible engine which works between the generator temperature T_1 and the surrounding temperature T_2. Then the amount of work that can be obtained from such an engine is given by

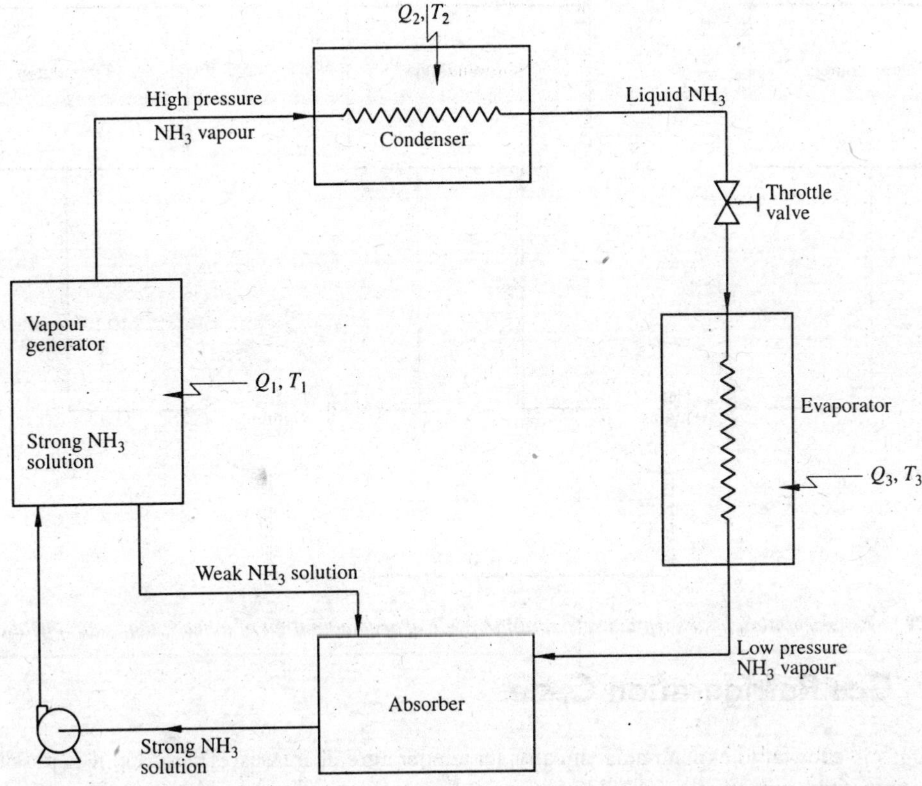

Fig. 9.28 *Schematic diagram of ammonia absorption refrigeration system*

$$W = Q_1 \frac{T_1 - T_2}{T_1} \tag{9.60}$$

Let this work be made use of to run a heat pump between the evaporator at T_3 and the surroundings at T_2. Let Q_3 represent the amount of energy extracted from the evaporator. Then for the heat pump

$$\frac{W}{Q_3} = \frac{T_2 - T_3}{T_3} \quad \text{or} \quad W = Q_3 \frac{T_2 - T_3}{T_3} \tag{9.61}$$

Therefore

$$W = Q_3 \frac{T_2 - T_3}{T_3} = Q_1 \frac{T_1 - T_2}{T_1} \tag{9.62}$$

The coefficient of performance of the absorption refrigeration system (neglecting the small amount of energy supplied to the pump) is given by

$$\frac{Q_3}{Q_1} = \frac{T_3(T_1 - T_2)}{T_1(T_2 - T_3)} \tag{9.63}$$

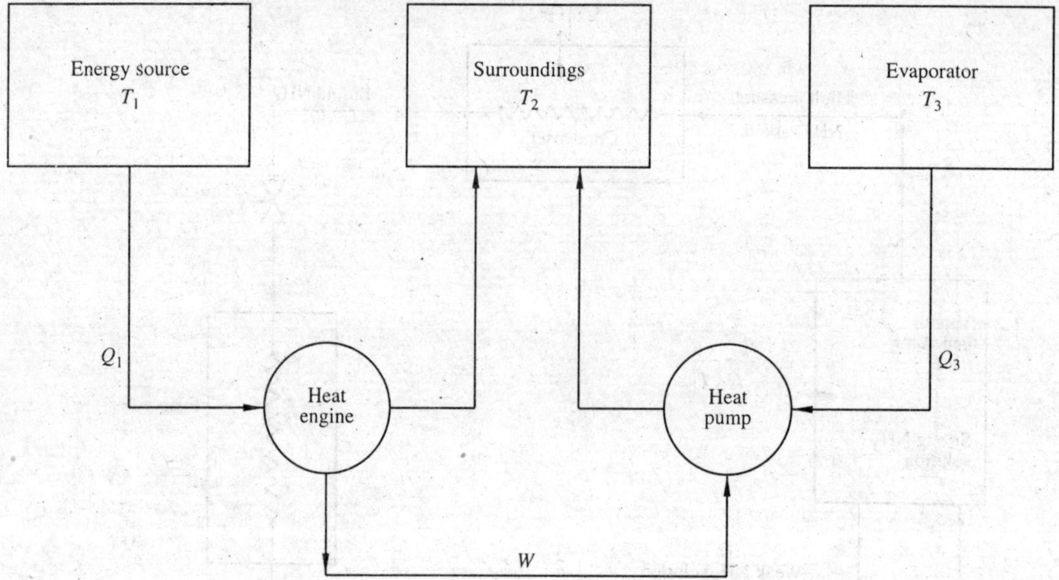

Fig. 9.29 *Representation of absorption refrigeration cycle as a combination of a heat pump and a heat engine*

9.18 Gas Refrigeration Cycle

When a gas is allowed to expand adiabatically, its temperature decreases. Therefore, it is possible to produce cooling by expanding a high pressure gas to low pressure. The simplest gas refrigeration cycle is the reversed Brayton cycle shown in Fig. 9.30.

The atmospheric air at state 1 is compressed isentropically to state 2. The compressed air is cooled at constant pressure to state 3 by exchanging energy with the surroundings. Then the air is allowed to expand isentropically to state 4. Because of the adiabatic expansion, the temperature of the gas is reduced. The work obtained during the expansion process can be used to run the compressor.

$$\text{Work done on compressor, } W_C = h_2 - h_1 = C_P(T_2 - T_1) \tag{9.64}$$

$$\text{Work delivered by the expander } = W_e = h_3 - h_4 = C_P(T_3 - T_4) \tag{9.65}$$

Therefore, the net work required is given by

$$W = C_P(T_2 - T_1 - T_3 + T_4) \tag{9.66}$$

This amount of work is supplied to the compressor from an external source. The cold air available at T_4 is then fed to the cooler where it absorbs energy as heat at constant pressure from the body to be refrigerated. The air leaving the cooler at state 1 is recycled. The coefficient of performance of this refrigeration system is given by

$$\text{COP} = \frac{Q_L}{W} = \frac{T_1 - T_4}{T_2 - T_1 - T_3 + T_4} \tag{9.67}$$

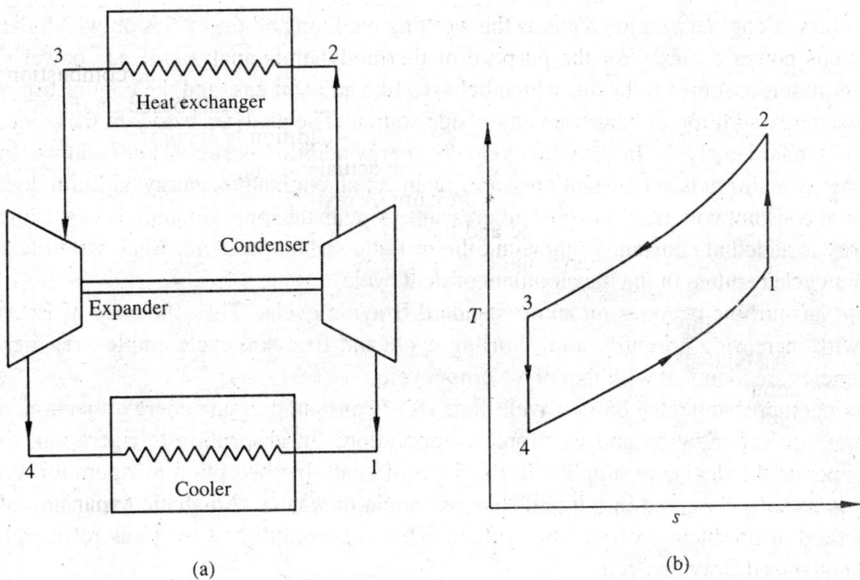

Fig. 9.30 *(a) Schematic diagram (b) T–s diagram of a gas refrigeration cycle*

9.19 Summary

Analysis of power cycles and refrigeration cycles provides valuable information regarding the design of new cycles and for improving the existing cycles. The power cycles and refrigeration cycles are classified into two groups as: Vapor power cycles/Refrigeration cycles in which the working fluid undergoes a phase change during the cyclic process; and Gas power cycles/Refrigeration cycles in which the working fluid does not undergo any phase change.

The ideal cycle for thermal power plants is the Rankine cycle. The ideal Rankine cycle is a modification of the Carnot cycle in which the isentropic compression of wet steam is replaced by isentropic compression of saturated liquid by employing a pump. In the Rankine cycle the energy addition as heat to water takes place in a steam generator at constant pressure. Then the steam undergoes isentropic expansion in a turbine, followed by a constant pressure energy rejection as heat in a condenser. The cycle is completed by isentropic compression of saturated liquid to steam generator pressure. In a practical Rankine cycle the performance is improved by superheating the steam before it is fed to a turbine. Objectionable moisture content in the turbine exhaust is avoided by introducing the reheat modification in the Rankine cycle. In the reheat modification, the steam is extracted from the turbine at an intermediate pressure and reheated before it is fed to another turbine for further expansion. The average temperature at which energy addition takes place can also be increased by preheating the feed water before it enters the steam generator. For this purpose the regenerative Rankine cycle was devised in which part of the steam is withdrawn at some intermediate stage in the turbine and used to preheat the feed water. In modern thermal power plants several stages of reheating and regeneration are employed to improve the efficiency of the power plant. With a view of improving the energy conversion efficiency, supercritical and binary vapor power cycles were developed.

A large class of engines employ a gas as the working medium and the cycles on which they operate are termed gas power cycles. For the purpose of thermodynamic analysis of gas power cycles the working medium is assumed to be air, which behaves like an ideal gas, and the combustion process is replaced by energy addition as heat from an outside source. The analysis based on these idealizations is called air-standard analysis. In the Otto cycle the energy addition is at constant volume. In a Diesel cycle the energy addition is at constant pressure. In an actual engine the energy addition does not take place either at constant volume or at constant pressure. A realistic approximation is to assume that part of the energy is added at constant volume and the rest at constant pressure. Such modification in the analysis of a cycle resulted in the development of dual cycle.

A simple gas turbine operates on an air-standard Brayton cycle. The efficiency of Brayton cycle increases with increasing pressure ratio. Stirling cycle and Ericsson cycle employ regenerators and their efficiencies are identical with that of a Carnot cycle.

The vapor compression refrigeration cycle consists of constant pressure energy rejection, throttling, constant pressure vaporization and isentropic compression. In absorption refrigeration, the energy needed to operate the device is supplied in the form of heat. In absorption refrigeration system, the refrigerant is usually dissolved in a liquid (like ammonia in water). Adiabatic expansion of gases is another method of producing refrigeration effect. This is accomplished by a gas refrigeration cycle which is the reversed Brayton cycle.

9.20 Review Questions

1. What are the main objectives in a power cycle and in a refrigeration cycle?
2. Classify the power cycles based on the combustion of fuel.
3. Classify the power cycles based on the nature of the working medium.
4. Why the Carnot cycle cannot be used for steam power plants?
5. What modifications have been made in the Carnot vapor power cycle while developing the Rankine cycle?
6. Sketch an ideal Rankine cycle on a T–s diagram.
7. Why the thermal efficiency of an ideal Rankine cycle is lower than that of a Carnot cycle operating between the same temperature levels?
8. How does a practical Rankine cycle differ from an ideal Rankine cycle?
9. Compare the thermal efficiencies of ideal Rankine cycle and practical Rankine cycle. Why should they differ from each other?
10. Define the isentropic efficiency of a turbine.
11. Define the isentropic efficiency of a pump.
12. Why is it undesirable to have excessive moisture content in steam turbine? What is the maximum allowed moisture content in a steam turbine?
13. Does the pressure and temperature of steam increase in a steam generator?
14. What are the methods to increase the thermal efficiency of a power cycle? Which of these methods are not practical for a Rankine cycle and why?
15. Sketch three Rankine cycles, on T–s diagram, operating between the same maximum and minimum temperature levels, but with different steam generator pressures.
16. Discuss the drawbacks associated with an increase in the pressure of steam generator.

17. Describe a reheat Rankine cycle and explain how it is likely to increase the thermal efficiency of the cycle.
18. Does the efficiency of a Rankine cycle change by reheat modification?
19. Describe a regenerative Rankine cycle and explain how it is likely to increase the thermal efficiency of the cycle.
20. What is a supercritical vapor power cycle? What are the advantages of this cycle?
21. What is a binary vapor power cycle? What advantages it has over a simple Rankine cycle?
22. What is an internal combustion engine and how does it differ from Rankine cycle?
23. What is an air-standard cycle?
24. How the combustion and exhaust processes are approximated in air-standard cycles?
25. What is the difference between spark ignition and compression engines?
26. How do the thermal efficiencies of air-standard Otto cycle and Carnot cycle compare for the same temperature limits? Explain.
27. Why is it not possible to use high compression ratios in Otto cycle?
28. Sketch an air-standard Otto cycle on P–v and T–s diagrams.
29. Define the compression ratio.
30. How does the thermal efficiency of an Otto cycle depend on the compression ratio?
31. What is knocking in an IC engine?
32. Does the thermal efficiency of an Otto engine depend on whether the gas or working fluid is monatomic or diatomic? Explain.
33. Is it possible to increase the compression ratio indefinitely in an Otto engine with a view of increasing its thermal efficiency? If not specify the reasons.
34. Define the mean effective pressure.
35. What is a compression ignition engine?
36. Sketch an air-standard Diesel cycle on P–v and T–s diagrams.
37. How does the Diesel cycle differ from Otto cycle?
38. What are the advantages associated with the Diesel cycle as compared to the Otto cycle?
39. Define the cut-off ratio for an air-standard Diesel cycle and what does it signify?
40. For a fixed value of compression ratio, how does the cut-off ratio affect the thermal efficiency of a Diesel cycle?
41. For fixed values of compression ratio and energy rejection, compare the thermal efficiencies of Otto and Diesel cycles.
42. Sketch the Otto and Diesel cycles, having the same maximum temperature and pressure and the same energy rejection, on P–v and T–s diagrams and compare their thermal efficiencies.
43. Sketch an air-standard dual cycle on P–v and T–s diagrams.
44. How does the dual cycle differ from Otto and Diesel cycles?
45. Define the constant-volume pressure ratio for a dual cycle.
46. Define the pressure ratio for a Brayton cycle.
47. Sketch an air-standard Brayton cycle on P–v and T–s diagrams and derive a relation to estimate the thermal efficiency of a Brayton cycle in terms of pressure ratio.
48. Derive a relation to estimate the optimum pressure ratio in a Brayton cycle (for obtaining maximum work output) for fixed values of maximum and minimum temperatures in the cycle.
49. Sketch the air-standard Stirling cycle on P–v and T–s diagrams and derive a relation to estimate the thermal efficiency of the cycle.
50. How does the Ericsson cycle differ from Stirling cycle?

51. What is meant by refrigeration?
52. How the refrigeration cycles are classified based on the working medium?
53. Which refrigeration cycle is taken as a standard of perfection against which the other refrigeration cycles are compared?
54. Does the COP of a refrigerator increase or decrease if the temperature difference between the high temperature body and low temperature body, is increased?
55. Explain why the Carnot vapor compression refrigeration cycle is not practical.
56. What modifications have been made in the Carnot vapor compression refrigeration cycle to obtain a practical vapor compression refrigeration cycle?
57. Suppose a refrigerator and heat pump operate between the same temperature levels, for which device will the COP be higher?
58. Sketch an ideal vapor compression refrigeration cycle on a *T–s* diagram.
59. Why a throttle valve is used in place of an isentropic turbine in the vapor compression refrigeration cycle?
60. In a practical vapor compression refrigeration cycle, the saturated liquid leaving the condenser is throttled to low pressure instead of subjecting it to isentropic expansion in a turbine. Does this modification lead to any loss in the refrigeration capacity? Explain with the help of a *T–s* diagram.
61. What is a ton of refrigeration?
62. What are the important considerations in the selection of a refrigerant?
63. Explain the ammonia absorption refrigeration system.
64. Derive a relation to estimate the COP of an absorption refrigeration system.
65. What is the advantage of using absorption refrigeration system over the vapor compression refrigeration system?
66. Sketch a gas refrigeration cycle on *T–s* diagram and derive a relation to estimate the COP of this cycle.
67. What is the advantage of employing a reversed Brayton cycle for cooling an air–craft?

9.21 Problems

9.1 In a power plant operating on a Rankine cycle, steam at 100 bar enters a turbine and expands to 0.1 bar. Determine the minimum temperature of the steam which enters the turbine to ensure a quality of 0.9 at the exit of the turbine.

9.2 In a thermal power plant steam enters a turbine at 500°C and the condenser is maintained at 0.1 bar. It is required that the quality of steam at the turbine exit should be at least 0.9. Determine the pressure at which steam should be supplied to the turbine.

9.3 Determine the efficiency of a steam power plant operating on a Carnot cycle with isothermal energy addition as heat at 3 MPa and an isothermal energy rejection as heat at 5 kPa. At the beginning of the isothermal energy addition the fluid is a saturated liquid and at the end of the isothermal energy addition it is in a saturated vapor state.

9.4 In the test of a turbine in a thermal power plant, the following data has been recorded. Superheated steam at 3 MPa and 300°C enters the turbine. The steam leaves the turbine at 5 kPa with a moisture content of 0.15. Determine the isentropic efficiency of the turbine.

9.5 The test results for a small steam turbine show that steam enters the turbine at 5 MPa and 500°C, and leaves the turbine at 100 kPa and 150°C. Determine the isentropic efficiency of the turbine.

9.6 A thermal power plant operates on a Rankine cycle. Superheated steam at 5 MPa and 400°C enters the turbine where it expands to the condenser pressure of 5 kPa. The saturated liquid leaving the condenser is fed back to the steam generator. Determine the thermal efficiency of the power plant and the steam production rate if the power output of the plant is 1 MW.

9.7 In an ideal Rankine cycle steam at 40 bar and 400°C enters a turbine and leaves at 10 kPa. If the power output of the turbine is 10 MW, determine

(a) The mass flow rate of steam through the turbine
(b) The thermal efficiency of the cycle and
(c) The quality of steam leaving the turbine.

9.8 In a thermal power plant, operating on a Rankine cycle, steam at 50 bar and 500°C enters a turbine, the isentropic efficiency of which is 0.85. The condenser operates at 10 kPa pressure. Neglecting the workdone on the pump, calculate the thermal efficiency of the cycle.

9.9 In an ideal Rankine cycle steam at 150 bar and 500°C enters a turbine and the condensers operate at 10 kPa. If the power output of the cycle is 100 MW, determine

(a) The mass flow rate of steam
(b) The thermal efficiency of the cycle and
(c) The mass flow rate of cooling water through the condensers, if the water enters the condensers at 30°C and leaves at 40°C.

9.10 A thermal power plant operates on an ideal Rankine cycle between the pressure limits 90 bar and 10 kPa. The moisture content of the steam at the turbine exit should not exceed 10 percent. If the flow rate of steam through the cycle is 100 kg/s, determine

(a) The minimum temperature at which the steam should enter the turbine and
(b) The thermal efficiency of the cycle.

9.11 It is desired to study the effect of the condenser pressure on the efficiency of a Rankine cycle. Superheated steam at 3 MPa and 300°C enters the turbine. Calculate the thermal efficiency of the Rankine cycle if the exhaust pressure of the turbine is adjusted to 5 kPa, 10 kPa, 50 kPa and 100 kPa. Discuss the results from the available energy point of view.

9.12 Superheated steam at 400°C enters the turbine of a thermal power plant operating on a Rankine cycle. The exhaust steam leaves the turbine at 5 kPa. It is desired to study the effect of pressure at the turbine inlet on the efficiency of the Rankine cycle and the quality of steam leaving the turbine. Calculate the Rankine cycle thermal efficiency and the moisture content of the exhaust steam if the turbine inlet pressure is 1 MPa, 2 MPa, 5 MPa and 8 MPa.

9.13 Superheated steam at 3 MPa enters the turbine of a thermal power plant operating on a Rankine cycle. The exhaust steam leaves the turbine at 5 kPa. It is desired to study the effect of superheat of the steam on the Rankine cycle efficiency. Calculate the thermal efficiency of the cycle and the moisture content of the exhaust steam, if the superheated steam enters the turbine at 300°C, 400°C, 500°C and 600°C.

9.14 A thermal power plant operates on a Rankine cycle. Superheated steam at 3 MPa and 350°C enters the turbine and leaves at 5 kPa. The isentropic efficiency of the turbine is 0.85. Determine the thermal efficiency of the power plant and the steam production rate if the power output of the plant is 1 MW.

9.15 In a thermal power plant operating on a Rankine cycle, superheated steam at 3 MPa and 350°C enters a turbine and leaves at 5 kPa. The saturated liquid leaving the condenser is compressed to the steam generator pressure by a pump whose isentropic efficiency is 0.75. The isentropic efficiency of the turbine is 0.85. Determine the thermal efficiency of the power plant and the steam circulation rate if the power output of the plant is 1 MW.

9.16 In a power plant employing Rankine cycle with reheat, superheated steam at 150 bar and 500°C enters the first stage of the turbine. The condenser is maintained at 0.1 bar. The exit steam from the first stage of the turbine is reheated to 500°C before it is fed to the second stage of the turbine. Calculate the thermal efficiency of the power plant if steam expands to (a) 90 bar in the first stage of the turbine, and (b) 60 bar in the first stage of the turbine.

9.17 A thermal power plant operates on a Rankine cycle with reheat modification. Superheated steam enters the turbine at 3.0 MPa and 350°C. After expansion to 500 kPa, the steam is reheated to 350°C and expanded in a second turbine to the condenser pressure of 5 kPa. Calculate the thermal efficiency of the plant if the isentropic efficiency of each turbine is 0.85, and the isentropic efficiency of the pump is 0.75.

9.18 In a thermal power plant, operating on a reheat Rankine cycle, superheated steam enters the turbine at 3.0 MPa and 600°C. After expansion to 1 MPa, the steam is reheated to 600°C and expanded in a second turbine to the condenser pressure of 5 kPa. Calculate the thermal efficiency of the cycle if the isentropic efficiency of each turbine is 0.85 and the isentropic efficiency of the pump is 0.75.

9.19 A thermal power plant operates on a regenerative Rankine cycle. Superheated steam at 3.5 MPa and 500°C enters the turbine. A suitable fraction of the steam is extracted from the turbine at 0.6 MPa and the remaining steam in the turbine is allowed to expand to the condenser pressure of 10 kPa. The extracted steam is mixed with the condensate after it has been raised to 0.6 MPa and the mixture is fed to the feed pump. Calculate the thermal efficiency of the power plant.

9.20 In a supercritical steam power plant, the feed pump raises the pressure of water from the condenser pressure of 50 kPa to the steam generator pressure of 25 MPa. Steam at 25 MPa and 600°C enters the turbine and is expanded to the condenser pressure of 50 kPa. If the isentropic efficiency of the turbine is 0.85 and that of the pump is 0.8, calculate the thermal efficiency of the plant.

9.21 Consider an air-standard Otto cycle with a compression ratio of 8. At the beginning of the compression stroke the air is at 100 kPa and 300 K. The thermal energy addition during the combustion process is 2500 kJ/kg of air. Determine

(a) The temperature and pressure at the terminal points of all the processes.
(b) The thermal efficiency of the cycle.
(c) The work done per kg air and
(d) The mean effective pressure.

9.22 An air-standard Otto cycle has a compression ratio of 8.5. At the beginning of the compression $P_1 = 1$ bar and $T_1 = 300$ K. The energy addition in the cycle is 1500 kJ/kg air. Determine the thermal efficiency of the cycle, the net work done per cycle and the mean effective pressure.

9.23 An air-standard Otto cycle has a compression ratio of 6.5. The pressure and temperature at the beginning of the compression stroke are 1 bar and 320 K, respectively. During a cycle 1000 kJ energy is added per kg air. Determine the thermal efficiency of the cycle, the maximum temperature in the cycle and the maximum pressure in the cycle.

9.24 An engine operates on an air-standard Otto cycle with a compression ratio of 8. The pressure and temperature at the end of the compression stroke are 1.5 MPa and 700 K, respectively. The pressure at the end of the energy addition process is 4.5 MPa. Determine the thermal efficiency of the cycle and the net work per kg of air.

9.25 In an air-standard Otto cycle, the net work done is 1000 kJ/kg and the maximum temperature in the cycle is 3200 K. The temperature at the end of the isentropic compression is 800 K. Determine the compression ratio of the engine.

9.26 In an air-standard Otto cycle, the temperature and pressure at the beginning of the compression stroke are 100 kPa and 300 K, respectively. The maximum pressure and temperature in the cycle are 7.5 MPa and 3200 K, respectively. Determine the thermal efficiency and the net work done per kg of air.

9.27 An air-standard Otto cycle has a compression ratio of 7.5. The conditions of air at the beginning of the compression stroke are 1 bar and 300 K. The maximum temperature in the cycle is 1100 K and during a cycle 2 kJ of energy is added to the air. Determine the net work done in one cycle and the mean effective pressure.

9.28 In an air-standard Otto cycle with a compression ratio of 8, the temperature, pressure and volume at the beginning of the compression stroke are 300 K, 100 kPa and 0.25 m^3 respectively. The maximum temperature of the cycle is 3000 K. Determine the thermal efficiency of the cycle.

9.29 For an air-standard Otto cycle determine the compression ratio corresponding to maximum work per unit mass between the temperature T_1 (minimum) and T_3 (maximum).

9.30 In an air-standard Otto cycle, the conditions of air at the beginning of the compression stroke are 1 bar and 27°C. If the compression ratio is 8, the cylinder volume is 250 cm^3 and 5 kJ energy is added during the energy addition process determine

(a) The thermal efficiency
(b) The work output of the cycle and
(c) The mean effective pressure.

9.31 An air-standard Otto cycle with a compression ratio of 8.5 takes in air at 1 bar and 300 K before the compression stroke begins. If the energy added is 1500 kJ/kg air, determine

(a) The work output of the cycle
(b) The maximum pressure and temperature of the cycle, and
(c) The mean effective pressure.

9.32 A four cylinder spark ignition engine (Otto cycle) has a compression ratio of 8 and each cylinder has a volume of 250 cm^3. At the beginning of the compression stroke the air is at 1 bar and 27°C and the maximum temperature in the cycle is 2000 K. Determine

(a) The thermal efficiency
(b) The energy added per cycle in each cylinder, and
(c) The number of revolutions per minute required for a power output of 30 kW.

9.33 An engine operates on an air-standard Diesel cycle. The conditions at the start of the compression stroke are given by 100 kPa and 300 K. The compression ratio is 15 and the maximum temperature in the cycle is 2000 K. Determine the thermal efficiency and the mean effective pressure.

9.34 An air-standard Diesel cycle has a compression ratio of 18 and a cut-off ratio of 2. The conditions of air at the beginning of the compression are 1 bar and 300 K. Determine the thermal efficiency, the maximum pressure in the cycle and the mean effective pressure.

9.35 In an air-standard Diesel cycle, the conditions at the start of the compression stroke are $V = 0.05\ m^3$, $P = 100$ kPa, and $T = 300$ K. The pressure at the end of the compression stroke is 4 MPa. The energy addition is 500 kJ/kg air. Determine the compression ratio, the cut-off ratio, the work done/kg air and the thermal efficiency of the cycle.

9.36 In an air-standard Diesel cycle, the minimum temperature and pressure are 300 K and 100 kPa, respectively. The maximum temperature of the cycle is 1500 K. The energy supplied is 600 kJ/kg air. Determine the thermal efficiency of the cycle.

9.37 In an air-standard Diesel cycle, the conditions of air at the beginning of the compression stroke are 1 bar and 300 K. The compression ratio is 16 and the energy addition is 1000 kJ/kg air. Determine the maximum temperature and pressure in the cycle, the cut-off ratio and the thermal efficiency of the cycle.

9.38 In an air-standard Diesel cycle with a compression ratio of 18, the maximum temperature in the cycle is 1800 K. The conditions of air at the beginning of compression are 0.95 bar and 310 K. If the mass of air which undergoes the cyclic change is 0.05 kg, determine the thermal efficiency of the cycle, the net work done per cycle and the cut-off ratio.

9.39 An air-standard Diesel cycle has a compression ratio of 18 and a cut off ratio of 3. At the beginning of the compression stroke the pressure and temperature are 100 kPa and 320 K. Determine the energy added per kg air, the net work done per kg air and the thermal efficiency of the cycle.

9.40 In an air-standard Diesel cycle the temperatures before and after the compression stroke are 300 K and 900 K, respectively. The energy supplied is 900 kJ/kg air. Determine the compression ratio, the cut-off ratio, and the maximum cycle temperature.

9.41 In an air-standard Diesel cycle with a compression ratio of 16, the conditions at the start of the compression stroke are 100 kPa and 300 K. The entropy change during the energy rejection process is −0.7 kJ/kg K. Determine the energy added per kg air, the maximum temperature of the cycle, the temperature of the air before the energy rejection process, and the thermal efficiency of the cycle.

9.42 Show that the mean effective pressure of an air-standard Diesel cycle is

$$P_m = P_1 \frac{\gamma r_0^\gamma (r_c - 1) - r_0 (r_c^\gamma - 1)}{(\gamma - 1)(r_0 - 1)}$$

9.43 An air-standard Diesel cycle operates with a compression ratio of 16 and a cut-off ratio of 2. At the beginning of the compression stroke, the air is at 1 bar and 300 K. Determine

(a) The maximum temperature and pressure of the cycle
(b) The energy added per kg air, and
(c) The mean effective pressure of the cycle.

9.44 In an air-standard Diesel cycle with a compression ratio of 15, the air at the beginning of the compression stroke is at 1 bar and 27°C. The volume of the cylinder is 350 cm^3 and 0.5 kJ of energy is added during the constant pressure process. Calculate

(a) The thermal efficiency of the cycle
(b) The maximum pressure and temperature of the cycle, and
(c) The mean effective pressure.

9.45 A four-cylinder Diesel engine has a compression ratio of 16 and a cut-off ratio of 2. The volume of each cylinder is 250 cm^3 and at the beginning of the compression stroke the air is at 1 bar and 300 K. Determine the power output of the engine at 1500 rpm.

is 0.5 kg/s, determine the COP, the power consumption of the refrigerator and the refrigeration capacity (in tons) of the unit.

9.62 An ideal vapor compression refrigeration system with Freon-12 as the refrigerant has a capacity of 100 kW. The evaporator pressure is 120 kPa and the condenser pressure is 960 kPa. Determine the power required by the unit and its COP.

9.63 A vapor compression refrigeration cycle, with Freon-12 as the refrigerant, has an evaporator temperature of −20°C and a condenser pressure of 10 bar. Saturated vapor exits the compressor and saturated liquid leaves the condenser. The mass flow rate of the refrigerant is 0.5 kg/s. Determine the COP and the refrigeration capacity (expressed in tons) of the unit.

9.64 A vapor compression refrigeration cycle, with Freon-12 as the refrigerant, has an evaporator temperature of −20°C. Saturated vapor enters the compressor and saturated liquid leaves the condenser at 40°C. If the mass flow rate of the refrigerant is 1 kg/s, determine the COP, the power consumption and the refrigeration capacity.
Properties of superheated Freon-12 at 960.255 kPa are as follows:

T(°C)	s (kJ/kg K)	h (kJ/kg)
50	0.7066	210.7246
60	0.7456	218.2097

9.65 A vapor compression refrigeration cycle employs Freon-12 as the refrigerant and operates between the temperatures −25°C and 35°C. If the vapor leaving the compressor and the liquid leaving the condenser are in their respective saturated states, and assuming that the compressor is reversible and adiabatic, calculate the COP, the power required and the mass flow rate of the refrigerant for a 50 kW refrigeration capacity.

9.66 The average rate of energy flow as heat to the atmosphere from the condenser of a domestic refrigerator is 1 MJ/h. The refrigerator uses a 200 W motor of 80% efficiency for the compressor operation and the compressor runs 30% of the time. Calculate the COP and the rate of refrigeration.

9.67 A Carnot engine operates between two reservoirs with temperatures 735°C and 35°C and rejects 30 kW to the low temperature reservoir. The Carnot engine drives the compressor of an ideal vapor compression refrigerator employing Freon-12 as the refrigerant. The refrigerator operates between the temperature levels −30°C and 35°C. Determine the COP, the capacity of the refrigerator, and the refrigerant flow rate.

9.68 In an ideal vapor compression refrigeration cycle Freon-12 enters the compressor at 180 kPa and the saturated liquid enters the throttle valve at 10 bar. Determine

(a) The temperature of Freon-12 leaving the compressor
(b) The COP of the refrigeration cycle, and
(c) The power input to the compressor for a 5 ton refrigeration unit.

9.69 A vapor compression refrigerator employing Freon-12 as the refrigerant operates between the temperature levels 40°C and −20°C. If the quality of the refrigerant at the beginning and at the end of the heat absorption process are 0.2 and 0.85, respectively, determine

(a) The Coefficient of Performance
(b) The evaporator and condenser pressures, and
(c) The power consumption for a 2.5 ton refrigerator unit.

9.70 Consider two vapor compression refrigeration cycles. In one refrigeration cycle, the refrigerant (Freon-12) enters the throttling valve as saturated liquid at 40°C, while in the other refrigeration cycle Freon-12 enters as subcooled liquid at 30°C. For both the cycles the evaporator pressures and the condenser pressures are the same and the refrigerant leaves the compressor as saturated vapor.

(a) Sketch both the cycles on *T–s* diagram, and
(b) Which cycle will have higher COP?

9.71 An ideal vapor compression refrigerator employing Freon-12 as the refrigerant operates between the pressure levels 1.5 bar and 9.6 bar. If the mass flow rate of the refrigerant is 0.25 kg/s, determine

(a) The COP
(b) The power input, and
(c) The refrigeration capacity in tons of refrigeration.

9.72 If the throttling valve in the refrigerator of Problem 9.71 is replaced by an isentropic turbine, determine the percentage increase in the refrigeration capacity.

9.73 Consider a 10 ton refrigerator system which operates on an ideal vapor compression refrigeration cycle with Freon-12 as the refrigerant. The refrigerant exits the compressor as saturated vapor at 42°C. The refrigerant enters the throttle valve as saturated liquid at 42°C. The evaporator pressure is 1.3 bar. Calculate

(a) The COP
(b) The quality of the refrigerant at the end of throttling process
(c) The mass flow rate of the refrigerant, and
(d) The power consumption.

9.74 Consider a refrigeration system using Freon-12 as the refrigerant. If the refrigerator is to operate in an environment at 40°C, what is the minimum pressure to which the refrigerant should be compressed? Why?

9.75 A refrigeration system using Freon-12 as the refrigerant should maintain the refrigerated space at −20°C. At what pressure the refrigerant should be maintained in the evaporator? Why?

9.76 A vapor compression refrigeration system using Freon-12 as the refrigerant is required to maintain the refrigerated space at −24°C while the ambient atmosphere is at 38°C. Select reasonable pressures for the evaporator and condenser and specify the reasons why you selected those values.

9.77 Air at 1 bar and 27°C enters the compressor of an ideal gas refrigeration (reversed Brayton) cycle. The pressure ratio of the cycle is 3 and the compressed air enters the turbine at 50°C. Determine

(a) The maximum and minimum temperatures in the cycle
(b) The Coefficient of Performance, and
(c) The mass flow rate of air for a 2 ton refrigeration unit.

10 Gas–Vapor Mixtures and Psychrometry

Learning Objectives

This chapter enables the readers to:

- Understand why the thermodynamic analysis of non-reactive mixtures is easy compared to the analysis of reactive mixtures.
- Know the gravimetric analysis and molar analysis of a mixture.
- Define mass fraction and mole fraction of a component in a mixture.
- Calculate the molar mass of a mixture.
- Convert the gravimetric analysis data into molar analysis data and *vice-versa*.
- State the Dalton's law of additive pressures and the Amagat's law of additive volumes.
- Define the component pressure, partial pressure and component volume of a component in a gas mixture.
- Predict the properties of an ideal gas mixture from a knowledge of the properties of the constituent species.
- Estimate the values of C_P and C_v of an ideal gas mixture.
- Distinguish between dry air and atmospheric air.
- Know the criterion of equilibrium between liquid water and air–water vapor mixture.
- Calculate the dew point temperature of an air–water vapor mixture.
- Define the terms specific humidity and relative humidity.
- Know about adiabatic saturation temperature, dry-bulb temperature and wet-bulb temperature.
- Calculate the specific humidity and relative humidity of air–water vapor mixtures.
- Understand psychrometric charts.
- Use the psychrometric chart in the analysis of processes dealing with air–water vapor mixtures.
- Understand the various processes used in the field of air-conditioning.
- Analyze the adiabatic mixing process of different streams of air–water vapor mixtures.
- Analyze the processes of dehumidification with cooling, evaporative cooling and humidification.
- Analyze the performance of cooling towers.

So far in our discussion attention was focused on a single component, simple system, or pure substances, such as water. In practice, however, we also deal with mixtures. For example, the air we breathe in is a mixture of oxygen and nitrogen; and the fuel oil used in our automobiles is a mixture of several hydrocarbons. The gas inside the automobile cylinder is a mixture of CO_2, CO, H_2O, O_2, N_2, SO_2, oxides of nitrogen, and unburned hydrocarbons. However, in the analysis of air-standard cycles, for simplicity we have assumed the working fluid to be air which is treated as an ideal gas.

This approximation is valid only at low temperatures where there is no possibility of chemical reaction between O_2 and N_2 (present in the air), or when the chemical composition does not change. At sufficiently low temperature the air is available in two phases. When the air is present in both the phases, the composition of the liquid phase is different from the composition of the vapor phase. Then, it cannot be treated as a pure substance. In the thermodynamic analysis of processes a knowledge of the thermodynamic properties of mixtures is essential.

Consider a mixture of N_2, H_2 and NH_3 at a temperature and pressure where the chemical reaction between N_2 and H_2 is feasible. In such a mixture the mole numbers of the species or constituents present cannot be varied independently. On the other hand, in a mixture of O_2 and N_2 at ambient temperature and pressure, where there is no possibility of the formation of oxides of nitrogen, the mole numbers of the components O_2 and N_2 can be varied independently. Therefore, the non-reacting mixtures present less difficulty in the thermodynamic analysis compared to reactive mixtures. In this chapter we concentrate on the study of non-reacting mixtures. If a mixture is homogeneous throughout both in chemical composition as well as in physical state, the mixture is said to be in a phase. The mixture of alcohol and water is an example of a liquid phase and a mixture of gases (like O_2 and N_2) is an example of a gas phase.

10.1 Mixtures of Ideal Gases

Suppose the molecules that constitute the system are independent and that there are no attractive forces between them. That is, the behavior of one molecule is not affected by the presence of the other molecules. Then, the properties of the mixture can be determined from the properties of the constituent species. Such an approximation is valid for mixtures of gases at sufficiently low densities.

A complete description of a multicomponent system requires the specification of the composition in addition to pressure and temperature. The composition of a mixture is usually specified in terms of either the mass (m_i) of each component or the number of moles (N_i) of each component. The total mass (m) of a mixture is the sum of the masses of all the constituent species. That is,

$$m = \sum_i m_i \tag{10.1}$$

The mass fraction (ϕ_i) of the component i is given by

$$\phi_i = \frac{m_i}{m} \tag{10.2}$$

An analysis of a mixture made on mass basis is called *gravimetric analysis*. If the analysis of a mixture is based on the number of moles of the constituent species then, it is called *molar analysis*. The total number of moles (N) of a mixture is given by the sum of the number of moles of the constituent species. That is,

$$N = \sum_i N_i \tag{10.3}$$

and the mole fraction (y_i) of component i is given by

$$y_i = \frac{N_i}{N} \tag{10.4}$$

where N_i is the number of moles of component i. All the mass fractions ϕ_i and mole fractions y_i are not independent and there exist relationships between them.

$$\sum \phi_i = 1 \quad \text{and} \quad \sum y_i = 1 \tag{10.5}$$

The mass (m_i) of a component i is related to the number of moles N_i and the molar mass M_i of the component i.

$$m_i = N_i M_i \tag{10.6}$$

Similarly, the molar mass of a mixture can be defined as the ratio of the mass m of the mixture to the total number of moles N of the mixture. That is,

$$M = \frac{m}{N} = \frac{\sum_i m_i}{\sum_i N_i} = \frac{\sum_i N_i M_i}{\sum_i N_i} = \sum_i y_i M_i \tag{10.7}$$

If dry air is treated as a mixture of nitrogen and oxygen only (ignoring the presence of other species), the composition of air is given by $y_{O_2} = 0.21$ and $y_{N_2} = 0.79$. Then, the molar mass of air is given by

$$M = 0.21 \times 32 + 0.79 \times 28 = 28.84 \text{ kg/kmol}$$

The atmospheric air contains traces of argon and other gases also. When these gases are included in the calculation the molar mass of air is 28.97 kg/kmol.

***Example* 10.1** The molar analysis of a gaseous fuel indicates that it contains 40% CH_4; 20% C_2H_6; 25% H_2 and 15% N_2. Determine (a) the molar mass of the fuel and (b) the gravimetric analysis.

Solution (a) Choose 100 moles of the fuel as the basis for calculation.

Component	y_i	N_i	M_i	m_i	ϕ_i
CH_4	0.4	40	16	0.64	0.374
C_2H_6	0.2	20	30	0.60	0.351
H_2	0.25	25	2	0.05	0.029
N_2	0.15	15	28	0.42	0.246

Molar mass of the mixture, M, is given by:

$$M = \sum y_i M_i = 0.4(16) + 0.2(30) + 0.25(2) + 0.15(28) = 17.1 \text{ kg/kmol}$$

(b) The sum of column 5 in the above table gives the mass of the mixture.

$$m = \sum m_i = 0.64 + 0.60 + 0.05 + 0.42 = 1.71 \text{ kg}$$

The mass fraction of CH_4

$$y_{CH_4} = \frac{0.64}{1.71} = 0.374$$

Similarly. the mass fractions of other species are calculated and they are shown in column 6 of the above table.

10.2 P–v–T Relationships for Ideal Gas Mixtures

The P–v–T behavior of an ideal gas is expressed by the relation $Pv = RT$. When two or more ideal gases are mixed, the mixture also behaves like an ideal gas, because the behavior of a gas molecule is not influenced by the presence of the other similar or dissimilar molecules. The P–v–T relationships of a mixture of gases is usually based on two models: Dalton's law of additive pressures and Amagat's law of additive volumes.

The Dalton's law of additive pressures states that the pressure P exerted by a gas mixture is equal to the sum of the component pressures, p_i. The component pressure p_i of component i is the pressure the component i would exert if it existed alone at the mixture temperature and volume. The Dalton's law gives:

$$P = \sum_i p_i \tag{10.8}$$

A schematic representation of the Dalton's law of additive pressures is shown in Fig. 10.1. The component pressure p_i or the pressure exerted by the component i if it existed alone at the mixture temperature T and volume V is given by

$$p_i = \frac{N_i RT}{V} \tag{10.9}$$

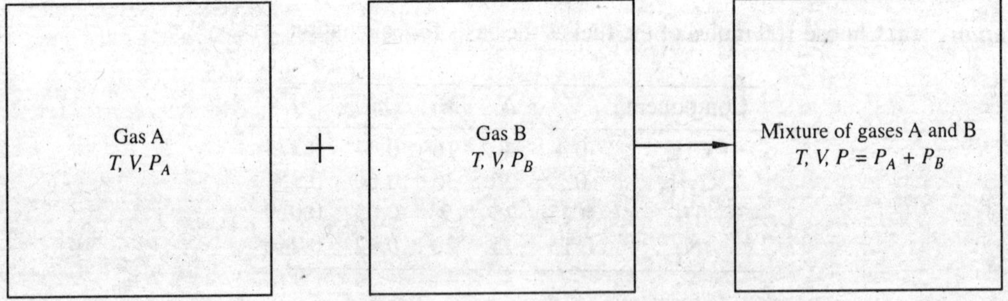

Fig. 10.1 *Schematic representation of Dalton's law of additive pressures*

Then the Dalton's law of additive pressures gives

$$P = \sum_i p_i = \frac{\sum N_i RT}{V} = \frac{RT \sum N_i}{V} = \frac{NRT}{V} \tag{10.10}$$

From Eqs. (10.9) and (10.10), we get

$$\frac{p_i}{P} = \frac{N_i RT/V}{NRT/V} = \frac{N_i}{N} = y_i$$

or

$$p_i = y_i P \tag{10.11}$$

The product $y_i P$ is frequently called the partial pressure of component i and the ratio p_i/P is called the pressure fraction of component i.

The Amagat's law of additive volumes states that the volume of a gas mixture is equal to the sum of the volumes each gas would occupy if it existed alone at the mixture temperature and pressure. That is,

$$V = \sum_i V_i \tag{10.12}$$

where V_i is the component volume of component i. That is, V_i is the volume occupied by component i if it existed alone at the mixture temperature and pressure and is given by

$$V_i = \frac{N_i RT}{P} \tag{10.13}$$

The Amagat's law of additive volumes gives

$$V = \sum_i V_i = \frac{\sum_i N_i RT}{P} = \frac{RT}{P}\sum N_i = \frac{NRT}{P} \tag{10.14}$$

From Eqs. (10.13) and (10.14), we get

$$\frac{V_i}{V} = \frac{N_i RT/P}{NRT/P} = \frac{N_i}{N} = y_i \tag{10.15}$$

The ratio V_i/V is called the volume fraction of component i. Thus for ideal gas mixtures, the mole fraction, the pressure fraction and the volume fraction of a component are identical. The composition of a gas mixture is usually determined by means of a volumetric analysis. Therefore, a volumetric analysis is identical to the molar analysis for an ideal gas mixture.

***Example* 10.2** A particular gas mixture has the following analysis by volume. O_2 = 20% ; N_2 = 50%; H_2O = 10% and CO_2 = 20%. If the gas mixture is available at 300 K and 100 kPa, calculate the partial pressures of each constituent and express the composition in mass fractions. If the gas mixture held in a closed vessel is treated with diethylamine solution, which absorbs the CO_2, and the gases are maintained at constant temperature, calculate the final pressure of the gas and express its composition by volume as well as by mass fraction.

Solution Since the volume fraction is equal to the mole fraction, the initial gas composition is $y(O_2)$ = 0.2, $y(N_2)$ = 0.5; $y(H_2O)$ = 0.1 and $y(CO_2)$ = 0.2. Consider one mole of the mixture as the basis for calculation.

Component	y_i	p_i (kPa)	M_i (kg/kmol)	m (kg$\times 10^3$)	ϕ_i
O_2	0.2	20	32	6.4	0.2065
N_2	0.5	50	28	14.0	0.4516
H_2O	0.1	10	18	1.8	0.0580
CO_2	0.2	20	44	8.8	0.2839

Molar mass of the mixture, $M = \sum y_i M_i = 31.0$ kg/kmol

Once the gas mixture is treated with diethylamine, all the CO_2 present in the gas mixture is absorbed. Then in the final gas

$p(O_2) = 20$ kPa; $p(H_2O) = 10$ kPa and $p(N_2) = 50$ kPa

Therefore final pressure $= P = \sum p_i = 80$ kPa.

Then, for the final gas mixture, we have

Component	Mole fraction (y_i)	volume %	M_i (kg/kmol)	Mass $M_i y_i$	Mass fraction
O_2	0.250	25.0	32	8	0.2883
N_2	0.625	62.5	28	17.5	0.6306
H_2O	0.125	12.5	18	2.25	0.0811

Molar mass $= M = \sum y_i M_i = 27.75$

Mole fraction $= y_i =$ Pressure fraction $= \dfrac{p_i}{P} =$ Volume fraction $= \dfrac{V_i}{V}$

10.3 Properties of Ideal Gas Mixtures

In a mixture of ideal gases, the molecules are independent of each other and there are no attraction forces between them. The properties of such a mixture can be determined from a knowledge of the properties of the constituent species and the composition. Since there are no attraction forces between the molecules, the total energy of the mixture is equal to the sum of the energies of the independent species that constitute the mixture. Therefore, the extensive properties like internal energy, enthalpy and entropy of the mixture are given by

$$U = \sum_i U_i = \sum_i N_i u_i \quad \text{or} \quad \sum_i m_i u_i \tag{10.16}$$

$$H = \sum_i H_i = \sum_i N_i h_i \quad \text{or} \quad \sum_i m_i h_i \tag{10.17}$$

$$S = \sum_i S_i = \sum_i N_i s_i \quad \text{or} \quad \sum_i m_i s_i \tag{10.18}$$

The molar or specific properties of the mixture are given by:

$$u = \sum_i y_i u_i \quad \text{or} \quad \sum_i \phi_i u_i \tag{10.19}$$

$$h = \sum_i y_i h_i \quad \text{or} \quad \sum_i \phi_i h_i \tag{10.20}$$

$$s = \sum_i y_i s_i \quad \text{or} \quad \sum_i \phi_i s_i \tag{10.21}$$

The changes in internal energy, enthalpy and entropy of a gas mixture are given by:

$$\Delta U = \sum_i \Delta U_i = \sum_i N_i \Delta u_i \quad \text{or} \quad \sum_i m_l \Delta u_i \tag{10.22}$$

$$\Delta H = \sum_i \Delta H_i = \sum_i N_i \Delta h_i \quad \text{or} \quad \sum_i m_i \Delta h_i \tag{10.23}$$

$$\Delta S = \sum_i \Delta S_i = \sum_i N_i \Delta s_i \quad \text{or} \quad \sum_i m_i \Delta s_i \tag{10.24}$$

We know that for an ideal gas

$$\Delta u = C_v \Delta T \quad \text{and} \quad \Delta h = C_P \Delta T$$

Therefore, for a mixture of ideal gases, we get

$$\Delta U = \sum N_i \Delta u_i = \sum N_i C_{vi} \Delta T$$

or

$$\frac{\Delta U}{N} = \Delta u = \frac{\sum N_i C_{vi} \Delta T}{N} = \sum y_i C_{vi} \Delta T = C_v \Delta T \tag{10.25}$$

where the molar heat capacity at constant volume for the mixture is given by:

$$C_v = \sum_i y_i C_{vi} \tag{10.26}$$

Similarly, we get the molar heat capacity at constant pressure for the mixture as

$$C_P = \sum_i y_i C_{Pi} \tag{10.27}$$

Similarly on a mass basis, the specific heats of the mixture are given by:

$$C_v = \sum_i \phi_i C_{vi} \quad \text{and} \quad C_P = \sum_i \phi_i C_{Pi} \tag{10.28}$$

Estimation of Δu and Δh for an ideal gas mixture is relatively easy since it requires a knowledge of the initial and final temperatures. However, care should be exercised in the estimation of Δs for an ideal gas mixture since it depends on both temperature and component pressures of the constituent species. The entropy change of individual ideal gases in a mixture is given by

$$\Delta s_i = C_{Pi} \ln \frac{T_{2i}}{T_{1i}} - R \ln \frac{P_{2i}}{P_{1i}} = C_{Pi} \ln \frac{T_{2i}}{T_{1i}} - R \ln \frac{y_{2i} P_2}{y_{1i} P_1} \tag{10.29}$$

where $y_{2i}P_2$ and $y_{1i}P_1$ denote the final and initial partial pressures of component i. The calculation of entropy change for a mixture is illustrated in the following example.

***Example* 10.3** Consider an isolated system separated into two compartments by a partition as shown in Fig. 10.2. One compartment of volume V_A contains an ideal gas A at temperature T and pressure P.

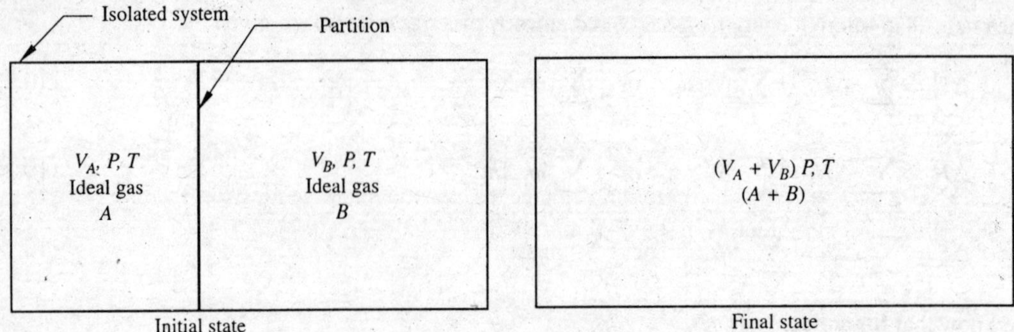

Fig. 10.2 *Sketch for Example 10.3*

The second compartment of volume V_B contains an ideal gas B at the same temperature and pressure. The partition is removed and the gases are allowed to mix. Determine the change in the internal energy, enthalpy and entropy associated with the mixing process.

Solution

For an isolated system: $Q = 0$ and $W = 0$. Therefore, $U_f - U_i = 0$

$$U_i = N_A u_{Ai} + N_B u_{Bi}$$

where N_A and N_B are the mole numbers of the gases A and B, respectively.

$$U_f = N \sum y_i u_{if} = (N_A + N_B)(y_A u_{Af} + y_B u_{Bf})$$

$$= (N_A + N_B)\left(\frac{N_A}{N_A + N_B} u_{Af} + \frac{N_B}{N_A + N_B} u_{Bf}\right) = N_A u_{Af} + N_B u_{Bf}$$

Since A and B are ideal gases, $u_A = u_A(T)$ and $u_B = u_B(T)$

Hence,

$$N_A u_{Af} + N_B u_{Bf} - N_A u_{Ai} - N_B u_{Bi} = 0$$

or,

$$N_A(u_{Af} - u_{Ai}) + N_B(u_{Bf} - u_{Bi}) = 0$$

We know that $u(T)$ is a monotonically increasing function of temperature; that is, if T increases u increases; or if T decreases, u decreases. Therefore, both the terms in the parenthesis are either positive or negative, but not opposite in sign, since the gases are initially at the same temperature. Therefore, the above equation is true only when the terms are independently equal to zero. Hence, $(u_{Af} - u_{Ai}) = 0$ and $(u_{Bf} - u_{Bi}) = 0$. Or,

$$T_f - T_i = 0$$

That is, there is no change either in the temperature or in the internal energy upon mixing ideal gases.

We know that $h = h(T)$ only, for an ideal gas. Hence,

$$H_i = N_A h_{Ai} + N_B h_{Bi}$$

$$H_f = (N_A + N_B)\sum y_i h_{if} = N_A h_{Af} + N_B h_{Bf}$$

Therefore, $H_f - H_i = 0$; that is, there is no change in the enthalpy on mixing ideal gases.

The partial pressure of gas A in the final mixture is given by

$$p_A = \frac{N_A RT}{V_A + V_B}$$

Similarly, the partial pressure of gas B in the mixture is given by

$$p_B = \frac{N_B RT}{V_A + V_B}$$

Therefore, the total pressure of the mixture is given by

$$P_f = \sum p_i = \frac{N_A RT}{V_A + V_B} + \frac{N_B RT}{V_A + V_B}$$

Before mixing: $PV_A = N_A RT$ and $PV_B = N_B RT$

Therefore, $P(V_A + V_B) = (N_A + N_B)RT$ or,

$$P = \frac{(N_A + N_B)RT}{V_A + V_B}$$

Therefore, $P_f = P$; that is, there is no change in the pressure in this case.

Initial entropy $= S_i = N_A s_{Ai} + N_B s_{Bi}$

Final entropy after mixing, S_f is given by

$$S_f = (N_A + N_B)s_f = (N_A + N_B)(y_A s_{Af} + y_B s_{Bf})$$

$$S_f = N_A s_{Af} + N_B s_{Bf}$$

Hence, the change in entropy ΔS is given by

$$\Delta S = N_A(s_{Af} - s_{Ai}) + N_B(s_{Bf} - s_{Bi})$$

We know that the change in entropy of an ideal gas is given by:

$$s_f - s_i = C_P \ln\frac{T_f}{T_i} - R\ln\frac{P_f}{P_i}$$

In this case $T_f = T_i$. P_i is the initial pressure and P_f is the final pressure of the component, which is the final partial pressure.

Therefore,

$$s_{Af} - s_{Ai} = -R\ln\frac{p_{Af}}{P} = -R\ln y_A$$

$$s_{Bf} - s_{Bi} = -R \ln \frac{p_{Bf}}{P} = -R \ln y_B$$

$$\Delta S = -N_A R \ln y_A - N_B R \ln y_B$$

or,

$$\Delta s = \frac{\Delta S}{N_A + N_B} = -R y_A \ln y_A - R y_B \ln y_B$$

Since $y_i < 1$, $\Delta s > 0$. Therefore, there is an increase in the entropy on mixing the ideal gases at the same temperature and pressure.

***Example* 10.4** A mixture of 25% nitrogen and 75% hydrogen by volume is compressed isentropically from 300 K and 100 kPa to 500 kPa in the first stage of a multistage compressor in a fertilizer plant. The compression to still higher pressures is achieved in subsequent stages after the gas mixture is passed through intercoolers. It is desired to predict the temperature of the gas mixture after compression as well as the work required per unit mass of the mixture. Also evaluate the entropy change for each gas. Assume that the mixture behaves like an ideal gas.

Solution Applying the first law of thermodynamics to the compressor, we get:

$$h_2 - h_1 = C_P(T_2 - T_1) = -\dot{W}_s$$

Also for isentropic compression $s_2 = s_1$

Therefore, the final temperature of the gas mixture can be obtained from the equation:

$$T_2 = T_1 \left(\frac{P_2}{P_1}\right)^{\frac{\gamma-1}{\gamma}}$$

For an ideal gas mixture, the volume fraction is equal to the mole fraction. Therefore, $y_{H_2} = 0.75$ and $y_{N_2} = 0.25$

$$C_P = \sum y_i C_{Pi} = 0.75(28.6455) + 0.25(29.1783) = 28.7787 \text{ kJ/kmol K}$$

$$C_v = \sum y_i C_{vi} = 0.75(20.3311) + 0.25(20.8641) = 20.4644 \text{ kJ/kmol K}$$

Therefore,

$$\gamma = \frac{28.7787}{20.4644} = 1.4063$$

$$T_2 = 300(5)^{\frac{0.4063}{1.4063}} = 477.6 \text{ K}$$

$$-\dot{w}_s = 28.7787(477.6 - 300) = 5111.1 \text{ kJ/kmol}$$

Molar mass of the mixture $= M = \sum y_i M_i$

$$= 0.75(2.016) + 0.25(28.013) = 8.5153 \text{ kg/kmol}$$

$$-\dot{W}_s = \frac{5111.1}{8.5153} = 600.2 \text{ kJ/kg of mixture}$$

Now the entropy change of an ideal gas can be estimated by making use of the relation:

$$\Delta s = C_P \ln \frac{T_2}{T_1} - R \ln \frac{P_2}{P_1}$$

For Hydrogen:

$$s_2 - s_1 = 28.6455 \ln \frac{477.6}{300} - 8.314 \ln 5 = -0.060\,96 \text{ kJ/kmol K}$$

For Nitrogen:

$$s_2 - s_1 = 29.1783 \ln \frac{477.6}{300} - 8.314 \ln 5 = 0.186\,78 \text{ kJ/kmol K}$$

For mixture:

$$s_2 - s_1 = \sum y_i (s_2 - s_1)_i = 0.75(-0.060\,96) + 0.25(0.186\,78) = 0 \text{ kJ/kmol K}$$

***Example* 10.5** A 5 m³ tank contains 60% H_2 and 40% methane by volume at 100 kPa and 300 K. Determine the amount of methane to be added at 300 K to change the composition to 50% methane by volume. Also determine the final pressure of the mixture in the tank.

Solution Initial composition: $y_{H_2} = 0.6, y_{CH_4} = 0.4$
Therefore, $p_{H_2} = 60$ kPa and $p_{CH_4} = 40$ kPa

$$N_{H_2} = \frac{p_{H_2} V}{RT} = \frac{60 \times 10^3 \times 5}{8.314 \times 10^3 \times 300} = 0.1203 \text{ kmol}$$

$$N_{CH_4} = \frac{p_{CH_4} V}{RT} = \frac{40 \times 10^3 \times 5}{8.314 \times 10^3 \times 300} = 0.0802 \text{ kmol}$$

Final composition: $y_{CH_4} = 0.5, y_{H_2} = 0.5$, or, $N_{CH_4} = N_{H_2}$
Therefore methane to be added = 0.1203 – 0.0802 = 0.0401 kmol
Final pressure, P_f is given by:

$$P_f = \frac{NRT}{V} = \frac{(0.1203 + 0.1203)8.314 \times 10^3(300)}{5} = 120.02 \text{ kPa}$$

***Example* 10.6** A rigid and insulated tank of volume 2 m³ contains 50% air and 50% helium (by volume) at 100 kPa and 300 K. The tank is connected to a high pressure line carrying helium at 4 MPa and 600 K. The valve is opened and the helium enters the tank till the pressure inside the tank is 4 MPa. Then, the valve is closed and the tank is isolated. Determine: (a) the final temperature of the gas mixture in the tank, (b) the composition of the mixture in the tank, and (c) the amount of helium that enters the tank.

Solution

$$N_i = \frac{P_i V}{RT_i} = \frac{100 \times 10^3 \times 2}{8.314 \times 10^3 \times 300} = 0.0802 \text{ kmol}$$

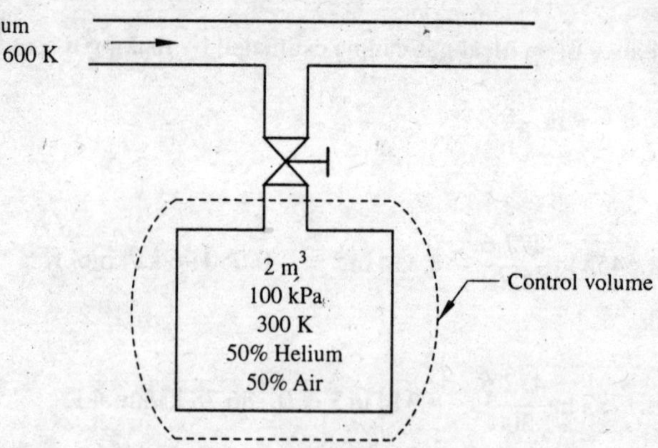

Fig. 10.3 *Sketch for Example 10.6*

Therefore, the quantity of air initially present in the tank $N_{air,0} = 0.0401$ kmol, $N_{He} = 0.0401$ kmol

Application of the first law of thermodynamics to the control volume shown in Fig. 10.3 gives:

$$(N_f - N_0)h_{He} = U_f - U_0 \tag{A}$$

where,

$$U_f = N_f u_f = N_f \left(\sum y_i u_i\right)_f = (N_{He} C_{vHe} + N_{air} C_{vair})_f T_f \tag{B}$$

$$U_0 = N_0 u_0 = N_0 \left(\sum y_i u_i\right)_0 = (N_{He} C_{vHe} + N_{air} C_{vair})_0 T_0$$

$$= \{0.0401 \times 12.4717 + 0.0401 \times 20.7889\}300 = 400.13 \text{ kJ}$$

$$T_f = \frac{P_f V}{RN_f} = \frac{4 \times 10^6 \times 2}{8.314 \times 10^3 (N_{He} + N_{air})_f} = \frac{962.23}{(N_{He} + N_{air})_f} \tag{C}$$

Substituting the value of U_0 and Eqs. (B) and (C) in Eq. (A), we obtain:

$$\{(N_f - N_0)C_P T\}_{He} = \frac{962.23(N_{He} C_{vHe} + N_{air} C_{vair})_f}{(N_{He} + N_{air})_f} - 400.13$$

or,

$$(N_{Hef} - 0.0401)20.786(600) = \frac{962.23\{12.4717 N_{Hef} + 0.0401 \times 20.7889\}}{N_{Hef} + 0.0401} - 400.13$$

or, $N_{Hef} = 0.995$ kmol

Therefore,

$$T_f = \frac{962.23}{0.995 + 0.0401} = 929.58 \text{ K}$$

Final gas mixture contains: He = 0.995 kmol, Air = 0.0401 kmol

Hence, the final composition : $y_{He} = 0.9613$; $y_{air} = 0.0387$

Helium that enters the tank is given by:

$$0.995 - 0.0401 = 0.9549 \text{ kmol}$$

10.4 Ideal Gas–Vapor Mixtures

Quite often an engineer encounters systems which are mixtures of condensable vapor and gas. The combustion products in the cylinder of an internal combustion engine contain water vapor, CO_2, CO and other components. In chemical process industries many of the gas streams contain condensable vapors and gases. In the field of air-conditioning the engineer deals with a mixture of air and water vapor. Therefore, a study of gas and condensable vapor mixtures is important in the thermodynamic analysis of processes. At low pressures, below the atmospheric pressure, the vapor behavior may be approximated as that of an ideal gas. The theory of ideal gas mixtures can be easily extended to the condensable vapor–gas mixtures at low pressures. If the temperature of the mixture is reduced, some of the vapor may condense. Therefore, the composition of the mixture changes during a given process. The theory of air–water vapor mixtures is presented in this section.

Air is a mixture of oxygen, nitrogen, argon and small amounts of other gases. The atmospheric air contains some water vapor also. The air which does not contain any water vapor is usually called dry air. The composition of dry air remains constant but the amount of water vapor present in the atmospheric air changes with temperature. The component pressure or the partial pressure of water vapor in the atmospheric air cannot be greater than the saturation pressure or vapor pressure of water at the temperature of the air. It can be observed from the saturated steam tables that the saturation pressure of water at 50°C is 12.335 kPa. At this low pressure, the water vapor behaves like an ideal gas. Hence, mixtures of air and water vapor can be treated as mixtures of ideal gases.

A mixture of air and water vapor is said to be saturated if the partial pressure of the water vapor in the mixture is equal to the saturation pressure of water at the temperature of the mixture.

When a mixture of air and water vapor is in equilibrium with liquid water, the criteria for equilibrium between the phases are: (i) the temperature of both the phases should be the same, and (ii) the chemical potential of a component is the same in both the phases. Since the solubility of air in liquid water is negligible, it is reasonable to consider the liquid phase as pure water. Then water (H_2O) is the only component which is present in both the phases. Therefore, at equilibrium:

$$\mu_{f,\,H_2O} = \mu_{g,\,H_2O} \tag{10.30}$$

The chemical potential of a component i is equal to the partial molar Gibbs free energy. For a pure substance at constant temperature, the change in the Gibbs free energy is given by:

$$d\mu = dg = vdp \tag{10.31}$$

If the substance is treated as an ideal gas then we can write:

$$d\mu = dg = \frac{RT}{P}dP = RTd\ln P \tag{10.32}$$

Thus, the chemical potential of water vapor in the air–water vapor mixture is related to the partial pressure of water vapor in the mixture. Hence, if the saturation pressure of water at the mixture temperature is equal to the partial pressure of water vapor in the air–water vapor mixture, the two phases—liquid water and air–vapor mixture—are in a state of thermodynamic equilibrium. That is, there is no possibility for further transfer of one component (water) from one phase to the other. Hence, air is saturated with water vapor. In fact, the total pressure has some effect on the saturation pressure, that is, the presence of inert gases like helium, argon in the mixture affects the saturation pressure. However, the change in the saturation pressure is so small that it can be neglected for all practical purposes.

Suppose air and liquid water are brought into contact with each other and are allowed to remain that way for a long time. Then the two phases attain a state of equilibrium and the air gets saturated with water vapor. Suppose the mixture is initially saturated at a given temperature and the temperature of the mixture is raised by heating. To maintain a state of equilibrium between the two phases more water evaporates to raise the partial pressure of water vapor in the mixture.

10.4.1 Dew point temperature

If the air–water vapor mixture which is initially not saturated, is cooled at constant pressure, the partial pressure of water vapor in the mixture remains constant till it is equal to the saturation pressure of water. Further cooling results in condensation of the water vapor. The temperature at which the vapor condenses when the air–water vapor mixture is cooled at constant pressure, is called the Dew point. In Fig. 10.4 if the initial state of the mixture is 1, a constant pressure cooling leads the mixture along the path 1–2. The partial pressure of water vapor in the mixture does not change till the mixture reaches state 2. Further cooling at constant pressure results in condensation of water vapor. At state 2, the partial pressure of water vapor in the mixture is equal to the saturation pressure of water. The temperature at state 2 is called the dew point temperature since it is associated with the formation of dew or water droplets.

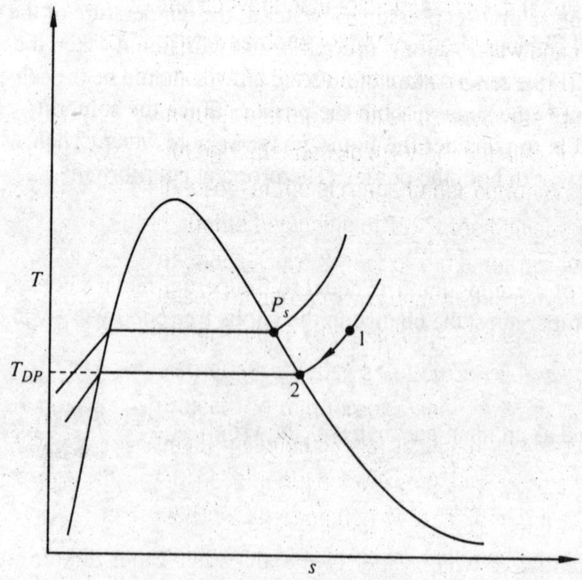

Fig. 10.4 *T–s diagram to show the dew point*

10.4.2 Specific humidity and relative humidity

The specific humidity or humidity ratio is defined as the ratio of the mass of water vapor to the mass of dry air in the mixture.

Specific humidity (or humidity ratio) is defined as:

$$\omega = \frac{m_w}{m_a} \tag{10.33}$$

Assuming that the water vapor follows the ideal gas law, we can write:

$$\omega = \frac{m_w}{m_a} = \frac{(p_w V/RT)M_w}{(p_a V/RT)M_a} = 0.622\frac{p_w}{p_a} \tag{10.34}$$

where, p_a is the partial pressure of air, and p_w is the partial pressure of water vapor. Then

$$\omega = 0.622\frac{p_w}{P - p_w} \tag{10.35}$$

The relative humidity, ϕ, is defined as the ratio of the amount of water vapor, m_w, present in the air to the maximum amount of water vapor, $m_{w,\,\max}$, the air can hold at the same temperature. At a given temperature, air can hold the maximum amount of water vapor when the air is saturated with water vapor. In that state, the partial pressure of water vapor in the air is equal to the saturation pressure P_s of water at the temperature of the air. Thus, the relative humidity ϕ is given by

$$\phi = \frac{m_w}{m_{\max}} = \frac{M_w p_w V/RT}{M_w P_s V/RT} = \frac{p_w}{P_s} \tag{10.36}$$

where M_w is the molar mass of water. Hence, the relative humidity ϕ can be redefined as the ratio of the partial pressure of water vapor in the air–water vapor mixture to the saturation pressure of water at the mixture temperature. The relative humidity varies from zero for dry air to one for saturated air. Equation (10.35) can be rewritten as:

$$\omega = 0.622\frac{p_w}{P - p_w} = 0.622\frac{p_w}{P_s}\frac{P_s}{P - p_w} = 0.622\phi\frac{P_s}{P - \phi P_s} \tag{10.37}$$

$$\phi = \frac{\omega P}{(0.622 + \omega)P_s} \tag{10.38}$$

***Example* 10.7** An air–water vapor mixture has a relative humidity of 60% at 1 atm and 30°C. Determine the specific humidity, the dew point, the mass of water vapor and the mass of dry air per 100 m^3 of the mixture, and the partial pressures of water vapor and air.

Solution Relative humidity $\phi = 0.60 = \dfrac{p_w}{P_s}$

At 30°C, the saturation pressure of water $P_s = 4.241$ kPa

$$\phi = 0.60 = \frac{p_w}{4.241}$$

Therefore, partial pressure of water vapor $= p_w = 2.5446$ kPa.

Total pressure $P = p_w + p_a = 1$ atm. $= 101.325$ kPa

or, partial pressure of air is given by

$$p_a = 101.325 - 2.5446 = 98.78 \text{ kPa}$$

Specific humidity ω is given by

$$\omega = 0.622\frac{p_w}{p_a} = 0.622\frac{2.5446}{98.78} = 0.016\,02 \text{ kg water/kg dry air}$$

The dew point is the temperature at which the saturation pressure is equal to the partial pressure of water ($P_s = p_w = 2.5446$ kPa).

From steam tables, the temperature corresponding to this pressure is read as 21.25°C. Hence $T_{DP} =$ 21.25°C. Therefore, dew will form if the ambient temperature falls below 21.25°C.

Mass of water vapor per 100 m^3 of mixture

$$= \frac{p_w V}{RT} M_w = \frac{2.5446 \times 10^3 \times 100 \times 18}{8.314 \times 10^3 \times 303} = 1.817 \, kg$$

Mass of dry air per 100 m^3 of mixture

$$= \frac{P_a V}{RT} M_a = \frac{98.78 \times 10^3 \times 100 \times 28.97}{8.314 \times 10^3 \times 303} = 113.54 \text{ kg}$$

In the coastal areas, during the summer months it is not unusual for the temperature to be as high as 42°C, with the relative humidity of air approaching 90%. Let us estimate the dew point temperature of this air.

$$\phi = 0.90 = \frac{p_w}{P_s} = \frac{p_w}{8.2578} \quad \text{or} \quad p_w = 7.432 \text{ kPa}$$

(Note that P_s at 42°C = 8.2578 kPa.)

Hence the Dew point temperature ($p_w = P_s = 7.432$ kPa) = 40.13°C. If the temperature of the mixture falls below 40.13°C (> human body temperature) the dew will form. This clearly indicates that if one moves around in an atmosphere which is hot and humid his body will be covered with water droplets which are at a higher temperature than that of the body. This explains the discomfort one feels in hot and humid atmosphere.

10.5 Adiabatic Saturation

Specific humidity or relative humidity of an air–water vapor mixture can be measured in principle with the help of a device called an adiabatic saturator. Consider the device shown in Fig. 10.5 into which the air-water vapor mixture whose specific humidity is to be measured, flows steadily.

Let us assume that the device is sufficiently long and provides enough time of contact for the air–water vapor mixture to come to equilibrium with the water. Then the air–water vapor mixture leaves the adiabatic saturator as saturated. Let the device be insulated so that there are no energy losses. Let us also assume that water is added continuously at the temperature of the outgoing air at a rate which

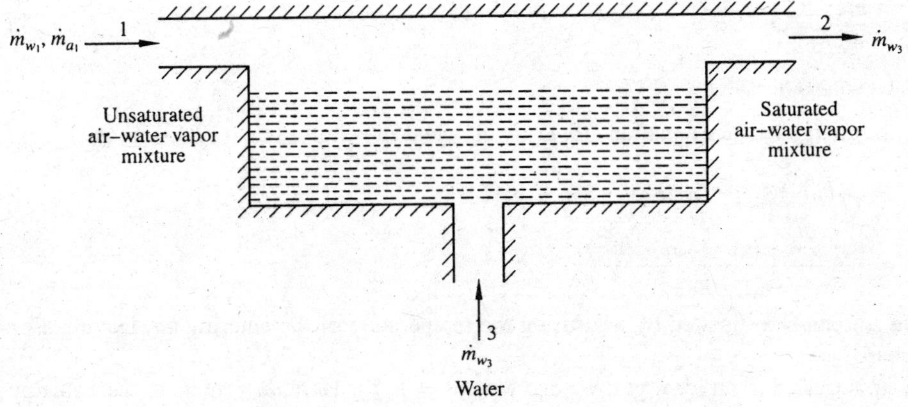

Fig. 10.5 *Schematic diagram of adiabatic saturator*

is equal to the rate of evaporation of water. Since the unsaturated air–water vapor mixture is sweeping over a layer of liquid water, some water evaporates. The energy needed for the evaporation comes from the air mixture. Hence, the air–water vapor mixture leaves the adiabatic saturator at a temperature lower than that of the entering air. As the air leaving the adiabatic saturator is in equilibrium with the liquid water, the temperature of the liquid water is equal to the temperature of the saturated air–water vapor mixture. Considering the unit as the control volume, the mass and energy balance equations can be written as:

Mass balance for air:

$$\dot{m}_{a1} = \dot{m}_{a2} \tag{10.39}$$

Mass balance for water:

$$\dot{m}_{w1} + \dot{m}_{w3} = \dot{m}_{w2} \tag{10.40}$$

Energy balance:

$$\dot{m}_{a1}h_{a1} + \dot{m}_{w1}h_{w1} + \dot{m}_{w3}h_{w3} = \dot{m}_{a2}h_{a2} + \dot{m}_{w2}h_{w2} \tag{10.41}$$

where,

$\dot{m}_{a1}$ = mass flow rate of dry air into the device
$\dot{m}_{w1}$ = mass flow rate of water vapor into the device
$\dot{m}_{w3}$ = mass flow rate of liquid water entering the device at temperature T_2
$\dot{m}_{a2}$ = mass flow rate of dry air leaving the device
$\dot{m}_{w2}$ = mass flow rate of water vapor leaving the device
h_{a1}, h_{a2} = specific enthalpy of dry air at the inlet and outlet of the device, respectively
h_{w1}, h_{w2} = specific enthalpy of saturated water vapor at the inlet and exit of the device, respectively
h_{w3} = specific enthalpy of the liquid water entering the device.

Dividing Eq. (10.39) by $\dot{m}_{a1} = \dot{m}_{a2}$, we get:

$$\frac{\dot{m}_{w1}}{\dot{m}_{a1}} + \frac{\dot{m}_{w3}}{\dot{m}_{a1}} = \frac{\dot{m}_{w2}}{\dot{m}_{a2}}$$

$$\omega_1 + \frac{\dot{m}_{w3}}{\dot{m}_{a1}} = \omega_2 \tag{10.42}$$

$$\dot{m}_{w3} = \dot{m}_{a1}(\omega_2 - \omega_1) \tag{10.43}$$

Dividing the energy balance equation (10.41) by $\dot{m}_{a1}$ and substituting Eq. (10.43) we get:

$$h_{a1} + \omega_1 h_{w1} + (\omega_2 - \omega_1)h_{w3} = h_{a2} + \omega_2 h_{w2} \tag{10.44}$$

$$\omega_1 = \frac{h_{a2} - h_{a1} + \omega_2(h_{w2} - h_{w3})}{h_{w1} - h_{w3}} \tag{10.45}$$

Therefore, ω_1 can be estimated by measuring the temperature of the entering air, leaving air and the total pressure.

If the air is treated as an ideal gas, we can write $h = h(T)$. Here, h_{w2} refers to the enthalpy of the saturated water vapor at temperature T_2 and h_{w3} refers to the enthalpy of the saturated liquid water at temperature T_2. Therefore,

$$h_{w2} - h_{w3} = h_{fg2} \tag{10.46}$$

where, h_{fg2} = the latent heat of vaporization of water at the temperature of the air that leaves the adiabatic saturator. Then

$$\omega_1 = \frac{C_P(T_2 - T_1) + \omega_2 h_{fg2}}{h_{g1} - h_{f2}} \tag{10.47}$$

where, h_{w1} is replaced by h_{g1} to avoid confusion, and h_{g1} represents the enthalpy of the saturated water vapor at state 1. Similarly h_{w3} is replaced by h_{f2} which stands for the enthalpy of the saturated liquid water at state 2.

In deriving a relation for the specific humidity, the temperature of the liquid water entering the adiabatic saturator is assumed to be equal to that of the saturated air–water vapor mixture leaving the adiabatic saturator. Without this assumption it is not possible to replace $(h_{w2} - h_{w3})$ by h_{fg2}. Then, the value of $(h_{w2} - h_{w3})$ is different from (h_{fg2}). In Eq. (10.45), $(h_{w2} - h_{w3})$ is weighted by ω_2 which is usually a small quantity. Therefore, if the temperature of the liquid water entering the adiabatic saturator is marginally different from that of the outgoing air, the error in the value of ω_1 calculated from Eq. (10.45) will be less than 1%. For all practical purposes, the adiabatic saturation temperature of air–water vapor mixtures is independent of the make-up water temperature, and depends only on the specific humidity and temperature of the entering air.

10.6 Psychrometer

The adiabatic saturation process discussed above can be used to determine the specific humidity and relative humidity of unsaturated air–water vapor mixture. However, in practice it is very difficult to attain saturated state by this method without employing an extremely long flow device or a spray mechanism to achieve saturation condition at the exit.

The principle of adiabatic saturation is used in measuring the specific humidity of air–water vapor mixtures in a laboratory using a simple instrument called the wet-bulb thermometer or a psychrometer.

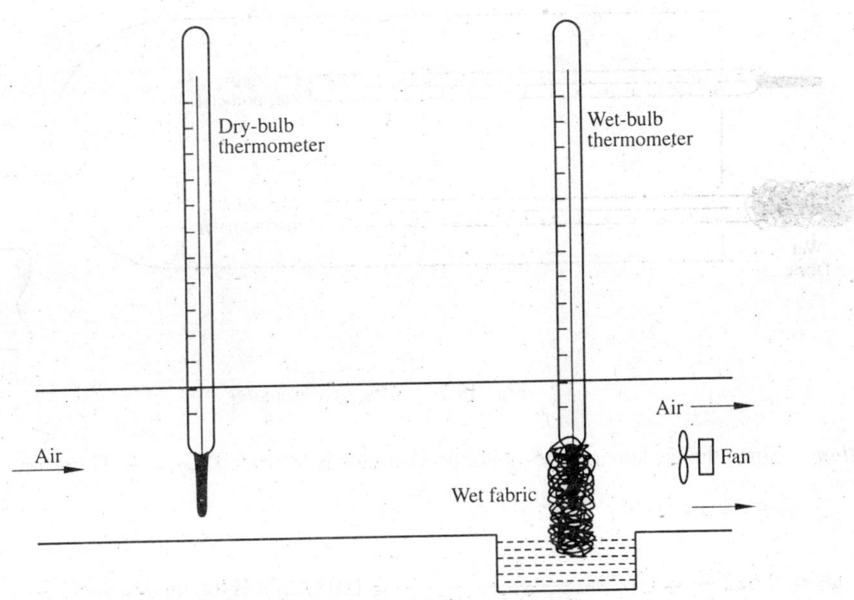

Fig. 10.6 *A simple psychrometer*

A schematic diagram of a simple psychrometer (dry-bulb and wet-bulb thermometer) is shown in Fig. (10.6).

A wick which is always maintained as wet surrounds one of the thermometers. The water is drawn up the wick by capillary action. When the air flows through the instrument, the first thermometer reads the temperature of the entering air and it is called the dry bulb temperature (T_{DB}). The air gets saturated while passing over the wet wick, and its temperature reduces, which is recorded by the second thermometer called the wet-bulb thermometer. The temperature recorded by the wet-bulb thermometer is called the wet-bulb temperature (T_{WB}).

The dry-bulb temperature is simply the temperature of the mixture. The wet-bulb temperature depends on the rates of energy, and mass transfer from the water to the air, energy transfer by radiation from the surroundings, air velocity and other factors. Therefore, the wet-bulb temperature depends upon the type of instrument also and is not just a property of the mixture. On the other hand, the adiabatic saturation temperature is governed by the equilibrium between the two phases and is a property of the system. It has been experimentally found that for air–water vapor mixtures at atmospheric pressures, the wet-bulb temperature is very close to the adiabatic saturation temperature. For other mixtures like alcohol–air, this need not be true.

A sling psychrometer shown in Fig. (10.7) is also commonly used to determine the dry-bulb and wet-bulb temperatures. The sling psychrometer containing the dry-bulb and wet-bulb thermometers, can be whirled around in the atmospheric air to provide the required air flow over the wet wick. Then the readings on the thermometers can be recorded after they attain a steady state.

***Example* 10.8** A mixture of air and water vapor enters an adiabatic saturator, shown in Fig. 10.5, at 40°C and 1 atm. pressure (101.325 kPa) and leaves at 30°C and 1 atm. pressure. Determine the specific humidity and relative humidity of the inlet air.

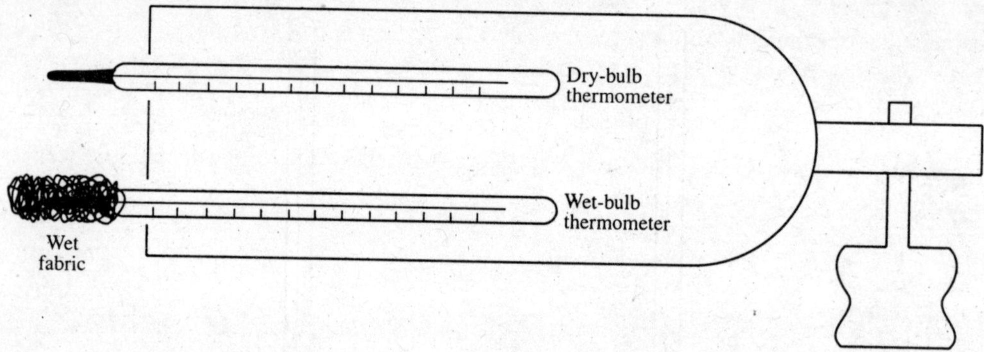

Fig. 10.7 *Sling psychrometer*

Solution Since the air leaving the adiabatic saturator is saturated, $\omega_2 = 1$. Therefore,

$$p_{w2} = P_s = 4.241 \text{ kPa} \quad (\text{at } 30°\text{C})$$

$$\omega_2 = 0.622\frac{p_w}{p_a} = 0.622\frac{4.241}{(101.325 - 4.241)} = 0.0272 \text{ kg H}_2\text{O/kg dry air}$$

$$h_{fg2} = 2430.74 \text{ kJ/kg} \quad h_{g1} = 2574.4 \text{ kJ/kg}; \quad h_{f2} = 125.66 \text{ kJ/kg}$$

$$C_P(air) = 1.005 \text{ kJ/kg K}$$

The specific humidity of the inlet air can be estimated by making use of the adiabatic saturation relation:

$$\omega_1 = \frac{C_P(T_2 - T_1) + \omega_2 h_{fg2}}{h_{g1} - h_{f2}} = \frac{1.005(-10) + 0.0272 \times 2430.74}{2574.4 - 125.66}$$

$$\omega_1 = 0.0229 \text{ kg H}_2\text{O/kg dry air}$$

$$\omega_1 = 0.0229 = 0.622\frac{p_w}{p_a} = 0.622\frac{p_w}{101.325 - p_w}$$

Hence, $p_w = 3.598$ kPa

Relative humidity ϕ_1 is given by:

$$\phi_1 = \frac{p_w}{P_s} = \frac{3.598}{7.375} = 0.4879$$

(Saturation pressure at 40°C = 7.375 kPa)

10.7 Psychrometric Chart

The properties of air–water vapor mixtures can be evaluated by making use of the adiabatic saturation relation. Since the air–water vapor mixtures are encountered in the field of air-conditioning, necessitating the use of the adiabatic saturation relation several times, its solution has been presented in a convenient and ready to use graphical form. Such a graphical representation of the solution of the adiabatic saturation relation is called a psychrometric chart. The dry-bulb temperature, the wet-bulb temperature, the specific humidity, the relative humidity and the enthalpy of the mixture are the usual properties which are shown on a psychrometric chart. The psychrometric charts are usually based on a total pressure of one standard atmosphere (101.325 kPa). If the total pressure of the mixture is fixed, lines of constant relative humidity and wet-bulb temperature can be drawn on the chart.

Usually the air–water vapor mixture enthalpy is expressed on the basis of the dry air, that is, in terms of kJ/kg dry air, and is denoted by $h^\star$.

$$h^\star = h_a + \omega h_w = h_a + \omega h_g \tag{10.48}$$

That is, $h^\star$ represents the enthalpy of 1 kg of dry air and the enthalpy of the accompanying water vapor. The enthalpy of the air is measured relative to 0°C whereas h_g is taken from steam tables (the enthalpy of the saturated liquid at 0°C is taken as zero). The adiabatic saturation relation (Eq. 10.44) can be rewritten in terms of the enthalpy of the air as:

$$h_1^\star + (\omega_2 - \omega_1)h_{w3} = h_2^\star \tag{10.49}$$

The value of the second term on the left-hand side of this equation is very small compared to the values of the other terms. Hence, for many engineering applications, it is reasonable to neglect the term $(\omega_2 - \omega_1)h_{w3}$. Then the above equation reduces to

$$h_1^\star = h_2^\star \tag{10.50}$$

That is, in an adiabatic saturator $h^\star$ remains constant. The value of $h_2^\star$ depends on the adiabatic saturation temperature only, because once T_2 is fixed, p_w is fixed and hence ω_2 is fixed. Therefore, the lines of constant $h^\star$ coincide with the lines of constant wet-bulb temperature. A schematic diagram of the psychrometric chart is shown in Fig. 10.8. The psychrometric chart, at a pressure of 101.325 kPa, presented in Appendix Fig. A.3 can be used to solve the problems discussed in this chapter.

***Example* 10.9** A sample of moist air at 1 atm. pressure when tested with a psychrometer gives 40°C and 30°C as the dry-bulb temperature and wet-bulb temperature, respectively. Determine the specific humidity, relative humidity, partial pressure of water vapor and the enthalpy of the mixture. Compare the results read from the psychrometric chart with the values calculated by making use of the adiabatic saturation relation.

Solution Application of the adiabatic saturation relation:

The adiabatic saturation relation is :

$$\omega_1 = \frac{C_P(T_2 - T_1) + \omega_2 h_{fg2}}{h_{g1} - h_{f2}}$$

In the previous problem we estimated:

$$\omega_1 = 0.0229 \text{ kg } H_2O/\text{kg dry air}$$

$$\phi_1 = 0.4879 \quad \text{and} \quad p_w = 3.598 \text{ kPa}$$

$$h_1^\star = h_2^\star = h_{a1} + \omega_1 h_{g1}$$
$$= 1.005 \times 40 + 0.0229 \times 2574.4 = 99.15 \text{ kJ/kg dry air}$$

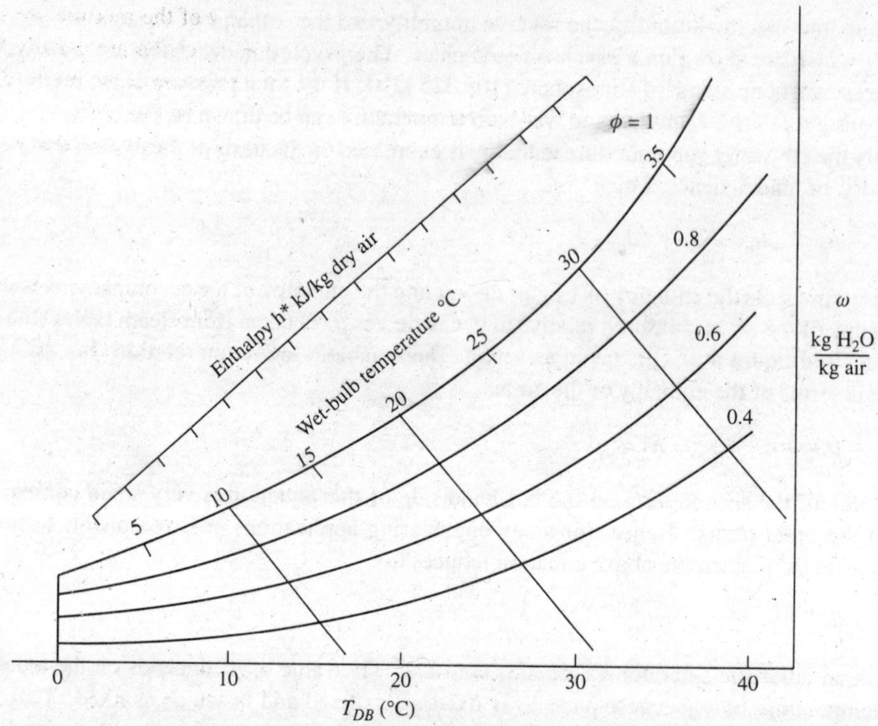

Fig. 10.8 *Schematic diagram of psychrometric chart*

Use of psychrometric chart

Locate 40°C on T_{DB} axis, and the 30°C constant wet-bulb temperature line. The intersection of the $T_{WB} = 30$°C with $T_{DB} = 40$°C fixes the state of the gas mixture, (shown as point *A* in Fig. 10.9). From the chart read:

$$\phi = 50\%; \quad h^* = 101 \text{ kJ/kg dry air}$$

$$\omega = 0.0230 \, kg \text{ H}_2\text{O/kg dry air}$$

10.8 Air-Conditioning

Human beings would like to live in an environment which is neither hot nor cold: neither very humid nor dry. Modern air-conditioning systems, which can cool, heat, humidify and dehumidify the air—in other words, to condition the air to the desired levels—have been developed to provide comfortable environment for living. The human body generates heat which is to be rejected to the surroundings. The rate of heat generation depends on the activity. It varies from 90 W while sleeping to 440 W when doing heavy physical work. The rate of heat transfer from the body depends on the temperature difference between the human body and the surroundings. The average temperature of a human body

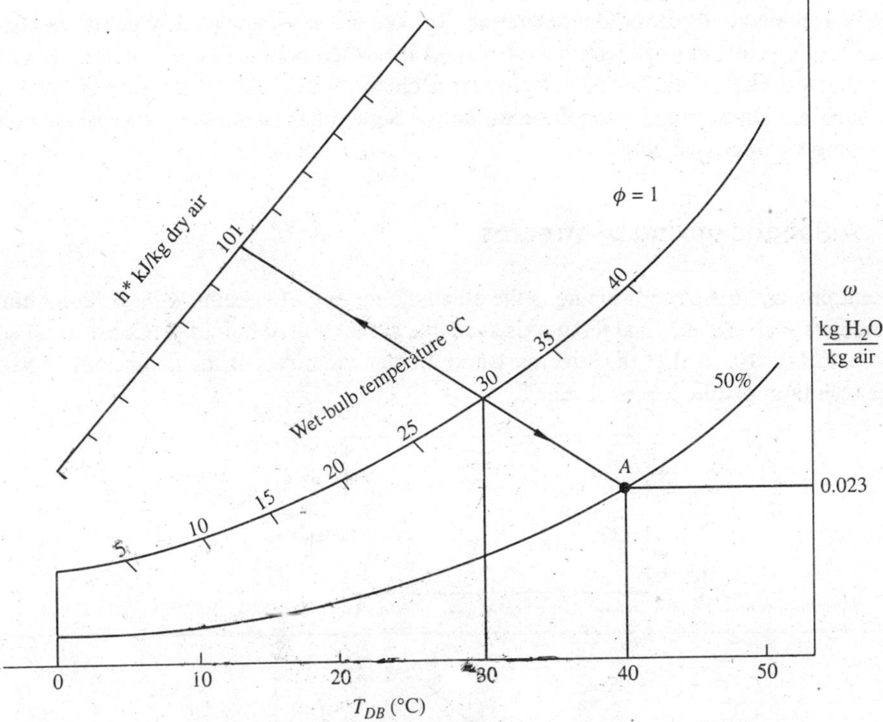

Fig. 10.9 *Sketch for Example 10.9*

is 37°C. In cold environment, that is in winter, the body looses more heat than it normally generates which results in discomfort. In hot environment, that is in summer, the heat transfer from the body to the surroundings at a temperature higher than the body temperature is not possible by normal heat transfer modes. In such hot environment the body helps out by perspiring or sweating more. As this sweat evaporates, it absorbs heat from the body and cools it. If the relative humidity of the surrounding air is very high, the rate of evaporation of the sweat from the body will be low. Therefore, the two important factors which are to controlled, for comfortable living, are temperature and relative humidity of the surrounding air. For comfortable living the surrounding air temperature should be in the range 22°C to 27°C and the relative humidity of the air should be in the range 40 to 60 percent.

The theory of air–water vapor mixtures is extensively used in the field of air-conditioning. Heating, cooling, humidifying and dehumidifying, and adiabatic mixing of different streams of air are the various important processes which are normally encountered in the design of air-conditioning systems. These processes can be conveniently represented on a psychrometric chart. The design or operation of the air-conditioning equipment is not discussed here; but, the thermodynamic analysis of the streams with particular reference to the estimation of humidity is discussed in this section.

Suppose we are interested in dehumidifying the air by removing part of the moisture. The humidity of the air can be reduced by bringing it into contact with silica gel, lithium chloride solution or some other hygroscopic material which absorbs water. These methods are usually expensive and cannot be adopted when a large quantity of air is to be dehumidified. Hence, an alternative and less expensive method has to be found. In general, the air can be cooled below its dew point reducing its humidity, and then it can be reheated to the desired temperature. To cool the air below its dew point, the air may be passed over cooling coils through which a refrigerant at low temperatures is circulated. Alternatively, the air can be cooled below its dew point by spraying chilled water into the air. One of these methods is usually used in commercial air-conditioning units. Some of the common processes employed in air-conditioning are discussed below.

10.8.1 Adiabatic mixing of streams

An important process in air-conditioning is the adiabatic mixing of streams with different humidities and temperatures such that the final mixture is available at the desired humidity. Consider an adiabatic mixer shown in Fig. 10.10. Let the streams 1 and 2 enter the mixer at the conditions shown in the figure. The resulting mixture leaves at state 3.

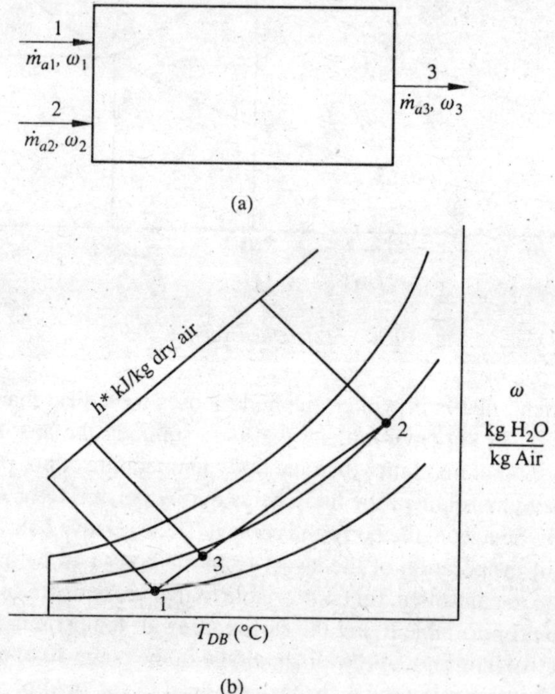

Fig. 10.10 *(a) Adiabatic mixer (b) Psychrometric chart showing adiabatic mixing process*

Considering the mixer as the control volume, the material and energy balance equations can be written as:

Mass balance for air:

$$\dot{m}_{a1} + \dot{m}_{a2} = \dot{m}_{a3} \tag{10.51}$$

Mass balance for water:

$$\dot{m}_{a1}\omega_1 + \dot{m}_{a2}\omega_2 = \dot{m}_{a3}\omega_3 \tag{10.52}$$

Energy balance:

$$\dot{m}_{a1}h_1^\star + \dot{m}_{a2}h_2^\star = \dot{m}_{a3}h_3^\star \tag{10.53}$$

Eliminating $\dot{m}_{a3}$ from the above relations, we get:

$$\frac{\dot{m}_{a1}}{\dot{m}_{a2}} = \frac{\omega_2 - \omega_3}{\omega_3 - \omega_1} = \frac{h_2^\star - h_3^\star}{h_3^\star - h_1^\star} \tag{10.54}$$

That is, the resulting stream 3 lies on a straight line joining the two streams 1 and 2. The line is divided into two segments at point 3 in inverse proportion to the masses of the two streams. We can use Eq. (10.54) to determine the state of the stream 3. In Fig. 10.10(b) points 1 and 2 represent the initial states of the streams and point 3 represents the state of the resultant mixture.

***Example* 10.10** If 10 m^3/min of air at 1 atm and 20°C with $\phi_1 = 90\%$ is mixed with 30 m^3/min of air at 1 atm. and 40°C with $\phi_2 = 20\%$, determine the state of the resulting mixture.

Solution Stream 1:

$$\phi_1 = 0.90 = \frac{p_w}{P_s} = \frac{p_w}{2.337}$$

$$p_w = 2.1033 \text{ kPa}$$

$$\omega_1 = 0.622\frac{p_w}{P_a} = 0.622\frac{2.1033}{(101.325 - 2.1033)}$$
$$= 0.0132\ kg\ H_2O/\text{kg dry air}$$

From the psychrometric chart:

$$\omega_1 = 0.0132 \text{ kg } H_2O/\text{kg dry air}$$

which is same as the calculated value. In view of this observation the required data for the other streams is directly read from the chart.

Mole fraction of water

$$= \frac{p_w}{P} = \frac{2.1033}{101.325} = 0.0208$$

Assuming the mixture to be an ideal gas, we get:

$$v_1 = \frac{RT}{P} = \frac{8.314 \times 10^3 \times 293}{101.325 \times 10^3} = 24.054 \text{ m}^3/\text{kmol}$$

Then,

$$10 \text{ m}^3/\text{min} = \frac{10}{24.054} = 0.4157 \text{ kmol/min}$$

Flow rate of dry air = 0.4157(0.9792) = 0.407 kmol/min, or

$$\dot{m}_{a1} = 0.407 \times 28.97 = 11.791 \text{ kg/min}$$

Stream 2:

$$T_{DB} = 40°C; \quad \phi 2 = 0.2$$

From psychrometric chart:

$$\omega_2 = 0.009 \text{ kg } H_2O/\text{kg dry air}$$

$$p_w = 1.3 \text{ kPa}$$

Mole fraction of water vapor

$$= \frac{1.3}{101.325} = 0.0128$$

$$v_2 = \frac{8.314 \times 10^3 \times 313}{101.325 \times 10^3} = 25.683 \text{ m}^3/\text{kmol}$$

Therefore,

$$30 \text{ m}^3/\text{min} = \frac{30}{25.683} = 1.168 \text{ kmol/min}$$

Flow rate of dry air = 1.168(1 − 0.0128) = 0.985 kmol/min, or

$$\dot{m}_{a2} = 0.985 \times 28.97 = 28.535 \text{ kg/min}$$

$$\frac{\dot{m}_{a1}}{\dot{m}_{a2}} = \frac{\omega_2 - \omega_3}{\omega_3 - \omega_1}$$

$$\frac{11.791}{28.535} = \frac{0.009 - \omega_3}{\omega_3 - 0.0132}$$

Hence $\omega_3 = 0.01$ kg H_2O/kg dry air.

From chart we read $h_2^\star = 64$ kJ/kg dry air and $h_1^\star = 54$ kJ/kg dry air.

$$\frac{\dot{m}_{a1}}{\dot{m}_{a2}} = \frac{h_2^\star - h_3^\star}{h_3^\star - h_1^\star}$$

$$\frac{11.791}{28.535} = \frac{64 - h_3^\star}{h_3^\star - 54} \quad or \quad h_3^\star = 61 \text{ kJ/kg dry air}$$

Streams 1 and 2 are located on the psychrometric chart as shown in Fig. 10.11. Stream 3 lies on the line joining 1 and 2. The values of ω_3 and $h_3^\star$ are calculated. Knowing the value of ω_3, the relative humidity is read as 29% and T_{DB} of the mixture is read as 35.5°C from the chart as shown in Fig. 10.11.

***Example* 10.11** It is required to supply air at 25°C into a room. For this purpose cold and humid air from a desert cooler can be mixed with atmospheric air and the resultant mixture can be supplied into the room. A desert cooler gives 10 m³/min of air at 20°C and 80% relative humidity. The atmospheric air is available at 40°C and 20% relative humidity. Determine the volume of atmospheric air that can be mixed with the cold air.

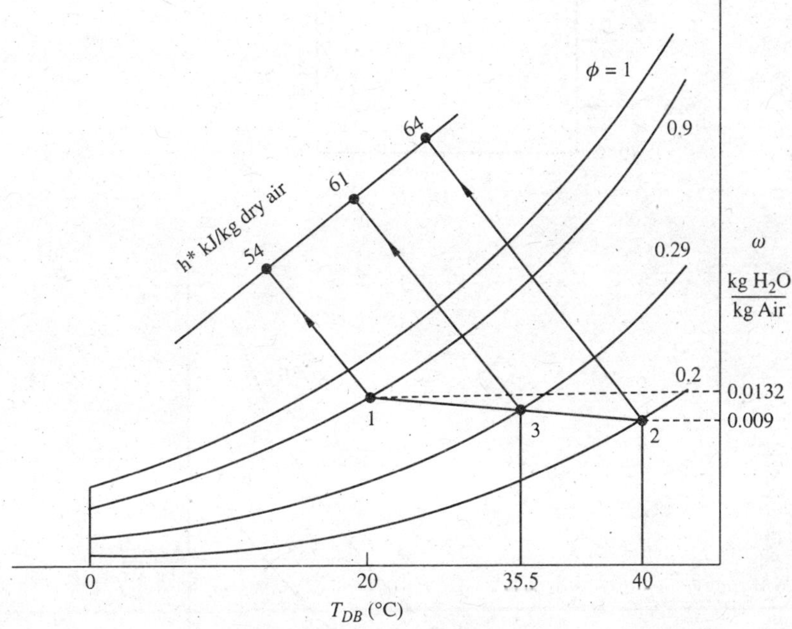

Fig. 10.11 *Sketch for Example 10.10*

Solution $\omega_1 = 0.0118$ kg H_2O/kg air; $p_{w1} = 1.8$ kPa; $\omega_2 = 0.009$ kg H_2O/kg air; $p_{w2} = 1.3$ kPa

Locate (1) and (2) as shown in Fig. 10.12. The intersection of the $T_{DB} = 25°C$ line with the straight line joining (1) and (2) gives the state (3).

$$\omega_3 = 0.011 \text{ kg } H_2O/\text{kg dry air}$$

$$\frac{\dot{m}_{a1}}{\dot{m}_{a2}} = \frac{\omega_2 - \omega_3}{\omega_3 - \omega_1} = \frac{0.009 - 0.011}{0.011 - 0.0118} = 2.5$$

$$\frac{\dot{m}_{a1}}{\dot{m}_{a2}} = \frac{p_{a1}V_1T_2}{p_{a2}V_2T_1} = 2.5 \quad \text{or} \quad V_2 = \frac{p_{a1}V_1T_2}{2.5p_{a2}T_1}$$

$$V_2 = \frac{(101.325 - 1.8)10(313.15)}{2.5(101.325 - 1.3)293.15} = 4.25 \text{ m}^3/\text{min}$$

Hence 4.25 m^3/min of atmospheric air can be mixed with 10 m^3/min of cold air.

10.8.2 Dehumidification

Suppose a mixture of air–water vapor at state B (see Fig. 10.13) is heated at constant pressure. The specific humidity of the mixture does not undergo any change, but relative humidity decreases. If the mixture in state *i* is cooled at constant pressure, its specific humidity does not change until it reaches the saturation temperature (state A); the relative humidity, however, increases. Further cooling results in condensation of the water vapor and the specific humidity decreases. The constant pressure heating and cooling processes are shown on a psychrometric chart in Fig. 10.13.

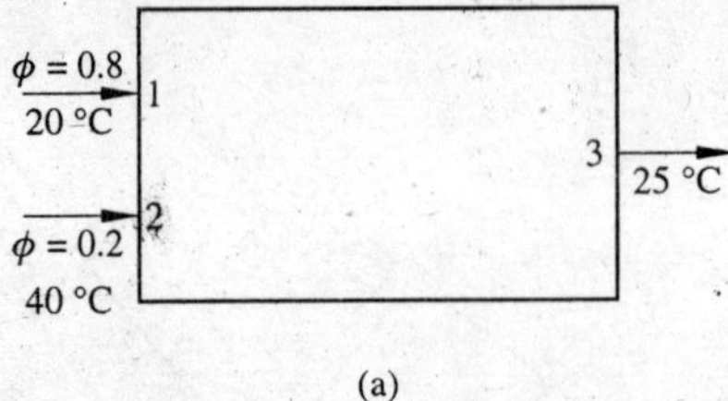

(a)

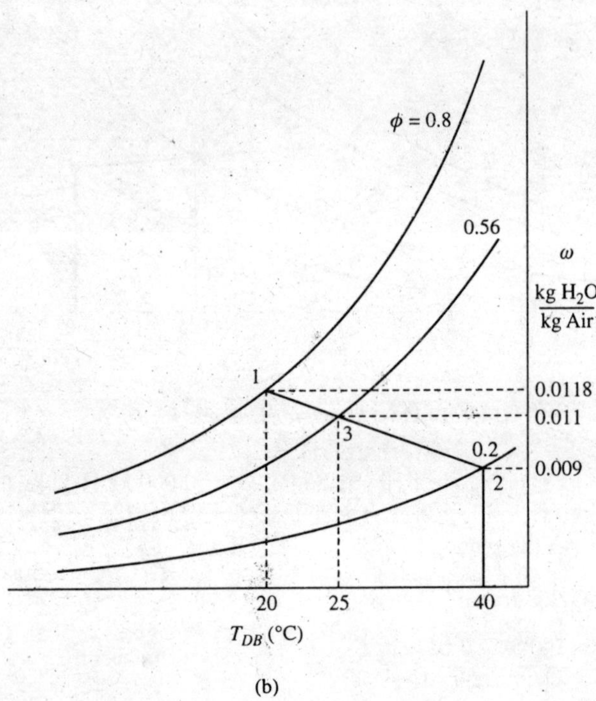

(b)

Fig. 10.12 *Sketch for Example 10.11 (a) Adiabatic mixer (b) Psychrometric chart*

A close observation of the process (*iABf*) suggests that dehumidification of an air–water vapor mixture can be achieved by cooling the mixture below its dew point temperature, allowing some water vapor to condense, and then reheating the mixture to the desired temperature. A schematic diagram of such an air conditioning unit is shown in Fig. 10.14.

In some commercial air-conditioning plants, chilled water below the dew point temperature of the mixture is sprayed into the air to be dehumidified instead of employing the cooling coils. The fine spray provides adequate heat transfer area and the air leaves with less humidity at the temperature of the chilled water. Then the air is reheated to the desired temperature. A schematic diagram of such a dehumidifier is shown in Fig. 10.15.

***Example* 10.12** Moist air at 1 atm. pressure and 60 percent relative humidity enters a dehumidifier unit at 40°C. The air is required to leave the unit at 20°C with 40% relative humidity. If the flow rate of

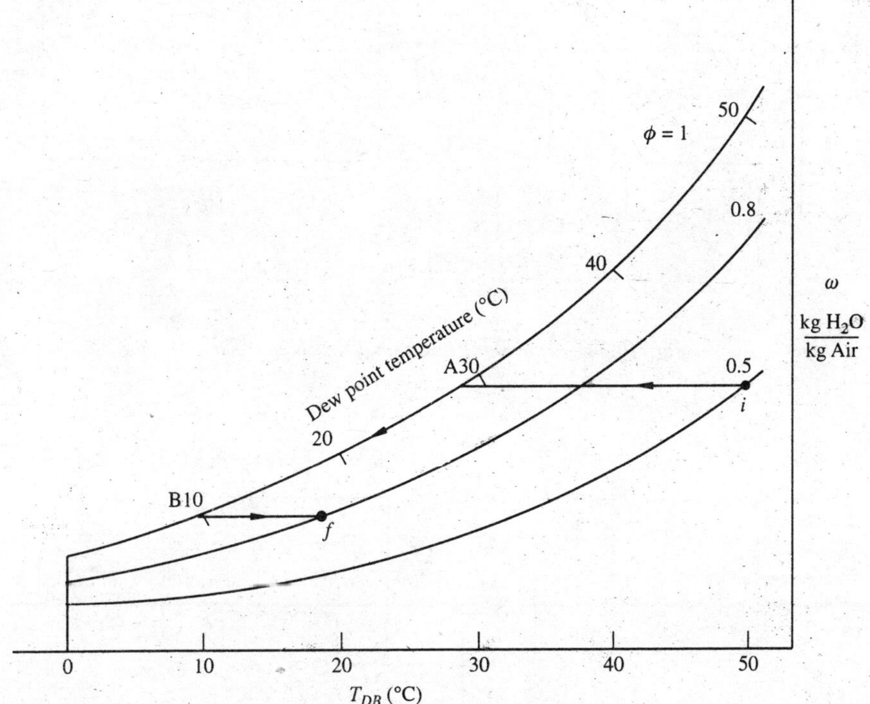

Fig. 10.13 *Psychrometric chart showing constant pressure cooling and heating processes*

the inlet air is 1 m^3/s, determine the rates of cooling and heating required to operate the dehumidifier. Also calculate the amount of water condensed in the refrigeration unit.

Solution $\phi_1 = 0.8$; $T_{DB} = 40°C$.

Hence, $\omega_1 = 0.0284$ kg H_2O/kg dry air;

$p_{w1} = 4.4$ kPa.

$h_1^* = 114$ kJ/kg dry air (from psychrometric chart).

$$\dot{m}_{a1} = \frac{p_{a1}VM}{RT_1} = \frac{(101.325 - 4.4) \times 10^3 \times 28.97}{8.314 \times 10^3 313.15} = 1.0785 \text{ kg/s dry air}$$

Conditioned air:

$$T_{DB} = 20°C; \quad \phi_4 = 0.4$$

Therefore $\omega_4 = 0.0056$ kg H_2O/kg dry air and $h_4^* = 34.5$ kJ/kg dry air.

The air enters the refrigeration unit at state 1, gets cooled below its dew point and leaves as saturated air at state 2. Then the air enters the heating unit and leaves at state 4. The process is shown on the psychrometric chart as 1A24 in Fig. 10.16.

Therefore $\omega_2 = \omega_4 = 0.0056$ kg H_2O/kg dry air.

T_{WB} (at state 2) = 6°C; $\phi_2 = 1.0$; and $h_2^* = 20.5$ kJ/kg dry air

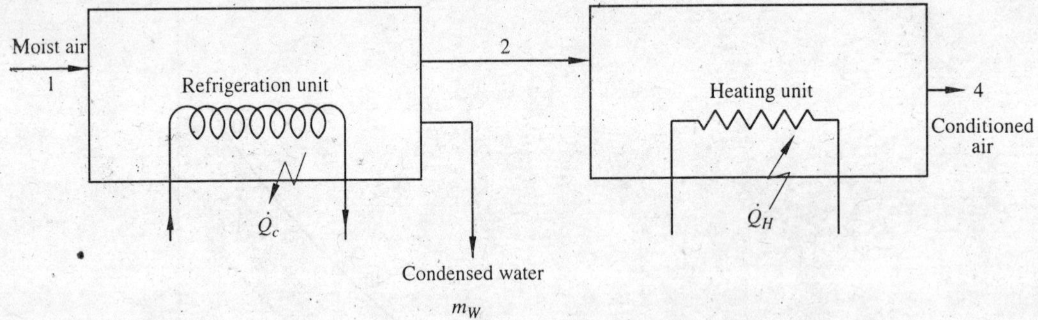

Fig. 10.14 *Schematic diagram of a dehumidifier*

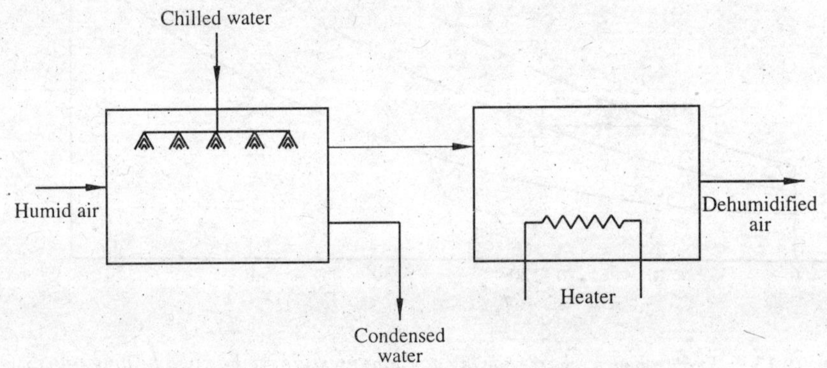

Fig. 10.15 *Schematic diagram of a spray dehumidifier*

Now choose the heating unit as the control volume.
Then energy balance gives

$$\dot{m}_{a2}h_2^{\star} + \dot{Q} = \dot{m}_{a4}h_4^{\star}$$

Mass balance for air:

$$\dot{m}_{a2} = \dot{m}_{a4} \quad \text{or} \quad \dot{Q} = \dot{m}_{a2}(h_4^{\star} - h_2^{\star})$$

Now choose the refrigeration unit as the control Volume. Then,
Mass balance for air:

$$\dot{m}_{a1} = \dot{m}_{a2}$$

Mass balance for water:

$$\dot{m}_{a1}\omega_1 = \dot{m}_{a2}\omega_2 + \dot{m}_w$$

or

$$\dot{m}_w = \dot{m}_{a1}(\omega_1 - \omega_2) = 1.0785(0.0284 - 0.0056) = 0.0246 \text{ kg/s}$$

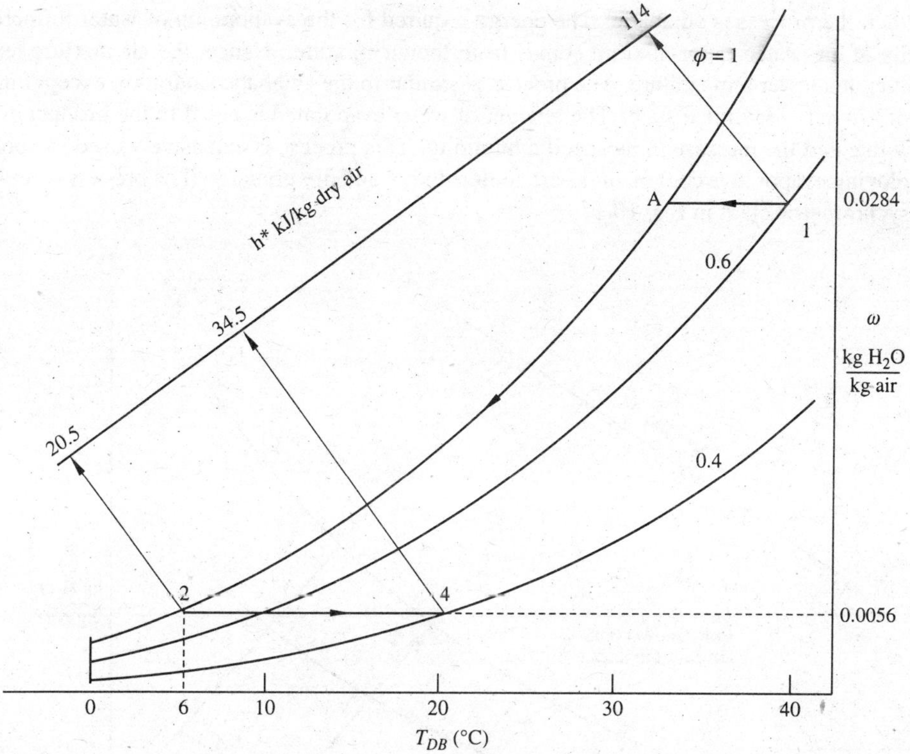

Fig. 10.16 *Sketch for Example 10.12*

Therefore, water condensed in the refrigeration unit = 0.0246 kg/s

Energy balance:

$$\dot{m}_{a1}h_1^* = \dot{m}_{a2}h_2^* + \dot{m}_w h_{w3} + \dot{Q}_c$$

$$\dot{Q}_c = \dot{m}_{a1}(h_1^* - h_2^*) - \dot{m}_w h_{w3}$$

$$= 1.0785(114 - 20.5) - 0.0246(25.21) = 100.2 \text{ kJ/s}$$

($h_w = 25.21$ kJ/kg $= h_f$ at 6°C from steam tables)

Hence Refrigeration capacity = 100.2 kW.

Then the energy balance equation for the heating unit gives

$$\dot{Q} = \dot{m}_{a2}(h_4^* - h_2^*) = 1.0785(34.5 - 20.5) = 15.1 \text{ kJ/s}$$

or the rate of heating required = 15.1 kW.

10.8.3 Humidification with cooling

If an air–water vapor mixture is passed through porous pads soaked in water, the mixture leaves in a saturated or near saturated state. Since no energy transfer occurs between the surroundings and the humidifier, the process is adiabatic. The energy required for the evaporation of water to increase the humidity of air–water vapor mixture comes from the air mixture. Hence, the air mixture leaves the humidifier at a lower temperature. The process is similar to the adiabatic saturation except that the air may not leave in a saturated state. The amount of water evaporated is equal to the product of the dry air flow rate and the increase in its specific humidity. This process is extensively used to cool homes by employing evaporative coolers or desert coolers in hot and dry climates. The process is represented on a psychrometric chart in Fig. 10.17.

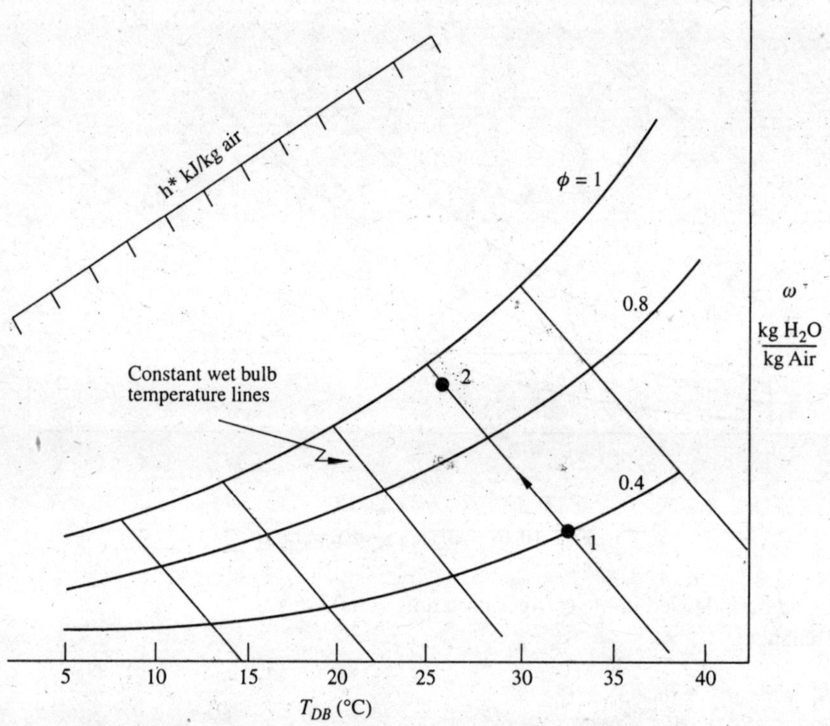

Fig. 10.17 *Psychrometric chart showing humidification with cooling*

***Example* 10.13** On a hot summer day the atmospheric air is at 40°C with a relative humidity of 20%. Suppose a desert cooler is used to increase the relative humidity to 80%, estimate the temperature of the exit air and the minimum temperature to which the air can be cooled by employing a well designed desert cooler.

Solution Atmospheric air:

$$T_{DB} = 40°\text{C}; \quad \phi_1 = 0.20$$

The evaporative cooling process is shown on a psychrometric chart in Fig. 10.18. Locate the initial state 1 at the intersection of $T_{DB} = 40°\text{C}$ and $\phi = 0.20$ and follow the constant wet-bulb temperature line till it meets $\phi = 0.80$ curve (shown as point 2). Then read the temperature at point 2 as 24.8°C.

When ϕ is raised to 80%, $T_{DB} = 24.8°C$ = Temperature of the air leaving the desert cooler.

Now follow the constant wet-bulb temperature line from point 1 till it intersects $\phi = 1$ curve and read the temperature of state 3 a 22°C. Therefore, the minimum temperature that can be achieved is 22°C.

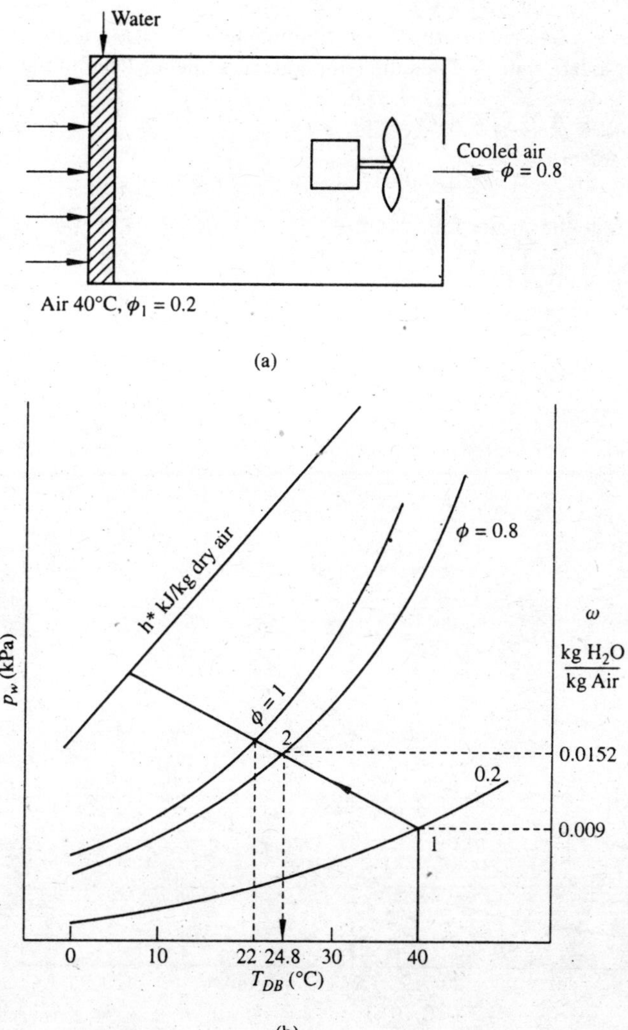

Fig. 10.18 *Sketch for Example 10.13 (a) evaporative cooler (b) representation of evaporative cooling process on a psychrometric chart*

***Example* 10.14** On a hot summer day the conditions of the atmospheric air are 49°C and 20% relative humidity. If an evaporative cooler is used, which delivers the air with 90% relative humidity, determine the temperature of the cold air that leaves the cooler. If the cooler handles 1 m^3/s of dry air, estimate the amount of water required to operate the cooler for a period of 5 hours.

Solution Atmospheric air:

$$T_{DB} = 49°\text{C}; \quad \phi_1 = 0.2$$

$$\omega_1 = 0.0147\ kg\ \text{H}_2\text{O/kg dry air}$$

Locate the initial state 1 as shown in Fig. 10.19. Follow the $h^{\star}$ = constant line from 1 till it intersects with $\phi = 0.9$ line and locate state 2. Then the temperature of the air leaving the cooler is 29°C.

$$T_2 = 29°\text{C}; \quad \omega_2 = 0.023\ kg\ \text{H}_2\text{O/kg dry air}$$

$$\dot{m}_w = \dot{m}_{a1}(\omega_2 - \omega_1) = 1(0.023 - 0.0147) = 0.0083\ \text{kg/s}$$

Water required to operate the cooler for 5 hours

$$= 0.0083 \times 5 \times 3600 = 149.4\ \text{kg}.$$

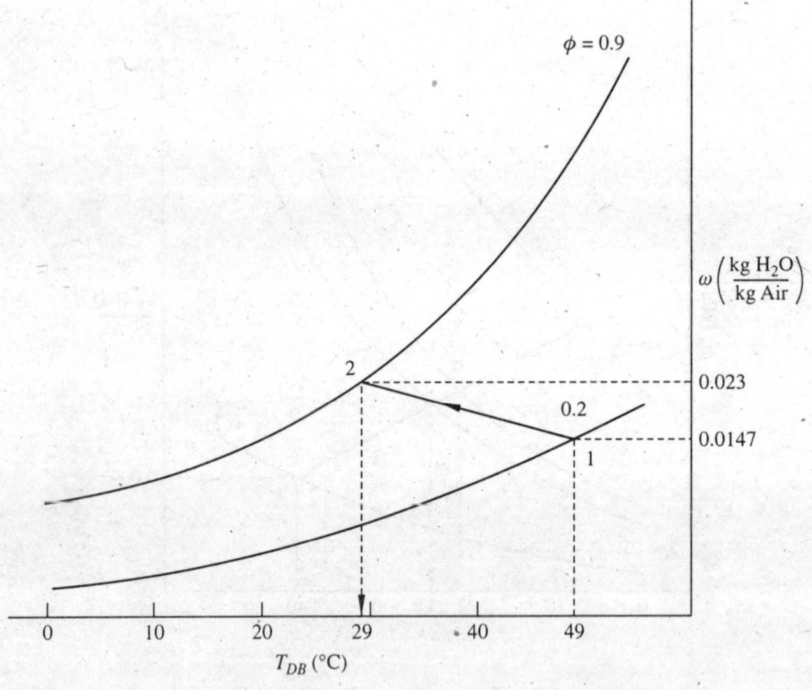

Fig. 10.19 *Sketch for Example 10.14*

10.9 Cooling Towers

A cooling tower is a device which uses the principle of evaporative cooling to cool warm water. Warm water is cooled by evaporation and in the process the humidity of the air increases. Since cooling is by evaporation, it is theoretically possible to have the cold water at the wet-bulb temperature of the air.

Condensers of thermal power plants, large scale refrigeration and air conditioning units require large quantity of cooling water. Hence, cooling towers are extensively used to provide large quantity of cold water at a temperature lower than the dry-bulb temperature of the ambient air.

There are two types of cooling towers used in industry: natural draft cooling towers and forced draft cooling towers. Warm water is sprayed at the top of the tower and the fine spray provides a large surface area for energy transfer and evaporation. At the same time atmospheric air is blown at the bottom which leaves at the top of the tower in nearly saturated state. Natural draft and forced draft cooling towers are shown in Fig. 10.20.

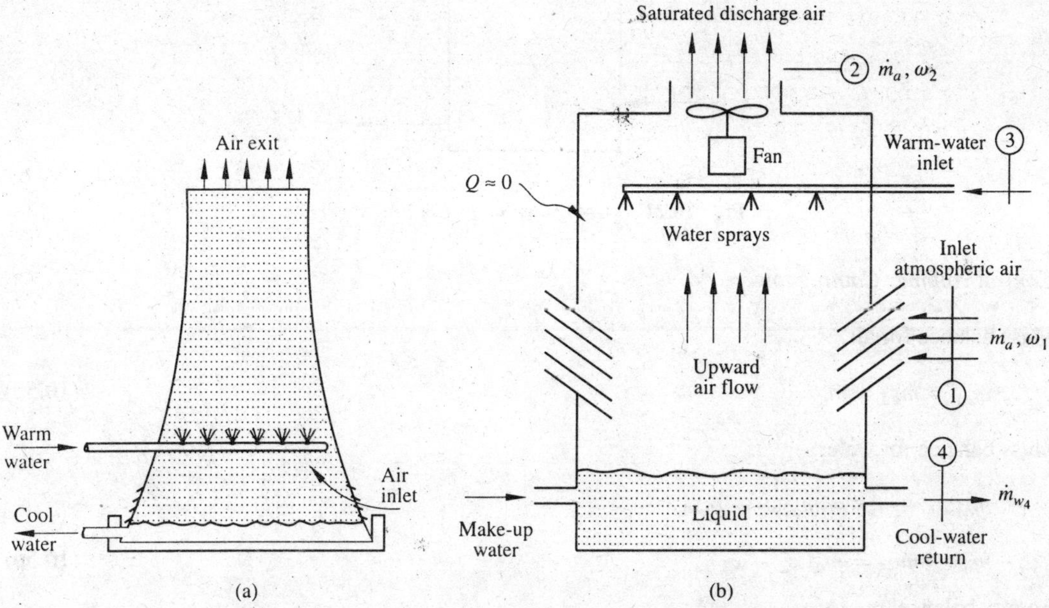

Fig. 10.20 *Schematic diagram of (a) Natural draft cooling tower (b) Forced draft cooling tower*

In cooling towers, since the process of cooling is by evaporation, the relative humidity of the entering atmospheric air is important. If the atmospheric air is saturated, cooling does not take place in the evaporative cooling towers. Depending upon the relative humidity of the entering air, the air flow rate is to be adjusted to cool the water to the desired temperature. As the air is humidified by evaporation of the water, make up water has to be added in the cooling towers. Since there is a temperature difference between the air and water, energy transfer as heat occurs from water to the air. However, in certain regions of the tower, the temperature of the water may fall below the air temperature and energy transfer occurs from air to the water. To analyze the complex processes that take place in a cooling tower, a knowledge of energy transfer through various modes like conduction, convection and radiation, and mass transfer and fluid mechanics is needed in addition to a knowledge of thermodynamics. However, we can obtain some information regarding the performance of cooling towers by applying the mass balance and energy balance. In such an analysis we treat the cooling tower as a control volume as shown in Fig. 10.21 and ignore the processes that take place in it.

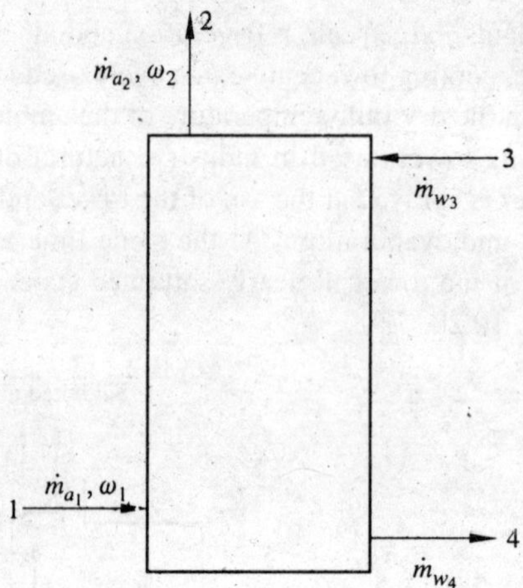

Fig. 10.21 *Control volume: Cooling tower*

Control volume: Cooling tower

Mass balance for air:

$$\dot{m}_{a1} = \dot{m}_{a2} = \dot{m}_a \tag{10.55}$$

Mass balance for water:

$$\dot{m}_{a1}\omega_1 + \dot{m}_{w3} = \dot{m}_{a2}\omega_2 + \dot{m}_{w4}$$

$$\dot{m}_{w3} - \dot{m}_{w4} = \dot{m}_a(\omega_2 - \omega_1) \tag{10.56}$$

Energy balance:

$$\dot{m}_{a1}h_1^\star + \dot{m}_{w3}h_3 = \dot{m}_{a2}h_2^\star + \dot{m}_{w4}h_4$$

$$\dot{m}_{w3}h_3 - \dot{m}_{w4}h_4 = \dot{m}_a(h_2^\star - h_1^\star) \tag{10.57}$$

Eqs. (10.56) and (10.57) can be solved to obtain:

$$\dot{m}_a = \frac{\dot{m}_{w3}(h_3 - h_4)}{(h_2^\star - h_1^\star) - (\omega_2 - \omega_1)h_4} \tag{10.58}$$

In the analysis of cooling towers, the value of $\dot{m}_{w3}$ is known in addition to the temperatures of the entering and leaving water. A knowledge of the conditions of the inlet and outlet air allows one to determine the mass flow rate of dry air required and the amount of make-up water.

The principle of evaporative cooling is also used in spray ponds, where warm water is sprayed into air and is cooled by the air as it falls into the pond. In spray ponds the water loss due to air drift is quite high and the water is unprotected from dust.

***Example* 10.15** An evaporative cooling tower is used in a large thermal power plant to provide cold water at a temperature of 30°C to the condensers. The warm water leaves the condensers at 40°C at a rate of 1000 kg/s. The atmospheric air is at 40°C with $\phi = 0.2$ and leaves the cooling tower at 35°C with 80% relative humidity. Determine the make up water required and the air flow rate.

Solution The conditions of the air entering the cooling tower and the air leaving the cooling tower are shown in Fig. 10.22.

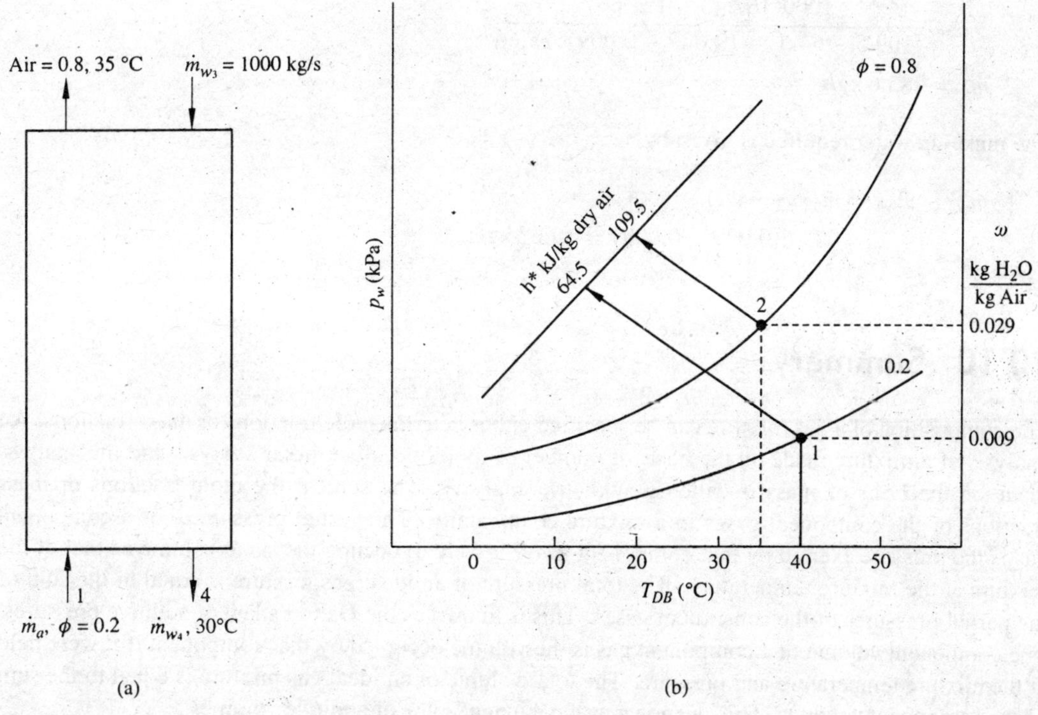

Fig. 10.22 *Sketch for Example 10.15 (a) Cooling tower (b) Psychrometric chart showing the conditions of air*

Entering air:

$$T_{DB} = 40°C; \quad \phi_1 = 0.2$$

$$\omega_1 = 0.009 \text{ kg H}_2\text{O/kg dry air}$$

$$h_1^\star = 64.5 \text{ kJ/kg dry air}$$

Leaving air:

$$T_{DB} = 35°C; \quad \phi_2 = 0.8$$

$$\omega_2 = 0.029 \; kg \text{ H}_2\text{O/kg dry air}$$

$$h_2^\star = 109.5 \text{ kJ/kg dry air}$$

From steam tables:

$$h_3(h_f \; at \; 40°C) = 167.45 \text{ kJ/kg}$$

$$h_4(h_f \text{ at } 30°\text{C}) = 125.66 \text{ kJ/kg}$$

Now, we can use Eq. (10.58) to calculate the flow rate of dry air.

$$\dot{m}_a = \frac{\dot{m}_{w3}(h_3 - h_4)}{(h_2^* - h_1^*) - (\omega_2 - \omega_1)h_4}$$

$$= \frac{1000(167.45 - 125.66)}{(109.5 - 64.5) - (0.029 - 0.009)125.66}$$

$$\dot{m}_a = 983.6 \text{ kg/s}$$

The make-up water required is given by

$$\dot{m}_{w3} - \dot{m}_{w4} = \dot{m}_a(\omega_2 - \omega_1)$$

$$= 983.6(0.029 - 0.009) = 19.67 \text{ kg/s}$$

10.10 Summary

The composition of a gas mixture can be specified either in terms mole fractions or mass fractions. An analysis of a mixture made on the basis of number of moles is called molar analysis and the analysis made on the basis of mass is called gravimetric analysis. The sum of the mole fractions or mass fractions of the component gases in a mixture equals unity. The partial pressure p_i of a component gas is the pressure exerted by that component gas if it were to occupy the same volume as that of the mixture at the mixture temperature. The total pressure of an ideal gas mixture is equal to the sum of the partial pressures of the constituent gases. This is known as the Dalton's law of additive pressures. The component volume of a component gas is the volume occupied by that component if it were held at the mixture temperature and pressure. The total volume of an ideal gas mixture is equal to the sum of the component volumes. This is known as the Amagat's law of additive volumes.

For an ideal gas mixture the volume fraction and the pressure fraction of a component gas is equal to its mole fraction. The properties of an ideal gas mixture can be evaluated in terms of the properties of the constituent gases and mole fractions or mass fractions of the constituent gases.

The temperature at which the vapor condenses when the air–water-vapor mixture is cooled at constant pressure is called the dew point temperature. The specific humidity or humidity ratio ω is defined as the ratio of the mass of water vapor to the mass of dry air in the mixture. The specific humidity is frequently expressed in terms of partial pressure of water vapor as $\omega = 0.622p_w/(P - p_w)$. The relative humidity ϕ is defined as the ratio of the partial pressure of water vapor in the mixture to the saturation pressure of water at the mixture temperature. The specific humidity or the relative humidity of an air–water-vapor mixture is measured with an adiabatic saturator or psychrometer. The temperature of a gas mixture is usually called the dry-bulb temperature. The wet-bulb temperature is the temperature attained by the air–water-vapor mixture when it is adiabatically saturated. A knowledge of wet-bulb temperature and dry-bulb temperature is sufficient to determine the properties of air–water-vapor mixture. In order to simplify calculations, the specific humidity, the relative humidity and the enthalpy of air–water-vapor mixtures are shown on a psychrometric chart. Usually the mixture enthalpy h^* is expressed on the basis of dry air. That is, h^* represents the sum of the enthalpies of one kg dry air and the accompanying water vapor.

Heating, cooling, humidifying, dehumidifying and adiabatic mixing of different streams of air are the various important processes which are normally encountered in the design of air-conditioning systems. These processes are discussed and the use of psychrometric chart is illustrated through several examples.

10.11 Review Questions

1. What are the different methods of expressing the composition of a gas mixture?
2. What is meant by gravimetric analysis?
3. What is meant by molar analysis?
4. How many variables are required for complete description of a thermodynamic system consisting of n constituents?
5. What is the relation between the mass of a component and its molar mass?
6. How does one estimates the molar mass of a mixture of gases?
7. What is the difference between dry air and atmospheric air?
8. What is component pressure and what is partial pressure? In which case the partial pressure is equal to component pressure?
9. State the Dalton's law of additive pressures.
10. What is Amagat's law of additive volumes?
11. What is meant by component volume?
12. What is pressure fraction of a component and what is volume fraction of a component, in a gas mixture?
13. What is the relation between pressure fraction, volume fraction and mole fraction of a component in an ideal gas mixture?
14. How is it possible to assume that water vapor behaves as an ideal gas in analyzing the mixtures of air and water vapor?
15. What is saturated air–water-vapor mixture?
16. Define the terms specific humidity and relative humidity.
17. What is dew point temperature?
18. How does one measures the specific humidity and relative humidity of a mixture of air–water vapor?
19. What is dry-bulb temperature and what is wet-bulb temperature?
20. What is adiabatic saturation temperature?
21. What is the difference between adiabatic saturation temperature and wet-bulb temperature?
22. For which air–water-vapor mixtures the wet-bulb temperature becomes equal to the dry-bulb temperature?
23. What is a psychrometer?
24. Sketch a psychrometric chart and indicate the lines of constant wet-bulb temperature and constant relative humidity on the chart.
25. How is the enthalpy of air–water-vapor mixture defined?
26. What is meant by air-conditioning?
27. Suppose an unsaturated air–water-vapor mixture is heated at constant pressure. Does the specific humidity increase or decrease? Does the relative humidity of the mixture change?

28. When does the moisture starts condensing if an unsaturated air–water-vapor mixture is cooled at constant pressure?
29. Suppose two streams of air–water-vapor mixtures of known specific humidity and temperatures are adiabatically mixed in a flow device. Is it possible to determine the state of the resultant mixture? Explain with the help of a psychrometric chart.
30. Sketch the process of dehumidification with cooling on a psychrometric chart.
31. What is evaporative cooling?
32. Which principle is made use of in the design of desert coolers?
33. What is a cooling tower and how does it work?
34. Why does the water kept in a porous earthen pot which is exposed to flowing air in shade gets cooled?
35. Is it possible to obtain saturated air from unsaturated air without adding any water? Explain.
36. It is usually observed that the outer surface of a glass filled with iced water is covered with a thin layer of water droplets. How do you explain this phenomenon?
37. Suppose a person wearing spectacles goes for a long walk in winter when the outside temperature is 10°C. After the walk, he enters a room which is maintained at 25°C and 40% relative humidity. Determine whether the spectacles glasses become fogged or not.
38. On a cold winter day if a person wearing spectacles moves out of a house which is held at 25°C and 40% relative humidity and goes for a walk in the open air, determine whether his spectacle glasses become fogged or not.

10.12 Problems

10.1 A mixture of gases has the following gravimetric analysis.

Constituent	CO_2	O_2	N_2	H_2	H_2O
Mass %	25	20	25	20	10

Determine:

(a) the composition of the mixture on a volume basis.
(b) the mole numbers of the mixture containing 100 kg of mixture at 100 kPa and 400 K.
(c) the specific heat at constant pressure for the mixture.

10.2 Suppose the mixture of Problem 10.1 is initially at 100 kPa and 400 K. Determine the partial pressure of each constituent in the mixture. The mixture is brought into contact with silica gel in a closed container which absorbs the water vapor. Assuming that the process takes place at constant temperature, determine the final pressure of the gas mixture and its composition by volume.

10.3 Suppose the mixture of Problem 10.1 is initially at 400 K and 100 kPa. The mixture is treated in a closed vessel with a solution of sodium hydroxide which preferentially absorbs CO_2. If the process takes place at constant temperature, determine the final pressure of the gas mixture and its molar composition.

10.4 The gas mixture of Problem 10.1 at 100 kPa and 400 K is first treated, in a closed vessel, with a solution of sodium hydroxide at constant temperature. Then it is brought into contact with anhydrous calcium carbonate which absorbs the water vapor in a closed vessel. The temperature of the mixture is maintained at 400 K. Determine the final pressure of the gas mixture, the composition by volume and the partial pressures of the constituents.

10.5 A natural gas having the molar analysis 40% propane (C_3H_8), 40% ethane (C_2H_6) and 20% methane (CH_4) enters a combustion chamber at 1 bar and 300 K. Determine the gravimetric analysis of the mixture and the mass flow rate, in kg/s, if the volumetric flow rate is 10 m^3/s.

10.6 A mixture of gases containing of 30 percent N_2, 50 percent He and 20 percent CO_2 by volume is used in a heat exchanger for cooling a highly inflammable oil from 250°C to 50°C. The specific heat of the oil is 1.2 kJ/kg K. Determine the flow rate of the gas mixture required to cool 3600 kg/h of the oil if the gas mixture enters the heat exchanger at 30°C and leaves at 50°C.

10.7 A gas mixture has the following gravimetric analysis: 50% N_2, 20% CO_2 and 30% He. If the mixture is compressed reversibly and adiabatically from 1 bar and 300 K to 5 bar, in a piston–cylinder assembly, determine the final temperature and the work done, in kJ/kg of mixture. Assume that the gas mixture behaves as an ideal gas.

10.8 An equimolar mixture of CO_2 and CH_4 enters an insulated nozzle at 3 bar and 400 K and expands isentropically to 1 bar. If the mixture enters the nozzle with negligible velocity, determine the final temperature of the mixture and the exit velocity.

10.9 The gravimetric analysis of a mixture of ideal gases is as follows: N_2 = 50%; O_2 = 10%; CO_2 = 20% and SO_2 = 20%. Determine:

(a) the molar mass of the mixture
(b) the mole fractions of CO_2 and SO_2, and
(c) the volume occupied by 10 kg of the mixture at 1 bar and 300 K.

10.10 The volumetric analysis of a gaseous fuel is given by CH_4 = 60%; C_2H_6 = 10%; CO_2 = 10% and the rest is N_2. Determine (a) the mass of 100 m^3 of this mixture at 1 bar and 30°C, and (b) the gravimetric analysis.

10.11 It is necessary to prepare a mixture of helium and carbon dioxide so that the molar mass of the mixture is 30 kg/kmol. Determine the volume ratio in which helium and carbon dioxide are to be mixed.

10.12 The volumetric analysis of a gas mixture is as follows: CO_2 = 30%; O_2 = 20%; N_2 = 30%; He = 10% and Ar = 10%. One kmol of this mixture is cooled from 227°C to 77°C by passing through a steady flow heat exchanger. Determine the heat interaction.

10.13 A rigid and insulated tank of volume 10 m^3 is divided into two equal compartments by a partition. One compartment is filled with Nitrogen at 10 bar and 500 K while the other compartment is filled with Helium at 2 bar and 300 K. The partition is removed and the gases are allowed to mix. Calculate:

(a) the final temperature and pressure of the mixture
(b) the molar composition of the mixture, and
(c) the entropy change associated with the mixing process.

10.14 A rigid and insulated tank of volume 3 m^3 is divided into two compartments. One compartment of volume 2 m^3 contains nitrogen at 5 bar and 500 K, while the second compartment of volume 1 m^3 contains helium at 1 bar and 600 K. The partition in between the two compartments is removed and the gases are allowed to mix until an equilibrium state is attained. Determine the final temperature, pressure and the entropy change.

10.15 An equimolar mixture of Helium and Nitrogen is compressed reversibly and adiabatically in a piston-cylinder assembly from the initial state of 1 bar and 300 K to a final pressure of 5 bar. Calculate:

(a) the final temperature
(b) the entropy change of Helium, and
(c) the entropy change of Nitrogen.

10.16 An equimolar mixture of Helium and Nitrogen at 3 bar and 400 K expands through an adiabatic nozzle to a pressure of 1 bar. If the gas mixture enters the nozzle with negligible velocity, determine the exit velocity.

10.17 An insulated and rigid tank of volume 1 m^3 is divided into two chambers by a partition. One chamber of volume 0.7 m^3 contains Helium at 20 bar and 400 K while the other chamber contains Nitrogen at 10 bar and 500 K. The partition is removed and the gases are allowed to mix. Calculate:

(a) the equilibrium temperature and pressure
(b) the entropy changes of Helium and Nitrogen, and
(c) the total entropy change.

10.18 A mixture of 50 percent propane and 50 percent oxygen by volume is compressed reversibly and adiabatically from 100 kPa and 300 K to 1 MPa. Determine the final temperature of the mixture and the work required to compress 100 moles of the gas mixture.

10.19 A rigid and insulated tank of volume 2 m^3 is divided into two equal compartments by a partition. One compartment contains He at 500 kPa and 300 K while the second compartment contains CO_2 at 1 MPa and 600 K. The partition is punctured and the gases are allowed to mix. Determine:

(a) the temperature and pressure of the mixture
(b) the composition of the mixture by volume
(c) the entropy change of He and CO_2, and
(d) the net entropy change associated with the process.

10.20 Atmospheric air contains approximately 79 percent N_2 and 21 percent O_2 by volume. Determine the minimum work required to separate one kmol of air at 300 K and 100 kPa into pure N_2 and O_2 at the same temperature and pressure.

10.21 The atomic mass of naturally available chlorine which is a mixture of Cl_2^{35} and Cl_2^{37} isotopes is 35.457. Determine the composition by volume of naturally available chlorine. Calculate the entropy of naturally available chlorine at 300 K and 100 kPa with reference to pure Cl_2^{35} and Cl_2^{37}, at the same temperature and pressure.

10.22 The specific heat ratio of the driver gas plays an important role in the operation of a shock-tube. For a particular purpose, a driver gas with a specific heat ratio of 1.5 is required. If a mixture of He and H_2 can be used as a driver gas, determine the composition by volume of the driver gas.

10.23 It is proposed to use methane as fuel in an internal combustion engine. During the suction stroke, 10 moles of air is taken for every mole of methane at 100 kPa and 300 K. If the compression ratio of the engine is 10, determine the pressure and temperature of the mixture at the end of the compression stroke.

10.24 A 5 m^3 tank contains 30% hydrogen, 30% argon and 40% methane by volume at 1 MPa and 300 K. It is necessary to change the composition of the gas mixture such that it contains 60%

methane by volume. Determine the amount of methane to be added. Also determine the final pressure of the mixture in the tank.

10.25 A rigid and insulated tank of volume 1 m^3 contains He at 100 kPa and 300 K. The tank is connected through a valve to a line carrying carbon dioxide at 400 kPa and 600 K. The valve is opened and the CO_2 enters the tank till the pressure inside the tank is 400 kPa. Then the valve is closed and the tank is isolated. Determine

(a) the final temperature of the mixture in the tank
(b) the amount of CO_2 that enters the tank; and
(c) the composition of the gas mixture in the tank.

10.26 A rigid and insulated tank of volume 1 m^3 contains CO_2 at 100 kPa and 300K. The tank is connected through a valve to a line carrying helium at 400 kPa and 600 K. The valve is opened and the helium flows into the tank till the pressure in the tank reaches 400 kPa. Then the valve is closed and the tank is isolated. Determine:

(a) the final temperature of the mixture in the tank
(b) the amount of helium that enters the tank
(c) the composition of the gas mixture in the tank; and
(d) the amount of energy to be removed from the tank to restore it to 300 K (neglect the heat capacity of the tank).

10.27 To study the explosive nature of premixed combustible gases, it is desired to prepare a mixture of propane and oxygen in a tank of volume 1 m^3. The tank is provided with two lines that carry propane and oxygen separately at 500 kPa and 300 K (see Fig. 10.23). The tank is initially evacuated. Oxygen is admitted rapidly until the tank pressure is 100 kPa. Then the oxygen supply valve is closed. The propane supply valve is opened allowing the propane to flow into the tank rapidly. The valve is closed when the pressure in the tank reaches 500 kPa. Estimate the composition of the gas mixture in the tank.

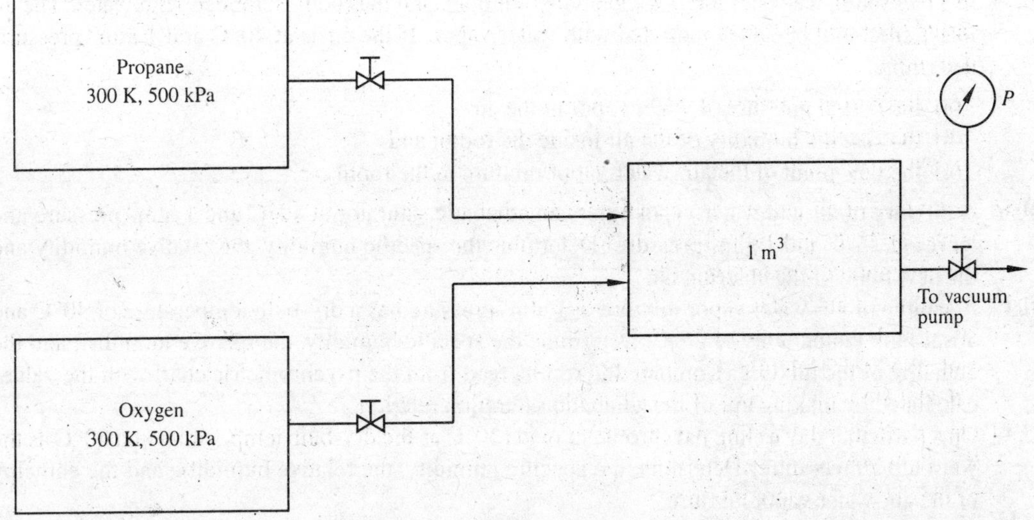

Fig. 10.23 *Sketch for Problem 10.27*

10.28 An auditorium of dimensions 20 m × 40 m × 6 m has air with a relative humidity of 60% at 1 atm. pressure and 30°C. Determine:

(a) the partial pressure of water vapor in the air.
(b) the specific humidity of the air.
(c) the mass of water vapor in the auditorium.
(d) the mass of dry air in the auditorium; and
(e) the dew point.

10.29 The temperature of a wall in a house is at 15°C. If the air in the house is at 20°C, what is the maximum relative humidity the air can have before condensation occurs on the wall.

10.30 An auditorium having a volume of 1×10^6 m^3 contains air at 30°C with a relative humidity of 70%. Determine the mass of water vapor in the auditorium and the dew point temperature.

10.31 Determine the mass of dry air and water vapor contained in a room of dimensions 5 m×7 m×4 m at 35°C and 1 bar. if the relative humidity of air is 70%.

10.32 A rigid tank of 1 m^3 volume contains dry air at 60°C and 3 bar. Water vapor is added until the air gets saturated at the same temperature. Determine (a) the mass of water vapor added, and (b) the final total pressure.

10.33 If the partial pressure of water vapor in the atmospheric air at 40°C is 5.5 kPa, determine:

(a) the specific humidity and relative humidity of the air
(b) the dew point temperature; and
(c) the enthalpy of the air.

10.34 A mixture of air and water vapor has a relative humidity of 40%. The total pressure of the mixture is 80 kPa. Calculate:

(a) the specific humidity
(b) the dew point temperature; and
(c) the enthalpy of the mixture.

10.35 In a bathroom, the water tap is accidentally left open and the room is flooded with water. The air inside the room becomes saturated with water vapor. If the air is at 40°C and 1 atm. pressure, determine:

(a) the partial pressure of water vapor in the air
(b) the specific humidity of the air inside the room; and
(c) the dew point of the air–water vapor mixture in the room.

10.36 A mixture of air and water vapor enters an adiabatic saturator at 40°C and 1 atm. pressure and leaves at 25°C and 1 atm. pressure. Determine the specific humidity, the relative humidity and the dew point of the entering air.

10.37 A sample of air–water vapor mixture at 1 atm. pressure has a dry-bulb temperature of 40°C and a wet-bulb temperature of 25°C. Determine the specific humidity, the relative humidity, and the enthalpy of the mixture. Compare the results read from the psychrometric chart with the values calculated by making use of the adiabatic saturation relation.

10.38 On a particular day a sling psychrometer reads 30°C as the dry-bulb temperature and 20°C as the wet-bulb temperature. Determine the specific humidity, the relative humidity, and the enthalpy of the air–water vapor mixture.

10.39 A psychrometer reads 35°C as the dry-bulb temperature and 25°C as the wet-bulb temperature. Determine the specific humidity, the relative humidity, and the enthalpy of the moist air without using the psychrometric chart. Read the corresponding values from the psychrometric chart and compare them with the calculated values.

10.40 The dry-bulb and wet-bulb temperatures of a mixture of air–water-vapor at one standard atmosphere pressure are 30°C and 20°C, respectively. Calculate: (a) the specific humidity, relative humidity and enthalpy of the mixture (b) the values of ω, ϕ and h^* from the psychrometric chart.

10.41 The following table gives the properties of a few air–water-vapor mixtures at a total pressure of one standard atmosphere. Using the psychrometric chart estimate the missing information in the table.

Mixture	T_{DB} (°C)	T_{WB} (°C)	ω	ϕ	Dew point (°C)
1	45	25			
2	35	20			
3	38			0.30	
4	30			0.60	
5	40		0.009		
6		18	0.008		

10.42 A mixture of nitrogen and water vapor at 100°C and 1 bar has the molar analysis 90% nitrogen and 10% water vapor. If the mixture is cooled at constant pressure, determine the temperature at which water vapor begins to condense.

10.43 The dry-bulb and wet-bulb temperatures of air in a room are 30°C and 20°C, respectively. If the air in the room is at 0.9 bar pressure, determine the specific humidity and relative humidity of the air in the room.

10.44 Using the psychrometric chart determine the dew point temperature of the humid air for which the dry-bulb and wet-bulb temperatures are 35°C and 20°C, respectively.

10.45 A mixture of air–water-vapor at 40°C with a relative humidity of 50% is cooled to 25°C while the pressure is held constant at 101.325 kPa. Determine:

(a) the relative humidity

(b) the dew point temperature in the final state; and

(c) the quantity of heat removed from the mixture containing 1 kg dry air.

Use the psychrometric chart to solve the problem.

10.46 An adiabatic mixing device operating at steady state receives 30 m^3/min of air at 40°C with a relative humidity of 60%. A second stream of air at 20°C with a relative humidity of 10% enters the mixing device at a rate of 10 m^3/min. If the mixing device is at 101.325 kPa pressure, determine the temperature and relative humidity of the exiting stream of mixture.

10.47 Two streams of moist air at 1 atm. pressure are adiabatically mixed in a steady flow mixer as shown in Fig. 10.24. Determine the specific humidity, the relative humidity, the temperature and the enthalpy of the resultant mixture.

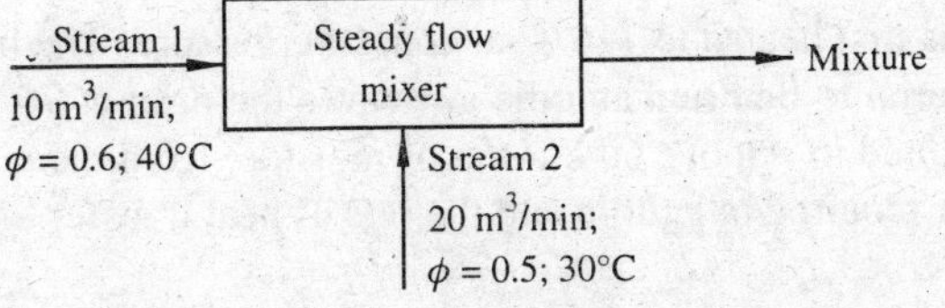

Fig. 10.24 *Sketch for Problem 10.47*

10.48 An air-conditioner delivers 1 m^3 of air at 5°C which is completely saturated with water vapor. If this air is mixed with 5 m^3 air at 35°C with a relative humidity of 60% in a room, determine the temperature, the specific humidity and the relative humidity of the air after mixing.

10.49 Two air-conditioners blow cold air into a common duct. The first air conditioner delivers 1 m^3/min of air at 20°C and 60% relative humidity while the second air-conditioner delivers 2 m^3/min of saturated air at 5°C. Determine the temperature and the relative humidity of the mixture.

10.50 It is required to blow air at 25°C into a room. For this purpose, chilled air can be mixed with an appropriate amount of atmospheric air and the mixture can be directed into the room. The atmospheric air is at 40°C with 40% relative humidity and the chilled air is available at 15°C with 80% relative humidity. Determine the ratio, in terms of dry air, in which the two air streams are to be mixed to obtain the required conditions. Also determine the relative humidity of the mixture.

10.51 A stream of air with a dry-bulb temperature of 40°C and a wet-bulb temperature of 25°C enters a mixing chamber at a rate of 3 kg dry air/s. A second stream of saturated air at 10°C enters the mixing chamber at a rate of 1 kg dry air/s. Assuming that the total pressure in the mixing chamber is 101.325 kPa; determine for the resultant mixture: (a) the specific humidity (b) the relative humidity; and (c) the temperature.

10.52 Moist air enters a device at a rate of 20 m^3/s at 1 standard atmosphere pressure with a dry-bulb temperature of 40°C and a wet-bulb temperature of 28°C. Liquid water at 10°C is sprayed into the air stream and the air leaves the device at 30°C. Determine the relative humidities of the entering air and exit air and the quantity of liquid water sprayed.

10.53 Moist air enters a cooling device (constant pressure) at 40°C with 60% relative humidity. The air is required to leave the device as saturated at 20°C. If the moist air enters the cooling device at 1 atm. pressure at a rate of 300 m^3/min, determine the amount of water condensed and the rate of cooling required in the device.

10.54 In an air-conditioning plant, saturated air at 10°C is heated at constant pressure to 25°C. Determine the specific humidity and the relative humidity of the air leaving the unit. Also, determine the energy required per kg of dry air for heating.

10.55 Atmospheric air at 37°C and 70% relative humidity is to be conditioned to 24°C and 40% relative humidity. If the pressure remains constant at 101.325 kPa, calculate:

(a) the amount of water to be removed
(b) the heat to be rejected in the cooling unit; and
(c) the quantity of heat to be added in the final process, if the flow rate of dry air through the unit is 10 kg/s.

10.56 Atmospheric air at 45°C with 20% relative humidity enters a desert cooler and leaves at 28°C. Determine the amount of water added per kg of dry air and the relative humidity of the cooled air. The total pressure of the air remains constant at one standard atmosphere in the desert cooler.

10.57 In a drier operating at steady state atmospheric air at 30°C and 101.325 kPa with a relative humidity of 40% is first heated to 110°C at constant pressure. The heated air is then allowed to pass over the material to be dried and the air leaves the drier at 70°C with a relative humidity of 0.5. If it is required to remove 60 kg/min of moisture from the material, determine the mass flow rate of dry air required in kg/min and the rate of heat transfer to the air as it passes through the heating unit.

10.58 Air enters a drier at 30°C with 20% relative humidity and leaves at 40°C with 60% relative humidity. Determine the flow rate of dry air at 1 atm. pressure if 10 kg/h of water is evaporated from the material in the drier.

10.59 Moist air at 1 atm pressure enters a dehumidifier unit at 35°C with 80% relative humidity. The air is required to leave the dehumidifier at 20°C with 40% relative humidity. If the flow rate of the dry air is 2 kg/s, determine the rate of cooling and the rate of heating required to operate the dehumidifier. Also determine the rate at which water is condensed in the refrigeration unit.

10.60 Atmospheric air at 40°C with 60% relative humidity is to be conditioned to 20°C with 50% relative humidity using a cooling coil, a condensate removal system and a heating section. If the air flow rate through the device is 10 kg/s of dry air, determine the energy transfer rates in the cooling section and heating section. Also determine the amount of water removed in the condensing unit.

10.61 To maintain comfortable conditions in an auditorium, 50 m^3/min of air at 35°C with 70% relative humidity is to be cooled and dehumidified to 20°C with 40% relative humidity by cooling and reheating in an air-conditioning plant. Determine the temperature to which the air is to be cooled. Also estimate the capacity of the refrigeration unit required for this purpose.

10.62 Air at 101.325 kPa, 10°C and 40% relative humidity enters a steady flow heating unit and leaves at 21°C. This air is then humidified, by injecting saturated steam at 101.325 kPa and finally the air leaves at 27°C and 50% relative humidity. If the flow rate of dry air through the unit is 60 kg/min determine (a) the amount of steam required for one hour, and (b) the power consumption of the heating unit.

10.63 On a summer day atmospheric air at 40°C with 30% relative humidity enters an evaporative cooler at a rate of 3 m^3/s. The air leaves the cooler with 80% relative humidity. Determine the temperature and the specific humidity of the air leaving the cooler. Also determine the amount of water required to operate the cooler for 10 hours.

10.64 A cooling system for an office building uses a desert cooler in summer months. If the atmospheric air is available at 42°C with 20% relative humidity, and the air leaves the desert cooler with 90% relative humidity, determine the temperature of the air leaving the cooler. If the cooler draws in 60 m^3/s of atmospheric air, estimate the amount of water required to operate the cooler continuously for a period of 8 hours.

10.65 Liquid water at 45°C from a power plant enters a cooling tower at a rate of 10000 kg/s. Atmospheric air at 25°C with 30% relative humidity is blown into the cooling tower. The air leaves the tower at 35°C with a relative humidity of 80%, while the water leaves the tower at 20°C. Determine the mass flow rate of dry air that enters the cooling tower and the quantity of make-up water required.

10.66 It is desired to cool water from 40°C to 25°C in a cooling tower. The atmospheric air is available at 30°C with a relative humidity of 20% and the air exits the cooling tower at 35°C with a relative humidity of 70%. If the flow rate of hot water into the cooling tower is 10^6 kg/min, determine the mass flow rate of dry air required and the amount of make-up water to be added.

10.67 A cooling tower is to be installed to supply 1000 kg/min of cooling water at 27°C. Water enters the cooling tower at a rate of 1000 kg/min at 45°C. The air enters the tower at 30°C with 30% relative humidity and leaves at 35°C with 80% relative humidity. Determine the make-up water required and the air flow rate.

10.68 A forced draft cooling tower is used in a large thermal power plant for providing cold water at a temperature of 30°C to the condensers. The water enters the cooling tower at 37°C at a rate of

1000 kg/s. The atmospheric air at 27°C with 40% relative humidity enters the tower and leaves at 33°C with 80% relative humidity. Determine the make-up water and the air flow rate required.

10.69 It is desired to cool 1000 kg/min of water from 40°C to 27°C. Atmospheric air at 40°C with a relative humidity of 20% enters the cooling tower and leaves at 30°C with 70% relative humidity. Calculate the flow rate of dry air through the cooling tower and the amount of make-up water required.

10.70 A small cooling tower is to be designed to cool 50 kg/s of water from 40°C to 24°C at a location where the atmospheric pressure is 90 kPa. Atmospheric air enters the cooling tower at 30°C with 30% relative humidity and leaves at 22°C with 80% relative humidity. Determine (a) the required flow rate of dry air, and (b) the flow rate of make-up water.

10.71 Suppose 100 kg of the gas mixture of Problem 10.1 is cooled at the constant pressure of 100 kPa. Determine the temperature at which water begins to condense. Determine the mass of water that condenses if the mixture is cooled 10°C below the condensation temperature.

10.72 Cold air is to be supplied for a library at the rate of 100 m^3/min. The air is to enter the building at 20°C with 40% relative humidity. Suppose the atmospheric air is available at 35°C with 70% relative humidity. Suggest a suitable air conditioning system and estimate the rates of cooling and heating required.

10.73 A gas mixture has the following composition by volume.

Constituent	CO_2	He	N_2	O_2	H_2O
Volume%	30	10	20	30	10

Suppose the mixture is available at 100 kPa and 400 K. It is desired to remove the water vapor by cooling the mixture. Suppose one kmol of the mixture is cooled at constant pressure. Determine the temperature at which water begins to condense. If the temperature of the mixture is reduced to 15°C below the condensation temperature, estimate the amount of water that is condensed.

Appendix A

Critical Constants of Some Selected Substances

Compound		Critical Temp. T_c (K)	Critical pressure P_c (bar)	Critical volume v_c (m^3/kmol)
Inorganic compounds				
Ammonia	NH_3	405.50	112.77	0.0725
Argon	A	151.20	48.64	0.075
Bromine	Br_2	584.00	103.35	0.114
Carbon dioxide	CO_2	304.20	73.87	0.094
Carbon disulphide	CS_2	552.00	79.03	0.170
Carbon monoxide	CO	133.00	34.96	0.093
Chlorine	Cl_2	417.00	77.11	0.124
Helium	He	5.26	22.90	0.058
Hydrogen	H_2	33.30	129.70	0.065
Hydrogen chloride	HCl	324.60	82.58	0.087
Hydrogen sulphide	H_2S	373.60	90.08	0.098
Neon	Ne	44.50	27.26	0.0417
Nitrogen	N_2	126.20	33.94	0.090
Nitric oxide	NO	179.20	65.86	0.058
Nitrous oxide	N_2O	309.70	72.65	0.0963
Oxygen	O_2	154.40	50.36	0.074
Sulphur dioxide	SO_2	430.70	78.83	0.122
Sulphur trioxide	SO_3	491.40	84.91	0.126
Water	H_2O	647.30	221.20	0.056

Contd.

Compound		Critical Temp. T_c (K)	Critical pressure P_c (bar)	Critical volume v_c (m^3/kmol)
Organic compounds				
Acetic acid	CH_3COOH	594.80	57.86	0.171
Acetone	CH_3COCH_3	508.70	47.22	0.213
Acetylene	C_2H_2	309.50	62.42	0.113
Aniline	C_6H_7N	698.80	52.99	0.274
Benzene	C_6H_6	562.10	49.24	0.260
Butane (iso)	C_4H_{10}	408.00	36.48	0.263
Butane (n)	C_4H_{10}	425.20	37.97	0.255
Butene (1)	C_4H_8	419.60	40.23	0.240
Carbon tetrachloride	CCl_4	556.40	45.60	0.276
Chlorobenzene	C_6H_5Cl	632.40	45.19	0.308
Chlorodifluoromethane	$CHClF_2$	369.60	49.14	0.165
Chloroform	$CHCl_3$	536.60	54.72	0.240
Cyclopentane	C_5H_{10}	511.76	21.08	0.260
Diethyl ether	$C_2H_5OC_2H_5$	467.00	39.62	0.281
Ethane	C_2H_6	305.43	48.84	0.148
Ethylamine	$C_2H_5NH_2$	456.40	56.24	0.181
Ethyl acetate	$C_2H_5COOCH_3$	523.30	38.30	0.286
Ethyl alcohol	C_2H_5OH	516.30	68.83	0.167
Ethylene	C_2H_4	283.10	51.17	0.124
Ethylene oxide	C_2H_4O	468.00	71.94	0.138
Freon 11	CCl_3F	471.20	43.77	0.248
Freon 12	CCl_2F_2	384.70	40.12	0.218
Freon 13	$CClF_3$	302.00	38.71	0.180
Freon 21	$CHCl_2F$	451.70	51.68	0.197
Freon 113	$C_2Cl_3F_3$	487.30	34.15	0.325
Heptane (n)	C_7H_{16}	540.16	27.37	0.426
Hexane (n)	C_6H_{14}	507.90	30.32	0.368
Methane	CH_4	190.70	46.41	0.099
Methyl acetate	CH_3COOCH_3	506.90	46.91	0.228
Methyl chloride	CH_3C1	416.30	66.77	0.143
Methyl alcohol	CH_3OH	513.20	75.54	0.118
Octane (n)	C_8H_{18}	569.40	24.97	0.486
Pentane (n)	C_5H_{12}	469.80	33.75	0.311
Propane	C_3H_8	369.90	42.57	0.200
Propyl alcohol (n)	C_3H_7OH	537.30	50.87	0.220
Propyl alcohol (iso)	C_3H_7OH	508.80	53.70	0.219
Propylene	C_3H_6	365.10	46.00	0.181
Toluene	$C_6H_5CH_3$	594.00	42.15	0.320

Source: Reid R.C., Prausnitz, J.M. and Poling, B.E., *The Properties of Gases and Liquids*, Fourth edition, McGraw-Hill, New York (1988)

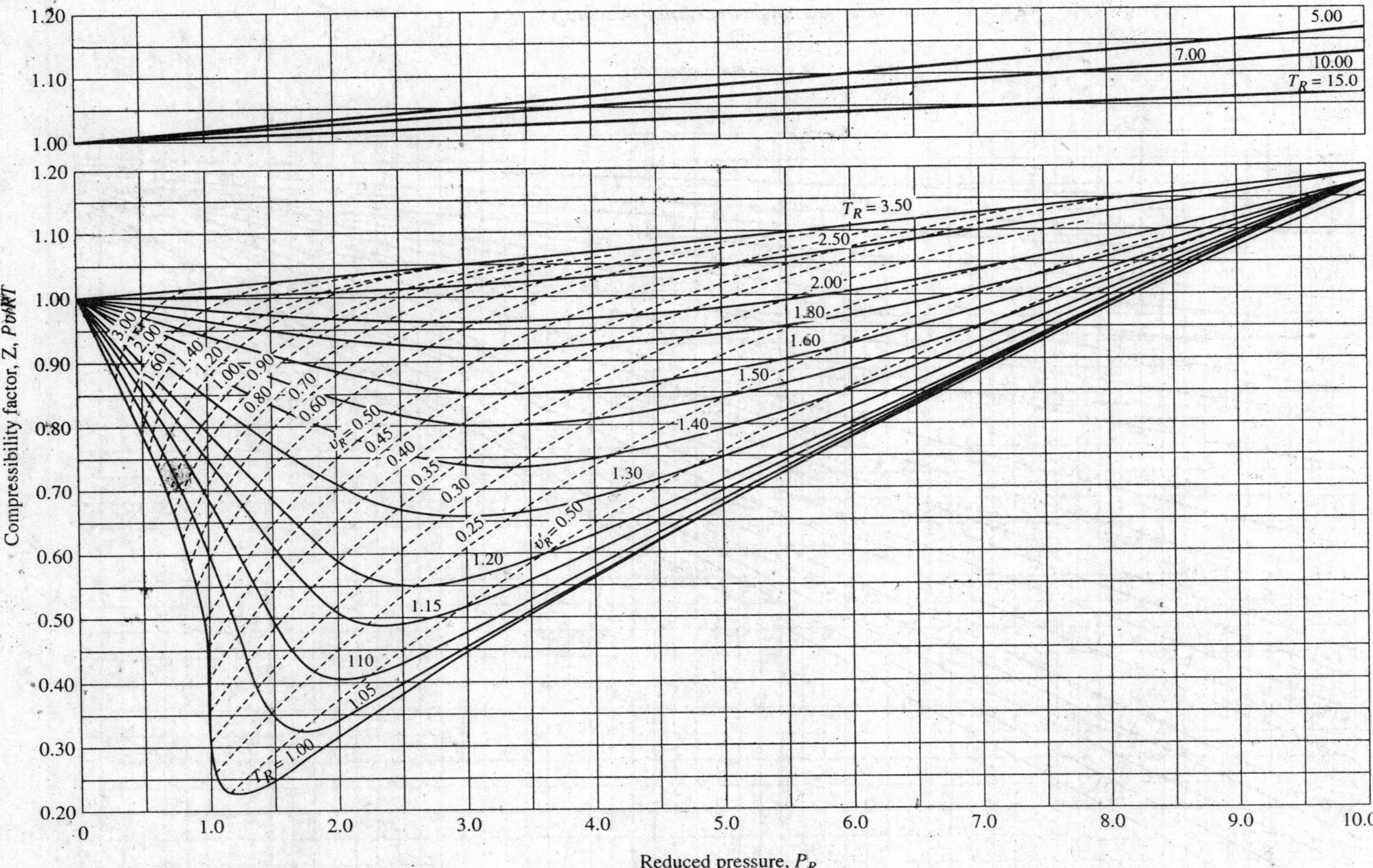

Fig. A.1 *Compressibility chart, low pressure range*

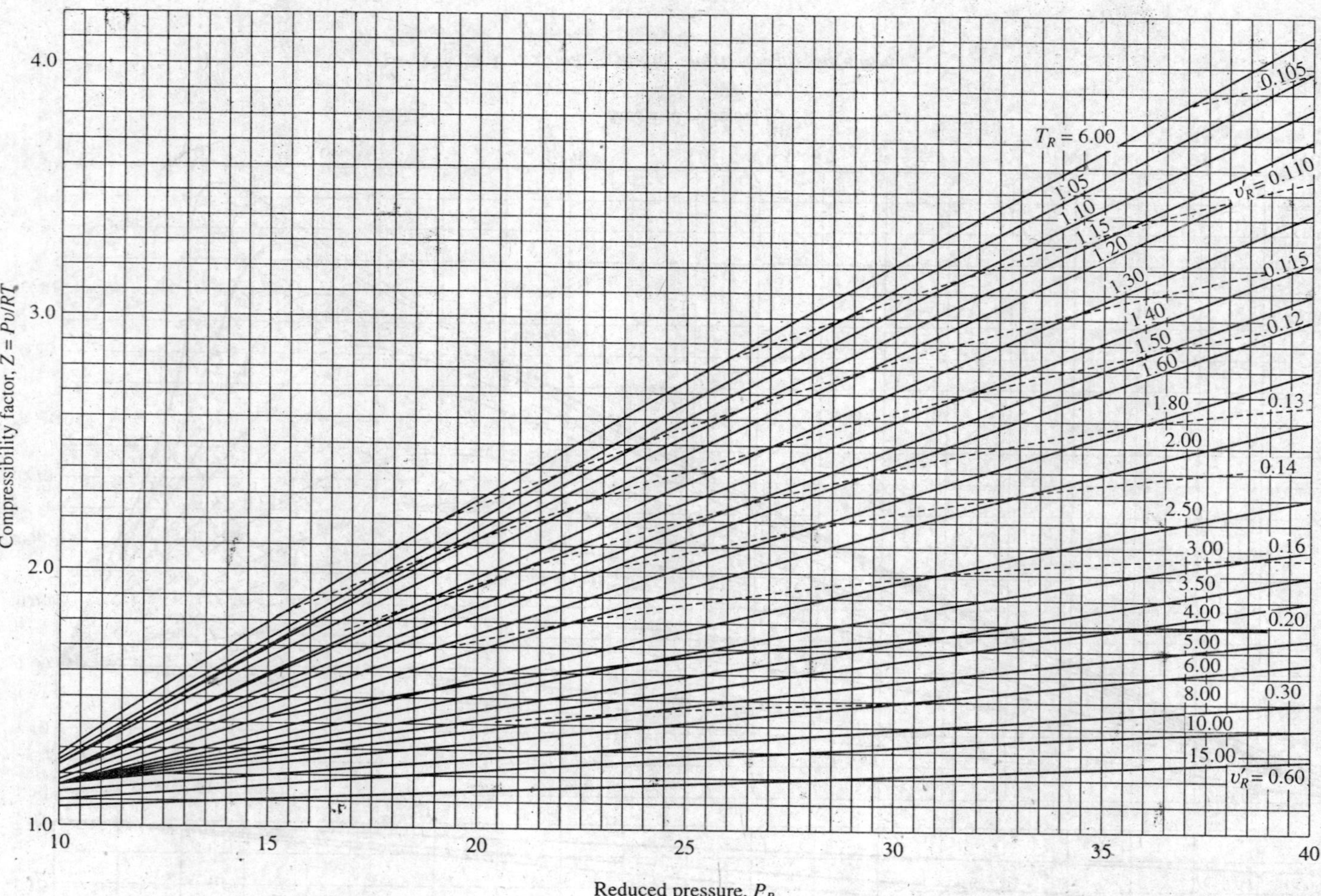

Fig. A.2 *Compressibility chart, high pressure range*

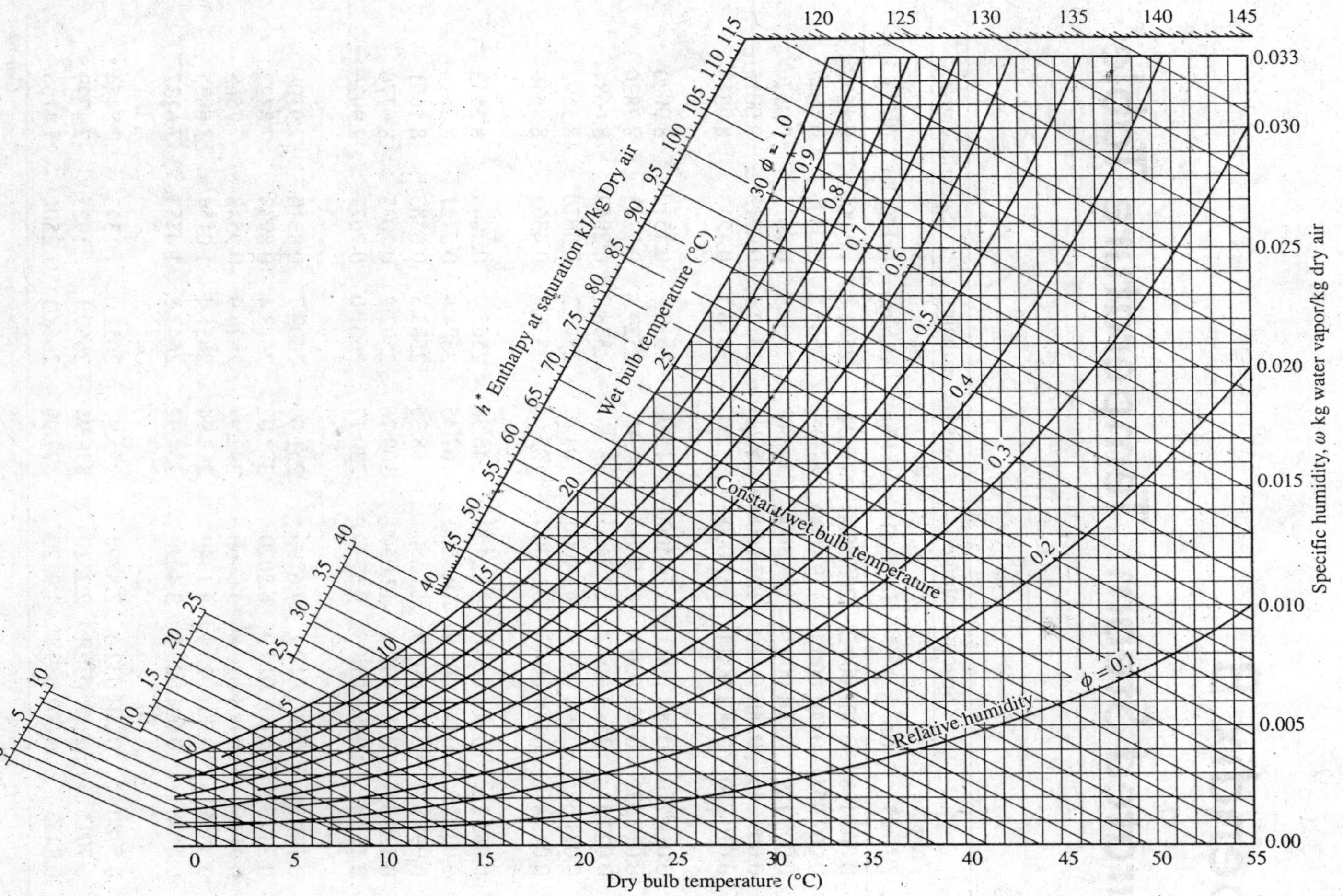

Fig. A.3 *Psychrometric chart at a pressure of 101.325 kPa*

Appendix B

Saturated Steam: Temperature Table

Temp. t (°C)	Pressure P (bar)	Specific volume v_f (m^3/kg)	v_g (m^3/kg)	Specific enthalpy h_f (kJ/kg)	h_g (kJ/kg)	Specific entropy s_f (kJ/kg K)	s_g (kJ/kg K)
0	0.006 108	0.001 000 2	206.3000	- 0.04	2501.6	- 0.0002	9.1577
2	0.007 055	0.001 000 1	179.9000	8.39	2505.2	0.0306	9.1047
4	0.008 129	0.001 000 0	157.3000	16.80	2508.9	0.0611	9.0526
6	0.009 345	0.001 000 0	137.8000	25.21	2512.6	0.0913	9.0015
8	0.010 720	0.001 000 1	121.0000	33.60	2516.2	0.1213	8.9513
10	0.012 270	0.001 0003	106.4000	41.99	2519.9	0.1510	8.9020
15	0.017 039	0.001 000 8	77.9800	62.94	2529.1	0.2243	8.7826
20	0.023 37	0.001 001 7	57.8400	83.86	2538.2	0.2963	8.6684
25	0.031 66	0.001 002 9	43.4000	104.77	2547.3	0.3670	8.5592
30	0.042 41	0.001 004 3	32.9300	125.66	2556.4	0.4365	8.4546
35	0.056 22	0.001 006 0	25.2400	146.56	2565.4	0.5049	8.3543
40	0.073 75	0.001 007 8	19.5500	167.45	2574.4	0.5721	8.2583
45	0.095 82	0.001 009 9	15.2800	188.35	2583.3	0.6383	8.1661
50	0.123 35	0.001 012 1	12.0500	209.26	2592.2	0.7035	8.0776
55	0.157 41	0.001 014 5	9.5790	230.17	2601.0	0.7677	7.9926
60	0.199 20	0.001 017 1	7.6790	251.09	2609.7	0.8310	7.9108
65	0.2501	0.001 019 9	6.2020	272.02	2618.4	0.8933	7.8322
70	0.3116	0.001 022 8	5.0460	292.97	2626.9	0.9548	7.7565
75	0.3855	0.001 025 9	4.1340	313.94	2635.4	1.0154	7.6835
80	0.4736	0.001 029 2	3.4090	334.92	2643.8	1.0753	7.6132
85	0.5780	0.001 032 6	2.8290	355.92	2652.0	1.1343	7.5454
90	0.7011	0.001 036 1	2.3610	376.94	2660.1	1.1925	7.4799
95	0.8453	0.001 039 9	1.9820	397.99	2668.1	1.2501	7.4166

Contd.

Temp. t (°C)	Pressure P (bar)	Specific volume v_f (m³/kg)	v_g (m³/kg)	Specific enthalpy h_f (kJ/kg)	h_g (kJ/kg)	Specific entropy s_f (kJ/kg K)	s_g (kJ/kg K)
100	1.0133	0.001 043 7	1.6730	419.06	2676.0	1.3069	7.3554
105	1.2080	0.001 047 7	1.4190	440.17	2683.7	1.3630	7.2962
110	1.4327	0.001 051 9	1.2100	461.32	2691.3	1.4185	7.2388
115	1.6906	0.001 056 2	1.0360	482.50	2698.7	1.4733	7.1832
120	1.9854	0.001 060 6	0.8915	503.72	2706.0	1.5276	7.1293
125	2.3210	0.001 065 2	0.7702	524.99	2713.0	1.5813	7.0769
130	2.7013	0.001 070 0	0.6681	546.31	2719.9	1.6344	7.0261
135	3.131	0.001 075 0	0.5818	567.68	2726.6	1.6869	6.9766
140	3.614	0.001 080 1	0.5085	589.10	2733.1	1.7390	6.9284
145	4.155	0.001 085 3	0.4460	610.60	2739.3	1.7906	6.8815
150	4.760	0.001 090 8	0.3924	632.15	2745.4	1.8416	6.8358
155	5.433	0.001 096 4	0.3464	653.78	2751.2	1.8923	6.7911
160	6.181	0.001 102 2	0.3068	675.47	2756.7	1.9425	6.7475
165	7.008	0.001 108 2	0.2724	697.25	2762.0	1.9923	6.7048
170	7.920	0.001 114 5	0.2426	719.12	2667.1	2.0416	6.6630
175	8.924	0.001 120 9	0.2165	741.07	2771.8	2.0906	6.6221
180	10.027	0.001 127 5	0.1938	763.12	2776.3	2.1393	6.5819
185	11.233	0.001 134 4	0.1739	785.26	2780.4	2.1876	6.5424
190	12.551	0.001 141 5	0.1563	807.52	2784.3	2.2356	6.5036
195	13.987	0.001 148 9	0.1408	829.88	2787.8	2.2830	6.4654
200	15.549	0.001 156 5	0.1272	852.37	2790.9	2.3307	6.4278
205	17.243	0.001 164 4	0.1150	874.99	2793.8	2.3778	6.3906
210	19.077	0.001 172 6	0.1042	897.74	2796.2	2.4247	6.3539
215	21.060	0.001 181 1	0.094 63	920.63	2798.3	2.4713	6.3176
220	23.198	0.001 190 0	0.086 04	943.67	2799.9	2.5178	6.2817
225	25.501	0.001 199 2	0.078 35	966.89	2801.2	2.5641	6.2461
230	27.976	0.001 208 7	0.071 45	990.26	2802.0	2.6102	6.2107
235	30.632	0.001 218 7	0.065 25	1013.80	2802.3	2.6562	6.1756
240	33.478	0.001 229 1	0.059 65	1037.60	2802.2	2.7020	6.1406
245	36.523	0.001 239 9	0.054 61	1061.60	2801.6	2.7478	6.1057
250	39.776	0.001 251 3	0.050 04	1085.80	2800.4	2.7935	6.0708
255	43.246	0.001 263 2	0.045 90	1110.20	2798.7	2.8392	6.0359
260	46.943	0.001 275 6	0.042 13	1134.90	2796.4	2.8848	6.0010
270	55.058	0.001 302 5	0.035 59	1185.20	2789.9	2.9763	5.9304

Contd.

Temp. t (°C)	Pressure P (bar)	Specific volume v_f (m³/kg)	Specific volume v_g (m³/kg)	Specific enthalpy h_f (kJ/kg)	Specific enthalpy h_g (kJ/kg)	Specific entropy s_f (kJ/kg K)	Specific entropy s_g (kJ/kg K)
280	64.202	0.001 332 4	0.030 13	1236.80	2780.4	3.0683	5.8586
290	74.461	0.001 365 9	0.025 54	1290.00	2767.6	3.1611	5.7848
300	85.927	0.001 404 1	0.021 65	1345.00	2751.0	3.2552	5.7081
310	98.700	0.001 448 0	0.018 33	1402.40	2730.0	3.3512	5.6278
320	112.890	0.001 499 5	0.015 48	1462.60	2703.7	3.4500	5.5423
330	128.630	0.001 561 5	0.012 99	1526.50	2670.2	3.5528	5.4490
340	146.050	0.001 638 7	0.010 78	1595.50	2626.2	3.6616	5.3427
350	165.350	0.001 741 1	0.008 799	1671.90	2567.7	3.7800	5.2177
360	186.750	0.001 895 9	0.006 940	1764.20	2485.4	3.9210	5.0600
370	210.540	0.002 213 6	0.004 973	1890.20	2342.8	4.1108	4.8114
374.15	221.200	0.003 17	0.003 17	2107.40	2107.4	4.4429	4.4429

Appendix C

Saturated Steam: Pressure Table

Pressure P (bar)	Temperature t (°C)	Specific volume v_f (m³/kg)	v_g (m³/kg)	Specific enthalpy h_f (kJ/kg)	h_g (kJ/kg)	Specific entropy s_f (kJ/kg K)	s_g (kJ/kg K)
0.01	6.98	0.001 000 1	129.20	29.34	2514.4	0.1060	8.9767
0.05	32.90	0.001 005 2	28.19	137.77	2561.6	0.4763	8.3960
0.10	45.83	0.001 010 2	14.67	191.83	2584.8	0.6493	8.1511
0.15	54.00	0.001 014 0	10.02	225.97	2599.2	0.7549	8.0093
0.20	60.09	0.001 017 2	7.650	251.45	2609.9	0.8321	7.9094
0.3	69.12	0.001 022 3	5.229	289.30	2625.4	0.9441	7.7695
0.4	75.89	0.001 026 5	3.993	317.65	2636.9	1.0261	7.6709
0.5	81.35	0.001 030 1	3.240	340.56	2646.0	1.0912	7.5947
0.6	85.95	0.001 033 3	2.732	359.93	2653.6	1.1454	7.5327
0.7	89.96	0.001 036 1	2.365	376.77	2660.1	1.1921	7.4804
0.8	93.51	0.001 038 7	2.087	391.72	2665.8	1.2330	7.4352
0.9	96.71	0.001 041 2	1.869	405.21	2670.9	1.2696	7.3954
1.0	99.63	0.001 043 4	1.694	417.54	2675.4	1.3027	7.3598
1.5	111.37	0.001 053 0	1.159	467.13	2693.4	1.4336	7.2234
2.0	120.23	0.001 060 8	0.8854	504.70	2706.3	1.5301	7.1268
2.5	127.43	0.001 067 5	0.7184	535.34	2716.4	1.6071	7.0520
3.0	133.54	0.001 073 5	0.6056	561.43	2724.7	1.6716	6.9909
3.5	138.87	0.001 078 9	0.5240	584.27	2731.6	1.7273	6.9392
4.0	143.62	0.001 083 9	0.4622	604.67	2737.6	1.7764	6.8943
5.0	151.84	0.001 092 8	0.3747	640.12	2747.5	1.8604	6.8192
6.0	158.84	0.001 100 9	0.3155	670.42	2755.5	1.9308	6.7575
7.0	164.96	0.001 108 9	0.2727	697.06	2762.0	1.9918	6.7052
8.0	170.41	0.001 115 0	0.2403	720.94	2767.5	2.0457	6.6192
9.0	175.36	0.001 121 3	0.2148	742.64	2772.1	2.0941	6.6192

Contd.

Pressure P (bar)	Temperature t (°C)	Specific volume v_f (m³/kg)	v_g (m³/kg)	Specific enthalpy h_f (kJ/kg)	h_g (kJ/kg)	Specific entropy s_f (kJ/kg K)	s_g (kJ/kg K)
10.0	179.88	0.001 127.4	0.1943	762.61	2776.2	2.1382	6.5862
11.0	184.07	0.001 133 1	0.1774	781.13	2779.7	2.1786	6.5497
12.0	187.96	0.001 138 6	0.1632	798.43	2782.7	2.2161	6.5194
13.0	191.61	0.001 143 8	0.1511	814.70	2785.4	2.2510	6.4913
14.0	195.04	0.001 148 9	0.1407	830.08	2787.8	2.2837	6.4651
15.0	198.29	0.001 153 9	0.1317	844.67	2789.9	2.3145	6.4406
16.0	201.37	0.001 158 6	0.1237	858.56	2791.7	2.3436	6.4175
17.0	204.31	0.001 163 3	0.1166	871.84	2793.4	2.3713	6.3957
18.0	207.11	0.001 167 8	0.1103	884.58	2794.8	2.3976	6.3751
19.0	209.80	0.001 172 3	0.1047	896.81	2796.1	2.4228	6.3554
20.0	212.37	0.001 176 6	0.099 54	908.59	2797.2	2.4469	6.3367
25	223.94	0.001 197 2	0.079 91	961.96	2800.9	2.5543	6.2536
30	233.84	0.001 216 3	0.066 36	1008.40	2802.3	2.6455	6.1837
35	242.54	0.001 234 5	0.057 03	1049.80	2802.0	2.7253	6.1228
40	250.33	0.001 252 1	0.049 75	1087.40	2800.3	2.7965	6.0685
45	257.41	0.001 269 1	0.044 04	1122.10	2797.7	2.8612	6.0191
50	263.91	0.001 285 8	0.039 43	1154.50	2794.2	2.9206	5.9735
60	275.55	0.001 318 7	0.032 44	1213.70	2785.0	3.0273	5.8908
70	285.79	0.001 351 3	0.027 37	1267.40	2773.5	3.1219	5.8162
80	294.97	0.001 384.2	0.023 53	1317.10	2759.9	3.2076	5.7471
90	303.31	0.001 417 9	0.020 50	1363.70	2744.6	3.2867	5.6820
100	310.96	0.001 452 6	0.018 04	1408.00	2727.7	3.3605	5.6198
120	324.65	0.001 526 8	0.014 28	1491.80	2689.2	3.4972	5.5002
140	336.64	0.001 610 6	0.011 50	1571.60	2642.4	3.6242	5.3803
160	347.33	0.001 710 3	0.009 308	1650.50	2584.9	3.7471	5.2531
180	356.96	0.001 839 9	0.007 498	1734.80	2513.9	3.8765	5.1128
200	365.70	0.002 037 0	0.005 877	1826.50	2418.4	4.0149	4.9412
220	373.69	0.002 671 4	0.003 728	2011.10	2195.6	4.2947	4.5799
221.20	374.15	0.003 17	0.003 17	2107.4.	2107.40	4.4429	4.4429

Reprinted from Chandrasekaran, K.D. and Venkateswarlu, D., *SI Units in Chemical Engineering and Technology*, Chemical Engineering Education Development Centre, IIT Madras (1974), with permission from Chairman, Centre for Continuing Education, IIT Madras.

Appendix D

Superheated Steam

Pressure (bar) (Saturation temperature)		Temperature (°C)							
		100	200	300	400	500	600	700	800
0.2	v	8.585	10.507	13.219	15.529	17.838	20.146	22.455	24.762
(60.09°C)	h	2686.3	2879.2	3076.4	3279.4	3489.0	3705.4	3928.7	4158.7
	s	8.1261	8.5839	8.9618	9.2882	9.5784	9.8416	10.0836	10.3085
0.4	v	4.2792	5.4478	6.6065	7.7625	8.9176	10.072	11.227	12.381
(75.89°C)	h	2683.8	2878.2	3075.9	3279.1	3488.8	3705.3	3928.6	4158.6
	s	7.8009	8.2625	8.6413	8.9680	9.2583	9.5216	9.7636	9.9885
0.6	v	2.8440	3.6280	4.4020	5.1740	5.9440	6.7140	7.4840	8.2540
(85.95°C)	h	2681.3	2877.3	3075.4	3278.8	3488.6	3705.1	3928.5	4158.5
	s	7.6085	8.0738	8.4536	8.7806	9.0710	9.3343	9.5764	9.8013
0.8	v	2.1260	2.7180	3.3000	3.8790	4.4570	5.0350	5.6130	6.1900
(93.51°C)	h	2678.8	2876.3	3075.0	3278.5	3488.4	3705.0	3928.4	4158.4
	s	7.4703	7.9395	8.3202	8.6475	8.9380	9.2014	9.4436	9.6685
1.0	v	1.6960	2.1720	2.6390	3.1020	3.5650	4.0280	4.4900	4.9520
(99.63°C)	h	2676.2	2875.4	3074.5	3278.2	3488.1	3704.8	3928.2	4158.3
	s	7.3618	7.8349	8.2166	8.5442	8.8348	9.0982	9.3405	9.5654
2	v		1.0804	1.3162	1.5492	1.7812	2.0129	2.2420	2.4754
(120.23°C)	h		2870.5	3072.1	3276.7	3487.0	3704.0	3927.6	4157.8
	s		7.5072	7.8937	8.2226	8.5139	8.7776	9.0201	9.2452
3	v		0.7164	0.8753	1.0314	1.1865	1.3412	1.4957	1.6499
(133.54°C)	h		2865.5	3069.7	3275.2	3486.0	3703.2	3927.0	4157.3
	s		7.3119	7.7034	8.0338	8.3257	8.5868	8.8325	9.0577

v in (m^3/kg); h in (kJ/kg); s in (kJ/kg K)

Contd.

Pressure (bar) (Saturation temperature)		Temperature (°C)						
		200	300	400	500	600	700	800
4	v	0.5343	0.6549	0.7725	0.8892	1.0054	1.1214	1.2372
(143.62°C)	h	2860.4	3067.2	3273.6	3484.9	3702.3	3926.4	4156.9
	s	7.1708	7.5675	7.8994	8.1919	8.4563	8.6992	8.9246
5	v	0.4250	0.5226	0.6172	0.7108	0.8039	0.8968	0.9896
(151.84°C)	h	2855.1	3064.8	3272.1	3483.8	3701.5	3925.8	4156.4
	s	7.0592	7.4614	7.7948	8.0879	8.3526	8.5957	8.8213
6	v	0.3520	0.4344	0.5136	0.5918	0.6696	0.7471	0.8245
(158.84°C)	h	2849.7	3062.3	3270.6	3482.7	3700.7	3925.1	4155.9
	s	6.9662	7.3740	7.7090	8.0027	8.2678	8.5111	8.7368
7	v	0.2999	0.3714	0.4396	0.5069	0.5737	0.6402	0.7066
(164.96°C)	h	2844.2	3059.8	3269.8	3481.6	3699.9	3924.5	4155.5
	s	6.8859	7.2997	7.6362	7.9305	8.1959	8.4395	8.6653
8	v	0.2608	0.3241	0.3842	0.4432	0.5017	0.5600	0.6181
(170.41°C)	h	2838.6	3057.3	3267.5	3480.5	3699.1	3923.9	4115.0
	s	6.8148	7.2348	7.5729	7.8678	8.1336	8.3773	8.6033
9	v	0.2303	0.2874	0.3410	0.3936	0.4458	0.4976	0.5493
(175.36°C)	h	2832.7	3054.7	3266.0	3479.4	3698.2	3923.3	4154.8
	s	6.7508	7.1771	7.5169	7.8124	8.0785	8.3225	8.5486
10	v	0.2059	0.2580	0.3065	0.3540	0.4010	0.4477	0.4943
(179.88°C)	h	2826.8	3052.1	3264.4	3478.3	3697.4	3922.7	4154.1
	s	6.6922	7.1251	7.4665	7.7627	8.0292	8.2734	8.4997
11	v	0.1859	0.2339	0.2782	0.3215	0.3643	0.4069	0.4493
(184.07°C)	h	2820.7	3049.6	3262.9	3477.2	3696.6	3922.0	4153.6
	s	6.6379	7.0778	7.4209	7.7177	7.9845	8.2289	8.4553
12	v	0.1692	0.2139	0.2547	0.2945	0.3338	0.3729	0.4118
(187.96°C)	h	2814.4	3046.9	3261.3	3476.1	3695.8	3921.4	4153.1
	s	6.5872	7.0342	7.3790	7.6765	7.9436	8.1882	8.4148
13	v	0.1551	0.1969	0.2348	0.2716	0.3080	0.3441	0.3800
(191.61°C)	h	2808.0	3044.3	3259.7	3475.0	3695.0	3920.8	4152.7
	s	6.5394	6.9938	7.3404	7.6385	7.9060	8.1508	8.3775

v in (m^3/kg); h in (kJ/kg); s in (kJ/kg K)

Contd.

Pressure (bar) (Saturation temperature)		Temperature (°C)						
		200	300	400	500	600	700	800
14	v	0.1429	0.1823	0.2177	0.2520	0.2859	0.3194	0.3528
(195.04°C)	h	2801.4	3041.6	3258.2	3473.9	3694.1	3920.2	4152.2
	s	6.4941	6.9561	7.3045	7.6032	7.8710	8.1161	8.3430
15	v	0.1324	0.1697	0.2029	0.2359	0.2667	0.2980	0.3292
(198.29°C)	h	2794.7	3038.9	3256.6	3472.8	3693.3	3919.6	4151.7
	s	6.4508	6.9207	7.2709	7.5703	7.8385	7.0838	8.3108
16	v		0.1587	0.1900	0.2202	0.2499	0.2793	0.3086
(201.37°C)	h		3036.2	3255.0	3471.7	3692.5	3918.9	4151.3
	s		6.8873	7.2394	7.5395	7.8080	8.0535	8.2807
17	v		0.1489	0.1785	0.2070	0.2351	0.2628	0.2904
(204.31°C)	h		3033.5	3253.5	3470.6	3691.7	3918.3	4150.8
	s		6.8557	7.2098	7.5105	7.7793	8.0250	8.2523
18	v		0.1402	0.1684	0.1954	0.2219	0.2481	0.2742
(207.11°C)	h		3030.7	3251.9	3469.5	3690.9	3917.7	4150.3
	s		6.8257	7.1816	7.4830	7.7522	7.9981	8.2256
19	v		0.1325	0.1593	0.1849	0.2101	0.2350	0.2597
(209.80°C)	h		3027.9	3250.3	3468.4	3690.0	3917.1	4149.9
	s		6.7970	7.1550	7.4570	7.7265	7.9727	8.2003
20	v		0.1255	0.1511	0.1756	0.1995	0.2232	0.2467
(212.37°C)	h		3025.0	3248.7	3467.2	3689.2	3916.5	4149.4
	s		6.7696	7.1296	7.4323	7.7022	7.9485	8.1763
22	v		0.113 43	0.137 00	0.159 34	0.181 19	0.202 76	0.224 15
(217.24°C)	h		3019.3	3245.5	3465.1	3687.6	3915.2	4148.4
	s		6.7179	7.0821	7.3862	7.6568	7.9035	8.1316
24	v		0.103 36	0.125 22	0.145 82	0.165 92	0.185 75	0.205 39
(221.78°C)	h		3013.4	3242.3	3462.9	3685.9	3914.0	4147.5
	s		6.6699	7.0384	7.3439	7.6152	7.8624	8.0908
26	v		0.094 83	0.115 26	0.134 38	0.153 01	0.171 35	0.189 52
(226.04°C)	h		3007.4	3239.0	3460.6	3684.3	3912.7	4146.6
	s		6.6249	6.9979	7.3048	7.5768	7.8245	8.0531

v in (m^3/kg); h in (kJ/kg); s in (kJ/kg K)

Contd.

Pressure (bar) (Saturation temperature)		Temperature (°C)					
		300	400	500	600	700	800
28	v	0.087 51	0.106 71	0.124 58	0.141 94	0.159 02	0.175 91
(230.05°C)	h	3001.3	3235.8	3458.4	3682.6	3911.5	4145.6
	s	6.5824	6.9601	7.2685	7.5412	7.7893	8.0183
30	v	0.081 16	0.099 31	0.116 08	0.132 34	0.148 32	0.164 12
(233.84°C)	h	2995.1	3232.5	3456.2	3681.0	3910.3	4144.7
	s	6.5422	6.9246	7.2345	7.5079	7.7564	7.9857
32	v	0.075 59	0.092 83	0.108 65	0.123 95	0.138 97	0.153 81
(237.45°C)	h	2988.7	3229.2	3454.0	3679.3	3909.0	4143.8
	s	6.5037	6.8912	7.2026	7.4767	7.7257	7.9552
34	v	0.070 68	0.087 11	0.102 09	0.116 57	0.130 71	0.144 70
(240.88°C)	h	2982.2	3225.9	3451.7	3677.7	3907.8	4142.8
	s	6.4669	6.8595	7.1724	7.4473	7.6967	7.9266
36	v	0.066 30	0.082 02	0.096 26	0.109 96	0.123 37	0.136 61
(244.16°C)	h	2975.6	3222.5	3449.5	3676.1	3906.5	4141.9
	s	6.4315	6.8294	7.1439	7.4195	7.6693	7.8995
38	v	0.062 37	0.077 47	0.091 04	0.104 06	0.116 81	0.129 37
(247.31°C)	h	2968.9	3219.1	3447.2	3674.4	3905.3	4141.0
	s	6.3973	6.8007	7.1168	7.3931	7.6434	7.8739
40	v	0.058 83	0.073 38	0.086 34	0.098 76	0.110 90	0.122 85
(250.33°C)	h	2962.0	3215.7	3445.0	3672.8	3904.1	4140.0
	s	6.3642	6.7733	7.0909	7.3680	7.6187	7.8495
45	v	0.051 34	0.064 72	0.076 43	0.087 43	0.098 43	0.109 10
(257.41°C)	h	2944.2	3207.1	3439.3	3668.6	3901.0	4137.7
	s	6.2852	6.7093	7.0311	7.3100	7.5619	7.7934
50	v	0.045 30	0.057 79	0.068 49	0.078 62	0.088 45	0.098 09
(263.91°C)	h	2925.5	3198.3	3433.7	3664.5	3897.9	4135.3
	s	6.2105	6.6508	6.9770	7.2578	7.5108	7.7431
60	v	0.036 14	0.047 38	0.056 59	0.065 18	0.073 48	0.081 59
(275.55°C)	h	2885.0	3180.1	3422.2	3656.2	3891.7	4130.7
	s	6.0692	6.5462	6.8818	7.1664	7.4217	7.6554

v in (m^3/kg); h in (kJ/kg); s in (kJ/kg K)

Contd.

Pressure (bar) (Saturation temperature)		Temperature (°C)					
		300	400	500	600	700	800
70 (285.79°C)	v	0.029 46	0.039 92	0.048 09	0.055 59	0.062 79	0.069 80
	h	2839.4	3161.2	3410.6	3647.9	3885.4	4126.0
	s	5.9327	6.4536	6.7993	7.0880	7.3456	7.5808
80 (294.97°C)	v	0.024 26	0.034 31	0.041 70	0.048 39	0.054 77	0.060 96
	h	2786.8	3141.6	3398.8	3639.5	3879.2	4121.3
	s	5.7942	6.3694	6.7262	7.0191	7.2790	7.5158
90 (303.31°C)	v		0.030 73	0.037 64	0.042 80	0.048 53	0.054 08
	h		3125.3	3389.2	3631.1	3873.0	4116.7
	s		6.3067	6.6728	6.9574	7.2196	7.4579
100 (310.96°C)	v		0.026 41	0.032 76	0.038 32	0.043 55	0.048 58
	h		3099.9	3374.6	3622.7	3866.8	4112.0
	s		6.2182	6.5994	6.9013	7.1660	7.4058
110 (318.05°C)	v		0.023 51	0.029 50	0.034 66	0.039 47	0.044 08
	h		3077.8	3362.2	3614.2	3860.5	4107.3
	s		6.1483	6.5432	6.8499	7.1170	7.3584
120 (324.65°C)	v		0.021 08	0.026 73	0.031 60	0.036 07	0.040 33
	h		3054.8	3349.6	3605.7	3854.3	4102.7
	s		6.0810	6.4906	6.8022	7.0718	7.3147
130 (330.83°C)	v		0.019 02	0.024 40	0.029 02	0.033 19	0.037 16
	h		3030.7	3336.8	3597.1	3848.0	4098.0
	s		6.0155	6.4409	6.7577	7.0298	7.2743
140 (336.64°C)	v		0.017 23	0.022 51	0.026 90	0.030 72	0.034 44
	h		3005.6	3323.8	3588.5	3841.7	4093.3
	s		5.9513	6.3937	6.7159	6.9906	7.2367
150 (342.13°C)	v		0.015 66	0.020 80	0.024 88	0.028 59	0.032 09
	h		2979.1	3310.6	3579.8	3835.4	4088.6
	s		5.8876	6.3487	6.6764	6.9536	7.2013
175 (354.64°C)	v		0.012 46	0.017 36	0.021 72	0.025 07	0.028 21
	h		2906.3	3276.5	3562.2	3822.8	4079.3
	s		5.7274	6.2432	6.6031	6.8857	7.1366

v in (m^3/kg); h in (kJ/kg); s in (kJ/kg K)

Contd.

Pressure (bar) (Saturation temperature)		Temperature (°C)				
		400	500	600	700	800
200	v	0.009 947	0.014 77	0.018 16	0.021 11	0.023 85
(365.70°C)	h	2820.5	3241.1	3535.5	3803.8	4065.3
	s	5.5585	6.1456	6.5043	6.7953	7.0511
220	v	0.008 251	0.013 12	0.016 33	0.019 07	0.021 60
(373.69°C)	h	2738.8	3211.7	3517.4	3791.1	4055.9
	s	5.4102	6.0716	6.4441	6.7410	7.0001
250	v	0.006 014	0.011 13	0.014 13	0.016 63	0.018 91
	h	2582.0	3165.9	3489.9	3771.9	4041.9
	s	5.1455	5.9655	6.3604	6.6664	6.9306
300	v	0.002 831	0.008 681	0.011 44	0.013 65	0.015 62
	h	2161.8	3085.0	3443.0	3739.7	4018.5
	s	4.4896	5.7975	6.2340	6.5560	6.8288
400	v	0.001 909	0.005 616	0.008 088	0.009 930	0.011 52
	h	1934.1	2906.8	3346.4	3674.8	3971.7
	s	4.1190	5.4762	6.0135	6.3701	6.6606
500	v	0.001 729	0.003 882	0.006 111	0.007 720	0.009 076
	h	1877.7	2723.0	3248.3	3610.2	3925.3
	s	4.0083	5.1782	5.8207	6.2138	6.5222
1000	v	0.001 446	0.001 893	0.002 668	0.003 536	0.004 341
	h	1797.6	2316.1	2857.5	3324.4	3714.3
	s	3.7738	4.4913	5.1505	5.6579	6.0397

v in (m^3/kg); h in (kJ/kg); s in (kJ/kg K)

Appendix E

Saturated Freon-12: Temperature Table

T (°C)	P^s (kPa)	v_f (m³/kg)	v_g (m³/kg)	h_f (kJ/kg)	h_g (kJ/kg)	s_f (kJ/kg K)	s_g (kJ/kg K)
−40.0	64.123	0.000 659	0.242 233	0.0000	169.7004	0.0000	0.7279
−38.0	70.376	0.000 662	0.222 125	1.7759	170.6318	0.0076	0.7256
−36.0	77.100	0.000 664	0.204 016	3.5560	171.5613	0.0151	0.7235
−34.0	84.319	0.000 667	0.187 683	5.3402	172.4887	0.0226	0.7215
−32.0	92.057	0.000 669	0.172 920	7.1287	173.4137	0.0300	0.7195
−30.0	100.341	0.000 672	0.159553	8.9215	174.3361	0.0374	0.7177
−28.0	109.193	0.000 675	0.147 433	10.7185	175.2557	0.0447	07159
−26.0	188.643	0.000 677	0.136420	12.5201	176.1723	0.0520	0.7142
−24.0	128.715	0.000 680	0.126 400	14.3261	177.0856	0.0592	0.7125
−22.0	139.436	0.000 683	0.117 268	16.1366	177.9955	0.0665	0.7109
−20.0	150.836	0.000 685	0.108 931	17.9517	178.9017	0.0736	0.7094
−18.0	162.940	0.000 688	0.101 311	19.7713	179.8040	0.0807	0.7080
−16.0	175.778	0.000 691	0.094 335	21.5957	180.7022	0.0878	0.7066
−14.0	189.379	0.000 694	0.087 938	23.4249	181.5959	0.0949	0.7052
−12.0	203.771	0.000 697	0.082 065	25.2589	182.4851	0.1019	0.7039
−10.0	218.985	0.000 700	0.076 667	27.0978	183.3695	0.1089	0.7027
−8.0	235.050	0.000 703	0.071 697	28.9417	184.2487	0.1158	0.7015
−6.0	251.998	0.000 706	0.067 115	30.7908	185.1227	0.1227	0.7004
−4.0	269.859	0.000 709	0.062 887	32.6450	185.9911	0.1296	0.6993
−2.0	288.661	0.000 712	0.058 981	34.5045	186.8538	0.1364	0.6983
0.0	308.440	0.000 716	0.055 366	36.3695	187.7104	0.1432	0.6973
2.0	329.222	0.000 719	0.052 019	38.2398	188.5608	0.1500	0.6963

Contd.

T (°C)	P^s (kPa)	v_f (m³/kg)	v_g (m³/kg)	h_f (kJ/kg)	h_g (kJ/kg)	s_f (kJ/kg K)	s_g (kJ/kg K)
4.0	351.047	0.000 723	0.048 914	40.1161	189.4095	0.1567	0.6954
6.0	373.944	0.000 726	0.046 033	41.9981	190.2414	0.1634	0.6945
8.0	397.942	0.000 730	0.043 355	43.8861	191.0712	0.1701	0.6936
10.0	423.083	0.000 733	0.040 863	45.7805	191.8935	0.1768	0.6928
12.0	449.392	0.000 737	0.038 542	47.6811	192.7082	0.1834	0.6920
14.0	476.905	0.000 741	0.036 379	49.5883	193.5149	0.1900	0.6912
16.0	505.660	0.000 744	0.034 361	51.5024	194.3132	0.1965	0.6904
18.0	535.684	0.000 748	0.032 476	53.4234	195.1030	0.2031	0.6897
20.0	567.019	0.000 752	0.030 713	55.3518	195.8836	0.2096	0.6890
22.0	599.698	0.000 756	0.029 063	57.2878	196.6550	0.2161	0.6883
24.0	633.756	0.000 761	0.027 518	59.2318	197.4165	0.2226	0.6876
26.0	669.233	0.000 765	0.026 069	61.1841	198.1679	0.2290	0.6869
28.0	706.156	0.000 769	0.024 709	63.1449	198.9088	0.2355	0.6863
30.0	744.567	0.000 774	0.023 432	65.1147	199.6387	0.2419	0.6856
32.0	784.508	0.000 778	0.022 232	67.0939	200.3571	0.2483	0.6850
34.0	826.003	0.000 783	0.021 103	69.0827	201.0637	0.2547	0.6844
36.0	869.108	0.000 788	0.020 040	71.0821	201.7577	0.2611	0.6838
38.0	913.842	0.000 793	0.019 038	73.0921	202.4388	0.2674	0.6831
40.0	960.255	0.000 798	0.018 093	75.1134	203.1063	0.2738	0.6825
42.0	1008.379	0.000 803	0.017 202	77.1464	203.7598	0.2801	0.6819
44.0	1058.266	0.000 808	0.016 359	79.1921	204.3982	0.2865	0.6813
46.0	1109.948	0.000 814	0.015 563	81.2510	205.0212	0.2928	0.6806
48.0	1163.453	0.000 820	0.014 810	83.3234	205.6281	0.2992	0.6800
50.0	1218.838	0.000 826	0.014 097	85.4106	206.2177	0.3055	0.6793
52.0	1276.143	0.000 832	0.013 422	87.5132	206.7894	0.3118	0.6787
54.0	1335.408	0.000 838	0.012 781	89.6322	207.3421	0.3182	0.6780
56.0	1396.661	0.000 844	0.012 173	91.7681	207.8752	0.3245	0.6773

The assistance of Mr. S.C. Gupta in generating Freon-12 tables is gratefully acknowledged.

Appendix F

Saturated Freon-12: Pressure Table

P^s (kPa)	T T(°C)	v_f (m^3/kg)	v_g (m^3/kg)	h_f (kJ/kg)	h_g (kJ/kg)	s_f (kJ/kg K)	s_g (kJ/kg K)
60	−41.4	0.000 658	0.257 692	−1.2435	169.0452	−0.0053	0.7295
65	−39.7	0.000 660	0.239 187	0.2570	169.8353	0.0011	0.7275
70	−38.1	0.000 662	0.223 235	1.6738	170.5776	0.0071	0.7258
75	−36.6	0.000 664	0.209 336	3.0126	171.2783	0.0128	0.7242
80	−35.2	0.000 665	0.197 114	4.2870	171.9419	0.0182	0.7227
85	−33.8	0.000 667	0.186 280	5.5033	172.5727	0.0232	0.7213
90	−32.5	0.000 669	0.176 606	6.6651	173.1741	0.0281	0.7200
95	−31.3	0.000 670	0.167 915	7.7781	173.7488	0.0327	0.7188
100	−30.1	0.000 672	0.160 061	8.8499	174.2994	0.0371	0.7177
110	−27.8	0.000 675	0.146 421	10.8768	175.3365	0.0453	0.7157
120	−25.7	0.000 678	0.134 975	12.7702	176.2990	0.0530	0.7139
130	−23.8	0.000 680	0.125 229	14.5484	177.1980	0.0601	0.7123
140	−21.9	0.000 683	0.116 825	16.2280	178.0418	0.0668	0.7108
150	−20.1	0.000 685	0.109 501	17.8234	178.8372	0.0731	0.7095
160	−18.5	0.000 688	0.103 060	19.3387	179.5901	0.0791	0.7083
170	−16.9	0.000 690	0.097 349	20.7888	180.3049	0.0847	0.7072
180	−15.4	0.000 692	0.092 249	22.1743	180.9855	0.0901	0.7061
190	−13.9	0.000 694	087.667	23.5064	181.6354	0.0952	0.7052
200	−12.5	0.000 696	0.083 526	24.7868	182.2572	0.1001	0.7043
215	−10.5	0.000 699	0.078 010	26.6246	183.1430	0.1071	0.7030
230	−8.6	0.000 702	0.073 187	28.3730	183.9779	0.1137	0.7019
245	−6.8	0.000 705	0.068 933	30.0393	184.7681	0.1199	0.7009
260	−5.1	0.000 708	0.065 152	31.6329	185.5182	0.1258	0.6999
275	−3.4	0.000 710	0.061 768	33.1636	186.2321	0.1315	0.6990

Contd.

P^s (kPa)	T T(°C)	v_f (m^3/kg)	v_g (m^3/kg)	h_f (kJ/kg)	h_g (kJ/kg)	s_f (kJ/kg K)	s_g (kJ/kg K)
290	−1.9	0.000 713	0.058 721	34.6338	186.9134	0.1369	0.6982
305	−0.3	0.000 715	0.055 962	36.0525	187.5649	0.1421	0.6974
320	−1.1	0.000 718	0.053 453	37.4205	188.1895	0.1470	0.6967
340	3.0	0.000 721	0.050 438	39.1775	188.9833	0.1534	0.6958
360	4.8	0.000 724	0.037 746	40.8629	189.7373	0.1594	0.6950
380	6.5	0.000 727	0.045 326	42.4823	190.4552	0.1651	0.6943
400	8.2	0.000 730	0.043 139	44.0433	191.1402	0.1707	0.6935
420	9.8	0.000 733	0.041 153	45.5527	191.7952	0.1760	0.6929
440	11.3	0.000 736	0.039 340	47.0118	192.4225	0.1810	0.6923
460	12.8	0.000 738	0.037 679	48.4253	193.0246	0.1859	0.6917
480	14.2	0.000 741	0.036 151	49.7983	193.6028	0.1907	0.6911
500	15.6	0.000 744	0.034 741	51.1321	194.1595	0.1953	0.6906
530	17.6	0.000 748	0.032 817	53.0672	194.9566	0.2019	0.6898
560	19.6	0.000 751	0.031 091	54.9270	195.7124	0.2082	0.6891
590	21.4	0.000 755	0.029 534	56.7215	196.4303	0.2142	0.6885
620	23.2	0.000 759	0.028 122	58.4559	197.1138	0.2200	0.6879
650	24.9	0.000 763	0.026 835	60.1346	197.7655	0.2256	0.6873
680	26.6	0.000 766	0.025 658	61.7636	198.3883	0.2309	0.6867
710	18.2	0.000 770	0.024 575	63.3452	198.9836	0.2361	0.6862
740	29.8	0.000 773	0.023 578	64.8837	199.5539	0.2411	0.6857
770	31.3	0.000 777	0.022 654	66.3828	200.1010	0.2460	0.6852
800	32.8	0.000 780	0.021 797	67.8449	200.6255	0.2507	0.6848
840	34.7	0.000 785	0.020 746	69.7396	201.2931	0.2568	0.6842
880	36.5	0.000 789	0.019 787	71.5775	201.9270	0.2626	0.6836
920	38.3	0.000 793	0.018 907	73.3639	202.5294	0.2683	0.6831
960	40.0	0.000 798	0.018 098	75.1013	203 1030	0.2737	0.6825
1000	41.7	0.000 802	0.017 351	76.7967	203.6486	0.2790	0.6820
1050	43.7	0.000 808	0.016 494	78.8585	204.2498	0.2854	0.6814
1100	45.6	0.000 813	0.015 711	80.8591	204.9043	0.2916	0.6808
1200	49.3	0.000 824	0.014 333	84.7076	206.0212	0.3034	0.6792
1300	52.8	0.000 834	0.013 157	88.3725	207.0166	0.3144	0.6784

The assistance of Mr. S.C. Gupta in generating Freon-12 tables is gratefully acknowledged.

Appendix G

Specific Heats of Some Gases at Low Pressure

Gas	Formula	C_P kJ/kg K	C_v kJ/kg K
Acetylene	C_2H_2	1.6947	1.3753
Air		1.0047	0.7176
Ammonia	NH_3	2.0890	1.5992
Argon	A	0.5208	0.3127
Butane	C_4H_{10}	1.7164	1.5734
Carbon dioxide	CO_2	0.8440	0.6552
Carbon monoxide	CO	1.0412	0.7444
Chlorine	Cl_2	0.4789	0.3617
Ethane	C_2H_6	1.7662	1.4897
Ethylene	C_2H_4	1.5482	1.2518
Helium	He	5.1926	3.1156
Hydrogen	H_2	14.2091	10.0849
Hydrazine	N_2H_4	1.6453	1.3815
Methane	CH_4	2.2537	1.7354
Neon	Ne	1.0299	0.6179
Nitrogen	N_2	1.0416	0.7448
Octane	C_8H_{18}	1.7113	1.6385
Oxygen	O_2	0.9216	0.6618
Propane	C_3H_8	1.6794	1.4909
Sulfur dioxide	SO_2	0.6225	0.4927
Water vapor	H_2O	1.8646	1.4033
Xenon	Xe	0.1582	0.0950

Appendix H

Standard Enthalpies of Formation and Gibbs Free Energies

Substance	Formula	State	ΔH_f^0 kJ/mol	ΔG_f^0 kJ/mol
Acetaldehyde	CH_3CHO	*g*	−164.4	−133.4
Acetic acid	CH_3COOH	*g*	−435.1	−376.9
Acetone	CH_3COCH_3	*g*	−217.7	−153.2
Acetylene	C_2H_2	*g*	226.9	209.3
Ammonia	NH_3	*g*	−45.72	−16.16
Benzene	C_6H_6	*g*	82.98	129.7
Carbon (diamond)	C	*s*	2.093	2.931
Carbon dioxide	CO_2	*g*	−393	−394.6
Carbon disulfide	CS_2	*g*	117.1	66.95
Carbon monoxide	CO	*g*	−110.6	−137.4
Carbon tetrachloride	CCl_4	*g*	−100.5	−58.28
Dodecane	$C_{12}H_{26}$	*g*	−291.1	50.07
Ethane	C_2H_6	*g*	−84.74	−32.95
Ethyl alcohol	C_2H_5OH	*g*	−235	−168.4
Ethylene	C_2H_4	*g*	52.34	68.16
Ethylene oxide	C_2H_4O	*g*	−52.67	−13.10
Formic acid	HCOOH	*g*	−378.9	−351.2
Formaldehyde	HCHO	*g*	−116	−110
Hydrogen chloride	HCl	*g*	−92.36	−95.33
Hydrogen sulfide	H_2S	*g*	−20.18	−33.08
Methane	CH_4	*g*	−74.9	−50.87
Methyl alcohol	CH_3OH	*g*	−201.3	−162.6

Contd.

Substance	Formula	State	ΔH_f^0 kJ/mol	ΔG_f^0 kJ/mol
Nitric oxide	NO	*g*	90.43	86.75
Nitrous oxide	N_2O	*g*	81.6	103.7
n-Butane	C_4H_{10}	*g*	−126.2	−16.10
n-Decane	$C_{10}H_{22}$	*g*	−249.8	33.24
n-Heptane	C_7H_{16}	*g*	−187.9	8.0
n-Hexane	C_6H_{14}	*g*	−167.3	−0.167
n-Nonane	C_9H_{20}	*g*	−229.2	24.83
n-Octane	C_8H_{18}	*g*	−208.6	16.40
n-Pentane	C_5H_{12}	*g*	−146.5	−8.37
Propane	C_3H_8	*g*	−103.9	−23.49
Sulfur dioxide	SO_2	*g*	−297.1	−300.4
Sulfur trioxide	SO_3	*g*	−396.0	−371.3
Water	H_2O	*g*	−242.0	−228.8

Source: Data adapted from R.C.Reid; J.M.Prausnitz and B.E.Poling, *The Properties of Gases and Liquids*, Fourth edition, McGraw Hill Book Company, New York (1988)

Appendix I

Standard Enthalpies of Combustion at 298.15 K

Compound	Formula (State)	$-\Delta H_c^0$ kJ/mol
Hydrocarbons		
Carbon (graphite)	$C(s)$	393.78
Carbon monoxide	$CO(g)$	283.18
Hydrogen	$H_2(g)$	286.03
Methane	$CH_4(g)$	890.94
Ethyne (acetylene)	$C_2H_2(g)$	1300.48
Ethene (ethylene)	$C_2H_4(g)$	1411.93
Ethane	$C_2H_6(g)$	1560.92
Propyne (allylene, methylacetylene)	$C_3H_4(g)$	1938.94
Propene (propylene)	$C_3H_6(g)$	2059.85
Propane	$C_3H_8(g)$	2221.54
n-Butane	$C_4H_{10}(g)$	2880.44
n-Pentane	$C_5H_{12}(g)$	3538.52
Benzene	$C_6H_6(g)$	3303.72
Cyclohexane	$C_6H_{12}(l)$	3932.53
n-Hexane	$C_6H_{14}(l)$	4165.91
Toluene (methylbenzene)	$C_7H_8(l)$	3912.56
n-Heptane	$C_7H_{14}(l)$	4820.14
n-Octane	$C_8H_{18}(l)$	5474.37
n-Decane	$C_{10}H_{22}(l)$	6782.87

Contd.

Compound	Formula (State)	$-\Delta H_c^0$ kJ/mol
Alcohols		
Methyl alcohol	$CH_4O(g)$	764.47
Ethyl alcohol	$C_2H_6O(g)$	1410.20
n-Propyl alcohol	$C_3H_8O(g)$	2069.37
n-Butyl alcohol	$C_4H_{10}O(g)$	2721.34
CHO Compounds		
Formaldehyde	$CCH_2O(g)$	563.84
Acetaldehyde	$C_2H_4O(g)$	1193.15
Acetone	$C_3H_6O(g)$	1822.60
Ethyl acetate	$C_4H_8O_2(g)$	2292.11
Phenol	$C_6H_6O(g)$	3129.84
Sulfur Compounds		
Carbonyl sulfide	$COS(g)$	553.54
Carbon disulfide	$CS_2(g)$	1103.31

The final products of combustion for Hydrocarbons, Alcohols and CHO compounds are $CO_2(g)$; $H_2O(l)$. For sulfur compounds, the final products of combustion are $CO_2(g)$; $SO_2(g)$ and $H_2O(l)$.

Adapted from Chandrasekaran, K.D and Venkateswarlu, D., *SI Units in Chemical Engineering and Technology*, Chemical Engineering Education Development Centre, IIT-Madras (1974), with permission from Chairman, Centre for Continuing Education, IIT Madras.

Appendix J

Isobaric Molar Heat Capacities in the Ideal Gas State

Compound	Formula	a	$b \times 10^2$	$c \times 10^5$	$d \times 10^8$
Acetaldehyde	C_2H_4O	7.716	18.23	−10.07	2.380
Acetic acid	CH_3COOH	4.84	25.49	−17.53	4.949
Acetone	C_3H_6O	6.301	26.06	12.53	2.038
Acetylene	C_2H_2	26.82	7.578	−5.007	1.412
Ammonia	NH_3	27.31	2.383	1.707	−1.185
Argon	Ar	20.8			
Benzene	C_6H_6	−33.92	47.39	−30.17	7.130
Carbon dioxide	CO_2	19.8	7.344	−5.602	1.715
Carbon monoxide	CO	30.87	−1.285	2.789	−1.272
Carbon disulfide	CS_2	27.44	8.127	−7.666	2.673
Carbon tetrachloride	CCl_4	40.72	20.49	−22.70	8.843
Chlorine	Cl_2	26.93	3.384	−3.869	1.547
Dodecane	$C_{12}H_{26}$	−9.328	0.01149	−63.47	13.59
Ethane	C_2H_6	5.409	17.81	−6.938	8.713
Ethanol	C_2H_5OH	9.014	21.41	−8.390	0.1373
Ethylene	C_2H_4	3.806	15.66	−8.348	1.755
Ethylene oxide	C_2H_4O	−7.519	22.22	−12.56	2.592
Helium	He	20.8			
Hydrogen	H_2	27.14	0.9274	−1.381	0.7645
Hydrogen chloride	HCl	30.67	−0.7201	0.1246	−0.3898
Methanol	CH_3OH	21.15	7.092	2.587	−2.852
Methane	CH_4	19.25	5.213	1.197	−1.132
Nitrogen	N_2	31.15	−1.357	2.680	−1.168

Contd.

Compound	Formula	a	$b \times 10^2$	$c \times 10^5$	$d \times 10^8$
Nitrogen dioxide	NO_2	24.32	4.836	−2.081	0.0293
Nitric oxide	NO	29.35	−93.78	0.9747	−0.4187
Nitrous oxide	N_2O	21.62	7.281	−5.778	1.830
n-Butane	C_4H_{10}	9.487	33.13	−1.108	−28.22
n-Pentane	C_5H_{12}	−3.626	48.73	−25.80	5.305
n-Hexane	C_6H_{14}	−4.413	58.20	−31.19	6.494
n-Heptane	C_7H_{16}	−5.146	67.62	−36.51	7.658
n-Octane	C_8H_{18}	−6.096	77.12	−41.95	8.855
n-Nonane	C_9H_{20}	−8.374	87.29	−48.23	10.31
Oxygen	O_2	28.11	−0.000 368	1.746	−1.065
Propane	C_3H_8	−4.224	30.63	−15.86	3.215
Sulfur dioxide	SO_2	23.85	6.699	−4.961	1.328
Sulfur trioxide	SO_3	19.21	0.1374	−11.76	3.70
Water	H_2O	32.24	0.1924	1.055	−0.3596

$$C_P^0 = a + bT + cT^2 + dT^3$$

C_P^0 is in J/mol K and T is in K.

Source: Data adapted from R.C.Reid; J.M.Prausnitz and B.E.Poling, *The Properties of Gases and Liquids* Fourth edition, McGraw-Hill Book Company, New York (1988).

Answers to Problems

Chapter 1

1.1 96.93 kPa; 0.9693 bar
1.2 5.876 bar
1.3 6.628 kPa
1.4 2.354 kPa
1.5 43.262 kPa
1.6 48.517 kg
1.7 60 N/m
1.8 $P = P_0 \left[1 + \left(\frac{1-\gamma}{\gamma}\right) \frac{gh}{P_0 v_0}\right]^{\frac{\gamma}{\gamma - 1}}$
1.9 0.71 bar
1.10 0.6979 bar
1.11 258.6 mol
1.12 500; 31.25; 62.5; 17.24
1.13 25.767 kmol
1.14 1.019 kg
1.15 5.76 MJ
1.16 207.36 MJ; Rs. 216/-
1.17 3.0264 MPa

Chapter 2

2.1 None
2.2 (a) only
2.3 Yes; $V = \pi R^2 h + \text{const}$
2.4 Second expression; $P = \dfrac{RT}{v - b} + \text{const}$
2.5 Extensive; Specific; Intensive; Intensive; Specific; Extensive; Extensive
2.6 12.5 kJ
2.7 964.5 MJ; 2.453 GJ
2.8 7.29 kJ
2.9 44.29 m/s
2.10 45.87 m
2.11 $(m - \rho V)gh$; Yes; Decreases

2.12 $\dfrac{P_1V_1 - P_2V_2}{\gamma - 1}$

2.13 6.545 MJ

2.14 Yes; 49.88 kJ

2.15 $RT\ln\left(\dfrac{v_2 - b}{v_1 - b}\right) - a\left(\dfrac{1}{v_1} - \dfrac{1}{v_2}\right)$

2.16 150 kJ

2.17 2.9894 MJ

2.18 96 kJ

2.19 (a) 115.13 kJ; (b) 34.187 kJ

2.20 (a) 4.5 J; (b) 6 J; (c) 4.5 J

2.21 0.2 m

2.22 4 kJ

2.23 1.21 J

2.24 580.8 kJ

2.25 (a) 1 MJ; (b) 0.6932 MJ; (c) 2.333 MJ

2.26 (a) $W = 0$; $Q = 0$; $\Delta U > 0$; $\Delta E = 0$ (b) $W = 0$; $Q = 0$; $\Delta U > 0$; $\Delta E < 0$

2.27 (a) $W < 0$; $Q = 0$; $\Delta E > 0$; $\Delta U > 0$ (b) $W = 0$; $Q = 0$; $\Delta E = 0$; $\Delta U > 0$
(c) (i) $W = 0$; $Q > 0$; $\Delta E > 0$; $\Delta U > 0$ (ii) $W < 0$; $Q = 0$; $\Delta E > 0$; $\Delta U > 0$

Chapter 3

3.1 (a) 83.14 m^3; (b) 82.6 m^3; (c) 82 m^3

3.2 (a) 0.52×10^{-3} m^3/mol; (b) 0.219×10^{-3} m^3/mol

3.3 (a) 6.544 MPa; (b) 6.5054 MPa

3.4 −5.8033 kJ; −2.3214 kJ; 5.7431 kJ

3.5 14.096 MPa; 10.02 MPa

3.7 $\dfrac{4R^2T_c^2}{P_ce^2}; \dfrac{RT_c}{P_ce^2}$

3.8 (a) $\dfrac{1}{T}; \dfrac{1}{P}$ (b) $\dfrac{Rv^2(v-b)}{RTv^3 - 2a(v-b)^2}; \dfrac{v^2(v-b)^2}{RTv^3 - 2a(v-b)^2}$

3.9 1.058 m^3

3.10 0.825; 0.8496

3.11 74.197 bar

3.12 0.597 m^3

3.13 (a) 19.9 MPa; (b) 18.298 MPa (c) 19.028 MPa

3.14 19.189 MPa

3.15 444.56 bar; 572.98 bar

3.16 357.4 K

3.17 (a) 11.418×10^{-3} m^3/kg; (b) 6.28×10^{-3} m^3/kg

3.18 76.63 bar

3.19 416.14 K

3.22 0.33 m^3/kg; 2540 kJ/kg; 6.34 kJ/kg K

3.23 (a) 0.784 (b) 0.847 (c) 0.955 (d) 0.875

3.24 0.6915 m^3

3.25	0.4364 m^3
3.26	15.07 kg; 45.21 kg
3.27	0.0448 kg; 0.9979
3.28	0.18 m^3/kg; 515 kJ/kg
3.29	91°C
3.30	0.94
3.31	0.949
3.32	815 kJ/kg; 400°C
3.33	70 kPa and $X = 0.74$; 700 kJ/kg
3.34	1.9854 bar; 0.4143
3.35	470 kJ/kg
3.36	3.5704 m^3/kg; 3.5704 m^3/kg; 3.5650 m^3/kg
3.37	7.141×10^{-3} m^3/kg; 3.928×10^{-3} m^3/kg; 6.111×10^{-3} m^3/kg
3.38	$v_{i+1} = \dfrac{RT}{P + \dfrac{a}{v_i^2}} + b$
3.39	14 kg; 11.79 kg; 14.43 kg; 22.08 kg
3.40	234.61°C
3.41	85.927 bar
3.42	1.692 956 6 m^3/kg 0.098 363 4 m^3/kg 0 m^3/kg
3.43	30.14 kg; 822.17 kg; 0.0354; 1061.34 kJ/kg; 1071.9 kJ/kg
3.44	9.9967 MJ
3.45	0.140 97 m^3/kg; 1081.34 kJ/kg
3.46	0°C with $X = 0{,}57$; −27.182 kJ
3.47	123.08 kg
3.48	1.9 kg
3.49	48.874 MPa
3.50	19.4 kg

Chapter 5

5.1	125 kJ; 75 kJ
5.2	−300; 700; −350; −150; 50; 600 kJ
5.3	180 kJ
5.4	(a) 0; 24.525 kJ/kg; −24.525 kJ/kg; 0 (b) 0; 0; −24.525 kJ/kg; 24.525 kJ/kg
5.5	Wet steam; $X = 0.986$ at 2.5 bar
5.6	Wet steam; $X = 0.81$ at 4 bar; −603.81 kJ
5.7	Wet steam $X = 7.847 \times 10^{-3}$ at 2 bar; 394.72 kJ
5.8	$X = 9.84 \times 10^{-2}$; 92.72 MJ
5.9	744.567 kPa; 1060.76 kJ
5.10	420.33 kJ
5.11	167.56 kJ; 567 kPa
5.12	308.44 kPa; $X = 0.5806$; 75.486 kJ
5.13	100.341 kPa; $X = 0.435$
5.14	16.87 kJ; 134.524 kJ

5.15	732.52 kJ; 461.8°C
5.16	Wet steam; $X = 0.933$ at 100 kPa
5.17	1.729 kJ; 1.729 kJ
5.18	Wet steam; $X = 0.811$ at 10 kPa; 832.24 kJ
5.19	100 kPa; −19.2°C; 22.68 kJ
5.20	87.297 kJ; 24.942 kJ
5.21	0; −3640.49 kJ/kmol; −5096.69 kJ/kmol; −1456.2 kJ/kmol; 4014.26 kJ/kmol; 4014.26 kJ/kmol; −1082.43 kJ; −1082.43 kJ
5.22	$-0.3069RT_1$
5.23	300 K; 50 kPa
5.24	−4.1324 kJ; −1.5492 kJ
5.25	544°C; 167.64 kJ; 91.18 kJ; 0.597 m above the base
5.26	462°C; 626.44 kJ
5.27	0.988 kJ
5.28	(a) Any cycle (b) Any process (c) Any reversible process
5.29	500 K; 0.1 MPa
5.30	(a) −5743.11 kJ (b) −2244.78 kJ; −9427.45 kJ
5.31	0.2243
5.32	25.497 MJ; 1.693 MJ
5.33	(a) 1620.16 K (b) 280.395 kJ
5.34	6 kJ; 280.395 kJ; 1620.16 K
5.35	(a) 1775.3 K (b) 3 kJ (c) 33.664 kJ

Chapter 6

6.1	148.17 kg
6.2	401.48 kW
6.3	637.73 kW; 0.93%; 0.0046%
6.4	56.394 kW
6.6	421.97 kW
6.7	3.02 kW
6.8	217.3°C
6.9	297.66 m/s; 0.1239 kg/s
6.10	607.12 m/s
6.11	0.97
6.12	0.96
6.13	0.3825
6.14	199.31 m/s
6.15	317.13 m/s
6.16	322.44 m/s
6.17	0.1866 kg/s
6.18	182.96°C; 0.4072 m^3/kg
6.19	711.86 K; 3.7334 kg
6.20	576.18°C; 5[illegible] 6 kg
6.21	10 ba[illegible]

6.22 3.603 MJ
6.23 3.366 kg
6.24 (a) 78.26 kg (b) 44.661 kg
6.25 396.23 K; 1.02 kg
6.26 7.24 kg
6.27 2 MPa; 362°C
6.28 4.509 kg
6.29 0.24 kg
6.30 0.0295
6.31 136.97
6.32 47.19 kg; 150.86 MJ
6.33 5.752 kg
6.34 41.942 MJ
6.35 (a) 11.655; 16.459 (b) 6.99% CO_2; 12.3% H_2O; 6.57% O_2; 74.14% N_2
6.36 7.7877
6.37 50.01
6.38 18.16
6.39 8.17% CO_2; 11.88% H_2O; 5.64% O_2; 74.31% N_2
6.40 17.25
6.41 (a) 4.5; 12.478 (b) 6.77% CO_2; 19.48% H_2O; 3.1% O_2; 70.65% N_2
6.42 (a) 15.125 (b) 10.54% CO_2; 11.85% H_2O; 3.29% O_2; 74.32% N_2
6.43 15.01; 8.66% CO_2; 9.38% H_2O; 6.67% O_2; 75.29% N_2
6.44 59.52; 140
6.45 17.98; 19.79
6.46 82.76% C; 17.24% H; 139.97
6.47 16.1; 9.4% CO_2; 5.24% O_2; 85.36% N_2
6.48 84.8% C; 15.2% H; 15.26
6.49 9.52; 117.86
6.50 9.83
6.51 871.38 kJ
6.52 598.034 kJ
6.53 −907.4 kJ; −518.82 kJ; −1256.5 kJ
6.54 −1171.196 kJ; −562.786 kJ; −1300.466 kJ
6.55 −1020.58 kJ
6.56 −3538.52 kJ; 206.364 kJ; −41.116 kJ
6.57 29105 kJ/kg
6.58 227.62 kJ; −31.837 kJ
6.59 $\Delta H_T^0 = -806 \times 10^3 + 8.81T + 1.2583 \times 10^{-2}T^2 - 2.727 \times 10^{-5}T^3 + 1.0645 \times 10^{-8}T^4$
where ΔH_T^0 is in Joule and T is in K
6.60 2090.2 K
6.61 2035.8 K
6.62 2011.4 K
6.63 1464.7 K
6.64 2047.3 K
6.65 1900.7 K

Chapter 7

7.1 (a) No (b) No

7.2 (a) Irreversible; Finite ΔT; (b) Irreversible; Conversion of W into U (c) Irreversible; Finite ΔP; (d) Irreversible; Finite ΔT (e) Irreversible; Mixing of non-identical gases (f) Irreversible; Mixing of matter at different states (g) Reversible

7.3 No

7.4 231.27 kJ; 231.27 kJ; No

7.8 13.084 kW; 3.084 kW

7.9 162°C; 150 kW

7.10 0.75

7.11 670.82 K; 552.79 kJ; 247.21 kJ

7.12 5.06

7.13 $\dfrac{1-\eta}{\eta}$

7.14 0.4 kJ

7.15 750 kW; 732.5 K

7.16 First possibility

7.17 0.894 kW

7.18 $\dfrac{T_2(T_1 - T_2)}{T_1(T_2 - T_L)}$

7.19 (a) $T_2 = \sqrt{T_1 T_3}$; (b) $T_2 = \dfrac{T_1 + T_3}{2}$

7.20 1800 kW rejected at 400 K and 850 kW absorbed at 300 K; 0.027

7.21 (a) 4.254 kW (b) 4.133

7.22 (a) 49.67 kW (b) 9.933

7.23 (a) 314.96 K (b) 4.704 (c) 24.8 kW

7.24 (a) 506.67 K (b) 893.2 kW

7.25 Not feasible

7.26 Feasible

7.27 No

7.28 19.76 kW

7.29 5.802 kW

7.30 0.341

7.31 No; Yes

7.32 (a) 547.72 K (b) 1.211 (c) 3.334 kW

7.33 8.571 kW

7.34 81.21 kW; 24.6 kW

7.35 252.97 K

7.36 6.068 MJ; 0.0137

7.37 11 MW; 4.0593×10^5

7.38 386.5 K; −1.2937 kJ/K; 1.2937 kJ/K; 0

7.39 Violates second law

7.40 1100 W; 2100 kW

7.41 114.19 kJ; 114.19 kJ; 0: 8.3733 kJ/kmol K; −5.9806 kJ/kmol K; −2.3918 kJ/kmol K

7.42 0.6747 kJ/K; 55.85 W

7.43 (a) 1.2268 kJ/kg K (b) 1.2268 kJ/kg K

7.44 6.0508 kJ/kg K; 4.0984 kJ/kg K. From steam tables: 6.0485 kJ/kg K; 4.0971 kJ/kg K

7.45 0.2345 kJ/kg K

7.46 3.035 bar

7.48 58.4452 kJ; 0.2064 kJ/kg K

7.49 1733.07 kJ; 4.2645 kJ/K

7.50 7.8344 kJ/kg K; 7.8349 kJ/kg K

7.51 (a) −111.941 J/K (b) 131.75 J/K (c) 19.809 J/K

7.52 (a) −4.9449 J/mol K (b) 11.0193 J/K (c) 6.0744 J/K

7.53 5.7628 J/mol K; No

7.54 15.84°C; 0.1181 kJ/K

7.55 (a) −8.915 kJ/K (b) 16.0312 kJ/K

7.56 −1.1286 kJ/K; 1.4224 kJ/K

7.57 −5.4076 kJ/K; 2.7608 kJ/K

7.58 11.61°C; 0.9611 kJ/K

7.59 430.3 m/s

7.60 1.11

7.62 1.0681 kJ/K

7.63 (a) −5.612 MJ/K; 6.484 MJ/K; 872 kJ/K (b) −4.126 MJ/K; 6.484 MJ/K; 2.358 MJ/K (c) −3.8899 MJ/K; 6.484 MJ/K; 2.5941 MJ/K

7.64 (a) 8.3012 kJ/kmol K; −6.311 kJ/kmol K; 1.9902 kJ/K (b) 8.3012 kJ/kmol K; −6.311 kJ/kmol K; 1.9902 kJ/K (c) 6.311 kJ/kmol K; −6.311 kJ/kmol K; 0

7.65 33.3497 kJ/K

7.66 $mC\left(\sqrt{T_1}-\sqrt{T_2}\right)^2$

7.67 4.349 MW

7.68 527.92 kW; 475.15 K

7.69 0.898 kg; 0.91 kg

7.70 $\gamma = \dfrac{\ln(P_0/P_1)}{\ln(P_2/P_1)}$

7.71 434.63 m/s; 2.833×10^{-3} m^2

7.72 10.83 kg/s

7.73 644.89 m/s; wet steam: $X = 0.9921$

7.74 0.8038 kg/s; 0.1962 kg/s; 1.1445 kJ/kg K

7.75 0.0216 kJ/kg K

7.76 (a) 0.2131 kJ/K; (b) −0.2007 kJ/K (c) 0.0124 kJ/K

7.77 −1.1469 kJ/kg K

7.78 6.5937 kJ/kg K; 6.5422 kJ/kg K

7.79 From B to A

7.80 0.5 kmol/s; 0.5 kmol/s; 14.2563 kJ/kmol K

7.81 Yes; Does not violate the laws of thermodynamics

7.82 (a) 0.354 kmol/s; 0.646 kmol/s (b) 4.146 kJ/kmol K (c) 218.44 K (d) 2373.9 kJ

7.83 (a) 3218.56 kJ (b) 0.5 kmol/s; 0.5 kmol/s (c) 9.1942 kJ/kmol K (d) 92.65 K; 728.51 K

7.84 (a) 189.42 K (b) 3218.56 kJ (c) 532.64 K; 67.36 K

7.85 (a) 1.1464×10^5 kg/h (b) 1.331 MW

7.86 Only the first claim is feasible

7.87 1273.4 kJ

7.88 4.4684 kJ/kmol K
7.89 2 MPa and 700°C; 71.2265 kJ/K
7.90 15.81 kg; 111.08 kJ/K
7.91 (a) 1714.03 kJ (b) 721 kJ (c) 993.03 kJ (d) 32.773 kJ
7.92 5741.29 kJ
7.93 1353.68 kJ
7.94 9337.6 kJ
7.95 (a) −1008.36 kJ/K; 1574.9 kJ/K (b) −293.82 MJ/h (c) 124.99 MJ/h (d) 168.83 MJ/h
7.96 (a) −272.67 kJ/K; 272.67 kJ/K (b) −48.203 MJ/min (c) 48.203 MJ/min (d) 0
7.97 (a) 332.42 kJ (b) 435.95 kJ
7.98 (a) 833.51 kJ (b) 0.2338
7.99 320.1 kJ

Chapter 8

8.4 (a) $Pv\kappa - Tv\beta$ (b) $\dfrac{T\beta}{\kappa} - P$
8.7 (a) $\dfrac{T\beta - 1}{\kappa}$ (b) $v(1 - T\beta)$ (c) $C_v \dfrac{\kappa}{\beta} + v$ (d) $C_v + \dfrac{v\beta}{\kappa}$
8.8 (a) $\dfrac{C_P}{\beta v}$ (b) C_P
8.21 $\dfrac{T_2}{T_1} = \left(\dfrac{P_2}{P_1}\right)^{\frac{\gamma - 1}{\gamma}}$
8.22 21.34 MPa
8.23 3.03°C
8.26 $\dfrac{1}{P}; \dfrac{v - b}{Pv}; \dfrac{v^2(v - b)^2}{RTv^3 - 2a(v - b)^2}$
8.27 $u_2 - u_1 = \int_{T_1}^{T_2} (C_P - Pv\beta)\, dT + \int_{P_1}^{P_2} (Pv\kappa - Tv\beta)\, dP$
8.28 $\Delta h = \int_{T_1}^{T_2} C_P dT + b\,(P_2 - P_1)$
8.30 $C_P - C_v = R$
8.31 $T(v - b)^{\gamma - 1} = \text{const}; P(v - b)^{\gamma} = \text{const}$
8.32 (b) $\dfrac{TR^2 v^3}{RTv^3 - 2a(v - b)^2}$
8.36 $(Z - 1)\dfrac{RT}{P}; -\dfrac{R}{P}$
8.37 $\dfrac{R}{v} + \dfrac{R}{v^2}\left(B + T\dfrac{dB}{dT}\right) + \dfrac{R}{v^3}\left(C + T\dfrac{dC}{dT}\right)$
8.39 $\dfrac{T}{C_P}; \dfrac{T}{C_v}$
8.40 $\infty; \dfrac{\gamma v}{\gamma - 1}; T; \gamma T;$
8.42 1509 m/s
8.43 5.13
8.44 1.49

8.45 6.725 MPa
8.46 −1.48°C; −3.71°C
8.47 Yes
8.48 8.226 kg
8.49 $\dfrac{Bv_{fg}P}{T}$; $\dfrac{Bv_{fg}P}{T^2}$
8.50 101.9°C
8.51 33.655 kJ/mol; Yes; Yes
8.52 3044 m
8.53 2898 m
8.54 2176.54 kJ/kg; 2173.59 kJ/kg
8.55 $F + P = C + 2$; 2
8.56 10.02 MJ
8.57 294.98 kJ; 179.88°C
8.58 1.963 MW
8.59 6 kW
8.60 1402.58 kW
8.61 4014.26 kJ
8.62 427.36 kJ

Chapter 9

9.1 798°C
9.2 20.64 bar
9.3 0.3965
9.4 0.7963
9.5 0.73
9.6 0.3816; 0.858 kg/s
9.7 (a) 9.377 kg/s (b) 0.3522 (c) 0.8163
9.8 0.3208
9.9 (a) 77.78 kg/s (b) 0.4143 (c) 3.3836×10^3 kg/s
9.10 (a) 776.1°C (b) 0.4417
9.11 0.3496; 0.3288; 0.2712; 0.2414
9.12 0.3156; 0.8826; 0.3451; 0.84; 0.3816; 0.7796; 0.398; 0.7441
9.13 0.3496; 0.7659; 0.3619; 0.8142; 0.3761; 0.8533; 0.3921; 0.8879
9.14 0.3024; 1.112 kg/s
9.15 0.3022; 1.114 kg/s
9.16 (a) 0.456 (b) 0.4766
9.17 0.3129
9.18 0.3567
9.19 0.3819
9.20 0.3544
9.21 (a) T = 689.2 K; 4173 K; 1816.4 K P = 1837.9 kPa; 11.13 MPa; 0.6056 MPa (b) 0.565 (c) 1412.5 kJ (d) 1874.83 kPa
9.22 0.5752; 862.8 kJ/kg; 11.357 bar

9.23	0.527; 2070.1 K; 42.05 bar
9.24	0.565; 567.6 kJ
9.25	8.8
9.26	0.5416; 989.3 kJ
9.27	1.1066 kJ; 2.279 bar
9.28	0.565
9.29	$\left(\frac{T_3}{T_1}\right)^{\frac{1}{2(\gamma-1)}}$
9.30	(a) 0.565 (b) 2.825 kJ (c) 12.914 MPa
9.31	(a) 862.8 kJ/kg air (b) 79.2 bar; 2796.44 K (c) 11.36 bar
9.32	(a) 0.565 (b) 0.2732 kJ (c) 3294
9.33	0.5911; 823.05 kPa
9.34	0.6316; 57.198 bar; 7.439 bar
9.35	13.942; 1.5783; 307.5 kJ; 0.615
9.36	0.6284
9.37	1904.75 K; 48.5 bar; 2.094; 0.609
9.38	0.6398; 26.192 kJ; 1.827
9.39	2043.4 kJ; 1203.77 kJ; 0.5981
9.40	15.5855; 1.9953; 1795.8 K
9.41	920.2 kJ; 1825.3 K; 795.7 K; 0.6135
9.43	(a) 1818.86 K; 48.5 bar (b) 913.7 kJ (c) 6.947 bar
9.44	(a) 0.5853 (b) 44.313 bar; 2110.56 K (c) 8.956 bar
9.45	8.1425 kW
9.46	(a) 2.112 (b) 0.6418 (c) 8.715 bar
9.47	1203 kJ; 1976.1 kJ; 1.497 MPa
9.48	659.69 kJ; 295.54 kJ; 0.448
9.49	175.97 kJ; 444.42 kJ; 0.3686
9.50	(a) 0.4 (b) 483.05 kJ (c) 281.56 kJ
9.51	315.56 kJ; 0.875
9.52	250 kJ/kg air; 1000 kJ/kg air; 0.75
9.54	$P' = \sqrt{P_1 P_2}$
9.55	(a) 222.28 kJ (b) 226.33 kJ
9.56	1.2857; 35 kW; 4.5
9.57	8.556 kW; 5.8544
9.58	First possibility
9.59	$\frac{mc}{T_2}(T_1 - T_2)^2$
9.60	136.3 kW
9.61	5.4769; 10.385 kW; 16.174
9.62	37 kW; 2.7025
9.63	3.004; 13.53
9.64	3.207; 32.362 kW; 29.513 tons
9.65	3.098; 16.14 kW; 0.504 kg/s
9.66	4.7871; 229.78 W
9.67	2.726; 185.83 kW; 1.934 kg/s

9.68 (a) 41.7°C (b) 3.3935 (c) 5.18 kW
9.69 (a) 3.8898 (b) 150.836 kPa; 960.255 kPa (c) 2.26 kW
9.70 (b) The cycle in which subcooled liquid enters the throttle valve.
9.71 (a) 3.1168 (b) 7.77 kW (c) 6.89
9.72 6.73
9.73 (a) 2.7085 (b) 0.3849 (c) 0.38 kg/s (d) 12.98 kW
9.74 1218.838 kPa
9.75 100.341 kPa
9.76 84.319 kPa; 1163.453 kPa
9.77 (a) 410.62 K; 235.99 K (b) 2.711 (c) 0.109 kg/s

Chapter 10

10.1 (a) CO_2 = 4.49%; O_2 = 4.94%; N_2 = 7.63%; H_2 = 79.10%; H_2O = 4.39% (b) 12.6417 kmol (c) 3.684 kJ/kg K
10.2 96.16 kPa; CO_2 = 4.67%; O_2 = 5.137%; N_2 = 7.935%; H_2 = 82.258%
10.3 96.06 kPa; O_2 = 5.143%; N_2 = 7.943%; H_2 = 82.344%; H_2O = 4.57%
10.4 91.67 kPa; O_2 = 5.389%; N_2 = 8.323%; H_2 = 86.288%; p_{O_2} = 4.94 kPa; p_{N_2} = 7.63 kPa; p_{H_2} = 79.10 kPa
10.5 53.66% C_3H_8; 36.58% C_2H_6; 9.76% CH_4; 13.15 kg/s
10.6 1.6256×10^3 kmol/h
10.7 535.6 K; 338.82 kJ/kg
10.8 311.65 K; 464.28 m/s
10.9 (a) 34.902 kg/kmol (b) y_{CO_2} = 0.1586; y_{SO_2} = 0.1091 (c) 7.1459 m^3
10.10 (a) 2.26 kg (b) CH_4 = 42.48%; C_2H_6 = 13.27%; CO_2 = 19.47%; N_2 = 24.78%
10.11 35:65
10.12 4.4973 MJ
10.13 (a) 466.78 K; 6.2237 bar (b) y_{N_2} = 0.75; y_{He} = 0.25 (c) 9.727 kJ/K
10.14 504.73 K; 3.6453 bar; 997.894 J/K
10.15 (a) 512.55 K (b) −2.2475 J/mol K (c) 2.2475 J/mol K
10.16 19.56 m/s
10.17 (a) 422.28 K; 17.314 bar (b) 1.5327 kJ/K; 0.4682 kJ/K (c) 2.0009 kJ/K
10.18 434.2 K; 583.25 kJ
10.19 (a) 509.42 K; 8.491 bar (b) 50% He (c) 2.4788 kJ/K; 0.2091 kJ/K (d) 2.6879 kJ/K
10.20 1.2819 MJ
10.21 4.4684 J/mol K
10.22 He = 47.11%; H_2 = 52.89%
10.23 2.443 MPa; 733.03 K
10.24 1.0023 kmol; 1.5 MPa
10.25 (a) 622.41 K (b) 0.0372 kmol (c) y_{He} = 0.5188; y_{CO_2} = 0.4812
10.26 (a) 541.19 K (b) 0.0488 kmol (c) y_{He} = 0.5489; y_{CO_2} = 0.4511 (d) 425.68 kJ
10.27 y_{O_2} = 0.1641; $y_{C_3H_8}$ = 0.8359
10.28 (a) 2.5446 kPa (b) 0.016 kg H_2O/kg dry air (c) 87.27 kg (d) 5452.65 kg (e) 21.25°C
10.29 0.729
10.30 212 12 kg; 23.81°C
10.31 154.25 kg; 3.873 kg

10.32 (a) 0.1296 kg (b) 3.1992 bar
10.33 (a) 0.0357 kg H_2O/kg dry air; 0.7458 (b) 34.56°C (c) 132.1 kJ/kg dry air
10.34 (a) 0.0238 kg H_2O/kg dry air (b) 28.7°C (c) 101.5 kJ/kg dry air
10.35 (a) 7.375 kPa (b) 0.0488 kg H_2O/kg dry air (c) 40°C
10.36 0.0138 kg H_2O/kg dry air; 0.298; 18.91°C
10.37 0.0138 kg H_2O/kg dry air; 0.298; 75.7 kJ/kg dry air. From chart: 0.0138 kg H_2O/kg dry air; 0.30; 76.5 kJ/kg dry air
10.38 0.0106 kg H_2O/kg dry air; 0.40; 57.5 kJ/kg dry air
10.39 0.0159 kg H_2O/kg dry air; 0.449; 76 kJ/kg dry air. From chart: 0.016 kg H_2O/kg dry air; 0.45; 76.5 kJ/kg dry air
10.40 (a) 0.0105 kg H_2O/kg dry air; 0.3966; 57 kJ/kg dry air (b) 0.0106 kg H_2O/kg dry air; 0.4; 57.5 kJ/kg dry air
10.41 0.0116; 0.195; 16.5; 0.0085; 0.25; 11.5; 23.7; 0.0125; 17.5; 23.8; 0.016; 21.4; 22; 0.195; 12.3; 30.2; 0.30; 10.5
10.42 45.83°C
10.43 0.012 39 kg H_2O/kg dry air; 0.4145
10.44 14.5°C
10.45 (a) 1 (b) 25°C (c) 24.16 kJ
10.46 35°C; 0.60
10.47 0.0182 kg H_2O/kg dry air; 0.56; 33.2°C; 80.5 kJ/kg dry air
10.48 25°C; 0.0114 kg H_2O/kg dry air; 0.56
10.49 10°C; 0.86
10.50 0.6833 kg dry atmospheric air to be mixed with 1 kg dry chilled air; 0.63
10.51 (a) 0.0121 kg H_2O/kg dry air (b) 0.4 (c) 32.5°C
10.52 0.415; 0.869; 0.094 kg/s
10.53 4.4 kg/min; 0.296 MW
10.54 0.0075 kg H_2O/kg dry air; 0.39; 15 kJ
10.55 (a) 0.206 kg/s (b) 791.35 kW (c) 140 kW
10.56 0.0072 kg; 0.8
10.57 585.94 kg/min; 797.17 kW
10.58 434.78 kg/h
10.59 173.9 kW; 29 kW; 0.0454 kg/s
10.60 857.1 kW; 115 kW; 0.21 kg/s
10.61 6°C; 72.1 kW
10.62 (a) 29.52 kg (b) 12 kW
10.63 28°C; 0.019 kg H_2O/kg dry air; 619.2 kg
10.64 24.8°C; 1.41 × 10^4 kg
10.65 154 68 kg/s; 358.9 kg/s
10.66 1.079 × 10^6 kg/min; 213 64 kg/min
10.67 28.15 kg/min; 1340.24 kg/min
10.68 10.5 kg/s; 617.77 kg/s
10.69 4202.6 kg/min; 39.92 kg/min
10.70 (a) 482.5 kg/s (b) 2.98 kg/s
10.71 30.5°C; 4.76 kg
10.72 157.26 kW; 28.86 kW
10.73 45.83°C; 1.04 kg

Index